INTRODUCTORY ALGEBRA

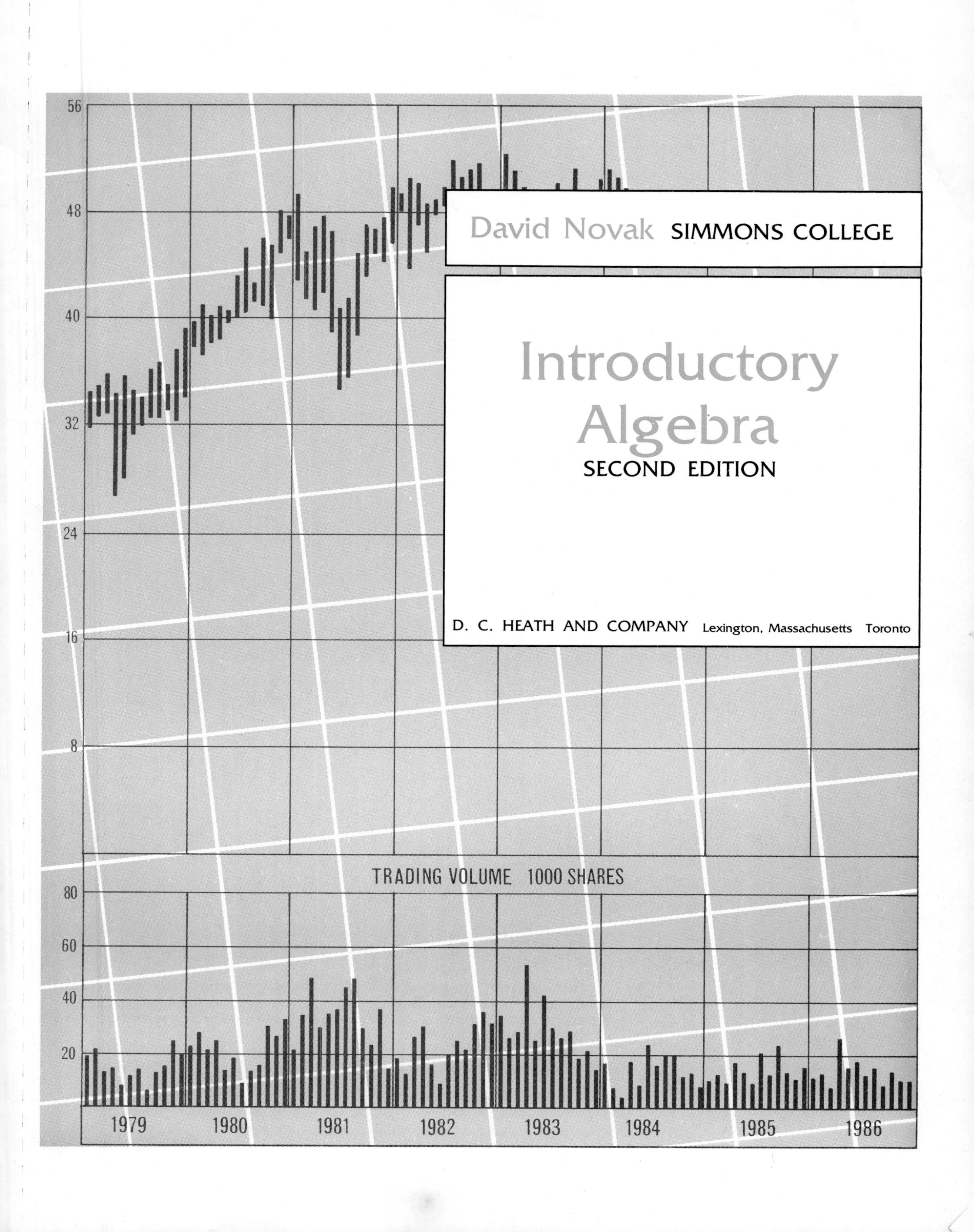
56
48
40
32
24
16
8
David Novak SIMMONS COLLEGE
Introductory Algebra
SECOND EDITION
D. C. HEATH AND COMPANY Lexington, Massachusetts Toronto
TRADING VOLUME 1000 SHARES
80
60
40
20
1979
1980
1981
1982
1983
1984
1985
1986

Cover art by Hamant Design.

Published simultaneously in Canada.

Printed in the United States of America.

International Standard Book Number: 0-669-12231-9

For Dena Marie
and
Jeannie Anne

PREFACE

Introductory Algebra, Second Edition reflects the comments and suggestions of numerous instructors and students who used the first edition, while continuing to exemplify my own beliefs about how students learn mathematics. Although students need constant practice to reinforce their skills, it is also important to provide them with an understanding of some of the "whys" of mathematics. In writing both editions it has been a challenge to determine in each section the amount of verbiage and the number of examples necessary to achieve my goal. Other textbooks I have used in the classroom either delve too heavily into the reasons behind mathematical manipulations or reduce mathematics to rote drill. With the help of current users of my text, I think this second edition has a better balance of explanation and problem-solving practice than the first edition.

Features

- *Introductory Algebra,* Second Edition is the second in a series of textbooks including *Basic Mathematics,* Second Edition and *Intermediate Algebra,* Second Edition which, combined with a large supplemental package, provide instructors with a complete, flexible system for teaching fundamental mathematics. A list of supplements follows the preface.
- Algorithms are given to aid students in solving many different kinds of equations. For example, Chapter 3 introduces a five-step algorithm for solving linear equations.
- Exponents are introduced in Chapter 1 and then treated in detail in Chapter 2.
- Analyzing and solving word problems is thoroughly discussed in Chapter 4. Numerous applications appear in examples and exercises throughout the remainder of the text.
- An entire chapter (Chapter 6) is devoted to graphing, a skill important for many subsequent courses.
- There are over 4000 problems in the text, many rewritten, and thus improved, from the first edition.
- Exercises are of three types: marginal problems for immediate reinforcement, section problems for homework assignments, and review problems and chapter tests for supplemental practice.
- Learning objectives at the beginning of each section concisely list the goals of the section.

Teaching Methods

In the Classroom: Since each section typically presents material that can be covered in one class period, the book is well-suited to many classroom situations. Marginal exercises can be done as preparation for class, they can function as in-class assignments with discussion to follow, or they can be assigned as out-of-class work. A self-paced individualized format can also be used, freeing the instructor to check each student's work in progress. Each section also provides plenty of homework problems, which appear on pages that can be torn out and handed in as homework assignments.

In a Learning Laboratory or Self-Study Setup: The text provides step-by-step detail throughout that enables students to master each skill at an individual pace. The available supplements are particularly helpful in either classroom or learning lab. Please see the listing of these after the preface.

Acknowledgments

This revision has benefited from the suggestions of many people. I would like to thank the following: C. J. Alexander, Des Moines Area Community College; Don Bohmont, Metropolitan Technical Community College; Janet Buck, Seattle Pacific University; Barbara Buhr, Fresno City College; Jonette Clark, Phillips University; Charles C. Edgar, Onondaga Community College; David Ellenbogen, Cape Cod Community College; Marilyn Foglesong, Michigan Christian College; Frank Garcia, North Seattle Community College; Helen Hancock, Shoreline Community College; Jane Keller, Metropolitan Technical Community College; Jerry Kerner, Nicolet College and Technical Institute; Lynne Kotrous, Central Community College; Marilynn Kruck, Shoreline Community College; Doty Latuszek, Kirtland Community College; Jay I. Lippmann, College of the Desert; Amy Lobdell, Edison Community College; Kenneth Lueder, Metropolitan Technical Community College; Ernest Madison, Metropolitan Technical Community College; Alice Madson, Kankakee Community College; James H. Mason, Belmont Technical College; Eric Morgan, Nash Technical College; Michelle A. Mosman, Des Moines Area Community College; Adelaida Roque, Columbia College (Illinois); Micky Smith, Livingston University; Myrna Sorensen, Santa Ana College; Deborah Trytten, William Penn College; Karen Viliska, Metropolitan Technical Community College; Michael White, Jefferson Community College; Jeff Yoder, Urbana College; Roberta Gansman, Guilford Technical Community College.

The staff at D. C. Heath has also contributed considerably to the book as you see it here, especially Mary Lu Walsh, Senior Mathematics Editor; Carolyn Johnson, Editorial Assistant; Cathy Brooks, Production Editor; and Susan Hamant, Cover Designer.

A very special thank-you to all the sales representatives at D. C. Heath for their suggestions for the second edition. I would also like to thank my wife Marian, without whom none of this would have been possible.

David Novak

SUPPLEMENTS

Introductory Algebra is the second in a series of worktexts that also includes *Basic Mathematics* and *Intermediate Algebra.* These books have been developed with a number of supplements to enable instructors to use the text in a variety of teaching situations.

GENTEST chapter tests may be generated on the IBM PC®, Apple II® series computer, or TRS 80™ and printed out on a wide variety of printers. Below are six features of GENTEST found useful by college teachers.

1. Since this book may be used for individually paced courses either in a learning lab or in a regular classroom, many test versions for each chapter and final exam are necessary for test security.
2. Because students are often required to retest until they reach a certain proficiency level, it is useful to be able to generate a retest different on every item from the test(s) the student has failed previously.
3. All versions of a chapter test review the same concepts and are of roughly the same difficulty.
4. Answer keys for each version are available so instructors, learning lab staff, or students themselves can check the tests.
5. The tests leave space for working out problems, which allows instructors to check the sources of student errors. Also, students can make test corrections on the test paper and file them for their future review and yours. By customizing the instructions given to the printer, one can increase or decrease the space between problems on an individual test.
6. Only one microcomputer and printer are needed.

Videotapes A series of 19 lectures, each 20 minutes long, have been prepared by the author to cover the following important topics:

1. Introduction and Addition and Subtraction of Integers
2. Multiplication and Division of Integers
3. Rational Exponents
4. Exponents
5. Laws for Rational Numbers and Simplifying Algebraic Expressions
6. Linear Equations in One Variable—Part I
7. Linear Equations in One Variable—Part II

IBM is a registered trademark of International Business Machines, Inc.
Apple is a registered trademark of Apple Computer, Inc.
TRS-80 is a trademark of Radio Shack, a division of Tandy Corp.

8. Word Problems
9. Polynomials
10. Factoring
11. Graphing
12. Straight Lines
13. 2-by-2 Systems (Substitution Method)
14. 2-by-2 Systems (Addition Method)
15. Simplification, Multiplication, and Division of Rational Expressions
16. Addition and Subtraction of Rational Expressions
17. Radicals
18. Quadratic Equations
19. Inequalities and Absolute Value

These lectures are very similar to Dr. Novak's own class presentations. The topic is introduced, needed definitions are explained, and then examples (related to or sometimes the same as those in the textbook) are given. Many of these lectures include word problems as examples. Developing the skills needed to analyze and solve applications is very important and is emphasized in these lectures.

The tapes offer students several advantages. In a self-paced program students may, after reading the textbook, view the tape(s) pertaining to a concept to reinforce the self-learning process. Within a lecture format, students may use the tapes as a review of class work, as a second lecture on a difficult topic, or as a substitute for missing a class lecture.

Student Guide to Margin Exercises This supplement provides completely worked-out solutions to every margin exercise in the textbook.

Instructor's Guide with Tests A brief overview of each chapter as well as over 50 sample problems per chapter are contained in this supplement. These sample problems may be used by instructors as lecture examples, in class exercises, homework problems, or test questions. Further, this supplement includes five forms of chapter tests, one diagnostic test, and two final exams. Answer keys for each test are supplied.

Answer Key This supplement includes the answers to every problem in the textbook.

CONTENTS

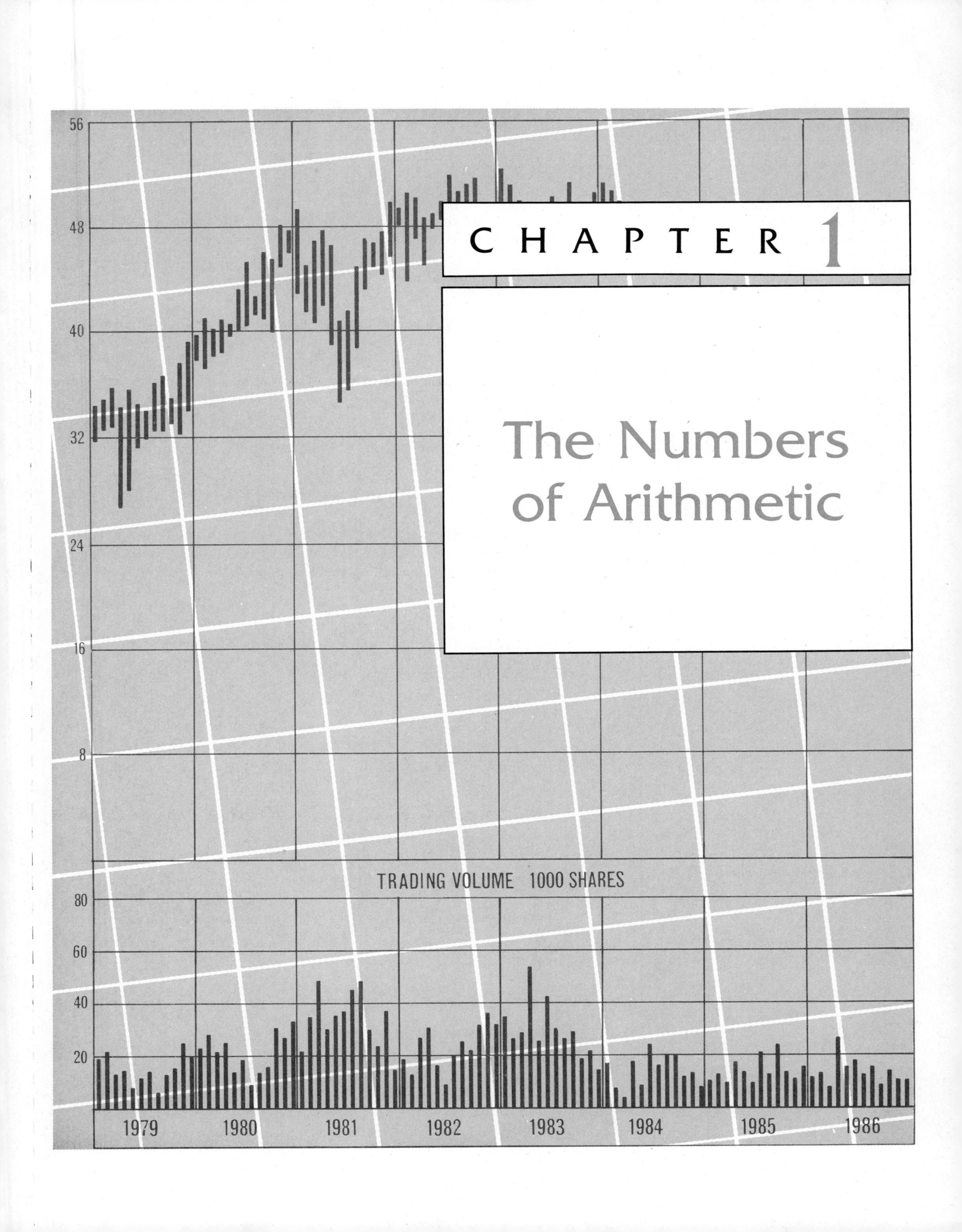

CHAPTER 1

The Numbers of Arithmetic

LEARNING OBJECTIVES

After finishing Section 1.1, you should be able to:

- Identify the set of natural numbers, the set of whole numbers, and the set of fractions.
- Represent fractions pictorially.

1.1

Introduction

Algebra is a generalization or an extension of arithmetic. Thus, before we start algebra, we are going to review some topics from arithmetic and present some of the terminology and notation that will be used throughout the book. We start by defining two sets of numbers.

DEFINITION 1.1 The set of *natural numbers* is the set

$$1, 2, 3, 4, 5, 6, \ldots$$

(where "..." means "and so on"). This set is also frequently called the set of *counting numbers.*

DEFINITION 1.2 If we include 0 along with the set of natural numbers, we obtain the set of *whole numbers*

$$0, 1, 2, 3, 4, \ldots$$

If we have a working knowledge of the set of whole numbers, we can solve a variety of problems. However, if a problem involves working with a part or a portion of a whole number, then we will need to know about the set of fractions.

DEFINITION 1.3 The set of *fractions* consists of all numbers of the form $\frac{m}{n}$, where m and n are whole numbers, except that n *cannot* be 0. The numbers

$$\frac{1}{2}, \frac{3}{17}, \frac{1}{4}, \frac{17}{3}, \frac{2}{5}, \frac{5}{1}, \quad \text{and} \quad \frac{0}{1}$$

are examples of fractions.

We can consider every whole number to be a fraction if we put the whole number over 1. For example, $4 = \frac{4}{1}$. Also, in the fraction $\frac{m}{n}$ the bar (—) may always be thought of as denoting division; that is, $\frac{m}{n}$ means m divided by n. Thus, $\frac{m}{n}$ may be written as $m \div n$. (We will sometimes see $\frac{m}{n}$ written in the form m/n.)

DEFINITION 1.4 In the fraction $\frac{m}{n}$, m is called the *numerator* of the fraction, and n is called the *denominator* of the fraction:

$$\frac{\text{numerator}}{\text{denominator}}$$

DO EXERCISES 1 THROUGH 3.

1. In the fraction $\frac{2}{7}$, which number is the numerator and which is the denominator?

2 = numerator
7 = denominator

2. Can the whole number 3 be thought of as a fraction? If so, how?

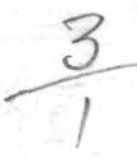

3. In the fraction $\frac{a}{b}$, is there any restriction on a or b? If so, what?

b cannot be 0.

If we want to represent a fraction pictorially, we use the denominator of the fraction to tell us how many equal parts to divide our whole unit into, and the numerator to tell us how many of these equal parts we need.

Example 1 Represent the fraction three-eighths $\left(\frac{3}{8}\right)$ pictorially. $\left(\text{Recall that } \frac{3}{8}\right.$ means 3 parts out of 8 parts.$\left.\right)$

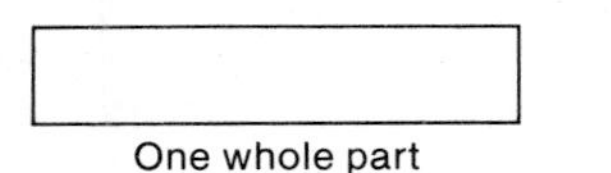

One whole part

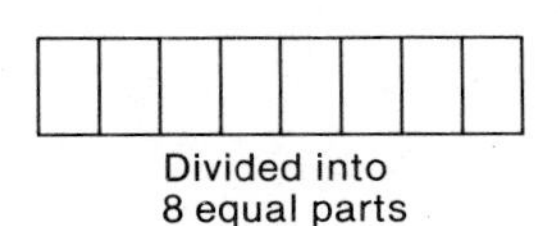

Divided into 8 equal parts

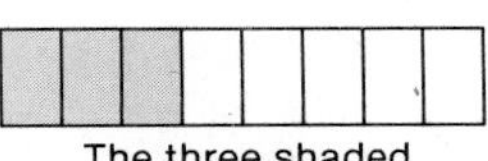

The three shaded parts out of 8 total parts represent the fraction $\frac{3}{8}$.

Example 2 Represent the fraction two-fifths $\left(\frac{2}{5}\right)$ pictorially.

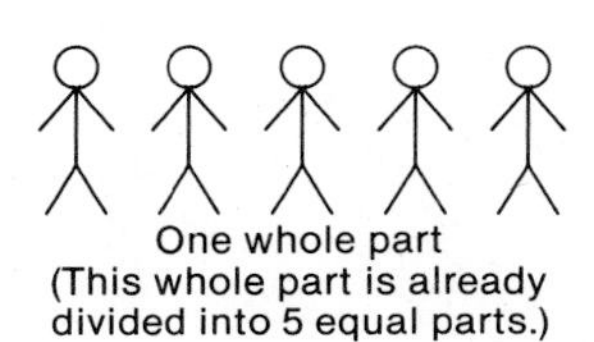

One whole part
(This whole part is already divided into 5 equal parts.)

Basketball Team

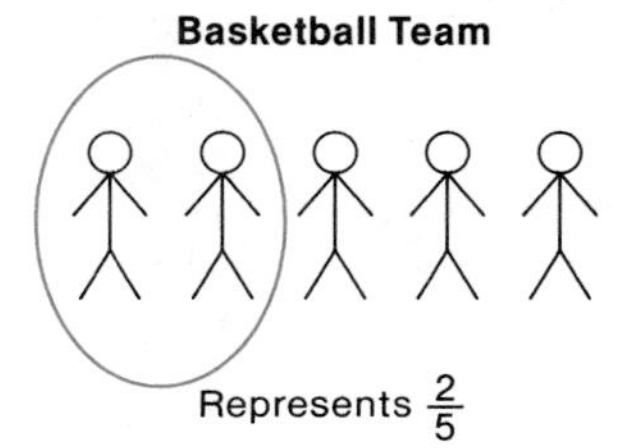

Represents $\frac{2}{5}$

DO EXERCISE 4.

Example 3 Represent the fraction five-thirds $\left(\frac{5}{3}\right)$ pictorially.

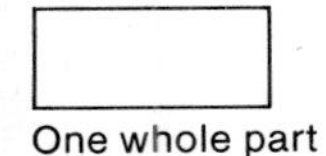

One whole part

Divided into 3 equal parts

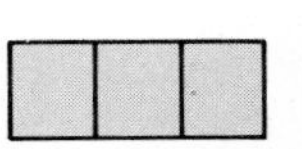

The shaded portion represents $\frac{5}{3}$.

DO EXERCISE 5.

The sets of natural numbers, whole numbers, fractions, and decimals (see Section 1.5) are the sets we work with in arithmetic. Our main goal in Chapter 2 will be to enlarge these sets of numbers to include negative numbers and still have all the rules and definitions used in arithmetic be valid for these larger sets of numbers used in algebra.

DO SECTION PROBLEMS 1.1.

4. Represent $\frac{2}{7}$ pictorially.

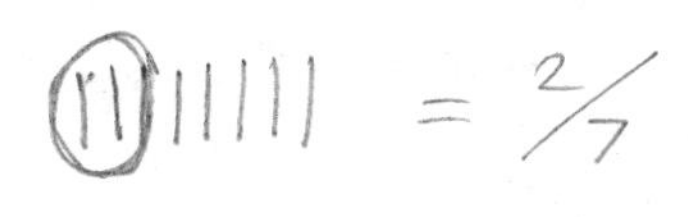

5. Represent seven-fourths pictorially.

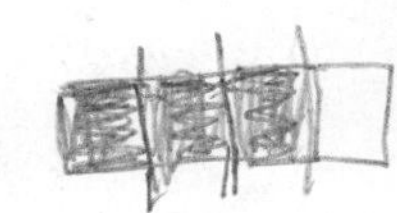

ANSWERS TO MARGINAL EXERCISES

1. 2 is the numerator and 7 is the denominator. **2.** Yes, write 3 as $\frac{3}{1}$.
3. Yes, b cannot be zero. **4.** 
5.

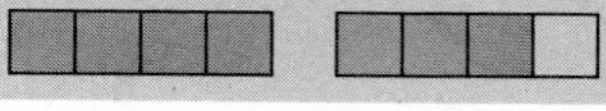

NAME COURSE/SECTION DATE

SECTION PROBLEMS 1.1

Classify the following as a natural number, whole number, and/or fraction.

1. 4 **2.** 0 **3.** $\frac{2}{3}$

4. 7 **5.** $\frac{5}{2}$ **6.** $\frac{1}{17}$

7. In the fraction $\frac{3}{8}$, which number is the numerator and which is the denominator?

8. In the fraction $\frac{15}{4}$ which number is the numerator and which is the denominator?

9. In the fraction 5 which number is the numerator and which is the denominator?

10. In the fraction $\frac{1}{39}$ which number is the numerator and which is the denominator?

11. In the fraction $\frac{m}{n}$ are there any restrictions on m and/or n?

12. Which of the following are well-defined fractions?
(a) $\frac{0}{5}$ (b) $\frac{1}{4}$ (c) $\frac{7}{0}$ (d) 3 (e) $\frac{0}{0}$

ANSWERS

1. natural
2. whole
3. fraction
4. natural
5. fraction
6. fraction
7. 3n, 8d
8. 15n, 4d
9. 5n, 2d
10. 1n, 39d
11. m/no, n/yes, not 0
12. b, d.

13. See figure.

14. See figure.

15. See figure.

16. See figure.

17. See figure.

18. See figure.

19. See figure.

20. See figure.

13. Represent $\frac{3}{5}$ pictorially.

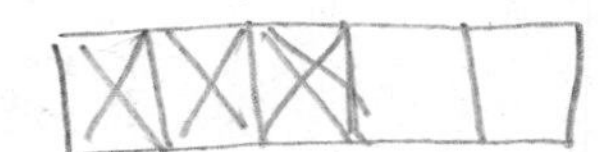

14. Represent $\frac{2}{3}$ pictorially.

15. Represent $\frac{7}{3}$ pictorially.

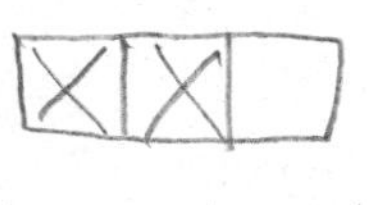

16. Represent $\frac{7}{6}$ pictorially.

17. Represent $\frac{2}{7}$ pictorially.

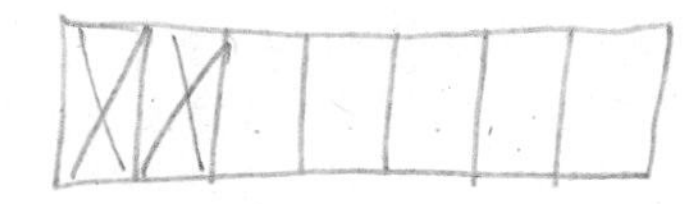

18. Represent $\frac{1}{4}$ pictorially.

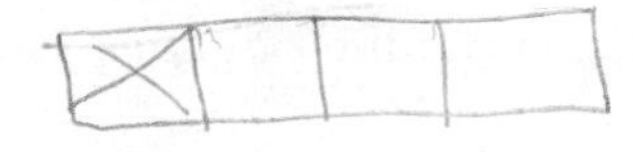

19. Represent $\frac{5}{2}$ pictorially.

20. Represent $\frac{12}{5}$ pictorially.

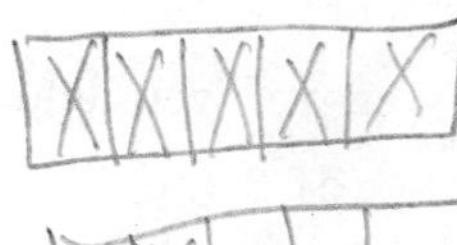

1.2

Multiplying, Reducing, and Dividing Fractions

LEARNING OBJECTIVES

After finishing Section 1.2, you should be able to:

- Multiply fractions.
- Reduce fractions to lowest terms.
- Divide fractions.

The first operation with fractions we consider is multiplication. However, before we give a rule for multiplication of fractions, we motivate the rule by an example.

Example 1 Consider the multiplication problem $\frac{1}{2} \cdot \frac{3}{4} = ?$ (The $\cdot$ indicates multiplication.) Notice that $\frac{1}{2} \cdot \frac{3}{4}$ means we want $\frac{1}{2}$ of $\frac{3}{4}$. Pictorially, we want to represent one-half of the fraction three-fourths.

(a) Represent the fraction $\frac{3}{4}$.

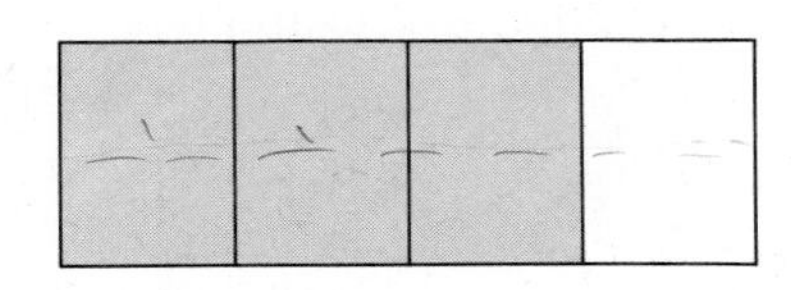

(b) We want one-half of the shaded area in (a).

(c) To do this, we divide the rectangle in (a) in half. (This has the effect of dividing our original whole unit into 8 equal parts.)

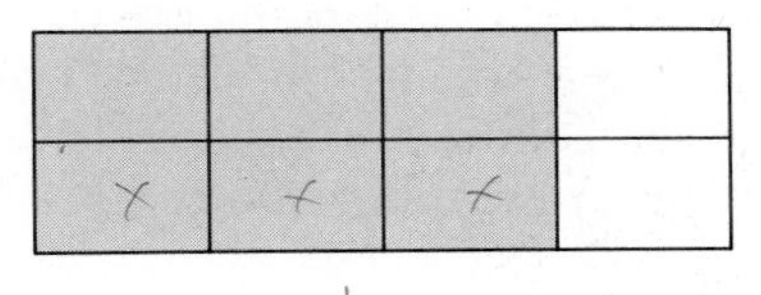

(d) Take $\frac{1}{2}$ of the shaded area,

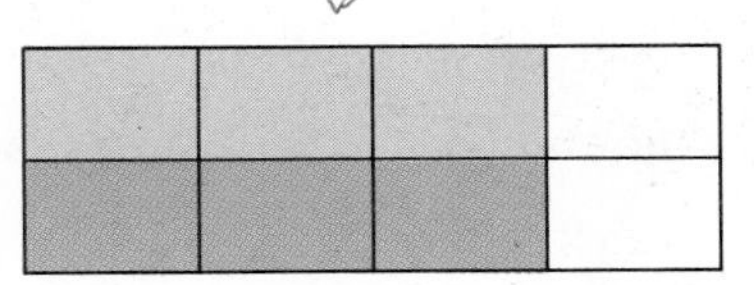

(e) and obtain three-eighths of our whole unit.

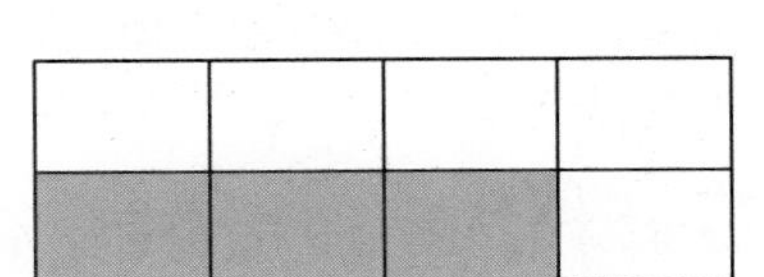

Thus $\frac{1}{2}$ of $\frac{3}{4} = \frac{1}{2} \cdot \frac{3}{4} = \frac{3}{8}$.

If we take a closer look at the result of multiplying $\frac{1}{2}$ by $\frac{3}{4}$, we obtain

$$\frac{1}{2} \cdot \frac{3}{4} = \frac{3}{8}$$

$3 = 1 \cdot 3$ = product of the numerators of the fractions $\frac{1}{2}$ and $\frac{3}{4}$

$8 = 2 \cdot 4$ = product of the denominators of the fractions $\frac{1}{2}$ and $\frac{3}{4}$

This example may be generalized to the following rule.

Calculate.

1. $\frac{3}{7} \cdot \frac{2}{5}$

2. $\frac{9}{4} \cdot \frac{2}{3}$

3. $1 \cdot \frac{7}{10}$

4. $\frac{1}{7} \cdot 2 \cdot \frac{3}{2}$

5. $\frac{4}{9} \times \frac{4}{5}$

6. Find all the factors of 20.

7. Find all the factors of 42.

8. Find all the factors of 17.

9. Is it true that if m is a factor of c, then m divides c with a remainder of zero; that is, m divides c evenly?

RULE 1.1 To multiply two (or more) fractions, multiply their numerators together and their denominators together; or, in symbols,

$$\frac{a}{b} \cdot \frac{c}{d} = \frac{a \cdot c}{b \cdot d}$$

Example 2 $\frac{2}{3} \cdot \frac{8}{9} = \frac{2 \cdot 8}{3 \cdot 9} = \frac{16}{27}$

Example 3 $\frac{5}{2} \cdot 3 = \frac{5}{2} \cdot \frac{3}{1} = \frac{5 \cdot 3}{2 \cdot 1} = \frac{15}{2}$

Example 4 $\frac{3}{5} \times 0 = \frac{3}{5} \times \frac{0}{1} = \frac{3 \times 0}{5 \times 1} = \frac{0}{5} = 0$

In Example 4 we used $\times$ for multiplication. However, in algebra x is often used to denote a variable, so in this book we will use different ways of indicating multiplication, for example:

$$2 \times 3 = 2 \cdot 3 = (2)(3) = 2(3) = (2)3$$

Example 5 $\frac{1}{7} \cdot \frac{3}{2} \cdot \frac{5}{4} = \frac{1 \cdot 3 \cdot 5}{7 \cdot 2 \cdot 4} = \frac{15}{56}$

DO EXERCISES 1 THROUGH 5.

Before we discuss reducing fractions to lowest terms, we must look at factors of whole numbers and the prime factorization of whole numbers.

Consider the whole number 12. We can write 12 as a product of two whole numbers in three different ways:

$$12 = 1 \cdot 12$$
$$12 = 2 \cdot 6$$
$$12 = 3 \cdot 4$$

The whole numbers 1, 2, 3, 4, 6, and 12 are called factors of 12.

DEFINITION 1.5 Given a whole number c, then a whole number m is a *factor* of c if there is another whole number n such that c is the product of m and n ($c = m \cdot n$). (This also says that n is a factor of c.)

Example 6 Find all the factors of 40.

$$40 = 1 \cdot 40$$
$$40 = 2 \cdot 20$$
$$40 = 4 \cdot 10$$
$$40 = 5 \cdot 8$$

Thus the factors of 40 are 1, 2, 4, 5, 8, 10, 20, and 40.

DO EXERCISES 6 THROUGH 9.

Every whole number greater than 1 has at least two factors, namely 1 and the number itself. However, some whole numbers have exactly two factors.

DEFINITION 1.6 A *prime number* is a whole number greater than 1 that has only 1 and itself as factors.

Example 7

(a) 2 is a prime since 1 and 2 are the only factors of 2.

(b) 3 is a prime since 1 and 3 are the only factors of 3.

(c) 4 is not a prime since 4 has a factor other than 1 and 4, namely, 2.

(d) 6 is not a prime since 1, 2, 3, and 6 are all factors of 6.

(e) 1 is not a prime since a prime must be strictly greater than 1.

DO EXERCISES 10 AND 11.

DEFINITION 1.7 A whole number greater than 1 that is not a prime is called a *composite number*.

We know that 12 is not a prime number. However, we can write 12 as a product of prime numbers—$12 = 2 \cdot 2 \cdot 3$. In fact, any composite number can be written as a product of prime numbers. The *prime factorization* of a whole number is the expression of the number as a product of prime numbers.

Example 8 Find the prime factorization of the following: (a) 15, (b) 42, (c) 90, (d) 11.

(a) $15 = 3 \cdot 5$ (Also, note that $15 = 1 \cdot 3 \cdot 5$, but this is not a prime factorization of 15 since 1 is not a prime.)

(b) $42 = 6 \cdot 7 = 2 \cdot 3 \cdot 7$

(c) $90 = 9 \cdot 10 = 3 \cdot 3 \cdot 2 \cdot 5$

(d) 11 is already prime and hence the prime factorization of 11 is 11.

DO EXERCISES 12 THROUGH 19.

Now we can return to reducing fractions.

If you have $\frac{1}{2}$ of a pie and your friend has $\frac{3}{6}$ of a pie, then you both have the same amount of pie. While the fractions $\frac{1}{2}$ and $\frac{3}{6}$ are not the same fractions, they are what we call equivalent fractions.

DEFINITION 1.8 The fractions $\frac{a}{b}$ and $\frac{c}{d}$ are *equivalent* if $a \cdot d = b \cdot c$, and if they are equivalent, we write $\frac{a}{b} = \frac{c}{d}$. Conversely, if $\frac{a}{b} = \frac{c}{d}$, then the fractions $\frac{a}{b}$ and $\frac{c}{d}$ are equivalent.

10. List the first ten prime numbers.

2, 3, 4, 5, 6, 7, 8, 9, 10

11. Why is 21 not a prime number?

Because of 1

Find the prime factorization of the following.

12. 15 $3 \cdot 5$

13. 6 $2 \cdot 3$

14. 30 $3 \cdot 5 \cdot 6 \cdot 10$

15. 28 $3 \cdot 4 \cdot 8 \cdot$

16. 17

17. 48

18. 37

19. 180

20. Are $\frac{3}{5}$ and $\frac{24}{40}$ equivalent?

21. Are $\frac{6}{8}$ and $\frac{14}{21}$ equivalent?

Example 9

(a) $\frac{1}{2}$ and $\frac{3}{6}$ are equivalent since $1 \cdot 6 = 2 \cdot 3$, and we write $\frac{1}{2} = \frac{3}{6}$.

(b) $\frac{1}{2}$ and $\frac{5}{10}$ are equivalent since $1 \cdot 10 = 2 \cdot 5$, and we write $\frac{1}{2} = \frac{5}{10}$.

Example 10

(a) $\frac{2}{3}$ and $\frac{5}{6}$ are not equivalent since $2 \cdot 6 \neq 3 \cdot 5$. (The symbol $\neq$ means "not equal to.")

(b) $\frac{2}{3}$ and $\frac{24}{36}$ are equivalent since $2 \cdot 36 = 3 \cdot 24$, and we write $\frac{2}{3} = \frac{24}{36}$.

In general, we have the *Fundamental Property of Fractions* which says that the fraction $\frac{m \cdot c}{n \cdot c}$ (where $c \neq 0$) is equivalent to the fraction $\frac{m}{n}$, since

$$\frac{m \cdot c}{n \cdot c} = \frac{m}{n} \cdot \frac{c}{c} = \frac{m}{n} \cdot 1 = \frac{m}{n}$$

DO EXERCISES 20 AND 21.

In Example 9, we had the three equivalent fractions $\frac{1}{2} = \frac{3}{6} = \frac{5}{10}$. The fraction $\frac{1}{2}$ is the reduced form of all fractions equivalent to $\frac{1}{2}$. Whenever a fraction is the final answer to a problem, we want it to be in reduced form.

DEFINITION 1.9 A fraction $\frac{m}{n}$ is in *reduced form* (also called *simplest form* or *lowest terms*) if the whole numbers m and n have no common factors larger than 1.

The following examples show how to reduce fractions to lowest terms. We use the property $\frac{m \cdot c}{n \cdot c} = \frac{m}{n}$ (where $c \neq 0$) to reduce fractions.

Example 11 Reduce the following fractions to lowest terms.

(a) $\frac{15}{21} = \frac{3 \cdot 5}{3 \cdot 7} = \frac{5}{7}$

Any time we can match a factor in the numerator with one in the denominator, we can reduce the fraction; that is,

$$\frac{m \cdot c}{n \cdot c} = \frac{m}{n} \text{ where } c \neq 0$$

(b) $\frac{12}{30} = \frac{3 \cdot 4}{3 \cdot 10} = \frac{4}{10} = \frac{2 \cdot 2}{2 \cdot 5} = \frac{2}{5}$ or $\frac{12}{30} = \frac{6 \cdot 2}{6 \cdot 5} = \frac{2}{5}$

(c) $\frac{24}{36} = \frac{4 \cdot 6}{6 \cdot 6} = \frac{4}{6} = \frac{2 \cdot 2}{2 \cdot 3} = \frac{2}{3}$ or $\frac{24}{36} = \frac{2 \cdot 12}{3 \cdot 12} = \frac{2}{3}$

(d) $\frac{15}{28} = \frac{3 \cdot 5}{4 \cdot 7} = \frac{3 \cdot 5}{2 \cdot 2 \cdot 7} = \frac{15}{28}$

The fraction $\frac{15}{28}$ is already reduced since there are no common factors in the numerator or denominator of $\frac{15}{28}$.

(e) To help in reducing fractions, we sometimes put a slash mark (/) through common factors in the numerator and denominator.

$$\frac{30}{42} = \frac{3 \cdot 10}{6 \cdot 7} = \frac{3 \cdot 2 \cdot 5}{2 \cdot 3 \cdot 7} = \frac{\cancel{3} \cdot \cancel{2} \cdot 5}{\cancel{2} \cdot \cancel{3} \cdot 7} = \frac{5}{7}$$

DO EXERCISES 22 THROUGH 27.

Reduce to lowest terms.

22. $\frac{18}{48}$

23. $\frac{32}{12}$

24. $\frac{6}{115}$

25. $\frac{33}{22}$

26. $\frac{96}{24}$

27. $\frac{39}{60}$

In multiplying fractions we always put our final answer in reduced form.

Example 12 Multiply. (Put the answer in reduced form.)

(a) $\frac{12}{5} \cdot \frac{10}{6} = \frac{12 \cdot 10}{5 \cdot 6} = \frac{6 \cdot 2 \cdot 5 \cdot 2}{5 \cdot 6} = \frac{6 \cdot 5 \cdot 2 \cdot 2}{6 \cdot 5 \cdot 1} = \frac{2 \cdot 2}{1} = \frac{4}{1} = 4$

(b) $\frac{21}{20} \cdot \frac{4}{9} \cdot \frac{5}{14} = \frac{21 \cdot 4 \cdot 5}{20 \cdot 9 \cdot 14} = \frac{3 \cdot 7 \cdot 4 \cdot 5 \cdot 1}{4 \cdot 5 \cdot 3 \cdot 3 \cdot 2 \cdot 7}$

$= \frac{3 \cdot 7 \cdot 4 \cdot 5 \cdot 1}{3 \cdot 7 \cdot 4 \cdot 5 \cdot 3 \cdot 2} = \frac{1}{6}$

We rework (b) using slash marks to help identify the common factors.

(c) $\frac{21}{20} \cdot \frac{4}{9} \cdot \frac{5}{14} = \frac{21 \cdot 4 \cdot 5}{20 \cdot 9 \cdot 14} = \frac{3 \cdot 7 \cdot 4 \cdot 5 \cdot 1}{4 \cdot 5 \cdot 3 \cdot 3 \cdot 2 \cdot 7}$

$= \frac{\cancel{3} \cdot \cancel{7} \cdot \cancel{4} \cdot \cancel{5} \cdot 1}{\cancel{4} \cdot \cancel{5} \cdot \cancel{3} \cdot 3 \cdot 2 \cdot \cancel{7}} = \frac{1}{6}$

DO EXERCISES 28 THROUGH 33.

Multiply. (Leave your final answer in reduced form.)

28. $\frac{3}{10} \cdot \frac{2}{15}$

29. $\frac{9}{2} \cdot \frac{4}{9}$

30. $\frac{4}{5} \cdot \frac{3}{7}$

31. $\frac{1}{6} \cdot \frac{3}{8} \cdot \frac{8}{12}$

32. $5 \cdot \frac{2}{25}$

33. $\frac{4}{5} \cdot \frac{3}{14} \cdot \frac{5}{8}$

Next we consider division of fractions. Consider the division problem $\frac{3}{4} \div \frac{2}{5}$. Remember that we may interpret the — in $\frac{m}{n}$ as division, so we have

$$\frac{3}{4} \div \frac{2}{5} = \frac{\frac{3}{4}}{\frac{2}{5}} = \frac{\frac{3}{4}}{\frac{2}{5}} \cdot 1$$

$$= \frac{\frac{3}{4} \cdot \frac{5}{2}}{\frac{2}{5} \cdot \frac{5}{2}} = \frac{\frac{3}{4} \cdot \frac{5}{2}}{\frac{2}{5} \cdot \frac{5}{2}} = \frac{\frac{3}{4} \cdot \frac{5}{2}}{\frac{10}{10}} = \frac{\frac{3}{4} \cdot \frac{5}{2}}{1}$$

$$= \frac{3}{4} \cdot \frac{5}{2}$$

Thus

$$\frac{3}{4} \div \frac{2}{5} = \frac{3}{4} \cdot \frac{5}{2}$$

34. $\frac{3}{4} \div \frac{1}{4}$

35. $\frac{9}{64} \div \frac{5}{32}$

36. $\frac{6}{7} \cdot 3 \cdot \frac{14}{5}$

37. $3 \div \frac{1}{3}$

38. $\frac{7}{2} \div \frac{5}{4}$

39. $\frac{2}{5} \div 4$

We see that division of fractions involves changing the operation of division to the operation of multiplication and at the same time the fraction $\frac{2}{5}$ that we are dividing by changes to $\frac{5}{2}$ in the multiplication problem. The general rule follows:

RULE 1.2 $$\frac{a}{b} \div \frac{c}{d} = \frac{a}{b} \cdot \frac{d}{c}$$

We see that $\div$ changes to $\cdot$; the fraction we are dividing by, $\frac{c}{d}$, changes (inverts) to $\frac{d}{c}$; and the fraction $\frac{a}{b}$ does not change.

Example 13 $\frac{3}{4} \div \frac{2}{5} = \frac{3}{4} \cdot \frac{5}{2} = \frac{3 \cdot 5}{4 \cdot 2} = \frac{15}{8}$

Example 14 $\frac{5}{7} \div 3 = \frac{5}{7} \div \frac{3}{1} = \frac{5}{7} \cdot \frac{1}{3} = \frac{5 \cdot 1}{7 \cdot 3} = \frac{5}{21}$

Example 15 $\frac{7}{6} \div \frac{21}{2} = \frac{7}{6} \cdot \frac{2}{21} = \frac{7 \cdot 2}{6 \cdot 21} = \frac{7 \cdot 2}{2 \cdot 3 \cdot 7 \cdot 3} = \frac{7 \cdot 2 \cdot 1}{7 \cdot 2 \cdot 3 \cdot 3} = \frac{1}{9}$

Example 16 $0 \div \frac{2}{3} = \frac{0}{1} \cdot \frac{3}{2} = \frac{0 \cdot 3}{1 \cdot 2} = \frac{0}{2} = 0$

Recall that 0 divided by a nonzero number is a valid operation, and the answer is always 0. However, we can never divide by 0.

DO EXERCISES 34 THROUGH 39.

DO SECTION PROBLEMS 1.2.

ANSWERS TO MARGINAL EXERCISES

1. $\frac{6}{35}$ **2.** $\frac{18}{12}$ **3.** $\frac{7}{10}$ **4.** $\frac{6}{14}$ **5.** $\frac{16}{45}$ **6.** 1, 2, 4, 5, 10, 20
7. 1, 2, 3, 6, 7, 14, 21, 42 **8.** 1, 17 **9.** Yes **10.** 2, 3, 5, 7, 11, 13, 17, 19, 23, 29 **11.** $21 = 3 \cdot 7$ **12.** $3 \cdot 5$ **13.** $2 \cdot 3$ **14.** $2 \cdot 3 \cdot 5$
15. $2 \cdot 2 \cdot 7$ **16.** 17 **17.** $2 \cdot 2 \cdot 2 \cdot 2 \cdot 3$ **18.** 37
19. $2 \cdot 2 \cdot 3 \cdot 3 \cdot 5$ **20.** Yes **21.** No **22.** $\frac{3}{8}$ **23.** $\frac{8}{3}$ **24.** $\frac{6}{115}$
25. $\frac{3}{2}$ **26.** 4 **27.** $\frac{13}{20}$ **28.** $\frac{1}{25}$ **29.** 2 **30.** $\frac{12}{35}$ **31.** $\frac{1}{24}$
32. $\frac{2}{5}$ **33.** $\frac{3}{28}$ **34.** 3 **35.** $\frac{9}{10}$ **36.** $\frac{36}{5}$ **37.** 9 **38.** $\frac{14}{5}$
39. $\frac{1}{10}$

NAME ______ COURSE/SECTION ______ DATE ______

SECTION PROBLEMS 1.2

Find all the factors of the following numbers.

1. 24 **2.** 40 **3.** 72 **4.** 84 **5.** 105

6. Give an example of a whole number that has exactly one factor. How many whole numbers have only one factor?

Give the prime factorization of the following numbers.

7. 14 **8.** 25 **9.** 9 **10.** 52

11. 17 **12.** 18 **13.** 24 **14.** 35

15. 8 **16.** 36 **17.** 53 **18.** 900

Reduce to simplest terms.

19. $\frac{3}{9}$ **20.** $\frac{15}{25}$ **21.** $\frac{15}{28}$ **22.** $\frac{6}{27}$ **23.** $\frac{14}{7}$

24. $\frac{15}{20}$ **25.** $\frac{60}{21}$ **26.** $\frac{10}{21}$ **27.** $\frac{24}{36}$ **28.** $\frac{14}{27}$

Perform the indicated operations. (Leave all answers in reduced form.)

29. $\frac{4}{5} \cdot \frac{7}{2}$ **30.** $6 \cdot \frac{1}{3}$ **31.** $1 \cdot \frac{10}{9}$

ANSWERS

1. ______
2. ______
3. ______
4. ______
5. ______
6. ______
7. ______
8. ______
9. ______
10. ______
11. ______
12. ______
13. ______
14. ______
15. ______
16. ______
17. ______
18. ______
19. ______
20. ______
21. ______
22. ______
23. ______
24. ______
25. ______
26. ______
27. ______
28. ______
29. ______
30. ______
31. ______

32. ______
33. ______
34. ______
35. ______
36. ______
37. ______
38. ______
39. ______
40. ______
41. ______
42. ______
43. ______
44. ______
45. ______
46. ______
47. ______
48. ______
49. ______
50. ______
51. ______
52. ______
53. ______
54. ______
55. ______
56. ______
57. ______
58. ______

32. $4 \cdot \frac{1}{3} \cdot 5$

33. $\frac{4}{9} \cdot \frac{3}{2} \cdot \frac{21}{10}$

34. $0 \cdot \frac{2}{3}$

35. $\frac{3}{10} \cdot \frac{2}{5} \cdot \frac{25}{9}$

36. $\frac{3}{8} \cdot \frac{2}{3} \cdot \frac{12}{15}$

37. $\frac{4}{9} \cdot \frac{3}{15} \cdot \frac{1}{8}$

38. $\frac{4}{5} \div \frac{2}{7}$

39. $2 \div \frac{2}{3}$

40. $0 \div \frac{1}{2}$

41. $\frac{10}{3} \div \frac{5}{6}$

42. $\frac{3}{5} \div \frac{7}{4}$

43. $\frac{2}{5} \div 1$

44. $\frac{7}{9} \div 21$

45. $5 \div \frac{10}{3}$

46. $\frac{3}{5} \div \frac{9}{10}$

47. $\frac{4}{7} \div \frac{5}{6}$

48. $\frac{2}{7} \div \frac{3}{14}$

49. $\frac{2}{5} \cdot \frac{10}{3} \cdot \frac{15}{4}$

50. $1 \div \frac{2}{3}$

51. $0 \div \frac{4}{9}$

52. $\frac{1}{3} \cdot \frac{9}{2} \cdot 0$

53. $\frac{3}{25} \div \frac{2}{5}$

54. $\frac{14}{3} \div 1$

55. $\frac{2}{3} \div \frac{1}{4}$

56. $\frac{8}{3} \cdot \frac{15}{4} \cdot \frac{5}{6}$

57. $\frac{2}{9} \cdot \frac{3}{4}$

58. $\frac{4}{9} \div \frac{2}{3}$

1.3

Addition and Subtraction of Fractions

LEARNING OBJECTIVES

After finishing Section 1.3, you should be able to:

- Find the least common denominator of two fractions.
- Add fractions.
- Subtract fractions.

Before we give the rule for adding two fractions, we look at two examples.

Example 1 $\frac{3}{8} + \frac{2}{8} = ?$

Let's use a ruler to help us with this.

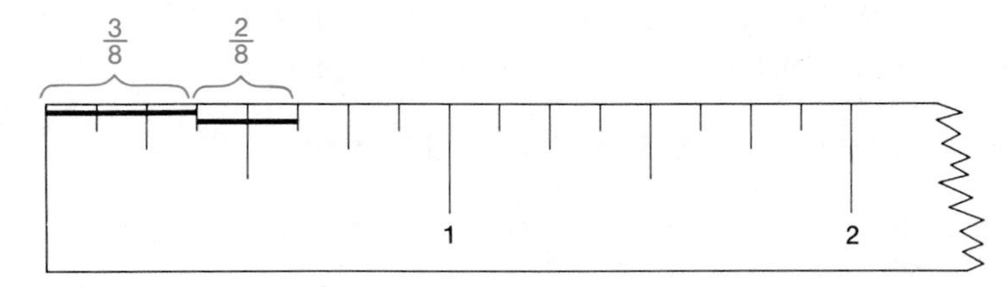

Thus $\frac{3}{8} + \frac{2}{8} = \frac{5}{8}$. To obtain the answer $\frac{5}{8}$, we added the numerators $(2 + 3)$ to get the new numerator 5, and we kept the same denominator (8).

Example 2 $\frac{1}{2} + \frac{3}{8} = ?$

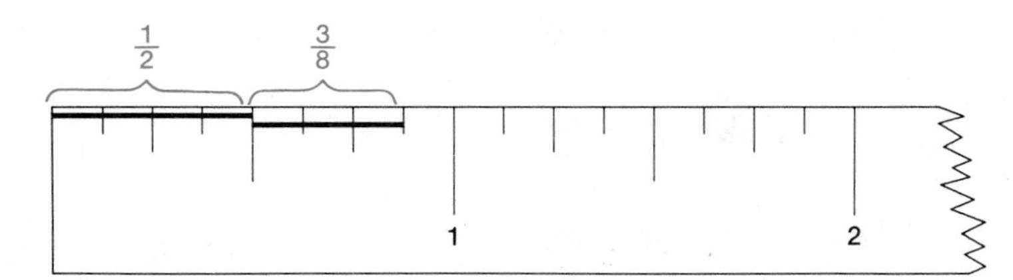

Thus $\frac{1}{2} + \frac{3}{8} = \frac{7}{8}$. What happened here? We didn't add the numerators (1 and 3) to get our new numerator 7; however, we did keep one of the denominators. Looking at the ruler we see that $\frac{1}{2} = \frac{4}{8}$. Thus, we can do the above problem as follows:

$$\frac{1}{2} + \frac{3}{8} = \frac{4}{8} + \frac{3}{8} = \frac{7}{8}$$

Notice here that the fractions we are adding have the same denominator (8) and that to obtain our answer, $\frac{7}{8}$, we add the numerators $(4 + 3)$ and keep the denominator (8).

The rule for addition of fractions follows.

> *RULE 1.3* To add two fractions that have the same denominator, we add the numerators and keep the denominator the same. In symbols,
>
> $$\frac{a}{b} + \frac{c}{b} = \frac{a + c}{b}$$

Add.

1. $\frac{1}{7} + \frac{2}{7}$

2. $\frac{1}{12} + \frac{2}{12} + \frac{1}{12}$

Find the LCD for Exercises 3 and 4.

3. $\frac{1}{3} + \frac{1}{7}$

4. $\frac{1}{2} + \frac{2}{3} + \frac{3}{4}$

Example 3 $\frac{4}{9} + \frac{1}{9} = \frac{4+1}{9} = \frac{5}{9}$

Example 4 $\frac{3}{10} + \frac{1}{10} + \frac{7}{10} = \frac{3+1+7}{10} = \frac{11}{10}$

DO EXERCISES 1 AND 2.

To add two fractions that do not have the same denominator, we must change the fractions to equivalent fractions that do have a common denominator. For this common denominator we will use what is called the least common denominator (LCD).

DEFINITION 1.10 The *least common denominator* (LCD) of two fractions is the smallest whole number that both denominators of the two fractions divide into with a remainder of zero; that is, it is the smallest whole number that both denominators divide into "evenly."

Very often it is possible to see what the LCD is going to be by inspection. Let's look at a few examples.

Example 5 The LCD for the problem $\frac{5}{12} + \frac{1}{4}$ is 12, since 12 is the smallest number that both 12 and 4 divide into evenly.

Example 6 The LCD for the problem $\frac{3}{5} + \frac{1}{3}$ is 15, since 15 is the smallest number that both 5 and 3 divide into evenly.

DO EXERCISES 3 AND 4.

At times the LCD is not so obvious, as, for example, in the problem $\frac{5}{18} + \frac{1}{15}$. Thus we need a method for finding the LCD in those instances where it is not easily recognizable.

RULE 1.4 To find the LCD of two (or more) denominators:

1. We write the prime factorization for each denominator.
2. In the LCD, we include each prime *the greatest* number of times it occurs in any *one of the individual factorizations.*

Example 7 Find the LCD for the problem $\frac{5}{18} + \frac{1}{30}$.

$$18 = 2 \cdot 9 = 2 \cdot 3 \cdot 3$$
$$30 = 6 \cdot 5 = 2 \quad \cdot 3 \cdot 5$$
$$\text{LCD} = 2 \cdot 3 \cdot 3 \cdot 5 = 90$$

In the LCD we included two 3's since twice was the greatest number of times 3 occurred in either factorization. We included one 2 and one 5 since once was the greatest number of times either number occurred in either factorization. We arranged identical primes in the same column so we could read downward to help find the LCD.

Example 8 Find the LCD for the problem $\frac{5}{12} + \frac{1}{10} + \frac{5}{6}$.

$$\begin{aligned} 12 &= 2 \cdot 6 = 2 \cdot 2 \cdot 3 \\ 10 &= 2 \cdot 5 = 2 \quad\quad \cdot 5 \\ 6 &= 2 \cdot 3 = 2 \quad \cdot 3 \\ \text{LCD} &= 2 \cdot 2 \cdot 3 \cdot 5 = 60 \end{aligned}$$

DO EXERCISES 5 AND 6.

Now we look at the problem of changing a fraction to an equivalent fraction having a denominator of our choice. This process is called *building up fractions*. Once again (as in reducing fractions) we will use the fact that $\frac{m \cdot c}{n \cdot c} = \frac{m}{n}$ (where $c \neq 0$); that is, the fact that $\frac{m \cdot c}{n \cdot c}$ and $\frac{m}{n}$ are equivalent fractions.

Example 9 Change $\frac{2}{3}$ to an equivalent fraction having 12 as its denominator.

$$\frac{2}{3} = \frac{?}{12}$$

To change a denominator of 3 to a denominator of 12, we have to multiply the 3 by 4. But if we multiply the denominator by 4, we must also multiply the numerator by 4 to assure that we obtain an equivalent fraction. Thus we have

$$\frac{2}{3} = \frac{2 \cdot 4}{3 \cdot 4} = \frac{8}{12}$$

Example 10 Change $\frac{3}{5}$ to an equivalent fraction having 30 for a denominator.

$$\frac{3}{5} = \frac{?}{30}$$

$$\frac{3}{5} = \frac{3 \cdot 6}{5 \cdot 6} = \frac{18}{30}$$

Example 11 Change 4 to an equivalent fraction having 3 for a denominator.

$$\frac{4}{1} = \frac{?}{3}$$

$$4 = \frac{4}{1} = \frac{4 \cdot 3}{1 \cdot 3} = \frac{12}{3}$$

DO EXERCISES 7 THROUGH 12.

Now we can add fractions that do not have the same denominator.

Example 12 $\frac{1}{4} + \frac{2}{3} = ?$ (The LCD is 12.)

$$\frac{1}{4} + \frac{2}{3} = \frac{?}{12} + \frac{?}{12} = \frac{1 \cdot 3}{4 \cdot 3} + \frac{2 \cdot 4}{3 \cdot 4} = \frac{3}{12} + \frac{8}{12} = \frac{11}{12}$$

Find the LCD for Exercises 5 and 6.

5. $\frac{1}{9} + \frac{3}{10} + \frac{1}{15}$

6. $\frac{1}{4} + \frac{9}{22}$

Fill in the numerators.

7. $\frac{3}{4} = \frac{}{16}$

8. $\frac{7}{3} = \frac{}{15}$

9. $\frac{4}{9} = \frac{}{18}$

10. $\frac{1}{2} = \frac{}{12}$

11. $3 = \frac{}{8}$

12. $\frac{3}{8} = \frac{}{128}$

Perform the indicated operation. (Leave answers in reduced form.)

13. $\frac{4}{5} + \frac{2}{15}$

14. $\frac{2}{3} + \frac{7}{4}$

15. $\frac{1}{2} + \frac{2}{3} + \frac{3}{4}$

16. $\frac{3}{4} + \frac{1}{8} + \frac{5}{12}$

17. $\frac{1}{30} + \frac{3}{28}$

18. $\frac{10}{3} \cdot \frac{4}{5}$

Perform the indicated operation.

19. $\frac{4}{5} - \frac{2}{5}$

20. $\frac{2}{3} - \frac{1}{4}$

21. $\frac{4}{7} - \frac{2}{5}$

22. $\frac{1}{2} \cdot \frac{2}{7}$

23. $\frac{1}{4} \div \frac{2}{3}$

24. $\frac{17}{20} - \frac{3}{4}$

25. $\frac{1}{5} + \frac{2}{3} + \frac{3}{10}$

26. $\frac{7}{8} - 0$

Example 13 $\frac{5}{18} + \frac{7}{30} = ?$ (The LCD is 90.)

$$\frac{5}{18} + \frac{7}{30} = \frac{?}{90} + \frac{?}{90} = \frac{5 \cdot 5}{18 \cdot 5} + \frac{7 \cdot 3}{30 \cdot 3} = \frac{25}{90} + \frac{21}{90} = \frac{46}{90} = \frac{23}{45}$$

Example 14 $3 + \frac{1}{5} + \frac{2}{3} = ?$ (The LCD is 15.)

$$3 + \frac{1}{5} + \frac{2}{3} = \frac{3 \cdot 15}{1 \cdot 15} + \frac{1 \cdot 3}{5 \cdot 3} + \frac{2 \cdot 5}{3 \cdot 5} = \frac{45}{15} + \frac{3}{15} + \frac{10}{15} = \frac{58}{15}$$

DO EXERCISES 13 THROUGH 18.

Subtraction of fractions is similar to addition except that we subtract the numerators instead of adding them.

Example 15 $\frac{2}{3} - \frac{1}{5} = ?$

$$\frac{2}{3} - \frac{1}{5} = \frac{2 \cdot 5}{3 \cdot 5} - \frac{1 \cdot 3}{5 \cdot 3} = \frac{10}{15} - \frac{3}{15} = \frac{10 - 3}{15} = \frac{7}{15}$$

Example 16 $\frac{3}{4} - \frac{2}{7} = ?$

$$\frac{3}{4} - \frac{2}{7} = \frac{21}{28} - \frac{8}{28} = \frac{21 - 8}{28} = \frac{13}{28}$$

In general, we have the following rule.

RULE 1.5 To subtract two fractions that have a common denominator, we subtract the numerators and keep the denominator the same; that is,

$$\frac{a}{b} - \frac{c}{b} = \frac{a - c}{b}$$

To subtract two fractions that have different denominators, we first change both fractions so they have a common denominator and then proceed as above.

DO EXERCISES 19 THROUGH 26.

DO SECTION PROBLEMS 1.3.

ANSWERS TO MARGINAL EXERCISES

1. $\frac{3}{7}$ **2.** $\frac{1}{3}$ **3.** 21 **4.** 12 **5.** 90 **6.** 44 **7.** 12

8. 35 **9.** 8 **10.** 6 **11.** 24 **12.** 48 **13.** $\frac{14}{15}$ **14.** $\frac{29}{12}$

15. $\frac{23}{12}$ **16.** $\frac{31}{24}$ **17.** $\frac{59}{420}$ **18.** $\frac{8}{3}$ **19.** $\frac{2}{5}$ **20.** $\frac{5}{12}$

21. $\frac{6}{35}$ **22.** $\frac{1}{7}$ **23.** $\frac{3}{8}$ **24.** $\frac{1}{10}$ **25.** $\frac{7}{6}$ **26.** $\frac{7}{8}$

NAME   COURSE/SECTION   DATE

SECTION PROBLEMS 1.3

Find the LCD for the following problems. Do not perform the addition.

1. $\frac{2}{5} + \frac{3}{10}$

2. $\frac{2}{3} + \frac{1}{7}$

3. $\frac{1}{18} + \frac{5}{27}$

4. $\frac{4}{15} + \frac{1}{21}$

5. $\frac{1}{5} + \frac{3}{10} + \frac{1}{4}$

6. $\frac{1}{6} + \frac{1}{28} + \frac{3}{10} + \frac{4}{45}$

Perform the indicated operation. (Leave all answers in reduced form.)

7. $\frac{4}{3} + \frac{3}{4}$

8. $\frac{13}{15} + \frac{2}{5}$

9. $\frac{1}{2} + \frac{3}{4} + \frac{7}{8}$

10. $\frac{5}{12} + \frac{1}{15}$

11. $\frac{2}{3} - \frac{4}{9}$

12. $\frac{3}{5} \cdot \frac{2}{9} \cdot \frac{15}{4}$

13. $\frac{7}{10} - \frac{3}{5}$

14. $\frac{3}{4} - \frac{2}{3}$

15. $\frac{7}{21} - \frac{1}{7}$

16. $5 - \frac{3}{4}$

17. $\frac{3}{10} - \frac{1}{5}$

18. $\frac{2}{9} + \frac{3}{4}$

19. $\frac{2}{3} + \frac{5}{7}$

20. $\frac{4}{15} + \frac{1}{21}$

21. $\frac{2}{5} + \frac{3}{10}$

ANSWERS

1. ______
2. ______
3. ______
4. ______
5. ______
6. ______
7. ______
8. ______
9. ______
10. ______
11. ______
12. ______
13. ______
14. ______
15. ______
16. ______
17. ______
18. ______
19. ______
20. ______
21. ______

22. ______________

23. ______________

24. ______________

25. ______________

26. ______________

27. ______________

28. ______________

29. ______________

30. ______________

31. ______________

32. ______________

33. ______________

34. ______________

35. ______________

36. ______________

37. ______________

38. ______________

39. ______________

40. ______________

41. ______________

42. ______________

43. ______________

44. ______________

45. ______________

22. $\frac{1}{18} + \frac{7}{15}$

23. $\frac{4}{5} + \frac{3}{10}$

24. $\frac{7}{12} - \frac{1}{18}$

25. $2 + \frac{1}{6} + \frac{3}{4}$

26. $\frac{1}{18} + \frac{5}{27}$

27. $\frac{7}{10} - \frac{1}{6}$

28. $\frac{1}{6} + \frac{3}{14} + \frac{1}{21}$

29. $\frac{1}{6} + \frac{1}{28} + \frac{3}{10} + \frac{4}{15}$

30. $\frac{4}{3} - \frac{1}{5}$

31. $\frac{6}{7} - \frac{4}{21}$

32. $3 \div \frac{2}{5}$

33. $\frac{3}{8} - \frac{1}{6}$

34. $\frac{3}{4} - \frac{1}{3}$

35. $\frac{2}{9} + \frac{1}{6} + \frac{3}{4}$

36. $\frac{3}{5} \div \frac{3}{10}$

37. $\frac{3}{7} \cdot \frac{21}{9}$

38. $\frac{1}{4} + \frac{5}{6} + \frac{1}{3}$

39. $\frac{13}{10} - \frac{3}{4}$

40. $3 \cdot \frac{2}{9} \cdot \frac{3}{5}$

41. $\frac{7}{10} \div \frac{14}{3}$

42. $\frac{2}{7} \div 3$

43. $\frac{3}{17} + \frac{4}{3}$

44. $\frac{17}{15} - \frac{3}{25}$

45. $2 + \frac{5}{6} + \frac{3}{20}$

1.4

Mixed Numbers

LEARNING OBJECTIVES

After finishing Section 1.4, you should be able to:

- Change mixed numbers to fractions and vice versa.
- Perform operations with mixed numbers.

Now we want to look at addition, subtraction, multiplication, and division with mixed numbers. First we need to define a mixed number. Then we will explore the connection between mixed numbers and fractions.

DEFINITION 1.11 A *mixed number* consists of a whole number and a fraction.

EXAMPLES: $1\frac{2}{3}$, $14\frac{1}{5}$, and $6\frac{2}{3}$

Note that $1\frac{2}{3}$ means $1 + \frac{2}{3}$ and is read "one and two-thirds."

Change to fractions.

1. $2\frac{2}{3}$

2. $7\frac{1}{4}$

3. $8\frac{2}{5}$

4. $12\frac{3}{4}$

Example 1 Change $2\frac{3}{5}$ to a fraction.

(a) *Long Method* (or, explanation of why Short Method works)

$$2\frac{3}{5} = 2 + \frac{3}{5} = \frac{2}{1} + \frac{3}{5} = \frac{10}{5} + \frac{3}{5} = \frac{13}{5}$$

(b) *Short Method*

Multiply the whole number (2) by the denominator (5) of the fraction.

$$2 \cdot 5 = 10$$

Add the result (10) to the numerator (3) of the fraction and place this result over the denominator (5) of the fraction.

$$10 + 3 = 13$$

Thus,

$$2\frac{3}{5} = \frac{13}{5}$$

Example 2 Change $4\frac{1}{6}$ to a fraction.

$$4\frac{1}{6} = \frac{(4 \cdot 6) + 1}{6} = \frac{24 + 1}{6} = \frac{25}{6}$$

The parentheses tell us first to multiply $4 \cdot 6$ and then add 1 to the result.

DO EXERCISES 1 THROUGH 4.

Example 3 Change $\frac{5}{3}$ to a mixed number.

(a) *Long Method*

$$\frac{5}{3} = \frac{3 + 2}{3} = \frac{3}{3} + \frac{2}{3} = 1 + \frac{2}{3} = 1\frac{2}{3}$$

Change to mixed numbers.

5. $\frac{7}{3}$

6. $\frac{15}{4}$

7. $\frac{28}{4}$

8. $\frac{173}{3}$

Perform the indicated operation. (Leave answers in reduced form.)

9. $3\frac{4}{5} \cdot \frac{10}{3}$

10. $1\frac{1}{8} \div 5$

11. $2\frac{2}{3} \cdot 1\frac{1}{2}$

12. $2\frac{3}{4} \div 1\frac{3}{5}$

(b) *Short Method*

Recall that $\frac{5}{3}$ means $5 \div 3$. Therefore, divide the numerator by the denominator:

$$\begin{array}{r} 1 \\ 3\overline{)5} \\ \underline{3} \\ 2 \end{array}$$

The remainder (2) becomes the numerator of the fraction in our mixed number. Thus $\frac{5}{3} = 1\frac{2}{3}$.

Example 4 Change $\frac{235}{4}$ to a mixed number.

$$\begin{array}{r} 58 \\ 4\overline{)235} \\ \underline{20} \\ 35 \\ \underline{32} \\ 3 \end{array}$$

Thus, $\frac{235}{4} = 58\frac{3}{4}$

DO EXERCISES 5 THROUGH 8.

To multiply or divide with mixed numbers, we *must* first change the mixed numbers to fractions.

Example 5 $1\frac{2}{3} \cdot 4\frac{1}{2} = \frac{5}{3} \cdot \frac{9}{2} = \frac{5 \cdot 9}{3 \cdot 2}$

$$= \frac{5 \cdot 3 \cdot 3}{3 \cdot 2} = \frac{3 \cdot 5 \cdot 3}{3 \cdot 2} = \frac{15}{2} \qquad \left(\text{or } 7\frac{1}{2}\right)$$

Example 6 $2\frac{2}{5} \div 3 = \frac{12}{5} \div \frac{3}{1} = \frac{12}{5} \cdot \frac{1}{3}$

$$= \frac{12 \cdot 1}{5 \cdot 3} = \frac{3 \cdot 4 \cdot 1}{3 \cdot 5} = \frac{4}{5}$$

DO EXERCISES 9 THROUGH 12.

To add or subtract with mixed numbers, we *may* change the mixed numbers into fractions. However, this step is *not necessary,* and we may add (subtract) the whole numbers and fractions separately.

Example 7 $1\frac{3}{4} + 2\frac{2}{3} + 7\frac{1}{6} = ?$ We may add the numbers horizontally or in a column. Here, we will use a column.

$$1\frac{3}{4} = 1\frac{9}{12}$$

$$2\frac{2}{3} = 2\frac{8}{12}$$

$$7\frac{1}{6} = 7\frac{2}{12}$$

$$10\frac{9+8+2}{12} = 10\frac{19}{12} = 10 + \frac{19}{12}$$

$$= 10 + 1\frac{7}{12} = 11\frac{7}{12}$$

Example 8 $5\frac{4}{5} - 1\frac{1}{15} = ?$

$$5\frac{4}{5} = 5\frac{12}{15}$$

$$-1\frac{1}{15} = -1\frac{1}{15}$$

$$4\frac{11}{15}$$

Example 9 $12\frac{1}{3} - 7\frac{3}{4} = ?$

$$12\frac{1}{3} = 12\frac{4}{12} = \overset{11}{\cancel{12}}\,1\frac{4}{12} = 11\frac{16}{12}$$

$$-7\frac{3}{4} = -7\frac{9}{12} = -7\frac{9}{12} = -7\frac{9}{12}$$

$$4\frac{7}{12}$$

Here we take 1 from the 12 and add it to the $\frac{4}{12}$, since we cannot subtract $\frac{9}{12}$ from only $\frac{4}{12}$.

DO EXERCISES 13 THROUGH 16.

DO SECTION PROBLEMS 1.4.

Perform the indicated operation. (Leave answers in reduced form.)

13. $4\frac{1}{3} - 2\frac{5}{8}$

14. $\frac{7}{8} + 5\frac{2}{3}$

15. $32\frac{5}{16} - 11\frac{7}{8}$

16. $3\frac{1}{4} + 2\frac{2}{3} + 14\frac{1}{6}$

ANSWERS TO MARGINAL EXERCISES

1. $\frac{8}{3}$ **2.** $\frac{29}{4}$ **3.** $\frac{42}{5}$ **4.** $\frac{51}{4}$ **5.** $2\frac{1}{3}$ **6.** $3\frac{3}{4}$ **7.** 7
8. $57\frac{2}{3}$ **9.** $\frac{38}{3}$ **10.** $\frac{9}{40}$ **11.** 4 **12.** $\frac{55}{32}$ **13.** $\frac{41}{24}$ **14.** $6\frac{13}{24}$
15. $20\frac{7}{16}$ **16.** $20\frac{1}{12}$

NAME COURSE/SECTION DATE

SECTION PROBLEMS 1.4

Change to a fraction.

1. $1\frac{3}{4}$
2. $6\frac{2}{5}$
3. $3\frac{2}{3}$
4. $4\frac{1}{5}$
5. $12\frac{3}{8}$
6. $2\frac{3}{7}$
7. $1\frac{5}{9}$
8. $15\frac{1}{6}$

Change to a mixed number.

9. $\frac{20}{3}$
10. $\frac{15}{2}$
11. $\frac{17}{4}$
12. $\frac{84}{7}$
13. $\frac{42}{9}$
14. $\frac{51}{3}$
15. $\frac{225}{11}$
16. $\frac{75}{4}$

Perform the indicated operation. (Leave all answers in reduced form.)

17. $1\frac{5}{7} \div 2\frac{2}{3}$
18. $4 \div 1\frac{3}{5}$
19. $2\frac{2}{5} \cdot 1\frac{1}{3}$
20. $1\frac{2}{5} \cdot \frac{1}{4} \cdot 3\frac{1}{2}$
21. $10\frac{1}{2} - 2\frac{5}{6}$
22. $17\frac{1}{5} - 2\frac{2}{3}$
23. $4\frac{2}{3} + 3\frac{2}{5}$
24. $1\frac{1}{6} + 4\frac{3}{10}$
25. $10\frac{1}{3} - 4\frac{3}{4}$

ANSWERS

1. ______
2. ______
3. ______
4. ______
5. ______
6. ______
7. ______
8. ______
9. ______
10. ______
11. ______
12. ______
13. ______
14. ______
15. ______
16. ______
17. ______
18. ______
19. ______
20. ______
21. ______
22. ______
23. ______
24. ______
25. ______

26. ______________

27. ______________

28. ______________

29. ______________

30. ______________

31. ______________

32. ______________

33. ______________

34. ______________

35. ______________

36. ______________

37. ______________

38. ______________

39. ______________

40. ______________

41. ______________

42. ______________

43. ______________

26. $4\frac{1}{5} + 3\frac{7}{10} + 1\frac{1}{2}$

27. $17 - 3\frac{3}{4}$

28. $0 \div 1\frac{1}{5}$

29. $14\frac{1}{3} + 16\frac{2}{5}$

30. $3\frac{1}{2} + 7\frac{2}{3} + \frac{7}{12}$

31. $1\frac{4}{5} \cdot 2\frac{2}{3}$

32. $1\frac{1}{5} \cdot \frac{2}{3}$

33. $1\frac{1}{5} \div 3$

34. $4\frac{1}{3} \div 1\frac{1}{6}$

35. $2\frac{3}{8} + 4 + 1\frac{1}{4}$

36. $12\frac{1}{3} - 3\frac{3}{5}$

37. $4\frac{3}{4} - 1\frac{1}{3}$

38. $17\frac{1}{5} + 12\frac{1}{2} + 3\frac{3}{10}$

39. $2\frac{7}{8} \div 3\frac{1}{2}$

40. $1 \div 2\frac{1}{5}$

41. $4\frac{2}{3} \cdot 1\frac{1}{5}$

42. $1\frac{1}{4} \cdot 2\frac{1}{3} \cdot 1\frac{1}{7}$

43. $2\frac{5}{6} + 1\frac{3}{4} + 3\frac{5}{9}$

1.5

Addition, Subtraction, and Rounding of Decimals

In this section we will look at decimals. In order to understand decimal notation, we need to know the names of the place values (or columns) that we use in our number system.

Decimal and Whole Number Place Values (Columns)

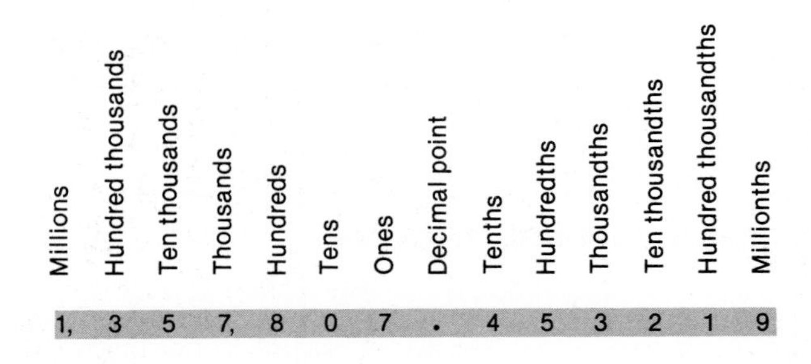

We see that all the names of the decimal columns end in *ths,* while the names of the whole number columns end in *s*. Note that we define the number of decimal places in a number to be the number of decimal columns to the right of the decimal point.

We can use the decimal columns to help change decimals to fractions and to help change certain fractions to decimals.

Example 1 Change the following decimals to fractions: (a) 0.5, (b) 0.04, (c) 2.12, (d) 0.316.

(a) 0.5 (five tenths) $= \frac{5}{10}$

(b) 0.04 (four hundredths) $= \frac{4}{100}$

(c) 2.12 (two and twelve hundredths) $= 2\frac{12}{100}$ $\left(\text{or } \frac{212}{100}\right)$

(d) 0.316 (three hundred sixteen thousandths) $= \frac{316}{1000}$

DO EXERCISES 1 THROUGH 4.

LEARNING OBJECTIVES

After finishing Section 1.5, you should be able to:

- Change decimals to fractions and vice versa.
- Add decimals.
- Subtract decimals.
- Round decimals.

Write as fractions.

1. 0.36

2. 0.0025

3. 253.174

4. 0.10

Write as decimals.

5. $\frac{37}{1000}$

6. $\frac{24}{10}$

7. $\frac{22}{10000}$

8. $\frac{285}{100}$

Perform the indicated operation.

9. $2.3 + 70.4 + 0.78$

10. $4.5 + 61.3 + 15.25 + 0.17$

11. $649.51 - 32.4$

12. Subtract 0.038 from 0.0562.

13. $\left(\frac{2}{3} \cdot \frac{5}{4}\right) + \frac{5}{3}$

Example 2 Change the following fractions to decimals: (a) $\frac{3}{10}$, (b) $\frac{3}{100}$, (c) $\frac{57}{1000}$, (d) $\frac{41}{10,000}$.

(a) $\frac{3}{10}$ (three tenths) $= 0.3$

(b) $\frac{3}{100}$ (three hundredths) $= 0.03$

(c) $\frac{57}{1000}$ (fifty-seven thousandths) $= 0.057$

(d) $\frac{41}{10,000}$ (forty-one ten thousandths) $= 0.0041$

Examples 1 and 2 show that the number of zeros in the denominator of the fraction is exactly the same as the number of decimal places in the decimal number. This is always the case.

DO EXERCISES 5 THROUGH 8.

If we add whole numbers in a column, recall that we must keep the respective columns of the numbers we are adding lined up so that we add ones to ones, tens to tens, hundreds to hundreds, and so on. The same is true when we add or subtract decimal numbers. We must keep the respective columns lined up. The way to do this is to keep the decimal points lined up.

Example 3 $12.2 + 7.64 + 3 = ?$ Note that for a whole number (such as 3) the decimal point is understood to be at the right of the number ($3 = 3.$).

Arrange in a column.	Insert zeros for convenience.	Add.
12.2	12.20	12.20
7.64	7.64	7.64
+ 3.	+ 3.00	+ 3.00
		22.84

Example 4 Subtract 1.756 from 3.48; that is, $3.48 - 1.756 = ?$

Arrange in a column.	We must insert a zero.	Subtract.
3.48	3.480	3.480
− 1.756	− 1.756	− 1.756
		1.724

DO EXERCISES 9 THROUGH 13.

Before we look at the rounding of decimals, we review the rounding of whole numbers.

RULE 1.6 To round a whole number to the nearest ten, locate the number in the column immediately to the right of the tens column—in this case, the ones column.

1. If the number in the ones column is 5 or greater, *add* one to the number in the tens column and place a zero in the ones column.
2. If the number in the ones column is less than 5, *do not change* the number in the tens column and place a zero in the ones column.

Example 5 Round 83 to the nearest ten.

Locate tens column.	Check the number immediately to the right.	Since it is a 3, which is less than 5, we round
↓ 83	83	83 to 80

Example 6 Round 248 to the nearest ten.

Locate tens column.	Check the number immediately to the right.	Since it is an 8, which is greater than 5, we round
↓ 248	248	248 to 250

DO EXERCISE 14.

To round whole numbers to the nearest hundred, thousand, ten thousand, and so forth, we use a rule similar to Rule 1.6.

Example 7 Round 3,763 to the nearest hundred.

Locate hundreds column.	Check to the right.	Since it is a 6, we round
↓ 3,763	3,763	3,763 to 3,800

Example 8 Round 48,359 to the nearest thousand.

Locate thousands column.	Check to the right.	Since it is a 3, we round
↓ 48,359	48,359	48,359 to 48,000

DO EXERCISE 15.

14. Round to the nearest ten.
(a) 33 (b) 426 (c) 1345

15. Round 15,538 to the nearest
(a) ten (b) hundred
(c) thousand
(d) ten thousand

16. Round 2.97541 to the nearest
(a) one (b) tenth
(c) hundredth
(d) thousandth
(e) ten thousandth

To round decimal numbers, we also use a rule similar to Rule 1.6.

Example 9 Round 0.3647 to the nearest hundredth.

Locate hundredths column.	Check to the right.	Since it is a 4, we round
0.3647	0.3647	0.3647 to 0.36

Note

We do not place a zero in the thousandths or ten thousandths column.

Example 10 Round 2.3718 to three decimal places, that is, to the nearest thousandth.

Locate thousandths column.	Check to the right.	Since it is an 8, we round
2.3718	2.3718	2.3718 to 2.372

Example 11 Round 2.951 to the nearest tenth.

Locate tenths column.	Check to the right.	Since it is a 5, we round
2.951	2.951	2.951 to 3.0

DO EXERCISE 16.

DO SECTION PROBLEMS 1.5.

ANSWERS TO MARGINAL EXERCISES

1. $\frac{36}{100}$ **2.** $\frac{25}{10000}$ **3.** $253\frac{174}{1000}$ or $\frac{253174}{1000}$ **4.** $\frac{10}{100}$
5. 0.037 **6.** 2.4 **7.** 0.0022 **8.** 2.85 **9.** 73.48
10. 81.22 **11.** 617.11 **12.** 0.0182 **13.** $\frac{5}{2}$ **14.** (a) 30
(b) 430 (c) 1350 **15.** (a) 15,540 (b) 15,500 (c) 16,000 (d) 20,000
16. (a) 3 (b) 3.0 (c) 2.98 (d) 2.975 (e) 2.9754

NAME | COURSE/SECTION | DATE

SECTION PROBLEMS 1.5

Change to a fraction.

1. 0.43 **2.** 1.005 **3.** 0.1234 **4.** 0.047 **5.** 2.1364

Change to a decimal.

6. $\frac{17}{100}$ **7.** $\frac{31}{10}$ **8.** $\frac{15}{100}$ **9.** $\frac{14}{1000}$ **10.** $\frac{306}{10{,}000}$

Perform the indicated operation.

11. $4.23 + 17.1 + 28$

12. $236.4 + 0.132 + 23.5$

13. $0.145 - 0.0129$

14. $13 - 0.005$

15. $1.419 + 0.003$

16. $4.16 + 25 + 0.139$

17. $4.032 - 1.019$

18. $12.3 - 9.148$

19.
$$\begin{array}{r} 23.4 \\ 1.59 \\ +\ 14.036 \\ \hline \end{array}$$

20.
$$\begin{array}{r} 4.6 \\ 0.19 \\ +12.084 \\ \hline \end{array}$$

21. Subtract 0.098 from 12.

22. Subtract 3.14 from 4.059.

ANSWERS

1. ______
2. ______
3. ______
4. ______
5. ______
6. ______
7. ______
8. ______
9. ______
10. ______
11. ______
12. ______
13. ______
14. ______
15. ______
16. ______
17. ______
18. ______
19. ______
20. ______
21. ______
22. ______

23. (a) ________
(b) ________
(c) ________
(d) ________
24. (a) ________
(b) ________
(c) ________
(d) ________
25. (a) ________
(b) ________
(c) ________
(d) ________
26. (a) ________
(b) ________
(c) ________
(d) ________
27. (a) ________
(b) ________
(c) ________
(d) ________
28. (a) ________
(b) ________
(c) ________
(d) ________
29. (a) ________
(b) ________
(c) ________
(d) ________
30. (a) ________
(b) ________
(c) ________
(d) ________
31. (a) ________
(b) ________
(c) ________
(d) ________
32. (a) ________
(b) ________
(c) ________

Round the following whole numbers to the nearest (a) ten, (b) hundred, (c) thousand, and (d) ten thousand.

23. 6,437 **24.** 77,640 **25.** 15,538 **26.** 8,496

Round the following decimal numbers to the nearest (a) tenth, (b) hundredth, (c) thousandth, and (d) ten thousandth.

27. 0.15471 **28.** 3.4976 **29.** 0.52654 **30.** 0.09183

31. Round 39,827 to the nearest

(a) ten
(b) hundred
(c) thousand
(d) ten thousand

32. Round 0.2538 to the nearest

(a) thousandth
(b) hundredth
(c) tenth

1.6

Multiplication and Division of Decimals

LEARNING OBJECTIVES

After finishing Section 1.6, you should be able to:

- Multiply decimals.
- Divide decimals.

For multiplication problems we often use the following terms:

DEFINITION 1.12 In the multiplication problem $a \cdot b = c$, a and b are called *factors* and c is called the *product.*

Multiplication of decimals is similar to multiplication of whole numbers except that we have to be careful in placing the decimal point in our answer (also called the product). Before we give a rule for this, we look at two examples of decimal multiplication.

1. $1.5 \times 0.03 = \frac{15}{10} \times \frac{3}{100}$

   $= \frac{45}{1000} = 0.045$

2. $10.22 \times 0.005 = \frac{1022}{100} \times \frac{5}{1000}$

   $= \frac{5110}{100{,}000} = 0.05110$

Note

The number of decimal places in the product is the same as the *sum* of the number of decimal places in the factors.

RULE 1.7 The number of decimal places in the product (answer) obtained by multiplying two decimal numbers is the *sum* of the number of decimal places in the factors.

Example 1 $0.325 \times 1.4 = ?$

Multiply.

Add number of decimal places in two factors.

$$
\begin{array}{r}
0.325 \\
\underline{1.4} \\
1300 \\
\underline{3250} \\
4550
\end{array}
\qquad
\begin{array}{r}
\rightarrow 3 \\
\rightarrow \underline{+1} \\
4
\end{array}
$$

Our answer requires 4 decimal places:

$$0.4550$$

Note that $0.4550 = .4550$ and the 0 in front of the .4550 is normally put in for convenience.

Perform the indicated operation.

1. 0.001×0.024

2. 21.459×1.3

3. $0.005 \times 0.12 \times 0.4$

Example 2 $0.0018 \times 0.024 = ?$

Multiply. Add number of decimal places in two factors.

$$\begin{array}{r} 0.0018 \\ \underline{0.024} \\ 72 \\ \underline{36} \\ 432 \end{array} \qquad \begin{array}{r} 4 \\ \underline{+\,3} \\ 7 \end{array}$$

Our answer requires 7 decimal places:

0.0000432

DO EXERCISES 1 THROUGH 3.

For division problems we often use the following terms:

DEFINITION 1.13 In the division problem $a \div b = c$, which can also be written

$$b\overline{)a}^{\,c}$$

a is called the *dividend*, b is called the *divisor*, and the answer, c, is called the *quotient*.

In the division of decimals we must take care in adjusting the decimal point if the divisor is not a whole number.

Example 3 Divide 35.1 by 15.

$15\overline{)35.1}$ Since the divisor is already a whole number, we do not have to adjust the decimal point.

$15\overline{)35.1}$ The decimal point moves straight up into the quotient.

Perform the division.

$$\begin{array}{r} 2.34 \\ 15\overline{)35.10} \\ \underline{30} \\ 5\,1 \\ \underline{4\,5} \\ 60 \\ \underline{60} \\ 0 \end{array}$$

Example 4 Divide 1.5 into 35.1.

$1.5.\overline{)35.1}$ Since the divisor is a decimal number, we have to move the decimal point one place to the right to make it a whole number.

$1.5\overline{)35.1}$ Now we must move the decimal point in the dividend the same number of places to the right.

$$\begin{array}{r} 23.4 \\ 15\overline{)351.0} \end{array}$$

Perform the division.

In Example 4, what has happened with the moving of the decimal points is as follows: 35.1 ÷ 1.5 is the same as 35.1/1.5, which is the same as

$$\frac{35.1 \times 10}{1.5 \times 10} = \frac{351}{15}$$

which is the same as 351 ÷ 15.

DO EXERCISES 4 AND 5.

Example 5 Divide 14 by 0.46. Round the answer to two decimal places, that is, to the nearest hundredth.

$.46\overline{)14.}$ Adjust the decimal point in the divisor and in the dividend.

$.46\overline{)14.00}$ We must insert two zeros in the dividend.

$$\begin{array}{r} 30.434 \\ 46\overline{)1400.000} \\ \underline{138} \\ 20\ 0 \\ \underline{18\ 4} \\ 1\ 60 \\ \underline{1\ 38} \\ 220 \\ \underline{184} \end{array}$$

Perform the division (we carry the division to three decimal places and then round back to two places).

Thus, our answer rounded to two decimal places is 30.43.

DO EXERCISES 6 THROUGH 8.

We have already changed certain fractions to decimal notation. For example,

$$\frac{3}{10} = 0.3 \qquad \frac{15}{1000} = 0.015 \qquad \frac{463}{10{,}000} = 0.0463 \qquad \frac{145}{100} = 1.45$$

We can change any fraction to decimal notation by recalling that the fractional bar means division $\left(\frac{a}{b} \text{ is the same as } a \div b\right)$. We divide the denominator of the fraction into the numerator to obtain one of two types of decimals.

1. A *terminating* (or finite) *decimal*. (That is, when we do the division, we will eventually have a remainder of zero.)
2. A *repeating decimal*. (When we do the division, we will never have a remainder of zero, but a series of numbers that will repeat over and over.)

Perform the indicated operation.

4. Divide 0.32 by 8.

5. Divide 0.32 by 0.8.

Perform the indicated operation.

6. Divide 59.96 by 25.6. Round to two decimal places.

7. Divide 0.02186 by 0.5. Round to the nearest thousandth.

8. Multiply 0.004 by 1.2.

Change the following fractions to decimals.

9. $\frac{5}{8}$

10. $\frac{2}{9}$

11. $\frac{1}{16}$

Example 6 Change $\frac{3}{8}$ to a decimal.

$$\begin{array}{r} 0.375 \\ 8\overline{)3.000} \\ \underline{2\,4} \\ 60 \\ \underline{56} \\ 40 \\ \underline{40} \\ 0 \end{array}$$

Thus $\frac{3}{8} = 0.375$ and is a finite decimal.

Example 7 Change $\frac{8}{11}$ to a decimal.

$$\begin{array}{r} 0.7272 \\ 11\overline{)8.0000} \\ \underline{7\,7} \\ 30 \\ \underline{22} \\ 80 \\ \underline{77} \\ 30 \\ \underline{22} \\ 8 \end{array}$$

Thus $\frac{8}{11} = 0.\overline{72}$ and is a repeating decimal.

Note

1. The bar over the 72 means the series of numbers 72 will repeat over and over again.
2. If we have nothing but zeros to bring down from the dividend, then once we have a remainder that repeats (in this case the remainder of 8) we know that we have a repeating decimal.

DO EXERCISES 9 THROUGH 11.

DO SECTION PROBLEMS 1.6.

ANSWERS TO MARGINAL EXERCISES

1. 0.000024 **2.** 27.8967 **3.** 0.00024 **4.** 0.04 **5.** 0.4 **6.** 2.34 **7.** 0.044 **8.** 0.0048 **9.** 0.625 **10.** $0.\overline{2}$ **11.** 0.0625

NAME ____ COURSE/SECTION ____ DATE ____

SECTION PROBLEMS 1.6

Perform the indicated operation. Round all answers to division problems to 3 decimal places.

1. 72×1.4

2. $(3.65) \cdot (1.9)$

3. 0.05×1.72

4. 6.04×0.003

5. $(1.324) \cdot (0.25)$

6. $(14.3) \cdot (0.02)$

7. $4.5 \times 1.32 \times 0.04$

8. $0.05 \times 1.34 \times 0.14$

9. $1.29 \div 0.08$

10. $17.1\overline{)536}$

11. $3.42 \div 15$

12. $23 \div 0.0034$

13. $6.5\overline{)0.981}$

14. $4 \div 0.025$

15. 7.8×0.012

ANSWERS

1. ____
2. ____
3. ____
4. ____
5. ____
6. ____
7. ____
8. ____
9. ____
10. ____
11. ____
12. ____
13. ____
14. ____
15. ____

16. ____________

17. ____________

18. ____________

19. ____________

20. ____________

21. ____________

22. ____________

23. ____________

24. ____________

25. ____________

26. ____________

27. ____________

28. ____________

29. ____________

30. ____________

16. 0.013×1.7

17. $1.56 \div 1.5$

18. $\frac{0.15}{2.36}$

19. Divide 4.3 into 231.

20. Divide 0.15 by 2.6.

21. $(1.6)(0.008)(20)$

22. $(0.1)(1.4)(0.07)(5.4)$

Change to a decimal.

23. $\frac{5}{12}$

24. $\frac{3}{17}$

25. $\frac{4}{25}$

26. $\frac{7}{8}$

27. $\frac{4}{3}$

28. $\frac{5}{33}$

29. $\frac{7}{4}$

30. $\frac{5}{16}$

1.7

Definition of Operations for Whole Numbers and Exponential Notation

LEARNING OBJECTIVES

After finishing Section 1.7, you should be able to:

- Use the definitions of multiplication, subtraction, division, and exponentiation for whole numbers.

In this section we want to recall how the operations of multiplication, division, and subtraction are defined for whole numbers. It is one of our objectives in the next chapter to extend these definitions to new sets of numbers that we will encounter in algebra. In the definitions in this section a, b, c, m, and n denote arbitrary whole numbers unless stated otherwise.

Consider the addition problem $3 + 3 + 3 + 3 + 3$. This is called *repeated addition* since the five terms we are adding are identical. We can also represent this problem with multiplication, that is, $3 + 3 + 3 + 3 + 3 = 5 \cdot 3$. In fact, repeated addition is how we define multiplication of two whole numbers. The general definition follows.

DEFINITION 1.14 Multiplication is actually repeated addition since $m \cdot n$ is n added to itself until we have m terms; that is,

$$m \cdot n = \underbrace{n + n + n + \cdots + n}_{m \text{ terms}}$$

Example 1

(a) $7 \cdot 2$ is defined as $2 + 2 + 2 + 2 + 2 + 2 + 2$.

(b) $4 \cdot 6$ is defined as $6 + 6 + 6 + 6$.

(c) $7 \cdot 8 = 8 + 8 + 8 + 8 + 8 + 8 + 8$.

DO EXERCISE 1.

1. Write as multiplication.
(a) $3 + 3 + 3 + 3 + 3$
(b) $7 + 7 + 7$

Subtraction is also defined in terms of addition. For example, $8 - 3$ is defined to be the number (in this case 5) that you must add to 3 to get 8; that is, $8 - 3 = c$ if and only if $c + 3 = 8$. The general definition follows.

DEFINITION 1.15 $a - b$ is the number you must add to b to get a; that is, $a - b = c$ if and only if $c + b = a$.

Example 2

(a) $10 - 7$ is 3 since $3 + 7 = 10$.

(b) $8 - 2$ is 6 since $6 + 2 = 8$.

DO EXERCISE 2.

2. Perform the following by using the definition of subtraction.
(a) $14 - 3$ (b) $5 - 1$

Division of whole numbers, while it can be thought of as repeated subtraction, is normally defined in terms of multiplication. For example, $20 \div 5 = \frac{20}{5}$

3. Perform the following by using the definition of division.
(a) $32 \div 8$ (b) $15 \div 3$

4. Find the multiplicative inverse of
(a) 7 (b) $\frac{3}{5}$ (c) 4.1

is the number (in this case 4) we must multiply by 5 to get 20; that is $20 \div 5 = c$ if and only if $c \cdot 5 = 20$. The general definition follows.

DEFINITION 1.16 $a \div b$ is the number we must multiply times b to get a; that is, $a \div b = c$ if and only if $c \cdot b = a$.

Example 3

(a) $20 \div 2$ is 10 since $10 \cdot 2 = 20$.

(b) $\frac{45}{9}$ is 5 since $5 \cdot 9 = 45$.

DO EXERCISE 3.

Another way of looking at division involves the use of multiplicative inverses. In what follows, a and b are arbitrary whole numbers, fractions, or decimals.

DEFINITION 1.17 Two numbers a and b are *multiplicative inverses* (or *reciprocals*) if $a \cdot b = 1$. Every number, other than 0, has a multiplicative inverse.

Example 4

(a) 5 and $\frac{1}{5}$ are multiplicative inverses since $5 \cdot \frac{1}{5} = 1$.

(b) $\frac{2}{3}$ and $\frac{3}{2}$ are multiplicative inverses since $\frac{2}{3} \cdot \frac{3}{2} = 1$.

(c) The multiplicative inverse of 2.3 is $\frac{1}{2.3}$.

RULE 1.8 The multiplicative inverse of a where $a \neq 0$ is $\frac{1}{a}$.

Recall that $a \div b$ is the number we have to multiply times b to get a. The number $a \cdot \frac{1}{b}$ fits this criterion:

$$a \cdot \frac{1}{b} \cdot b = \frac{a}{1} \cdot \frac{1}{b} \cdot \frac{b}{1} = \frac{a \cdot 1 \cdot b}{1 \cdot b \cdot 1} = \frac{a}{1} = a$$

Thus we can define $a \div b$ as a times the multiplicative inverse of b; that is, $a \div b = a \cdot \frac{1}{b}$. Note that this is precisely what we do when we divide with fractions; for example, $\frac{3}{4} \div \frac{2}{3} = \frac{3}{4} \cdot \frac{3}{2}$. In other words, $\frac{3}{4}$ divided by $\frac{2}{3}$ is the same as $\frac{3}{4}$ times the multiplicative inverse of $\frac{2}{3}$.

DO EXERCISE 4.

SOME USEFUL PROPERTIES OF WHOLE NUMBERS, FRACTIONS, AND DECIMALS

1. $a + 0 = a$
2. $a \cdot 1 = a$

3. $a \cdot 0 = 0$
4. $\frac{a}{1} = a \div 1 = a$
5. $\frac{0}{a} = 0 \div a = 0 \qquad (a \neq 0)$
6. We can *never* divide by zero. For example, $3 \div 0 = c$ if and only if $c \cdot 0 = 3$. But $c \cdot 0 = 0$. Thus if division by 0 were allowed, we would have the uncomfortable situation in which $0 = 3$.

We will see these properties again when we study algebra.

Just as multiplication is another way to represent repeated addition, we can also attach another meaning to repeated multiplication (such as $2 \cdot 2 \cdot 2 \cdot 2 \cdot 2$) if we use *exponential notation*. In exponential notation $2 \cdot 2 \cdot 2 \cdot 2 \cdot 2$ becomes 2^5, where 2 is called the *base* and 5 is called the *exponent* (or *power*). The base is the number we are multiplying times itself, and the exponent tells us how many factors of the base we have in our repeated multiplication.

$$2^5 = \underbrace{2 \cdot 2 \cdot 2 \cdot 2 \cdot 2}_{\text{5 factors of 2}}$$

(Base: 2; Exponent: 5)

In general, we have the following definition where a is an arbitrary whole number, fraction, or decimal, and n is a natural number.

DEFINITION 1.18

$$a^n = \underbrace{a \cdot a \cdot a \cdot \ldots \cdot a}_{n \text{ factors of } a}$$

Example 5 (1) Expand.

(a) $5^4 = 5 \cdot 5 \cdot 5 \cdot 5$

(b) $\left(\frac{1}{2}\right)^3 = \left(\frac{1}{2}\right)\left(\frac{1}{2}\right)\left(\frac{1}{2}\right)$

(c) $3^2 \cdot 7^6 = 3 \cdot 3 \cdot 7 \cdot 7 \cdot 7 \cdot 7 \cdot 7 \cdot 7$

(d) $9^1 = 9$ In fact, we have $a^1 = a$ for all a. Thus if a number has no exponent showing, the exponent is understood to be a positive 1.

(2) Write using exponential notation.

(a) $3 \cdot 3 \cdot 3 \cdot 3 \cdot 3 \cdot 3 \cdot 3 = 3^7$

(b) $\left(\frac{2}{5}\right)\left(\frac{2}{5}\right)\left(\frac{2}{5}\right)\left(\frac{2}{5}\right) = \left(\frac{2}{5}\right)^4$

(c) $(1.5)(1.5) + 4 = (1.5)^2 + 4^1$

DO EXERCISES 5 AND 6.

In the next chapter, after we have defined some new sets of numbers, we will expand our definition of exponential notation to include these new sets of numbers. We will also look at various exponential rules. For now, we will look at just two of these rules.

To motivate this first rule, note the following:

1. $2^3 \cdot 2^2 = \underbrace{2 \cdot 2 \cdot 2}_{2^3} \cdot \underbrace{2 \cdot 2}_{2^2} = 2^5$ Note that the sum of the two original exponents $(2 + 3)$ equals the final exponent (5).

2. $5^2 \cdot 5^4 = \underbrace{5 \cdot 5}_{5^2} \cdot \underbrace{5 \cdot 5 \cdot 5 \cdot 5}_{5^4} = 5^6$ Once again, we have $2 + 4 = 6$.

These examples lead to the following exponential rule where a is an arbitrary whole number, fraction, or decimal, and m and n are natural numbers.

5. Expand the following.
(a) 2^6
(b) $(4.3)^3$
(c) 7^1
(d) $2^4 + 3^2$

6. Write the following using exponential notation.
(a) $8 \cdot 8 \cdot 8 \cdot 8 \cdot 8$
(b) $\left(\frac{3}{4}\right)\left(\frac{3}{4}\right)\left(\frac{3}{4}\right)$
(c) $2 \cdot 2 \cdot 2 \cdot 7 \cdot 7 \cdot 7 \cdot 7$

RULE 1.8

$$a^m \cdot a^n = a^{m+n}$$

Note

1. In Rule 1.8 the bases must be the same.
2. The operation between a^m and a^n must be multiplication.
3. All the multiplication is done by adding the exponents. Hence do not multiply the bases.

Example 6 Simplify the following.

(a) $8^3 \cdot 8^5 = 8^{3+5} = 8^8$

(b) $\left(\frac{1}{2}\right)^2 \cdot \left(\frac{1}{2}\right)^4 = \left(\frac{1}{2}\right)^{2+4} = \left(\frac{1}{2}\right)^6$

(c) $2^3 + 2^4$ This cannot be simplified since the operation between 2^3 and 2^4 is not multiplication.

(d) $(2.7)^3(2.7) = (2.7)^3(2.7)^1 = (2.7)^{3+1} = (2.7)^4$

(e) $3^5 \cdot 7^4$ This cannot be simplified since the bases are not the same.

DO EXERCISE 7.

7. Simplify the following.
(a) $4^3 \cdot 4^2$
(b) $\left(\frac{1}{3}\right)^3 \cdot \left(\frac{1}{3}\right) \cdot \left(\frac{1}{3}\right)^2$
(c) $2^3 \cdot 3^2$

To motivate our second exponential rule, note the following:

1. $(5^2)^3 = 5^2 \cdot 5^2 \cdot 5^2 = 5^{2+2+2} = 5^6$
2. $(2^3)^4 = 2^3 \cdot 2^3 \cdot 2^3 \cdot 2^3 = 2^{3+3+3+3} = 2^{12}$

The relationship among the original exponents and the final exponent is $2 \cdot 3 = 6$.

Once again, we have $(3)(4) = 12$.

In general, when we have a base raised to an exponent and then raised to another exponent, we can use the following rule where a is an arbitrary whole number, fraction, or decimal, and m and n are natural numbers.

RULE 1.9

$$(a^m)^n = a^{m \cdot n}$$

Example 7 Simplify the following.

(a) $(2^4)^2 = 2^{4 \cdot 2} = 2^8$

(b) $2^4 \cdot 2^2 = 2^{4+2} = 2^6$

Note the difference between Example (a), where Rule 1.9 is used, and Example (b), where Rule 1.8 is used.

(c) $[(7.4)^2]^5 = (7.4)^{2 \cdot 5} = (7.4)^{10}$

DO EXERCISE 8.

8. Simplify the following.
(a) $(5^3)^4$
(b) $5^3 \cdot 5^4$
(c) $[(0.8)^2]^3$

ANSWERS TO MARGINAL EXERCISES

1. (a) $5 \cdot 3$ (b) $3 \cdot 7$ **2.** (a) $14 - 3 = 11$ since $11 + 3 = 14$ (b) $5 - 1 = 4$ since $4 + 1 = 5$ **3.** (a) $32 \div 8 = 4$ since $4 \cdot 8 = 32$ (b) $15 \div 3 = 5$ since $5 \cdot 3 = 15$ **4.** (a) $\frac{1}{7}$ (b) $\frac{5}{3}$ (c) $\frac{1}{4.1}$
5. (a) $2 \cdot 2 \cdot 2 \cdot 2 \cdot 2 \cdot 2$ (b) $(4.3)(4.3)(4.3)$ (c) 7 (d) $2 \cdot 2 \cdot 2 \cdot 2 + 3 \cdot 3$
6. (a) 8^5 (b) $\left(\frac{3}{4}\right)^3$ (c) $2^3 \cdot 7^4$ **7.** (a) 4^5 (b) $\left(\frac{1}{3}\right)^6$ (c) Cannot simplify
8. (a) 5^{12} (b) 5^7 (c) $(0.8)^6$

NAME

COURSE/SECTION

DATE

SECTION PROBLEMS 1.7

Write as multiplication.

1. $2 + 2 + 2 + 2$

2. $8 + 8 + 8 + 8 + 8 + 8 + 8$

3. $4 + 4$

4. $3 + 3 + 3 + 3$

5. $5 + 5 + 5 + 9 + 9$

6. $7 + 7 + 7 + 4 + 4 + 4 + 4 + 4$

Write as repeated addition.

7. $3 \cdot 5$

8. 4×7

9. $5 \cdot 1$

10. $6 \cdot 0$

11. $2 \cdot 3 + 4 \cdot 8$

12. $1 \cdot 9 + 3 \cdot 4$

Perform the following using the definition of subtraction.

13. $10 - 4$

14. $14 - 6$

15. $12 - 3$

16. $7 - 2$

17. $5 - 0$

18. $9 - 7$

Perform the following using the definition of division.

19. $24 \div 6$

20. $18 \div 3$

21. $\frac{50}{5}$

22. $\frac{20}{5}$

23. $14 \div 1$

24. $0 \div 8$

Find the multiplicative inverse (reciprocal).

25. 3

26. 5

27. $\frac{1}{6}$

28. $\frac{7}{2}$

29. $\frac{2}{3}$

30. $\frac{9}{5}$

31. 1.8

32. $\frac{1}{2.3}$

ANSWERS

1. ______
2. ______
3. ______
4. ______
5. ______
6. ______
7. ______
8. ______
9. ______
10. ______
11. ______
12. ______
13. ______
14. ______
15. ______
16. ______
17. ______
18. ______
19. ______
20. ______
21. ______
22. ______
23. ______
24. ______
25. ______
26. ______
27. ______
28. ______
29. ______
30. ______
31. ______
32. ______

33. ______
34. ______
35. ______
36. ______
37. ______
38. ______
39. ______
40. ______
41. ______
42. ______
43. ______
44. ______
45. ______
46. ______
47. ______
48. ______
49. ______
50. ______
51. ______
52. ______
53. ______
54. ______
55. ______
56. ______

Write using exponential notation.

33. $6 \cdot 6 \cdot 6 \cdot 6$

34. $(2.5)(2.5)(2.5)$

35. $\left(\frac{1}{3}\right)\left(\frac{1}{3}\right)\left(\frac{1}{3}\right)\left(\frac{1}{3}\right)\left(\frac{1}{3}\right)$

36. $1 \cdot 1 \cdot 1 \cdot 1 \cdot 1 \cdot 1 \cdot 1$

37. $2 \cdot 2 \cdot 2 \cdot 4 \cdot 4 \cdot 4 \cdot 4 \cdot 4$

38. $5 \cdot 5 \cdot 4 \cdot 4 \cdot 4 \cdot 2 \cdot 2 \cdot 2 \cdot 2 \cdot 2$

Expand.

39. 4^3

40. $(12)^2$

41. $\left(\frac{1}{5}\right)^4$

42. $(2.7)^2$

43. $2^3 \cdot 3^2$

44. $\left(\frac{3}{5}\right)^4\left(\frac{1}{2}\right)^2$

Simplify the following.

45. $3^4 \cdot 3^2$

46. $5 \cdot 5^4$

47. $\left(\frac{1}{2}\right)^4\left(\frac{1}{2}\right)^6$

48. $(1.4)^2(1.4)(1.4)^5$

49. $7^3 \cdot 6^2$

50. $4^3 + 5^2$

51. $(3^2)^4$

52. $\left[\left(\frac{2}{3}\right)^4\right]^3$

53. $(7^5)^2$

54. $7^5 \cdot 7^2$

55. $6^3 \cdot 6^5$

56. $(6^5)^3$

1.8

Properties of the Numbers of Arithmetic

LEARNING OBJECTIVES

After finishing Section 1.8, you should be able to:

- Identify the commutative, associative, and distributive laws.

The sets of whole numbers, fractions, and decimals satisfy certain laws and properties that we want to review in this section. In Chapter 2 we will extend these laws and properties to other sets of numbers that are needed for algebra.

As we are well aware, $2 + 4$ is the same as $4 + 2$, and $7 + \frac{1}{2}$ is the same as $\frac{1}{2} + 7$. These are examples of a general rule that is known as the commutative law for addition.

COMMUTATIVE LAW FOR ADDITION

$$m + n = n + m$$

where m and n are two arbitrary whole numbers, fractions, or decimals.

We have a similar situation in multiplication. For example, we know that $2 \cdot 3 = 3 \cdot 2$ and $\frac{3}{4} \cdot 12 = 12 \cdot \frac{3}{4}$. Once again, these are examples of a general rule that is known as the commutative law for multiplication.

COMMUTATIVE LAW FOR MULTIPLICATION

$$m \cdot n = n \cdot m$$

where m and n are two arbitrary whole numbers, fractions, or decimals. We could also write the commutative law as $mn = nm$, since mn means "m times n."

The commutative laws tell us that the *order* in which we add or multiply two numbers is immaterial. For example, if we want to add the two numbers 4 and 7, we may write this as $4 + 7$ or as $7 + 4$.

DO EXERCISES 1 THROUGH 3.

Which law is being demonstrated?

1. $2 \cdot 8 = 8 \cdot 2$

2. $8 + 10 = 10 + 8$

3. $3 + \frac{4}{5} = \frac{4}{5} + 3$

Consider the addition problem $2 + 5 + 4$, and the following two ways of performing the addition:

1. $(2 + 5) + 4 = 7 + 4 = 11$
2. $2 + (5 + 4) = 2 + 9 = 11$

We obtain the same answer in each case. Hence the way in which we *pair* the numbers to do the addition is immaterial. This is an example of the associative law for addition.

Which law is being demonstrated?

4. $\left(7 + \frac{1}{2}\right) + 5 = 7 + \left(\frac{1}{2} + 5\right)$

5. $(2 \cdot 4) \cdot \frac{3}{2} = 2 \cdot \left(4 \cdot \frac{3}{2}\right)$

6. $7 \cdot 13 = 13 \cdot 7$

7. $(4 + 5) + 6 = 4 + (6 + 5)$

ASSOCIATIVE LAW FOR ADDITION

$$(m + n) + r = m + (n + r)$$

where m, n, and r are three arbitrary whole numbers, fractions, or decimals.

As we might expect, there is a similar result for multiplication.

Example 1 Show that $(2 \cdot 3) \cdot \frac{1}{2}$ is the same number as $2 \cdot \left(3 \cdot \frac{1}{2}\right)$.

$(2 \cdot 3) \cdot \frac{1}{2}$	$2 \cdot \left(3 \cdot \frac{1}{2}\right)$
$(6) \cdot \frac{1}{2}$	$2 \cdot \left(\frac{3}{2}\right)$
$\frac{6}{2}$	$\frac{6}{2}$
3	3

The general rule is called the associative law for multiplication.

ASSOCIATIVE LAW FOR MULTIPLICATION

$$(m \cdot n) \cdot r = m \cdot (n \cdot r)$$

where m, n, and r are arbitrary whole numbers, fractions, or decimals. We could also write the associative law as $(mn)r = m(nr)$.

DO EXERCISES 4 THROUGH 7.

Example 2 The associative law for addition enables us to write the problem $2 + 5 + 4$ without using parentheses. Similarly, we can write the multiplication problem $2 \cdot 5 \cdot 4$ without using parentheses.

Example 3 To solve the problem $\frac{1}{3} + 4 + \frac{2}{3} + 2 = ?$ it is convenient to rearrange the terms and obtain the following.

$$\begin{aligned} \frac{1}{3} + 4 + \frac{2}{3} + 2 &= \frac{1}{3} + \frac{2}{3} + 4 + 2 \\ &= \frac{3}{3} + 4 + 2 \\ &= 1 + 4 + 2 \\ &= 5 + 2 \\ &= 7 \end{aligned}$$

The associative and commutative laws allow us to do the problem as above.

While the operations of addition and multiplication do satisfy the commutative and associative laws, the operations of division and subtraction do not. Division is not commutative since, for example, $20 \div 4$ (which equals 5) is not the same as $4 \div 20$ $\left(\text{which equals } \frac{1}{5}\right)$. Nor is division associative. Subtraction is not associative since, for example, $(10 - 5) - 2$ (which equals 3) is not the same as $10 - (5 - 2)$ (which equals 7). Nor is subtraction commutative.

DO EXERCISES 8 AND 9.

All the previous laws involved either the operation of addition or the operation of multiplication, but never both together. The distributive law does involve both. Before we state the distributive law, we look at two examples.

Example 4 Is $3 \cdot (4 + 2)$ the same as $3 \cdot 4 + 3 \cdot 2$?

$3 \cdot (4 + 2)$	$3 \cdot 4 + 3 \cdot 2$
$3(6)$	$12 + 6$
18	18

On the right-hand side we multiplied before we added.

Thus, they are the same.

Example 5 Show that $\frac{1}{2} \cdot \left(4 + \frac{1}{3}\right)$ is the same number as $\frac{1}{2} \cdot 4 + \frac{1}{2} \cdot \frac{1}{3}$.

$\frac{1}{2} \cdot \left(4 + \frac{1}{3}\right)$	$\frac{1}{2} \cdot 4 + \frac{1}{2} \cdot \frac{1}{3}$
$\frac{1}{2} \cdot \left(\frac{13}{3}\right)$	$2 + \frac{1}{6}$
$\frac{13}{6}$	$\frac{13}{6}$

Examples 4 and 5 are examples of a general rule known as the distributive law of multiplication over addition.

DISTRIBUTIVE LAW OF MULTIPLICATION OVER ADDITION

$$m \cdot (n + r) = m \cdot n + m \cdot r$$

where m, n, and r are arbitrary whole numbers, fractions, or decimals. We could also write the distributive law as $m(n + r) = mn + mr$.

DO EXERCISES 10 THROUGH 13.

8. Give an example to show that the operation of subtraction is not commutative.

9. Give an example to show that the operation of division is not associative.

Which law is being demonstrated?

10. $4 + 7 = 7 + 4$

11. $4 \cdot (3 + 2) = 4 \cdot 3 + 4 \cdot 2$

12. $\frac{2}{3} \cdot (3 \cdot 6) = \left(\frac{2}{3} \cdot 3\right) \cdot 6$

13. $\frac{1}{3} \cdot (2 + 5) = \frac{1}{3} \cdot 2 + \frac{1}{3} \cdot 5$

14. Fill in the blanks:
$3(\_\_\_ + 4) = 3 \cdot 2 + \_\_\_ \cdot 4$

15. Fill in the blanks:
$\_\_\_(\_\_\_ + 3) = 2 \cdot 4 + \_\_\_ \cdot 3$

Insert the symbol $>$, $<$, or $=$ between each pair of numbers.

16. 6, 7

17. $\frac{1}{3}$, 5

18. $\frac{4}{4}$, 1

19. $\frac{11}{13}$, $\frac{12}{13}$

20. $1\frac{1}{5}$, $\frac{6}{5}$

Example 6 Use the distributive law to help fill in the blanks.

(a) $2 \cdot (\_\_\_ + 3.1) = 2 \cdot 4 + \_\_\_ \cdot 3.1$

Answer: $2 \cdot (\underline{4} + 3.1) = 2 \cdot 4 + \underline{2} \cdot 3.1$

(b) $\_\_\_ \cdot (5 + \_\_\_) = 7 \cdot 5 + 7 \cdot \frac{2}{3}$

Answer: $\underline{7} \cdot \left(5 + \underline{\frac{2}{3}}\right) = 7 \cdot 5 + 7 \cdot \frac{2}{3}$

The distributive law will be very important for us in algebra. In fact we will see the commutative, associative, and distributive laws again in Chapter 2 after we have defined some new sets of numbers.

DO EXERCISES 14 AND 15.

Next we want to look at the *order property* of whole numbers, decimals, and fractions. First, though, it will be useful to introduce some new notation and terminology.

The whole numbers can be pictorially represented by using a number line.

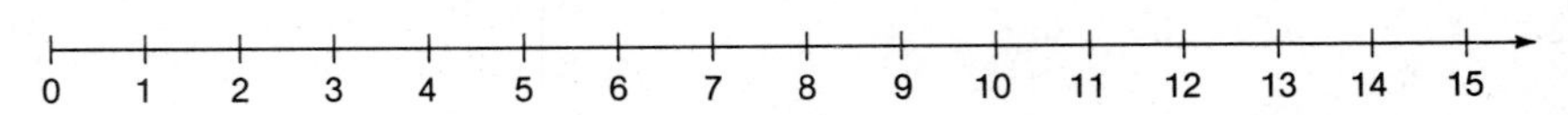

Now let us put a few fractions and mixed numbers on the number line.

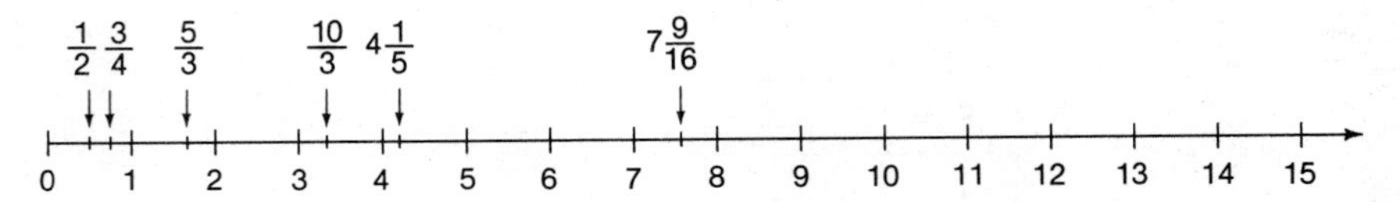

The following notation will be helpful to us. The symbol $<$ means "less than," so $3 < 14$ may be read "three is less than fourteen." The symbol $>$ means "greater than" so $5 > 0$ may be read "five is greater than 0." Intuitively, we know that 5 is greater than 3 and $\frac{2}{3}$ is greater than $\frac{1}{3}$. However, we need a mathematical definition of greater than (larger than).

> *DEFINITION 1.19* Given two whole numbers, fractions, or decimals m and n, $m > n$ if m is to the right of n on a number line.

Example 7 With the help of the following number line, insert the symbol $<$, $>$, or $=$ between the following pairs of numbers: (a) 5, 6; (b) $\frac{3}{2}$, $\frac{1}{3}$; (c) $\frac{4}{4}$, 1; (d) $\frac{3}{2}$, $\frac{4}{3}$.

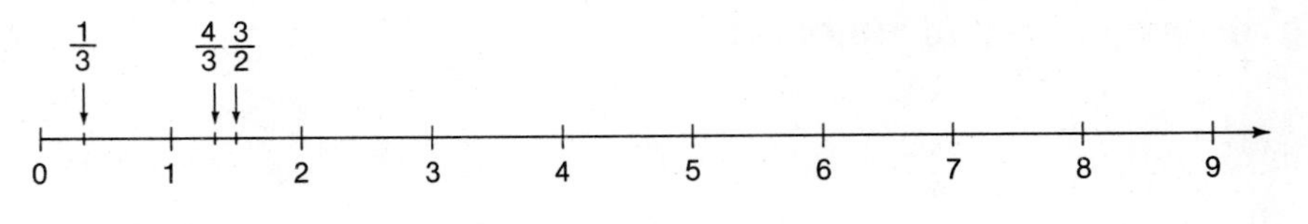

(a) $5 < 6$ (b) $\frac{3}{2} > \frac{1}{3}$ (c) $\frac{4}{4} = 1$ (d) $\frac{3}{2} > \frac{4}{3}$

DO EXERCISES 16 THROUGH 20.

The order property of whole numbers, fractions, and decimals says we can compare the size of any two whole numbers, fractions, or decimals.

ORDER PROPERTY Given two whole numbers, fractions, or decimals m and n, exactly one of the following is true.

1. $m > n$
2. $m < n$
3. $m = n$

Example 8 Which is larger, $\frac{4}{5}$ or $\frac{7}{9}$?

To compare these fractions, it helps to change them to equivalent fractions with a common denominator.

$$\frac{4}{5} = \frac{36}{45} \quad \text{and} \quad \frac{7}{9} = \frac{35}{45}$$

Thus, $\frac{4}{5} > \frac{7}{9}$ since $\frac{36}{45} > \frac{35}{45}$.

DO EXERCISES 21 AND 22.

DO SECTION PROBLEMS 1.8.

21. Which is larger, $\frac{3}{4}$ or $\frac{5}{7}$?

22. Which is larger, $\frac{14}{21}$ or $\frac{21}{28}$?

ANSWERS TO MARGINAL EXERCISES

1. Commutative (multiplication) **2.** Commutative (addition) **3.** Commutative (addition) **4.** Associative (addition) **5.** Associative (multiplication) **6.** Commutative (multiplication) **7.** Associative and Commutative (addition) **8.** Answers may vary. **9.** Answers may vary. **10.** Commutative (addition) **11.** Distributive **12.** Associative (multiplication) **13.** Distributive **14.** $3(\underline{2} + 4) = 3 \cdot 2 + \underline{3} \cdot 4$ **15.** $\underline{2}(\underline{4} + 3) = 2 \cdot 4 + \underline{2} \cdot 3$ **16.** $<$ **17.** $<$ **18.** $=$ **19.** $<$ **20.** $=$ **21.** $\frac{3}{4}$ is larger. **22.** $\frac{21}{28}$ is larger.

NAME COURSE/SECTION DATE

ANSWERS

1. ______
2. ______
3. ______
4. ______
5. ______
6. ______
7. ______
8. ______
9. ______
10. ______
11. ______
12. ______
13. ______

SECTION PROBLEMS 1.8

Which law is being demonstrated?

1. $3 \cdot (4 + 5) = 3 \cdot 4 + 3 \cdot 5$

2. $3 \cdot 5 + 3 \cdot 6 = 3 \cdot (5 + 6)$

3. $2 \cdot 7 = 7 \cdot 2$

4. $\frac{2}{5} + 3 = 3 + \frac{2}{5}$

5. $3 + (4 + 5) = (3 + 4) + 5$

6. $\frac{7}{4} \cdot \frac{2}{3} = \frac{2}{3} \cdot \frac{7}{4}$

7. $2 \cdot (7 + 3) = 2 \cdot 7 + 2 \cdot 3$

8. $(3 \cdot 4) \cdot \frac{1}{2} = 3 \cdot \left(4 \cdot \frac{1}{2}\right)$

9. $\frac{3}{5} \cdot \frac{7}{4} = \frac{7}{4} \cdot \frac{3}{5}$

10. $\frac{1}{5} \cdot (4 + 2) = \frac{1}{5} \cdot 4 + \frac{1}{5} \cdot 2$

11. Give an example to show that division is not commutative.

12. Given an example to show that division is not associative.

13. Give an example to show that subtraction is not commutative.

14. ________
15. ________
16. ________
17. ________
18. ________
19. ________
20. ________
21. ________
22. ________
23. ________
24. ________
25. ________
26. ________
27. ________
28. ________
29. ________

14. Give an example to show that subtraction is not associative.

Fill in the blanks.

15. $\underline{\qquad} \cdot (3 + 4) = 2 \cdot 3 + 2 \cdot \underline{\qquad}$

16. $3 \cdot (\underline{\qquad} + \underline{\qquad}) = 3 \cdot \frac{1}{2} + 3 \cdot 5$

17. $4 \cdot (2 + \underline{\qquad} + 5) = \underline{\qquad} \cdot 2 + 4 \cdot 3 + \underline{\qquad} \cdot 5$

18. $\underline{\qquad} \cdot (5 + \underline{\qquad}) = 3 \cdot 5 + 3 \cdot 6$

Insert $<$, $>$, or $=$ between each pair of numbers.

19. 4, 7

20. 3, 0

21. $\frac{12}{5}, \frac{11}{5}$

22. $\frac{22}{5}, 4\frac{2}{5}$

23. $\frac{5}{7}, \frac{3}{4}$

24. $\frac{3}{4}, \frac{2}{3}$

25. $1\frac{1}{9}, \frac{10}{9}$

26. $1, \frac{5}{5}$

27. $\frac{3}{10}, \frac{5}{13}$

28. Which law allows us to write the problem $2 + 5 + 7$ without using parentheses?

29. Why do we need parentheses in the problem $(20 - 12) - 5$?

NAME COURSE/SECTION DATE

CHAPTER 1 REVIEW PROBLEMS

SECTION 1.1

1. Represent $\frac{2}{3}$ pictorially.

2. Represent $\frac{7}{4}$ pictorially.

3. In the fraction $\frac{7}{4}$, which is the numerator and which is the denominator?

4. What is the difference between the set of whole numbers and the set of fractions?

5. Define a fraction.

SECTION 1.2

Find the prime factorization.

6. 21

7. 100

Reduce.

8. $\frac{6}{21}$

9. $\frac{48}{6}$

Perform the indicated operation.

10. $4 \div \frac{2}{3}$

11. $\frac{3}{4} \cdot \frac{2}{9} \cdot \frac{15}{17}$

12. $\frac{3}{7} \div 9$

13. $0 \cdot \frac{3}{5}$

ANSWERS

1. See figure.
2. See figure.
3. ______
4. ______
5. ______
6. ______
7. ______
8. ______
9. ______
10. ______
11. ______
12. ______
13. ______

14. ____________

15. ____________

16. ____________

17. ____________

18. ____________

19. ____________

20. ____________

21. ____________

22. ____________

23. ____________

24. ____________

25. ____________

26. ____________

27. ____________

28. ____________

29. ____________

SECTION 1.3

Find the LCD.

14. $\frac{1}{6} + \frac{5}{12}$

15. $\frac{1}{15} + \frac{1}{18} + \frac{1}{3}$

Perform the indicated operation.

16. $\frac{3}{4} + \frac{2}{3}$

17. $\frac{1}{4} + \frac{3}{5}$

18. $\frac{13}{15} - \frac{1}{3}$

19. $\frac{1}{2} + \frac{3}{4} + \frac{7}{8}$

SECTION 1.4

Change to a mixed number.

20. $\frac{19}{5}$

21. $\frac{132}{6}$

Change to a fraction.

22. $1\frac{1}{4}$

23. $12\frac{2}{3}$

Perform the indicated operation.

24. $7\frac{1}{5} - 4\frac{2}{3}$

25. $1\frac{1}{2} \div \frac{5}{6}$

26. $2\frac{1}{5} \cdot 2\frac{1}{7}$

27. $14\frac{1}{3} + 12\frac{1}{2} + 5\frac{1}{6}$

SECTION 1.5

Write as a fraction.

28. 0.014

29. 1.36

NAME ______ COURSE/SECTION ______ DATE ______

ANSWERS

30. ______
31. ______
32. ______
33. ______
34. ______
35. ______
36. ______
37. ______
38. ______
39. ______
40. ______
41. ______
42. ______
43. ______

Write as a decimal.

30. $\frac{6}{1000}$

31. $\frac{1421}{100}$

Perform the indicated operation.

32. $11.106 + 13 + 0.14 + 2.8$

33. $27.9 - 18.032$

34. $16 - 1.009$

35. $0.142 + 3.02 + 0.0005$

36. Round 48,735 to the nearest
(a) ten
(b) hundred
(c) thousand
(d) ten thousand

37. Round 0.1972 to the nearest
(a) thousandth
(b) hundredth
(c) tenth

SECTION 1.6

Perform the indicated operation. (Round all answers in division problems to 3 decimal places.)

38. 0.024×1.15

39. $2.56 \div 1.6$

40. $0.02 \times 1.5 \times 0.003$

41. $0.0141 \div 0.003$

Change to a decimal. (Round to the nearest thousandth.)

42. $\frac{3}{8}$

43. $\frac{5}{12}$

44. ____________

45. ____________

46. ____________

47. ____________

48. ____________

49. ____________

50. ____________

51. ____________

52. ____________

53. ____________

54. ____________

55. ____________

56. ____________

57. ____________

58. ____________

59. ____________

SECTION 1.7

Write as multiplication.

44. $4 + 4$

45. $8 + 8 + 8 + 8 + 8$

46. Calculate $15 - 4$ using the definition of subtraction for whole numbers.

47. Calculate $24 \div 8$ using the definition of division for whole numbers

Find the multiplicative inverse.

48. 12

49. $\frac{9}{2}$

Write using exponential notation.

50. $9 \cdot 9 \cdot 9 \cdot 9$

51. $5 \cdot 5 \cdot 6 \cdot 6 \cdot 6 \cdot 6 \cdot 6$

Simplify.

52. $9^2 \cdot 9^5$

53. $(2^3)^6$

54. $3^4 \cdot 3^2 \cdot 3$

SECTION 1.8

What law is being demonstrated?

55. $2 \cdot (3 + 7) = 2 \cdot 3 + 2 \cdot 7$

56. $4 + \frac{1}{3} = \frac{1}{3} + 4$

57. $2 \cdot \left(\frac{2}{3} \cdot 5\right) = \left(2 \cdot \frac{2}{3}\right) \cdot 5$

Insert $<$, $>$, or $=$ between each pair of numbers.

58. $\frac{2}{3}, \frac{4}{5}$

59. $6, 2\frac{1}{5}$

NAME ________ COURSE/SECTION ________ DATE ________

CHAPTER 1 TEST

Perform the indicated operation. (Leave all answers in reduced form.)

1. $\frac{3}{4} \div \frac{7}{8}$

2. $\frac{3}{10} + \frac{2}{15}$

3. $2\frac{2}{3} + 5\frac{4}{5}$

4. $\frac{2}{5} \cdot \frac{15}{8} \cdot \frac{1}{6}$

5. $10\frac{1}{3} - 2\frac{3}{4}$

6. $\frac{13}{18} - \frac{5}{12}$

7. $5 \div \frac{2}{3}$

8. $\left(3\frac{1}{2}\right)\left(2\frac{3}{7}\right)$

9. $15.5 - 1.026$

10. 1.4×0.05

11. $1.76 + 0.5 + 12 + 3.159$

12. $0.384 \div 1.6$

13. Find the LCD for the problem
$\frac{1}{6} + \frac{3}{20} + \frac{1}{45}$

ANSWERS

1. ________
2. ________
3. ________
4. ________
5. ________
6. ________
7. ________
8. ________
9. ________
10. ________
11. ________
12. ________
13. ________

14. (a) ________

(b) ________

15. See figure.

16. (a) ________

(b) ________

(c) ________

(d) ________

17. (a) ________

(b) ________

(c) ________

18. ________

19. ________

20. ________

14. Simplify.

(a) $(3^3)^4$ (b) $2^5 \cdot 2^2$

15. Represent $\frac{5}{6}$ pictorially.

16. Round 12.4815 to the nearest

(a) thousandth (b) hundredth (c) tenth (d) one

17. Insert $<$, $>$, or $=$ between each pair of numbers.

(a) 1.1, 1.09 (b) $2\frac{1}{3}, \frac{7}{3}$ (c) $\frac{5}{7}, \frac{2}{3}$

Which law is being demonstrated in each of Problems 18–20?

18. $2 \cdot 6 = 6 \cdot 2$

19. $3 + (4 + 6) = (3 + 4) + 6$

20. $4\left(\frac{2}{3} + 1\right) = 4 \cdot \frac{2}{3} + 4 \cdot 1$

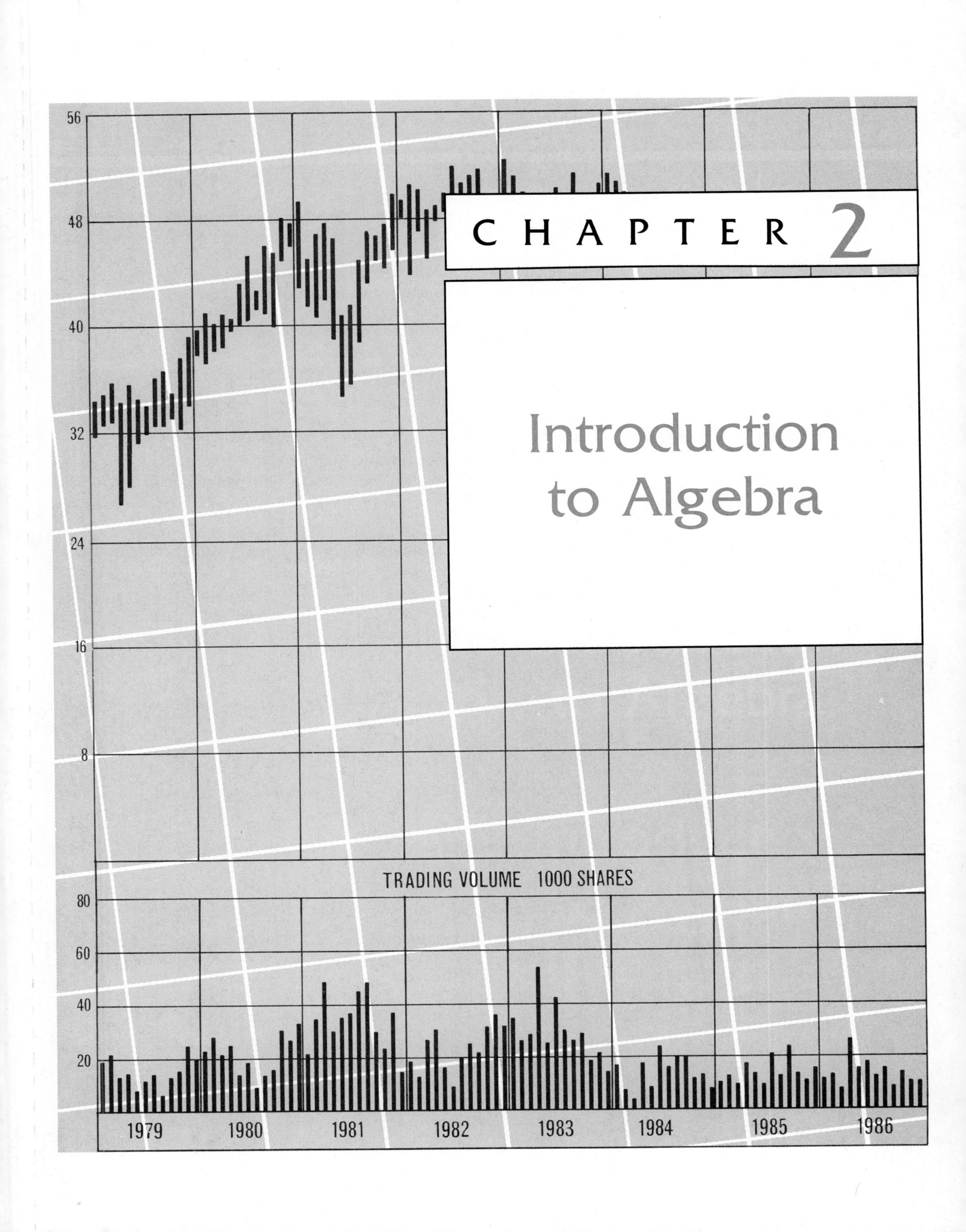

CHAPTER 2

Introduction to Algebra

2.1

Introduction

Algebra can be described as a generalization or an extension of arithmetic. This allows us to use algebra to solve problems that we cannot solve with just a knowledge of arithmetic.

In arithmetic, we normally work just with numbers. However, in algebra we also use letters. These letters represent unknown numbers and are called *variables*.

There are four common uses of variables in algebra.

1. Variables can represent an unknown numerical value.

   EXAMPLE: Consider the phrase

   three times a number plus four

   If we let the variable x stand for the unknown number, then the word phrase may be translated into the mathematical phrase

   $$3 \cdot x + 4$$

2. Variables can stand for an unknown number that possesses certain characteristics.

   EXAMPLE: Let m be any whole number that divides into 60 evenly (that is, m is a factor of 60). In this case m could be 1, 2, 3, 4, 5, 6, 10, 12, 15, 30, or 60, since these are all factors of 60.

3. Variables can be used to list a set of numbers.

   EXAMPLE: If we wanted to list the set of all whole numbers larger than 20

   $$\{21, 22, 23, 24, 25, \ldots\}$$

   we might also write all whole numbers y where

   $$y > 20$$

   Note

   The symbol { } is called roster notation and is one method used to indicate a set.

4. Variables can represent unknown quantities in a formula.

   EXAMPLE: The formula for the area of a triangle is given by

   $$A = \frac{b \cdot h}{2}$$

   where A = area, b = base, and h = height

Since algebra is an extension of arithmetic, the laws, properties, and operations of arithmetic that were discussed in Chapter 1 will be extended to the numbers and variables that we encounter in algebra. In fact, a large part of what we will learn in algebra is how to combine variables and numbers according to the rules of arithmetic as well as those of algebra.

2.2

Addition of Integers

In arithmetic we used only positive numbers and zero in our calculations. These numbers included the set of whole numbers {0, 1, 2, 3, 4, . . .}, decimals, and fractions, all of which we looked at in Chapter 1. Now we want to expand the number system to include negative numbers.

Recall the number line.

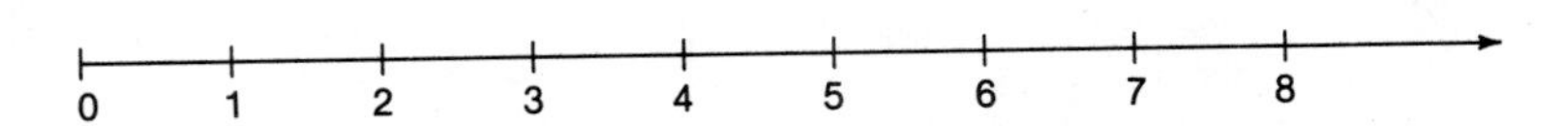

We extend the number line to the left and include some negative numbers.

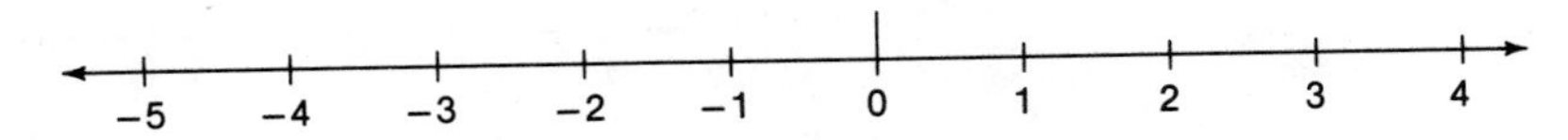

If you are wondering why we need negative numbers, note the following examples of situations in which negative numbers could be useful.

Example 1 Temperature. It is convenient to be able to write 20° above zero as +20° and 20° below zero as −20°.

Example 2 Business. A profit of \$600 may be written as +\$600 while a loss of \$600 may be written as −\$600.

Example 3 Distance. If 0 represents sea level, then +400 would represent 400 feet above sea level and −200 would represent 200 feet below sea level.

■ **DO EXERCISE 1.**

DEFINITION 2.1 The set of *integers* is the set

$$\{\ldots, -3, -2, -1, 0, 1, 2, 3, \ldots\}$$

Notice that every whole number is an integer but not every integer is a whole number.

The set of integers satisfies the commutative, associative, and distributive laws mentioned in Section 1.8.

■ **DO EXERCISE 2.**

The first operation we will look at involving the set of integers is addition. If we want to physically demonstrate the addition problem 2 + 3, we can use a number line.

LEARNING OBJECTIVES

After finishing Section 2.2, you should be able to:

- Add integers.
- Calculate the absolute value of integers.

1. Give an example of another situation where negative numbers might be useful.

2. Identify the following as a whole number, natural number, integer, or more than one of the above.

(a) 5 (b) −7 (c) 0
(d) 236 (e) −1

Demonstrate the following on a number line.

3. $2 + (-3)$

4. $(-4) + 2$

Example 4 Demonstrate $2 + 3$ using a number line.

(a) Draw a number line and start at 0 (the origin).

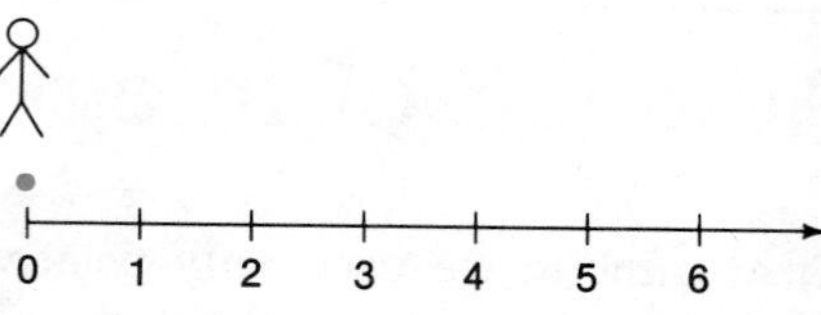

(b) The 2 indicates that we take two steps forward (or to the right).

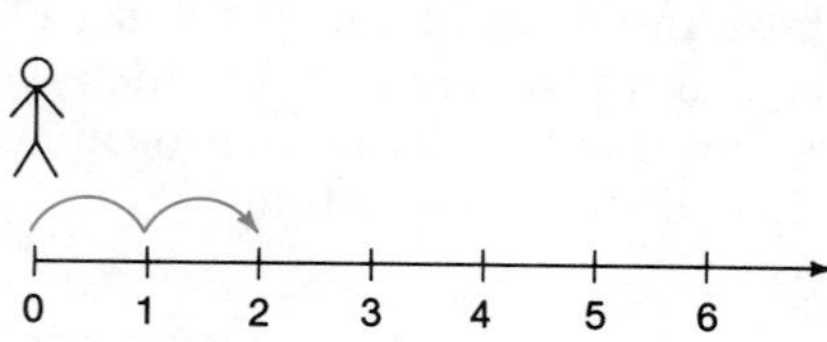

(c) The 3 indicates that we go forward three more steps.

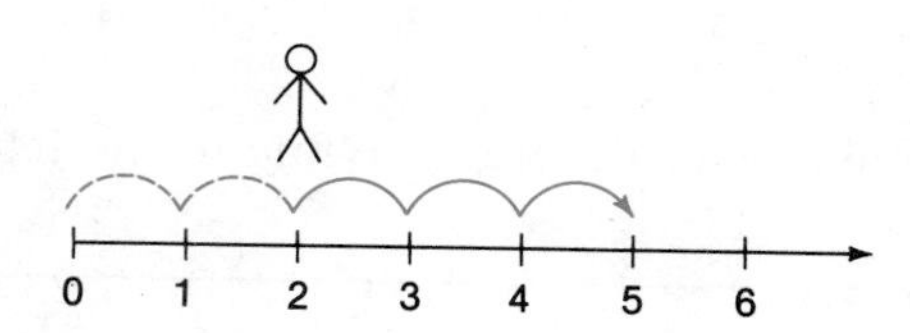

(d) We end up at 5, and thus, $2 + 3 = 5$.

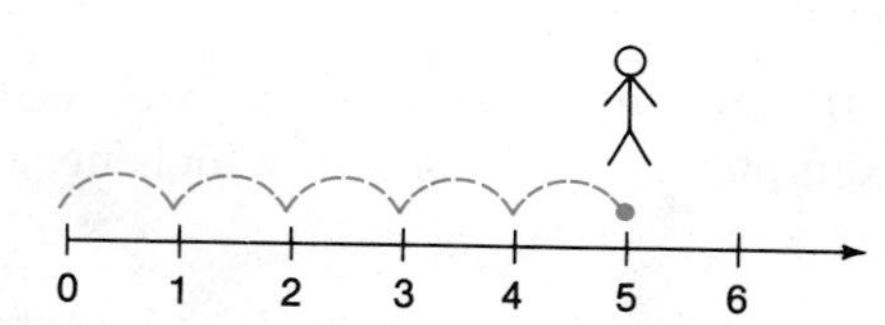

Well, if we can walk forward on a number line, we can also walk backward. Using a number line is one way to demonstrate addition of positive and negative integers, where a positive integer means we go forward (to the right) and a negative integer means we go backward (to the left).

Example 5 $3 + (-2) = ?$

(a) Positive 3 indicates 3 steps forward (always start at 0).

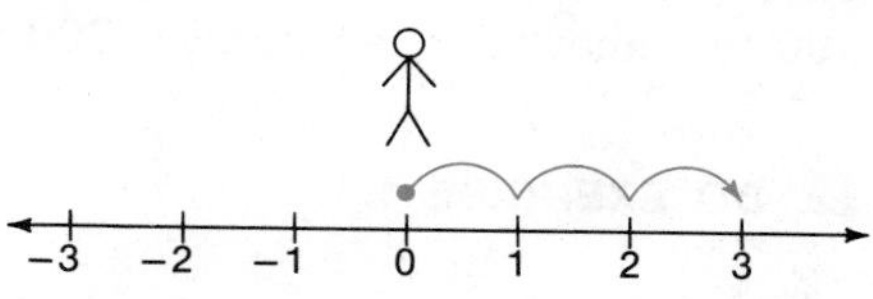

(b) Negative 2 indicates 2 steps backward.

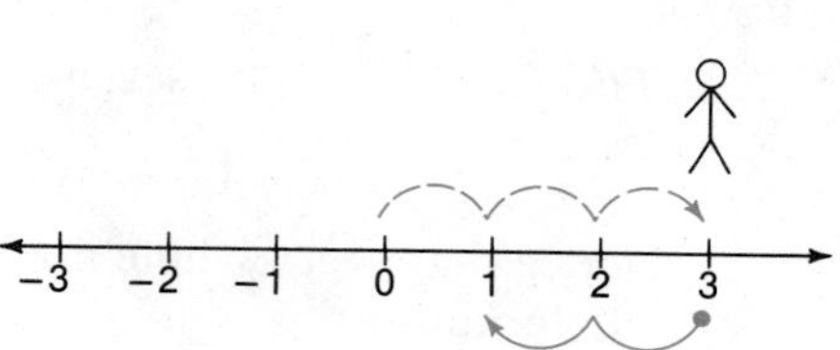

(c) We end up at 1. Thus $3 + (-2) = 1$.

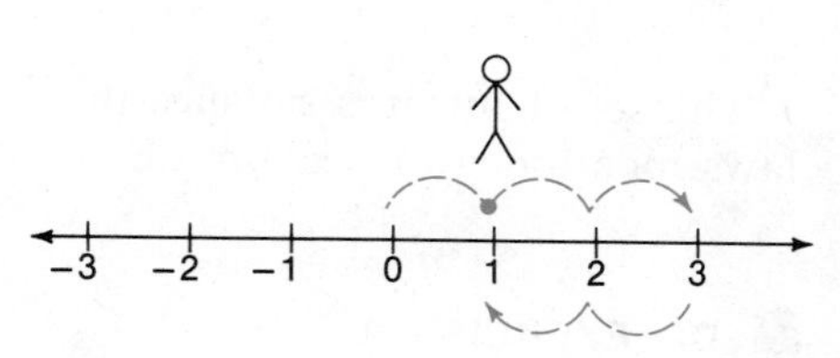

DO EXERCISES 3 AND 4.

Example 6 $4 + (-6) + 1 = ?$

(a) 4 indicates 4 steps forward.

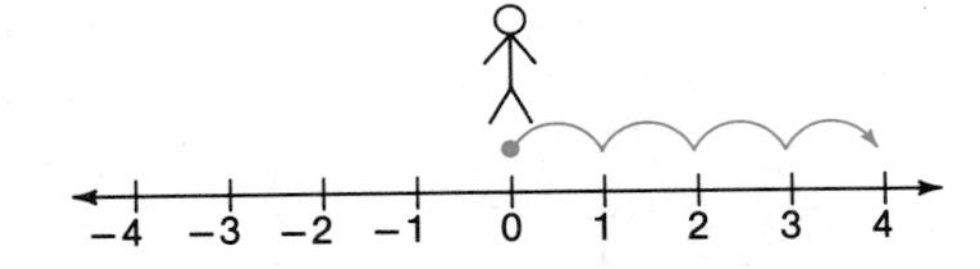

(b) -6 indicates 6 steps backward.

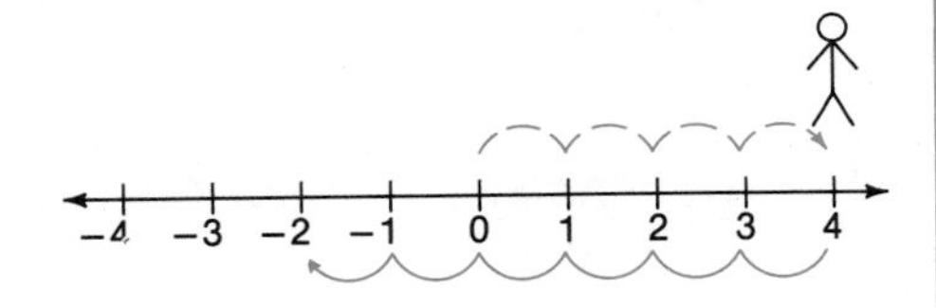

(c) 1 indicates 1 step forward.

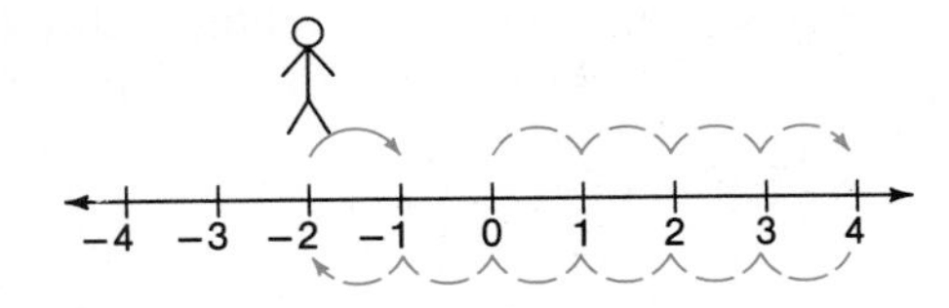

(d) Thus, $4 + (-6) + 1 = -1$.

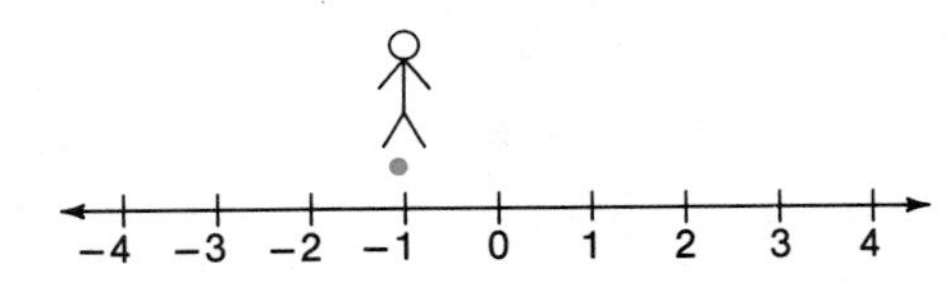

Example 7 $(-2) + (-3) = ?$

(a) 2 steps backward are followed by 3 steps backward.

(b) Thus, $(-2) + (-3) = -5$.

DO EXERCISES 5 THROUGH 7.

Using a number line to help with addition of integers is often useful. A problem such as $236 + (-250)$, however, is rather impractical to put on a number line even though we may mentally picture this problem as taking 236 steps forward followed by 250 steps backward. Thus, we will develop another method (or a general rule) for adding integers. But first we need the concept of absolute value.

DEFINITION 2.2 The *absolute value* of a number x, denoted $|x|$, is the distance the number x is from 0 on a number line.

Since a distance must be either positive or zero, the absolute value of any nonzero number is positive, and the absolute value of zero is zero.

Example 8 Find the absolute value of the following.

(a) $|3| = ?$
On a number line 3 is 3 units from 0. Thus, $|3| = 3$.

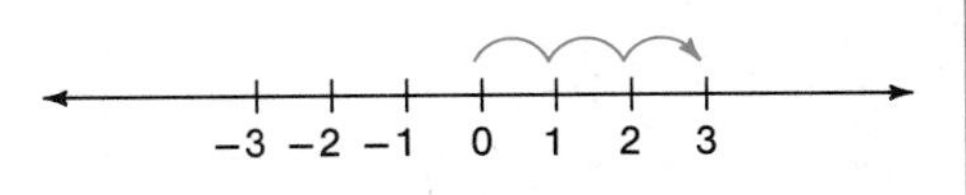

Demonstrate the following on a number line.

5. $(-1) + (-5)$

6. $4 + (-5) + 2 + (-3)$

7. Write out a few of the numbers in the following sets:

(a) The set of natural numbers
(b) The set of whole numbers
(c) The set of integers

Calculate.

8. $|-10|$

9. $|5|$

10. $-3 + (-4)$

11. $-10 + 2$

Calculate.

12. $|7 + (-10)|$

13. $|7| + |-10|$

14. $|(-3) + 5 + (-2)|$

15. $4 + (-1)$

16. $-2 + 5 + (-6)$

Calculate.

17. $416 + 512$

18. $-2240 + (-530)$

(b) $|-2| = ?$
Although -2 is a negative number, on a number line its distance from 0 is 2. Thus, $|-2| = 2$.

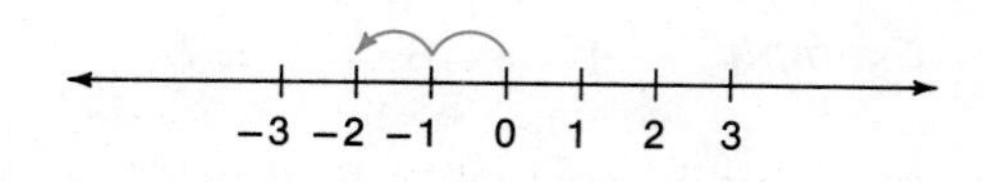

(c) $|72| = 72$

(d) $|-456| = 456$

DO EXERCISES 8 THROUGH 11.

Example 9 Find the absolute value of the following.
(a) $|-8 + 6|$ (b) $|-8| + |6|$

(a) $|-8 + 6| = |-2| = 2$

(b) $|-8| + |6| = 8 + 6 = 14$

DO EXERCISES 12 THROUGH 16.

The general rule for addition of integers is broken into two parts, Rule 2.1 and Rule 2.2.

> *RULE 2.1* To add two integers that have the same sign (that is, both are positive or both are negative):
> 1. Add the absolute values of the integers together.
> 2. Attach the common sign (+ or −) of the integers to the sum (answer).

Example 10 $452 + 78 = ?$

(a) $|452| + |78| = 452 + 78 = 530$

(b) Since 452 and 78 are both positive, the answer is positive. Thus, $452 + 78 = 530$.

Example 11 $-73 + (-214) = ?$

(a) $|-73| + |-214| = 73 + 214 = 287$

(b) Since we are adding two negative integers, the answer is negative. Thus, $-73 + (-214) = -287$. (We could think of this on a number line as taking 73 steps backward followed by an additional 214 steps backward.)

DO EXERCISES 17 AND 18.

RULE 2.2 To add two integers that have opposite signs (that is, one is negative and one is positive):

1. Find the difference of their absolute values.
2. Attach the sign of the integer with the greatest absolute value to the sum (answer).

Example 12 $-75 + 225 = ?$

(a) $|-75| = 75$, $|225| = 225$, and $225 - 75 = 150$

(b) The answer is positive since 225 has the greater absolute value and 225 is positive. Thus, $-75 + 225 = 150$. (Once again, we could think of this on a number line as taking 75 steps backward followed by 225 steps forward.)

Example 13 $417 + (-503) = ?$

(a) $|417| = 417$, $|-503| = 503$, and $503 - 417 = 86$

(b) The answer is negative since -503 has the greater absolute value and -503 is negative. Thus, $417 + (-503) = -86$.

DO EXERCISES 19 THROUGH 21.

Example 14 $-25 + 17 + (-43) + (-10) + 34 = ?$

We could add these numbers two at a time until we are finished. However, there is another way of doing the problem.

(a) Add all the positive numbers together. $17 + 34 = 51$

(b) Add all the negative numbers together. $-25 + (-43) + (-10) = -78$

(c) Thus, the problem reduces to $51 + (-78)$, and we proceed as in Examples 12 and 13 and obtain $51 + (-78) = -27$.

The associative and commutative laws justify adding the positive numbers together first and then adding the negative numbers together.

DO EXERCISES 22 THROUGH 24.

DO SECTION PROBLEMS 2.2.

Calculate.

19. $-74 + 36$

20. $216 + (-89)$

21. $-36 + (-50)$

Calculate.

22. $-17 + 23 + 45 + (-18)$

23. $-415 + (-2010) + 3514 + (-1058)$

24. $-216 + (-34) + (-746)$

ANSWERS TO MARGINAL EXERCISES

1. Answers may vary. **2.** (a) and (d) are natural numbers. (a), (c), and (d) are whole numbers. (a), (b), (c), (d), and (e) are integers.

3.

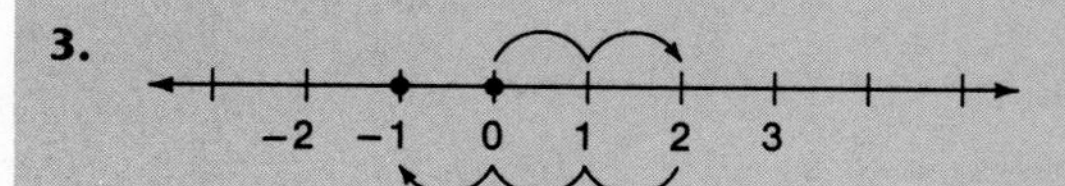

4.

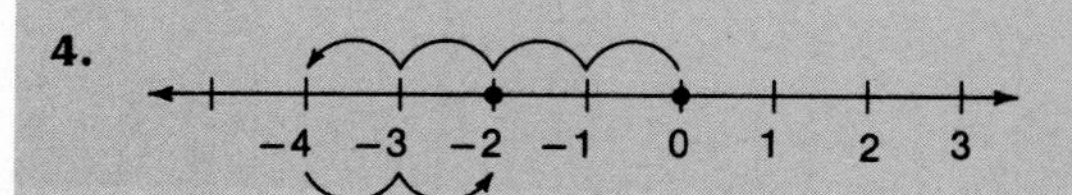

5.

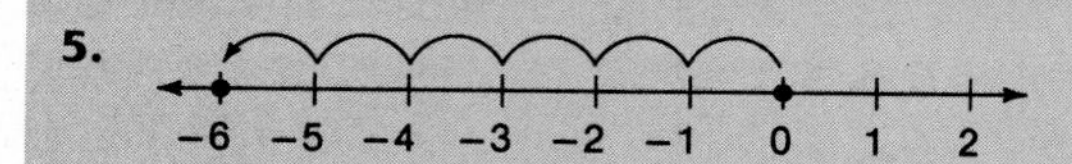

6.

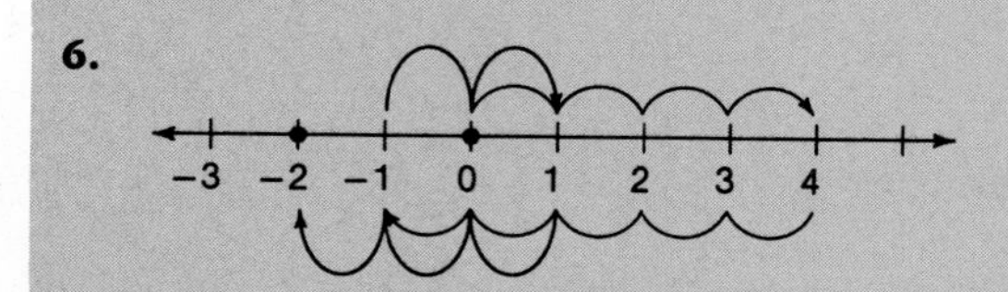

7. (a) $\{1, 2, 3, 4, \ldots\}$ (b) $\{0, 1, 2, 3, \ldots\}$ (c) $\{\ldots, -2, -1, 0, 1, 2, \ldots\}$
8. 10 **9.** 5 **10.** −7 **11.** −8 **12.** 3 **13.** 17 **14.** 0
15. 3 **16.** −3 **17.** 928 **18.** −2770 **19.** −38 **20.** 127
21. −86 **22.** 33 **23.** 31 **24.** −996

NAME COURSE/SECTION DATE

SECTION PROBLEMS 2.2

Work on a number line.

1. $4 + (-6)$
2. $-7 + 1$
3. $-2 + (-5)$
4. $-1 + (-4)$
5. $-2 + 5 + (-3)$
6. $4 + (-2) + 6$

Calculate.

7. $-5 + 7$
8. $6 + (-10)$
9. $-2 + (-3)$
10. $-12 + (-5)$
11. $8 + (-10) + (-3)$
12. $-8 + 4 + (-5)$
13. $-2 + (-5) + (-1)$
14. $-10 + 6 + (-2)$
15. $1 + (-4) + (-3) + 8$
16. $-17 + (-3) + 25 + (-4)$
17. $-3 + (-5) + (-4) + (-1)$
18. $12 + (-8) + 6 + (-10)$
19. $-2 + (-12) + 6$
20. $-5 + 11 + (-4)$
21. $|-7|$
22. $|4|$
23. $|8 + (-5)|$
24. $|8| + |-5|$
25. $|-6 + (-1)|$
26. $|-6| + |-1|$

ANSWERS

1. See figure.
2. See figure.
3. See figure.
4. See figure.
5. See figure.
6. See figure.
7. ______
8. ______
9. ______
10. ______
11. ______
12. ______
13. ______
14. ______
15. ______
16. ______
17. ______
18. ______
19. ______
20. ______
21. ______
22. ______
23. ______
24. ______
25. ______
26. ______

27. $713 + (-514)$

28. $214 + (-316)$

29. $-214 + (-89)$

30. $283 + 417$

31. $-417 + 215$

32. $-334 + (-275)$

33. $4134 + (-5216)$

34. $-513 + 416$

35. $27 + (-35) + (-74)$

36. $14 + (-38) + 64$

37. $-216 + 509 + (-204)$

38. $18 + (-14) + (-53)$

39. $273 + (-154) + (-240) + 75$

40. $216 + (-314) + (-216) + 79$

41. $-150 + (-260) + 430 + (-175)$

42. $-40 + 15 + 73 + (-170)$

43. $24 + 183 + (-340) + 265$

44. $453 + (-63) + (-750) + 300 + 515$

45. $-28 + (-140) + 263 + 114 + (-98)$

46. $650 + (-430) + 165 + (-270) + 245$

27. ______
28. ______
29. ______
30. ______
31. ______
32. ______
33. ______
34. ______
35. ______
36. ______
37. ______
38. ______
39. ______
40. ______
41. ______
42. ______
43. ______
44. ______
45. ______
46. ______

2.3

Subtraction of Integers

Recall the definition of subtraction for whole numbers in Chapter 1: $a - b$ is the number that we must add to b to get a (or in symbols, $a - b = c$ if and only if $c + b = a$) where a, b, and c are whole numbers. We are going to use the same definition for subtraction of integers. However, eventually we will derive from this definition a rule for subtraction that is much easier to work with than the definition of subtraction.

DEFINITION 2.3 $a - b = c$ if and only if $c + b = a$ where a, b, and c are integers.

Note

1. The operation of subtraction is defined in terms of addition.
2. The number we are subtracting, b, may be positive or negative.

Example 1 For the following subtraction problems, state whether the number we are subtracting is positive or negative.

(a) $4 - 2$. We are subtracting a positive 2.

(b) $3 - (-5)$. We are subtracting a negative 5.

(c) $-3 - (-1)$. We are subtracting a negative 1.

DO EXERCISES 1 THROUGH 5.

You may notice from these problems that we are currently using the sign $-$ for two different purposes: (1) to indicate the operation of subtraction (*minus*), and (2) to indicate the *negative* of a number. It is important to distinguish between these two uses for the sign $-$.

Example 2 Translate the following mathematical phrases into English.

(a) $-3 - 4$
Negative three **minus** four (or *negative* three **subtract** four)

(b) $7 - (-5)$
Seven **minus** *negative* five

(c) $-6 - (-1)$
Negative six **minus** *negative* one

(d) $-3 - 4 - (-7)$
Negative three **minus** four **minus** *negative* seven

DO EXERCISES 6 THROUGH 9.

In the following examples we use Definition 2.3 ($a - b = c$ if and only if $c + b = a$) to work the subtraction problems and find c.

Example 3 $a - b = c$ if and only if $c + b = a$.

(a) $7 - 5 = c$ if $c + 5 = 7$; thus $c = 2$.

(b) $2 - 6 = c$ if $c + 6 = 2$; thus $c = -4$.

LEARNING OBJECTIVES

After finishing Section 2.3, you should be able to:

- Subtract integers.
- Find the additive inverse of an integer.

For the following subtraction problems, tell whether the number we are subtracting is positive or negative.

1. $7 - 5$

2. $-3 - 10$

3. $4 - (-6)$

4. $1 - 5$

5. $-1 - (-3)$

Translate the following mathematical phrases into English.

6. $6 - 2$

7. $7 - (-5)$

8. $-2 - 1$

9. $-3 - (-4) - 6$

Use Definition 2.3 to work the following problems.

10. $7 - 3$

11. $-4 - 5$

12. $2 - (-5)$

13. $-6 - (-3)$

14. $-(-6)$ can be read "the negative of negative six." How else can it be read?

15. -7 can be read "negative seven." How else can it be read?

16. Find the additive inverse of 7.

17. Find the additive inverse of -12.

(c) $-2 - 4 = c$ if $c + 4 = -2$; thus $c = -6$.

(d) $5 - (-3) = c$ if $c + (-3) = 5$; thus $c = 8$.

(e) $-1 - (-4) = c$ if $c + (-4) = -1$; thus $c = 3$.

DO EXERCISES 10 THROUGH 13.

Using Definition 2.3 to work subtraction problems is somewhat tedious, so we want to develop a more practical method. As a first step, we define the additive inverse (or opposite) of a number. Recall that we defined multiplicative inverses (or reciprocals) in Section 1.8.

Note that the sum of 4 and -4 is 0; that is, $4 + (-4) = 0$. We say in this case that 4 and -4 are *additive inverses.* This also says that -4 is the additive inverse of 4, and 4 is the additive inverse of -4.

DEFINITION 2.4 Given a number x, the *additive inverse* (or *opposite*) of x is the number that added to x gives a sum of zero. The additive inverse of x will be denoted by $-x$. (This gives us a third use of the sign $-$. The other two uses are the operation of subtraction and the negative of a number.)

Example 4 What is the additive inverse of 3?

Since $3 + (-3) = 0$, the additive inverse of 3 is -3. Note that -3 can be read as "negative 3" or "the additive inverse of 3." Hence, we can actually interchange the words "negative" and "additive inverse."

Example 5 What is the additive inverse of -4?

Since $-4 + 4 = 0$, the additive inverse of -4 is 4. Recall that our notation for the additive inverse of x is $-x$. Thus, the additive inverse of -4 is denoted as

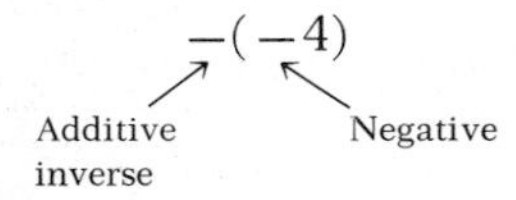

But the additive inverse of -4 is also 4 and, thus, $-(-4) = 4$. Also, $-(-4)$ can be read "the negative of a negative four."

DO EXERCISES 14 AND 15.

Example 6 What is the additive inverse of -10?

Since $-10 + 10 = 0$, the additive inverse of -10, denoted $-(-10)$, is 10. Once again, this also gives us $-(-10) = 10$.

In general, we have the following rule.

RULE 2.3 The additive inverse of negative x (or the negative of negative x) is x. In symbols we have $-(-x) = x$.

DO EXERCISES 16 AND 17.

Example 7 Calculate or simplify the following.
(a) $-(-3)$ (b) $4 + [-(-7)]$ (c) $-(-5) + (-8)$

(a) $-(-3) = 3$

(b) $4 + [-(-7)] = 4 + 7 = 11$

(c) $-(-5) + (-8) = 5 + (-8) = -3$

DO EXERCISES 18 AND 19.

Now we may give a rule for subtraction.

RULE 2.4 For integers a and b, we have $a - b = a + (-b)$.

Note

1. In Rule 2.4 we may regard $-b$ as the additive inverse of b or the negative of b.
2. Rule 2.4 says that a *minus* b is the same as a *plus* the *additive inverse* of b, or a *minus* b is the same as a *plus* the *negative* of b.
3. Rule 2.4 allows us to change the operation of subtraction to the operation of addition. Recall that we do something similar when we divide fractions. For example, $\frac{a}{b}$ *divided* by $\frac{c}{d}$ can be changed to $\frac{a}{b}$ *times* the *multiplicative inverse* of $\frac{c}{d}$. That is, $\frac{a}{b} \div \frac{c}{d} = \frac{a}{b} \cdot \frac{d}{c}$.
4. Rule 2.4 follows from the original definition of subtraction. Recall that $a - b$ is the number that added to b gives a. The number $a + (-b)$ fits this definition since $a + (-b) + b = a + 0 = a$. Thus, $a + (-b)$ is the same as $a - b$.

Example 8 Change the following subtraction problems to addition problems.

(a) $4 - 1 = 4 + (-1)$
(b) $2 - (-3) = 2 + [-(-3)]$
(c) $-4 - 6 = -4 + (-6)$
(d) $-1 - (-5) = -1 + [-(-5)]$

DO EXERCISES 20 THROUGH 25.

We will perform the subtraction problems in Example 9 by changing the subtraction to addition.

Example 9

(a) $4 - 1 = 4 + (-1) = 3$

(b) $2 - (-3) = 2 + [-(-3)] = 2 + 3 = 5$

(c) $-4 - 6 = -4 + (-6) = -10$

(d) $-1 - (-5) = -1 + [-(-5)] = -1 + 5 = 4$

(e) $3 - 10 = 3 + (-10) = -7$

DO EXERCISES 26 THROUGH 32.

Simplify and calculate.

18. $-(-3)$

19. $-4 + [-(-5)]$

Change the following subtraction problems to addition problems. (Do not calculate the final answer yet.)

20. $7 - 3$

21. $-2 - 4$

22. $5 - (-4)$

23. $-7 - 15$

24. $-3 - (-10)$

25. $2 - (-3)$

Perform the following subtraction problems by first changing the subtraction to addition.

26. $3 - 1$

27. $-2 - 1$

28. $4 - (-5)$

29. $-1 + (-15)$

30. $-(-3) - (-2)$

31. $-2 + (-4)$

32. $-4 - (-8)$

Calculate.

33. $-3 + 7 - 5$

34. $4 - (-3) - 7$

35. $-1 - (-4) + (-6)$

36. $-(-3) - (-4) - 5$

Calculate.

37. $-45 - 273$

38. $614 - (-212) - 345$

39. $-|-7| - |2|$

40. $-|-2| - |4 - 7|$

Example 10 Work the following problems.

(a) $-2 - (-4) + 7 = -2 + [-(-4)] + 7 = -2 + 4 + 7 = 9$

(b) $-2 - 4 - 6 = -2 + (-4) + (-6) = -12$

(c) $4 - 5 - (-8) + 6 = 4 + (-5) + [-(-8)] + 6 = 4 + (-5) + 8 + 6 = 13$

DO EXERCISES 33 THROUGH 36.

Example 11 Work the following problems.

(a) $418 - (-72) = 418 + [-(-72)] = 418 + 72 = 490$

(b) $-213 + 400 - 250 = -213 + 400 + (-250) = -463 + 400 = -63$

(c) $|-8| - |3| = 8 - 3 = 8 + (-3) = 5$

(d) $-|3| - |-4| = -3 - 4 = -3 + (-4) = -7$

Note that in parts (c) and (d) of Example 11 we must calculate the absolute values before we do the operation of subtraction.

DO EXERCISES 37 THROUGH 40.

For all the problems we have done in this section, we have changed subtraction to addition. We will continue to do this throughout the chapter and for most of Chapter 3. Eventually, as we gain confidence with the operation of subtraction, we can handle subtraction problems by doing the change to addition in our heads (for example, $-3 - 7 = -10$ instead of $-3 - 7 = -3 + (-7) = -10$). However, we may *always* change subtraction to addition if we feel more comfortable doing it that way.

DO SECTION PROBLEMS 2.3.

ANSWERS TO MARGINAL EXERCISES

1. Positive 5 **2.** Positive 10 **3.** Negative 6 **4.** Positive 5 **5.** Negative 3 **6.** Six minus two **7.** Seven minus negative five **8.** Negative two minus one **9.** Negative three minus negative four minus six **10.** $7 - 3 = 4$ since $4 + 3 = 7$ **11.** $-4 - 5 = -9$ since $(-9) + 5 = -4$ **12.** $2 - (-5) = 7$ since $7 + (-5) = 2$ **13.** $-6 - (-3) = -3$ since $-3 + (-3) = -6$ **14.** Additive inverse of negative 6 **15.** Additive inverse of 7 **16.** -7 **17.** 12 **18.** 3 **19.** 1 **20.** $7 + (-3)$ **21.** $-2 + (-4)$ **22.** $5 + [-(-4)]$ **23.** $-7 + (-15)$ **24.** $-3 + [-(-10)]$ **25.** $2 + [-(-3)]$ **26.** 2 **27.** -3 **28.** 9 **29.** -16 **30.** 5 **31.** -6 **32.** 4 **33.** -1 **34.** 0 **35.** -3 **36.** 2 **37.** -318 **38.** 481 **39.** -9 **40.** -5

NAME COURSE/SECTION DATE

ANSWERS

SECTION PROBLEMS 2.3

Use Definition 2.3 to work the following subtraction problems.

1. $6 - 14$ **2.** $-2 - 5$ **3.** $4 - (-6)$ **4.** $-2 - (-3)$

Calculate.

5. $4 - 5$ **6.** $7 - 4$ **7.** $2 - 7$ **8.** $-3 - 4$

9. $-12 - 5$ **10.** $-3 - (-4)$ **11.** $-1 - (-5)$ **12.** $-12 - 3$

13. $-2 - (-3)$ **14.** $-10 - (-2)$ **15.** $14 - (-2)$ **16.** $-5 - 7$

17. $7 - (-9)$ **18.** $-5 + 23$ **19.** $4 - 3 + 2$ **20.** $-1 + 6 - 10$

21. $5 - 6 - 7$ **22.** $-4 - 8$ **23.** $3 - (-6)$ **24.** $-10 - (-2)$

25. $12 - (-1)$ **26.** $-3 + 4 - 5$ **27.** $6 - 4 - (-1)$

28. $-3 - 2 - 6$ **29.** $4 - (-3) - 7$ **30.** $50 - (-80)$

31. $-10 - (-14) - 37$ **32.** $|3| - |-7|$ **33.** $-|-2| - |5|$

1. ______
2. ______
3. ______
4. ______
5. ______
6. ______
7. ______
8. ______
9. ______
10. ______
11. ______
12. ______
13. ______
14. ______
15. ______
16. ______
17. ______
18. ______
19. ______
20. ______
21. ______
22. ______
23. ______
24. ______
25. ______
26. ______
27. ______
28. ______
29. ______
30. ______
31. ______
32. ______
33. ______

34. ____________

35. ____________

36. ____________

37. ____________

38. ____________

39. ____________

40. ____________

41. ____________

42. ____________

43. ____________

44. ____________

45. ____________

46. ____________

47. ____________

48. ____________

49. ____________

50. ____________

51. ____________

52. ____________

53. ____________

54. ____________

55. ____________

34. $-306 - 489$

35. $16 - (-43) - 105$

36. $-3 - 4 + (-5)$

37. $-3 - 4 - 5$

38. $4 + (-3) - 7$

39. $-5 + 4 - (-6)$

40. $-(-3) - 5 - (-7)$

41. $4 + (-2) + (-7) - 5$

42. $-42 + 30 - 12$

43. $9 - (-3) - (-5) + 12$

44. $|7| - |-3|$

45. $-|4| + -|-6|$

46. $215 - (-613)$

47. $-333 - 201$

48. $-41 - (-63) + 74 - 89$

49. $-|6| - |-3|$

50. $|-4| - |7| - |-4|$

51. $-415 + 304$

52. $705 - (-143)$

53. $-206 + 304 - (-209)$

54. $153 - 214 - 612$

55. $-143 - (-205) - (-513)$

2.4

Multiplication and Division of Integers

As we already know, the product of two positive numbers is also positive. However, consider the problem of a positive number times a negative number, for example, $(4)(-3)$. Recall our definition of multiplication of whole numbers as repeated addition (Definition 1.14). For example,

$$4 \cdot 3 = 3 + 3 + 3 + 3 \qquad \text{or} \qquad 6 \cdot 4 = 4 + 4 + 4 + 4 + 4 + 4$$

We want to extend this definition to the set of integers and, in particular, to extend this definition to our problem of $(4)(-3)$. If we do this, we obtain

$$(4)(-3) = -3 + (-3) + (-3) + (-3) = -12$$

With the help of the commutative law for multiplication, the problem $(-4)(3)$ (a negative number times a positive number) becomes

$$(-4)(3) = (3)(-4) = -4 + (-4) + (-4) = -12$$

Thus in each of the above problems a negative number times a positive number results in a negative product (answer).

RULE 2.5 If we multiply a negative number and a positive number together, the product will be a negative number.

Example 1 Calculate the following.

(a) $(-2)(4) = -8$

(b) $(3)(-7) = -21$

(c) $(2)(-5)(4) = (-10)(4) = -40$

(d) $(-3)(5) + (2)(-4) = -15 + (-8) = -23$ Note that we multiply before we add.

DO EXERCISES 1 THROUGH 6.

Consider the problem $(-3)(-4)$, that is, a negative number times a negative number. With the help of the distributive law (see Chapter 1) we have

$$\begin{aligned} 0 = (-3)(0) &= (-3)[4 + (-4)] \\ &= (-3)(4) + (-3)(-4) \\ &= -12 + (-3)(-4) \end{aligned}$$

Thus, $0 = -12 + (-3)(-4)$, and $(-3)(-4)$ must be the number you add to -12 to get 0, or $(-3)(-4) = 12$. Thus, a negative number times a negative number gives us a positive product.

LEARNING OBJECTIVES

After finishing Section 2.4, you should be able to:

- Multiply integers.
- Divide integers.
- Apply the rules for the Order of Operations.

Perform the indicated operations.

1. $(4)(-6)$

2. $(-3)(5)$

3. $2 - 7$

4. $(-1)(5)(6)$

5. $-1 - (-10)$

6. $(5)(-2) + (-4)(3)$

Perform the indicated operations.

7. $(-4)(-4)$

8. $(6)(-2)$

9. $(-1)(-5)$

10. $-2 - 1$

11. $(-3)(-5)(4)$

12. $(5)(2) - (-2)(-3)$

Perform the indicated operations.

13. $\frac{-4}{2}$

14. $\frac{-15}{-3}$

15. $(-5)(4)$

16. $\frac{16}{4}$

17. $36 \div (-6)$

18. $-7 - (-10)$

RULE 2.6 If we multiply a negative number and a negative number together, the product will be a positive number.

Example 2 Calculate the following.

(a) $(-4)(-5) = 20$

(b) $(3)(-7) = -21$

(c) $(-2)(4)(-3) = (-8)(-3) = 24$

(d) $(-5)(-2)(-7) = (10)(-7) = -70$

(e) $(-3)(4) - (-5)(-2) = -12 - (10) = -12 + (-10) = -22$
We multiply before we subtract.

(f) $(-2)(-3)(-1)(-4) = (6)(-1)(-4) = (-6)(-4) = 24$

(g) $(-4)(0) = 0$ In general, the product of zero and any number will be zero.

Summarizing our rules for multiplying two numbers together, we have

1. If *both* numbers have the *same* sign, the product is *positive*.
2. If the *two* numbers have *different* signs, the product is *negative*.

DO EXERCISES 7 THROUGH 12.

Recall our definition of division of whole numbers (Definition 1.16): $\frac{a}{b} = c$ if and only if $c \cdot b = a$. We want to extend this definition to the set of integers.

Example 3 Note the following.

(a) $\frac{-12}{2} = -6$ since $(-6)(2) = -12$ A negative number divided by a positive number results in a negative quotient (answer).

(b) $\frac{16}{-2} = -8$ since $(-8)(-2) = 16$ A positive number divided by a negative number results in a negative quotient.

(c) $\frac{-20}{-5} = 4$ since $(4)(-5) = -20$ A negative number divided by a negative number results in a positive quotient.

RULE 2.7

1. If we divide two numbers that have the *same* sign, the quotient will be a *positive* number.
2. If we divide two numbers that have *different* signs, the quotient will be a *negative* number.
3. We can never divide by zero.

DO EXERCISES 13 THROUGH 18.

We want to look at problems that involve combinations of the four operations of addition, subtraction, multiplication, and division. When we have more than one operation in a problem, it appears we may have a choice as to which operation to perform first. For example, consider $2 + 3 \cdot 6$. If we add first and then multiply, we obtain $2 + 3 \cdot 6 = 5 \cdot 6 = 30$. If we multiply first and then add, we obtain $2 + 3 \cdot 6 = 2 + 18 = 20$. Since the answers are different, it obviously makes a difference which operation is performed first.

RULE 2.8 The following is the Order of Operations:

1. Perform operations within parentheses. (The symbols [ ] and { } have the same meaning in an algebraic expression as do parentheses, namely multiplication. The symbols ( ), [ ], { } are called *grouping symbols.*)
2. Multiply or divide proceeding in order from left to right.
3. Add or subtract proceeding in order from left to right.

Example 4 The correct order for doing the problem $2 + 3 \cdot 6$ is multiply and then add; that is,

$$2 + 3 \cdot 6 = 2 + 18 = 20$$

Example 5 Calculate the following.

(a) $12 + \frac{6}{3} = 12 + 2 = 14$ — Divide and then add.

(b) $3 + 4[6 + (-3)] = 3 + 4(3) = 3 + 12 = 15$ — Parentheses, multiply, and then add.

(c) $4 - 2(7 + 1) = 4 - 2(8) = 4 - 16 = 4 + (-16) = -12$ — Parentheses, multiply, and then subtract.

(d) $(-2)(3) + 4(-6) = -6 + (-24) = -30$ — Multiply and then add.

(e) $7 - 2 + 5 = 5 + 5 = 10$ — Add or subtract proceeding from left to right.

(f) $8 \div 4 \times 2 = 2 \times 2 = 4$ — Multiply or divide proceeding from left to right.

Later in this chapter we will add another operation to the Order of Operations.

DO EXERCISES 19 THROUGH 25.

DO SECTION PROBLEMS 2.4.

Calculate.

19. $3 + 2 \cdot 7$

20. $5 - 3(4)$

21. $-4 - \frac{10}{2}$

22. $-2 + 3(7 - 5)$

23. $6(-3) - 2$

24. $(4 + 5)8$

25. $6 \div 3 \times 2$

ANSWERS TO MARGINAL EXERCISES

1. -24 **2.** -15 **3.** -5 **4.** -30 **5.** 9 **6.** -22
7. 16 **8.** -12 **9.** 5 **10.** -3 **11.** 60 **12.** 4 **13.** -2
14. 5 **15.** -20 **16.** 4 **17.** -6 **18.** 3 **19.** 17
20. -7 **21.** -9 **22.** 4 **23.** -20 **24.** 72 **25.** 4

NAME COURSE/SECTION DATE

SECTION PROBLEMS 2.4

Calculate.

1. $(-3)(10)$
2. $(6)(5)$
3. $(-4)(-5)$
4. $(2)(-8)$
5. $(-2)(-3)(-4)$
6. $(-2)(3)(-1)$
7. $(4)(-2)(5)$
8. $(3)(-8)(-2)$
9. $\frac{4}{-2}$
10. $\frac{-16}{-2}$
11. $(-8) \div (4)$
12. $16 \div 8$
13. $0 \div (-1)$
14. $(-21) \div (-3)$
15. $\frac{-20}{4}$
16. $\frac{48}{8}$
17. $-7 - 3$
18. $(-7)(-3)$
19. $-7(-3)$
20. $-7 - (3)$
21. $(-5)(4)(0)$
22. $4 - (-6)$
23. $\frac{(4)(3)}{-2}$
24. $\frac{(-3)(-8)}{-4}$
25. $(-72)(-104)$
26. $(-3)(4)(-5)$
27. $\frac{24}{(-3)(2)}$

ANSWERS

1. ______
2. ______
3. ______
4. ______
5. ______
6. ______
7. ______
8. ______
9. ______
10. ______
11. ______
12. ______
13. ______
14. ______
15. ______
16. ______
17. ______
18. ______
19. ______
20. ______
21. ______
22. ______
23. ______
24. ______
25. ______
26. ______
27. ______

28. ____________

29. ____________

30. ____________

31. ____________

32. ____________

33. ____________

34. ____________

35. ____________

36. ____________

37. ____________

38. ____________

39. ____________

40. ____________

41. ____________

42. ____________

43. ____________

44. ____________

45. ____________

46. ____________

47. ____________

48. ____________

49. ____________

50. ____________

28. $2 + 3 \cdot 7$

29. $(-2)(3) - (-1)(-5)$

30. $-2 - \frac{15}{3}$

31. $-7 - (-3)(4)$

32. $4 - 2(3)$

33. $4 + \frac{(-2)(10)}{-4}$

34. $-3 - 5(2 - 7)$

35. $4 + 5 \cdot 7$

36. $6 - 4(2)$

37. $2 \cdot 4 + 3 \cdot 5$

38. $7 - 5(4 - 5)$

39. $4 \cdot 3 + 2$

40. $16 \div 8 \div 2$

41. $10 \div 2 \times 5$

42. $-5 - 3 \cdot 2$

43. $2 \cdot |3 - 7|$

44. $(-|-30|) \div (-6)$

45. $\frac{|5 - 15|}{-|-2|}$

46. $|-3| \cdot |4| - |-2|$

47. $\frac{15 - 5}{-2 - 3}$

48. $\frac{6 + 2 \cdot 4}{-2}$

49. $\frac{(-2 - 1)(2 - 6)}{3}$

50. $\frac{4 - 2(1 - 3)}{1 - 5}$

2.5

Rational Numbers

Once again, we are going to enlarge our number system by including additional negative numbers. Just as every integer has an additive inverse, every fraction (we looked at the set of fractions in Chapter 1) also has an additive inverse. For example, $\frac{3}{4} + \left(-\frac{3}{4}\right) = 0$ and thus $-\frac{3}{4}$ is the additive inverse of $\frac{3}{4}$ $\left(\text{also } \frac{3}{4} \text{ is the additive inverse of } -\frac{3}{4}\right)$. We have located a few fractions and their additive inverses on the following number line.

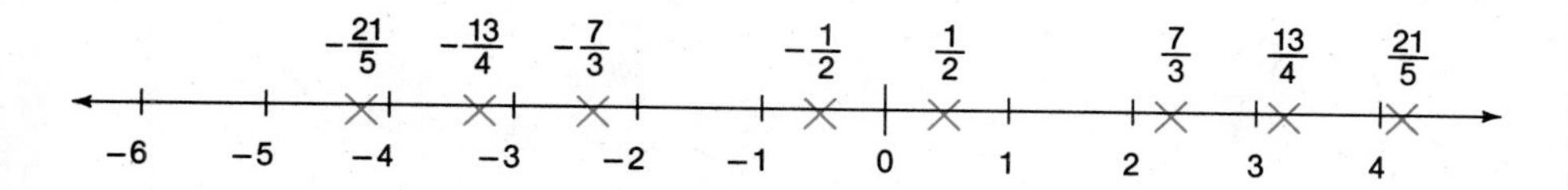

This leads us to a new set of numbers called the set of rational numbers. The set of rational numbers contains the set of fractions but also much more.

DEFINITION 2.5 The set of *rational numbers* is the set of all numbers of the form $\frac{m}{n}$ where m and n are arbitrary integers, except $n \neq 0$.

The rational numbers satisfy the commutative, associative, and distributive laws, which we will look at again in Section 2.8.

Up to this point, we have defined the following sets of numbers:

1. Set of natural numbers = $\{1, 2, 3, \ldots\}$
2. Set of whole numbers = $\{0, 1, 2, \ldots\}$
3. Set of integers = $\{\ldots, -3, -2, -1, 0, 1, 2, 3, \ldots\}$
4. Set of rational numbers = $\left\{\frac{m}{n} \,\middle|\, m \text{ and } n \text{ are integers, } n \neq 0\right\}$ (where the vertical bar in { | } is read "such that")

DO EXERCISES 1 AND 2.

Before we look at various operations on the set of rational numbers, note the following:

$$\frac{-20}{4} = -5 \qquad \frac{20}{-4} = -5 \qquad -\frac{20}{4} = -5 \qquad -\frac{-20}{-4} = -5$$

Thus,

$$\frac{-20}{4} = \frac{20}{-4} = -\frac{20}{4} = -\frac{-20}{-4}$$

and in general we have the following rule.

RULE 2.9 $$\frac{-a}{b} = \frac{a}{-b} = -\frac{a}{b} = -\frac{-a}{-b}$$

LEARNING OBJECTIVES

After finishing Section 2.5, you should be able to:

- Add rational numbers.
- Subtract rational numbers.
- Multiply rational numbers.
- Divide rational numbers.

1. Locate the following rational numbers on the given number line.

$A = -2\frac{1}{3}$ $\quad B = -\frac{4}{5}$

$C = \frac{3}{2}$ $\quad D = -1\frac{1}{6}$

$E = -3$ $\quad F = -3\frac{2}{5}$

−4 −3 −2 −1 0 1 2 3 4

2. Match the following numbers with the set(s) in which they belong.

$-\frac{2}{3}$	Natural numbers
0	Whole numbers
5	Integers
1.4	Rational numbers
−3	
$-4\frac{1}{5}$	

Express in standard form.

3. $-\frac{3}{4}$

4. $\frac{5}{-2}$

5. $\frac{-7}{-9}$

6. $-\frac{5}{-11}$

This rule allows us to move the negative sign in a fraction from the denominator to the numerator, from the denominator to in front of the fraction, from in front of the fraction to the numerator, and so forth.

DEFINITION 2.6 The fraction $\frac{-a}{b}$ is the *standard form* of the four equivalent fractions $\frac{-a}{b}, \frac{a}{-b}, -\frac{a}{b}, -\frac{-a}{-b}$.

Quite often when we work problems involving rational numbers, we find it convenient to express the rational number in standard form.

Example 1 Express the following fractions in standard form.

(a) $\frac{2}{-3}$ (b) $-\frac{4}{9}$ (c) $\frac{-3}{-5}$

(a) $\frac{2}{-3} = \frac{-2}{3}$

(b) $-\frac{4}{9} = \frac{-4}{9}$

(c) $\frac{-3}{-5} = \frac{-(-3)}{5} = \frac{3}{5}$

DO EXERCISES 3 THROUGH 6.

Now we want to extend the rules for adding, subtracting, multiplying, and dividing fractions that were developed in Chapter 1 to the set of all rational numbers.

Example 2 $\frac{2}{3} + \frac{-4}{3} = ?$

$$\frac{2}{3} + \frac{-4}{3} = \frac{2 + (-4)}{3} = \frac{-2}{3} \quad \left(\text{or } -\frac{2}{3}\right)$$

Example 3 $-\frac{1}{6} + \frac{-3}{12} = ?$

$$-\frac{1}{6} + \frac{-3}{12} = -\frac{2}{12} + \frac{-3}{12} = \frac{-2}{12} + \frac{-3}{12}$$

$$= \frac{-2 + (-3)}{12} = \frac{-5}{12} \quad \left(\text{or } -\frac{5}{12}\right)$$

Example 4 $2\frac{2}{3} + \left(-\frac{4}{5}\right) = ?$

$$2\frac{2}{3} + \left(-\frac{4}{5}\right) = \frac{8}{3} + \left(-\frac{4}{5}\right) = \frac{40}{15} + \frac{-12}{15}$$

$$= \frac{40 + (-12)}{15} = \frac{28}{15} \quad \left(\text{or } 1\frac{13}{15}\right)$$

DO EXERCISES 7 AND 8.

Example 5 $\frac{1}{2} - \frac{3}{4} = ?$

$$\frac{1}{2} - \frac{3}{4} = \frac{1}{2} + \left(-\frac{3}{4}\right) = \frac{2}{4} + \frac{-3}{4}$$

$$= \frac{2 + (-3)}{4} = \frac{-1}{4} \quad \left(\text{or } -\frac{1}{4}\right)$$

Example 6 $-\frac{3}{16} - \frac{1}{4} = ?$

$$-\frac{3}{16} - \frac{1}{4} = \frac{-3}{16} + \left(-\frac{1}{4}\right) = \frac{-3}{16} + \frac{-4}{16}$$

$$= \frac{-7}{16} \quad \left(\text{or } -\frac{7}{16}\right)$$

DO EXERCISES 9 AND 10.

Example 7 $-1\frac{1}{4} - 2\frac{1}{3} + 1\frac{1}{6} - 3 = ?$

$$-1\frac{1}{4} - 2\frac{1}{3} + 1\frac{1}{6} - 3 = -\frac{5}{4} + \left(-\frac{7}{3}\right) + \frac{7}{6} + (-3)$$

$$= \frac{-15}{12} + \frac{-28}{12} + \frac{14}{12} + \frac{-36}{12}$$

$$= \frac{-15 + (-28) + 14 + (-36)}{12}$$

$$= \frac{-65}{12} \quad \left(\text{or } -5\frac{5}{12}\right)$$

DO EXERCISE 11.

Perform the indicated operations.

7. $\frac{1}{5} + \frac{-3}{10}$

8. $-2\frac{1}{2} + \left(-\frac{1}{6}\right)$

Perform the indicated operations.

9. $-\frac{3}{5} - \frac{2}{3}$

10. $\frac{4}{-7} - \frac{-2}{5}$

Perform the indicated operations.

11. $2\frac{3}{4} - 1\frac{1}{16} - 3\frac{1}{8} + 2\frac{1}{4}$

The rules developed in Section 2.4 for multiplying or dividing positive and negative integers will also apply to positive and negative rational numbers. In other words,

1. The product (quotient) of two positive rational numbers is a positive number.
2. The product (quotient) of two negative rational numbers is a positive number.
3. The product (quotient) of one positive and one negative rational number is a negative number.

Example 8 $\left(\frac{3}{4}\right)\left(-\frac{1}{2}\right) = ?$

$$\left(\frac{3}{4}\right)\left(-\frac{1}{2}\right) = \left(\frac{3}{4}\right)\left(\frac{-1}{2}\right) = \frac{(3)(-1)}{(4)(2)} = \frac{-3}{8} \quad \left(\text{or } -\frac{3}{8}\right)$$

Example 9 $\left(-2\frac{1}{2}\right)\left(\frac{-4}{9}\right) = ?$

$$\left(-2\frac{1}{2}\right)\left(\frac{-4}{9}\right) = \left(\frac{-5}{2}\right)\left(\frac{-4}{9}\right) = \frac{(-5)(-4)}{(2)(9)} = \frac{20}{18} = \frac{10}{9}$$

DO EXERCISES 12 THROUGH 14.

Example 10 $\left(\frac{4}{5}\right) \div \left(-\frac{2}{3}\right) = ?$

$$\left(\frac{4}{5}\right) \div \left(-\frac{2}{3}\right) = \left(\frac{4}{5}\right)\left(\frac{-3}{2}\right) = \frac{(4)(-3)}{(5)(2)}$$

$$= \frac{-12}{10} = \frac{-6}{5} \quad \left(\text{or } -\frac{6}{5}\right)$$

Example 11 $\left(-3\frac{1}{8}\right) \div \left(-1\frac{1}{4}\right) = ?$

$$\left(-3\frac{1}{8}\right) \div \left(-1\frac{1}{4}\right) = \left(-\frac{25}{8}\right) \div \left(-\frac{5}{4}\right) = \left(\frac{-25}{8}\right)\left(\frac{-4}{5}\right)$$

$$= \frac{(-25)(-4)}{(8)(5)} = \frac{100}{40} = \frac{5}{2}$$

DO EXERCISES 15 THROUGH 17.

Perform the indicated operations.

12. $\left(-\frac{2}{3}\right) \cdot \left(\frac{7}{4}\right)$

13. $\left(-1\frac{3}{4}\right)\left(-\frac{3}{7}\right)$

14. $\left(-\frac{5}{6}\right)\left(\frac{3}{-10}\right)$

Perform the indicated operations.

15. $\left(-\frac{4}{9}\right) \div \left(\frac{1}{3}\right)$

16. $(-3) \div \left(-1\frac{1}{3}\right)$

17. $-\frac{1}{5} - \frac{2}{15}$

Finally in this section we consider problems involving positive and negative rational numbers expressed as decimals.

Example 12 $52.346 + (-45.14) + (-2.071) + 1.145 = ?$

$$52.346 + (-45.14) + (-2.071) + 1.145 = 53.491 + (-47.211) = 6.28$$

Note that in Example 12 we added the two positive decimals together and the two negative decimals together. Then we found the difference in the absolute values of 53.491 and −47.211 and attached the proper sign to our answer.

Example 13 $1.9 - 3.05 = ?$

$$1.9 - 3.05 = 1.9 + (-3.05) = -1.15$$

DO EXERCISES 18 THROUGH 20.

Example 14 $(3.046)(-1.15) = ?$

```
   3.046
 × 1.15
  15230
  3046
 3046
 3.50290
```

Thus $(3.046)(-1.15) = -3.5029$

Example 15 $(-25.6) \div (-0.08) = ?$

```
       3 20.
.08 )25.60
     24
      1 6
      1 6
```

Thus $(-25.6) \div (-0.08) = 320$

DO EXERCISES 21 THROUGH 24.

DO SECTION PROBLEMS 2.5.

Perform the indicated operations.

18. $(256.1)(-15.3)$

19. $73.1 - 104$

20. $65.4 + (-14.36) + (-75.1) + 6.42$

Perform the indicated operations.

21. $(9.42) \div (-0.3)$

22. $(-0.004)(-0.02)$

23. $(-9.8) \div (-0.04)$

24. $-4.26 - (-3.14) + 7.6 - 2.15$

ANSWERS TO MARGINAL EXERCISES

1.

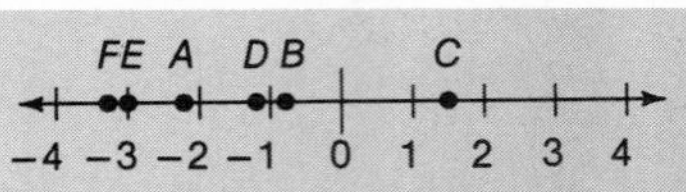

2. 5 is a natural number; 0, 5 are whole numbers; 0, 5, −3 are integers; $-\frac{2}{3}$, 0, 5, 1.4, −3, $-4\frac{1}{5}$ are rational numbers. **3.** $\frac{-3}{4}$ **4.** $\frac{-5}{2}$

5. $\frac{7}{9}$ **6.** $\frac{5}{11}$ **7.** $-\frac{1}{10}$ **8.** $-\frac{8}{3}$ **9.** $-\frac{19}{15}$ **10.** $-\frac{6}{35}$

11. $\frac{13}{16}$ **12.** $-\frac{7}{6}$ **13.** $\frac{3}{4}$ **14.** $\frac{1}{4}$ **15.** $-\frac{4}{3}$ **16.** $\frac{9}{4}$

17. $-\frac{1}{3}$ **18.** −3918.33 **19.** −30.9 **20.** −17.64

21. −31.4 **22.** 0.00008 **23.** 245 **24.** 4.33

NAME COURSE/SECTION DATE ANSWERS

SECTION PROBLEMS 2.5

Calculate.

1. $-6.5 + 4.7$
2. $3.41 - 10.9$
3. $-2.6 - 7.15$
4. $3.6 + (-1.9)$
5. $-\frac{3}{7} + \left(-\frac{5}{7}\right)$
6. $\frac{4}{9} - \frac{6}{9}$
7. $-\frac{3}{5} + \frac{7}{4}$
8. $\frac{3}{4} + \left(-\frac{2}{3}\right)$
9. $-\frac{5}{8} + \frac{1}{4}$
10. $\frac{3}{4} - \frac{5}{3}$
11. $-2.8 + (-5.3)$
12. $5.7 - (-7.2) + 6.6$
13. $-6.1 - (-3.8)$
14. $1.562 - (-3.5)$
15. $\frac{1}{3} - \frac{1}{2} + \left(\frac{-11}{6}\right)$
16. $-\frac{3}{5} - \left(-\frac{1}{2}\right)$
17. $\frac{2}{7} - \left(\frac{-3}{14}\right)$
18. $-\frac{1}{4} - \left(-\frac{2}{3}\right) - \frac{1}{6}$
19. $-\frac{5}{4} - 1\frac{1}{5} + 2\frac{1}{3}$
20. $2\frac{1}{2} - 3\frac{1}{4} - \left(-1\frac{1}{8}\right)$
21. $-1\frac{1}{5} - 2 - \left(-3\frac{1}{3}\right)$
22. $\frac{3}{4} - \left(\frac{-2}{3}\right)$
23. $(-2.5)(-1.3)$
24. $(-4.1)(9.5)$
25. $(-8.3)(-5)(0.01)$

1. ______
2. ______
3. ______
4. ______
5. ______
6. ______
7. ______
8. ______
9. ______
10. ______
11. ______
12. ______
13. ______
14. ______
15. ______
16. ______
17. ______
18. ______
19. ______
20. ______
21. ______
22. ______
23. ______
24. ______
25. ______

26. ____________
27. ____________
28. ____________
29. ____________
30. ____________
31. ____________
32. ____________
33. ____________
34. ____________
35. ____________
36. ____________
37. ____________
38. ____________
39. ____________
40. ____________
41. ____________
42. ____________
43. ____________
44. ____________
45. ____________
46. ____________
47. ____________
48. ____________
49. ____________
50. ____________

26. $(-1.3)(-0.004)(-45.1)$

27. $(7.2) \div (-1.2)$

28. $(-4.8) \div (-4)$

29. $\left(\frac{2}{3}\right)\left(-\frac{3}{5}\right)$

30. $\left(\frac{4}{9}\right)(0)$

31. $(4)\left(\frac{1}{-8}\right)$

32. $\left(-\frac{3}{4}\right) \div (1)$

33. $\left(\frac{-5}{4}\right) \div \left(\frac{-3}{4}\right)$

34. $\left(\frac{1}{5}\right) \div \left(-1\frac{2}{3}\right)$

35. $\left(-1\frac{2}{3}\right)\left(4\frac{1}{5}\right)$

36. $\left(-2\frac{1}{5}\right)\left(-1\frac{1}{4}\right)$

37. $\left(-1\frac{1}{3}\right) \div \left(-2\frac{1}{2}\right)$

38. $3 \div \left(-3\frac{1}{4}\right)$

39. $(-4.36)(-1.04)(0)$

40. $0 \div \left(-2\frac{1}{5}\right)$

41. $-2\frac{1}{4} \div 5$

42. $\left(-1\frac{1}{4}\right)\left(-\frac{2}{3}\right)$

43. $-\frac{1}{8} - \left(-\frac{2}{3}\right)$

44. $4 - 0.387$

45. $-2\frac{1}{4} - 3\frac{1}{2} + 2\frac{2}{3} - 1\frac{1}{3}$

46. $-3.4 + (-5.6)$

47. $52.3 - 104.15$

48. $(-4.2)(3.06)$

49. $(0.84) \div (-0.7)$

50. $(14.05)(-3.69)(0)$

2.6

Introduction to Exponents

We repeat Definition 1.18 from Chapter 1.

DEFINITION 1.18 For a whole number a and a natural number n

$$a^n = \underbrace{a \cdot a \cdot a \cdot \ldots \cdot a}_{n \text{ factors}}$$

The number a is called the *base* and n is called the *exponent* (or power).

We are going to use the same definition now, but we will allow the base a to be any positive or negative rational number. For now we will restrict n to natural numbers, but in the next section we will attach a meaning to exponents that are negative integers.

Note

1. The following are different ways of denoting multiplication of the number 2 and the variable x:

$$2 \cdot x = (2)(x) = (2)x = 2(x) = 2x$$

2. The following are different ways of denoting multiplication of the variable x and the variable y:

$$x \cdot y = (x)(y) = x(y) = (x)y = xy$$

3. In algebra we do not often use $\times$ to indicate multiplication since it can be confused with x, which is frequently used as a variable.

Example 1 Expand the following.

(a) $5^3 = 5 \cdot 5 \cdot 5$

(b) $x^4 = x \cdot x \cdot x \cdot x$

(c) $\left(\frac{1}{2}\right)^5 = \left(\frac{1}{2}\right)\left(\frac{1}{2}\right)\left(\frac{1}{2}\right)\left(\frac{1}{2}\right)\left(\frac{1}{2}\right)$

(d) $(-3)^2 = (-3)(-3)$

(e) $8^1 = 8$
In fact, we have $x^1 = x$ for all x. Thus if a number or variable has no exponent showing the exponent is understood to be a positive 1.

(f) $(2x)^3 = (2x)(2x)(2x)$

DO EXERCISES 1 THROUGH 3.

Example 2 Put the following in exponential notation.

(a) $4 \cdot 4 \cdot 4 \cdot 4 \cdot 4 = 4^5$

(b) $(-6)(-6)(-6)(-6) = (-6)^4$

(c) $\left(\frac{2}{3}\right)\left(\frac{2}{3}\right) = \left(\frac{2}{3}\right)^2$

(d) $-(6 \cdot 6 \cdot 6 \cdot 6) = -(6^4)$ [Note the difference between (b) and (d).]

(e) $\left(-\frac{1}{4}\right)\left(-\frac{1}{4}\right)(7)(7)(7) = \left(-\frac{1}{4}\right)^2 \cdot (7)^3$

DO EXERCISES 4 THROUGH 7.

LEARNING OBJECTIVES

After finishing Section 2.6, you should be able to:

- Use exponential notation.
- Apply the rules for the Order of Operations.

Expand the following.

1. 4^5

2. $(-2)^4$

3. $(xy)^3$

Write using exponents.

4. $8 \cdot 8 \cdot 8 \cdot 8 \cdot 8$

5. $(-4)(-4)(-4)$

6. $-(3 \cdot 3 \cdot 3 \cdot 3)$

7. $xxx + xx$

8. How would you write x^1 without using an exponent?

Calculate.

9. 6^2

10. 2^5

11. $(-3)^2$

12. $-(7)^2$

13. $-\left(-\frac{3}{4}\right)^2$

14. $\left(-\frac{1}{3}\right)^3$

Calculate.

15. $-3 \cdot 2^3$

16. $2(1 + 3) - 4$

17. $7 - 5(2 + 1)^2$

Example 3 Calculate.

(a) $(-4)^3 = (-4)(-4)(-4) = -64$

(b) $(-0.5)^2 = (-0.5)(-0.5) = 0.25$

(c) $-\left(\frac{2}{3}\right)^4 = -\left(\frac{2}{3}\right)\left(\frac{2}{3}\right)\left(\frac{2}{3}\right)\left(\frac{2}{3}\right) = -\frac{16}{81}$

(d) $2^3 \cdot 3^2 = 2 \cdot 2 \cdot 2 \cdot 3 \cdot 3 = 8 \cdot 9 = 72$

(e) $\left(\frac{1}{2}\right)^5 = \frac{1}{2} \cdot \frac{1}{2} \cdot \frac{1}{2} \cdot \frac{1}{2} \cdot \frac{1}{2} = \frac{1}{32}$

DO EXERCISES 8 THROUGH 14.

Consider the problem $4 \cdot 3^2$. There appear to be two ways of doing this problem.

1. Multiply first and then square: $4 \cdot 3^2 = 12^2 = 144$
2. Square first and then multiply: $4 \cdot 3^2 = 4 \cdot 9 = 36$

The answers are different, so it obviously matters which operation (multiplication or exponentiation) comes first. We can now revise Rule 2.8 (Order of Operations) to include exponentiation.

> *RULE 2.8 (REVISED)* The following is the Order of Operations:
>
> 1. Perform operations within parentheses.
> 2. Perform operations involving exponents.
> 3. Multiply or divide proceeding in order from left to right.
> 4. Add or subtract proceeding in order from left to right.

Example 4 Calculate.

(a) $4 \cdot 3^2 = 4 \cdot 9 = 36$ — Exponent, multiply

(b) $3 + 4(6 - 3) = 3 + 4(3) = 3 + 12 = 15$ — Parentheses, multiply, add

(c) $-2 \cdot 4^2 + 3 = -2 \cdot 16 + 3 = -32 + 3 = -29$ — Exponent, multiply, add

(d) $(2 - 5)^2 \cdot 5 + 1 = (-3)^2 \cdot 5 + 1 = 9 \cdot 5 + 1 = 45 + 1 = 46$ — Parentheses, exponent, multiply, add

DO EXERCISES 15 THROUGH 17.

DO SECTION PROBLEMS 2.6.

ANSWERS TO MARGINAL EXERCISES

1. $4 \cdot 4 \cdot 4 \cdot 4 \cdot 4$ **2.** $(-2)(-2)(-2)(-2)$ **3.** $xyxyxy$ **4.** 8^5 **5.** $(-4)^3$ **6.** $-(3^4)$ **7.** $x^3 + x^2$ **8.** x **9.** 36 **10.** 32 **11.** 9 **12.** -49 **13.** $-\frac{9}{16}$ **14.** $-\frac{1}{27}$ **15.** -24 **16.** 4 **17.** -38

NAME ______ COURSE/SECTION ______ DATE ______

ANSWERS

1. ______
2. ______
3. ______
4. ______
5. ______
6. ______
7. ______
8. ______
9. ______
10. ______
11. ______
12. ______
13. ______
14. ______

SECTION PROBLEMS 2.6

Expand.

1. 3^6

2. $(1.4)^3$

3. $\left(-\frac{2}{3}\right)^4$

4. $(2x)^5$

Write using exponents.

5. $7 \cdot 7 \cdot 7$

6. $(-3)(-3)(-3)(-3)(-3)$

7. $(1.2)(1.2)(1.2)(1.2)$

8. $-\left(\frac{2}{3} \cdot \frac{2}{3} \cdot \frac{2}{3}\right)$

9. $xxx + yy$

10. 5

Calculate.

11. 5^3

12. $(-7)^2$

13. $\left(\frac{3}{4}\right)^2$

14. $-(7)^2$

15. ____________

16. ____________

17. ____________

18. ____________

19. ____________

20. ____________

21. ____________

22. ____________

23. ____________

24. ____________

25. ____________

26. ____________

27. ____________

28. ____________

15. $\frac{(-4)^2}{5}$

16. $-(-3)^2$

17. $(1.3)^4$

18. $(-2)^3$

19. $-\left(\frac{1}{3}\right)^3$

20. $\left(-\frac{1}{2}\right)^5$

21. $5 \cdot 2^3$

22. $6 \cdot 3^2$

23. $6 - 2^2 \cdot 3$

24. $4 + 3 \cdot 4^2$

25. $16 \div 2^3$

26. $(5 - 2)^2 \cdot 3$

27. $(4 - 6)^3 \cdot 4 - 3$

28. $3 + 2^3(2^2 \cdot 3 - 2)$

2.7

The Exponential Rules

There are numerous exponential rules that are useful in algebra. The first one we have already mentioned. (In the following rules the variable x will be an arbitrary rational number.)

RULE 2.10

$$x^1 = x$$

for all rational numbers.

The second rule we first saw in Chapter 1 (Rule 1.8). We state the rule again where now the base x is a rational number and m and n are positive integers.

RULE 2.11

$$x^n \cdot x^m = x^{n+m}$$

Note

1. In Rule 2.11 the bases must be the same.
2. All the multiplying is done by adding the exponents. Do not multiply the bases.
3. The operation between x^n and x^m must be multiplication.

Example 1 Simplify the following.

(a) $10^4 \cdot 10^7 = 10^{4+7} = 10^{11}$

(b) $(-2)^3(-2)^2(-2)^4 = (-2)^{3+2+4} = (-2)^9$

(c) $x^5 \cdot x = x^5 \cdot x^1 = x^{5+1} = x^6$

(d) $x^3 + x^2$
This cannot be simplified since the operation between the two bases is addition and not multiplication.

(e) $(3x)^2(3x)^7 = (3x)^{2+7} = (3x)^9$

(f) $2^3 \cdot 3^4$
This cannot be simplified since the bases are not the same.

DO EXERCISES 1 THROUGH 10.

LEARNING OBJECTIVES

After finishing Section 2.7, you should be able to:

- Use the exponential rules to simplify expressions involving exponents.

Simplify.

1. $3^6 \cdot 3^2$

2. $(-4)^7(-4)^3$

3. $x^8 \cdot x^2$

4. $y^5 \cdot y$

5. $(-13)^{12}(-13)^{15}$

6. $x^2 \cdot y^2$

7. $a^2 \cdot a^5 \cdot a^4$

8. $(xy)^4(xy)^2$

9. $\left(\frac{1}{2}\right)^3\left(\frac{1}{2}\right)^2$

10. $a^2ba^3b^2$

We would like to extend our definition of exponents to include negative integers. Note the pattern in the following:

$$10^4 = 10{,}000$$
$$10^3 = 1000$$
$$10^2 = 100$$
$$10^1 = 10$$

As the exponent decreases by 1, we move the decimal point in the numbers on the right one place to the left. If we continue this pattern, we obtain the following:

$$10^4 = 10{,}000$$
$$10^3 = 1000$$
$$10^2 = 100$$
$$10^1 = 10$$
$$10^0 = 1$$

We will talk more about zero exponents later.

$$10^{-1} = 0.1 = \frac{1}{10} = \frac{1}{10^1}$$

Thus $10^{-1} = \frac{1}{10^1}$.

$$10^{-2} = 0.01 = \frac{1}{100} = \frac{1}{10^2}$$

Thus $10^{-2} = \frac{1}{10^2}$.

$$10^{-3} = 0.001 = \frac{1}{1000} = \frac{1}{10^3}$$

Thus $10^{-3} = \frac{1}{10^3}$.

$$\vdots$$

Thus, if we have 10 raised to a negative exponent, we can make the exponent positive by taking the 10 from the numerator and putting it into the denominator; that is, $10^{-n} = 1/10^n$. That this situation occurs for other bases is our next exponential rule.

RULE 2.12

$$x^{-n} = \frac{1}{x^n}$$

where n is a positive integer and $x \neq 0$. (Remember that $\neq$ means "not equal to.")

This rule is critical when we attempt to evaluate a number raised to a negative exponent, for example, 5^{-2}. Although 5^2 (5 to the positive 2 power) has a numerical meaning ($5^2 = 5 \cdot 5 = 25$), 5^{-2} (5 to the negative 2 power) has no numerical meaning until we change the negative exponent to a positive exponent, that is,

$$5^{-2} = \frac{1}{5^2} = \frac{1}{5 \cdot 5} = \frac{1}{25}$$

Thus $5^{-2} = \frac{1}{25}$. (Note that $(-5)^2 = (-5)(-5) = 25$, and, in particular, $(-5)^2 \neq 5^{-2}$.)

Example 2 Calculate and/or change negative exponents to positive exponents.

(a) $3^{-3} = \frac{1}{3^3} = \frac{1}{27}$

(b) $x^{-2} = \frac{1}{x^2} \quad (x \neq 0)$

(c) $(-3)^{-2} = \frac{1}{(-3)^2} = \frac{1}{9}$

(d) $-(3)^{-2} = -\frac{1}{(3)^2} = -\frac{1}{9}$

(e) $\left(\frac{1}{2}\right)^{-5} = \frac{1}{\left(\frac{1}{2}\right)^5} = 1 \div \left(\frac{1}{2}\right)^5 = 1 \div \frac{1}{32} = 1 \cdot \frac{32}{1} = 32$

(f) $x^{-3}y^2 = \frac{1}{x^3} \cdot \frac{y^2}{1} = \frac{y^2}{x^3} \quad (x \neq 0)$

DO EXERCISES 11 THROUGH 20.

We know what happens if a base raised to a negative exponent starts out in the numerator, but what if it starts out in the denominator, as in $1/5^{-2} = ?$ We can use Rule 2.12 to solve such problems, as the following examples show.

Example 3 Change negative exponents to positive exponents.

(a) $\frac{1}{5^{-2}} = \frac{1}{\frac{1}{5^2}} = 1 \div \frac{1}{5^2} = 1 \cdot \frac{5^2}{1} = 5^2$

(b) $\frac{1}{3^{-4}} = \frac{1}{\frac{1}{3^4}} = 1 \div \frac{1}{3^4} = 1 \cdot \frac{3^4}{1} = 3^4$

(c) $\frac{1}{x^{-3}} = \frac{1}{\frac{1}{x^3}} = 1 \div \frac{1}{x^3} = 1 \cdot \frac{x^3}{1} = x^3 \quad (x \neq 0)$

DO EXERCISES 21 AND 22.

Note the pattern in Example 3:

$$\frac{1}{5^{-2}} = 5^2, \qquad \frac{1}{x^{-3}} = x^3, \qquad \frac{1}{3^{-4}} = 3^4$$

In other words, to change the exponent from negative to positive, we brought the base from the denominator to the numerator and made the exponent positive.

Evaluate and/or change negative exponents to positive exponents (assume $x \neq 0$, $y \neq 0$).

11. 7^{-3}

12. 9^{-2}

13. $(-4)^{-2}$

14. $\left(\frac{1}{4}\right)^{-3}$

15. 6^{-1}

16. $\left(\frac{2}{3}\right)^{-4}$

17. y^{-3}

18. $(4x)^{-3}$

19. $4x^{-3}$

20. $2x^{-2}y^{-5}$

Write with positive exponents. Assume all variables are nonzero.

21. $\frac{1}{4^{-2}}$

22. $\frac{1}{x^{-5}}$

Write with positive exponents. Assume all variables are nonzero.

23. 3^{-3}

24. 7^{-3}

25. $\frac{1}{3^{-4}}$

26. $\frac{x^{-2}}{y^{-4}}$

27. $\left(\frac{2}{3}\right)^{-2}$

Calculate and simplify. (Leave no negative exponents in your final answer.)

28. $x^{-3} \cdot x^7$

29. $x^{-5} \cdot x \cdot x^{-2}$

30. $3^{-1} \cdot y^2 \cdot 3^3 \cdot y^{-4}$

Example 4 Change all negative exponents to positive exponents. Assume all variables are nonzero.

(a) $x^{-4} = \frac{1}{x^4}$

(b) $\frac{1}{4^{-3}} = 4^3$

(c) $\frac{x^{-2}}{y^3} = \frac{x^{-2}}{1} \cdot \frac{1}{y^3} = \frac{1}{x^2} \cdot \frac{1}{y^3} = \frac{1}{x^2 y^3}$

(d) $\frac{3^{-2}}{x^{-5}} = \frac{3^{-2}}{1} \cdot \frac{1}{x^{-5}} = \frac{1}{3^2} \cdot \frac{x^5}{1} = \frac{x^5}{3^2}$

(e) $7x^{-2} = \frac{7}{x^2}$

(f) $(7x)^{-2} = \frac{1}{(7x)^2}$

DO EXERCISES 23 THROUGH 27.

Now that we have a definition of negative exponents, we can remove the restriction that m and n must be positive in Rules 2.11 and 2.12. We obtain

RULE 2.11

$$x^n \cdot x^m = x^{n+m}$$

for all integers n and m and $x \neq 0$.

RULE 2.12

$$x^{-n} = \frac{1}{x^n}$$

for all integers n and $x \neq 0$.

Example 5 Calculate and/or simplify.

(a) $2^3 \cdot 2^{-2} = 2^{3+(-2)} = 2^1 = 2$

(b) $x^{-4} \cdot x^{-3} = x^{-4+(-3)} = x^{-7} = \frac{1}{x^7} \quad (x \neq 0)$

(c) $3^{-5} \cdot 3^3 \cdot 3^{-1} = 3^{-5+3+(-1)} = 3^{-3} = \frac{1}{3^3} = \frac{1}{27}$

In problems (b) and (c) of Example 5, note that we did not change negative exponents to positive exponents until after we had done all the simplifying we could. Then as a final step we changed x^{-7} to $1/x^7$ and 3^{-3} to $1/3^3$.

DO EXERCISES 28 THROUGH 30.

Our next exponential rule is for dividing when we have the same base.

RULE 2.13

$$\frac{x^n}{x^m} = x^{n-m}$$

where n and m are integers and $x \neq 0$.

Notice the following facts about Rule 2.13:

1. In the rule it is $n - m$ and not $m - n$. Since subtraction is not commutative, this makes a difference.
2. The bases must be the same.
3. All the division is done by subtracting the exponents. Do not divide the bases.
4. Rule 2.13 can be arrived at by using Rules 2.11 and 2.12:

$$\frac{x^n}{x^m} = \underbrace{x^n \cdot x^{-m}}_{\text{Rule 2.12}} = \underbrace{x^{n+(-m)}}_{\text{Rule 2.11}} = x^{n-m}$$

Example 6 Calculate and/or simplify. Assume all variables are nonzero.

(a) $\frac{x^7}{x^3} = x^{7-3} = x^4$

(b) $\frac{y^{-3}}{y^{-5}} = y^{-3-(-5)} = y^{-3+[-(-5)]}$

$= y^{-3+5} = y^2$

(c) $\frac{2^{-4}}{2^{-2}} = 2^{-4-(-2)} = 2^{-4+2}$

$= 2^{-2} = \frac{1}{2^2} = \frac{1}{4}$

(d) $\frac{3^5 \cdot 3^{-2}}{3^4} = \frac{3^{5+(-2)}}{3^4} = \frac{3^3}{3^4}$

$= 3^{3-4} = 3^{-1} = \frac{1}{3}$

DO EXERCISES 31 THROUGH 38.

Our next rule was presented first in Chapter 1 (Rule 1.9). It involves having a base raised to an exponent and then raised to another exponent.

RULE 2.14

$$(x^m)^n = x^{m \cdot n}$$

for all integers m and n and $x \neq 0$.

Simplify and/or calculate. Assume all variables are nonzero.

31. $\frac{y^{10}}{y^4}$

32. $\frac{x^{-3}}{x^2}$

33. $\frac{x^2}{x^{-5}}$

34. $\frac{a^4 \cdot a^{-3}}{a^{-2}}$

35. $\frac{3^5}{3^2}$

36. $\frac{10^{-2}}{10^{-4}}$

37. $\frac{4^3}{4^5}$

38. $\frac{2^{-2} \cdot 2^3}{2^{-1}}$

Simplify and/or calculate. Assume all variables are nonzero.

39. $(10^2)^3$

40. $(y^{-3})^5$

41. $y^{-3} \cdot y^5$

42. $(3^2)^{-2}$

43. $\frac{2^4}{2^{-1}}$

44. $\frac{(4^2)^3 \cdot 4^{-2}}{4^5}$

Be careful not to confuse Rule 2.11 ($x^n \cdot x^m = x^{n+m}$) where we add exponents and Rule 2.14 [$(x^m)^n = x^{m \cdot n}$] where we multiply exponents.

Example 7 Calculate and simplify. Assume all variables are nonzero.

(a) $(x^3)^{-2} = x^{(3)(-2)} = x^{-6} = \frac{1}{x^6}$

(b) $x^3 \cdot x^{-2} = x^{3+(-2)} = x^1 = x$

(c) $(2^2)^4 = 2^{(2)(4)} = 2^8 = 256$

(d) $$\frac{(2^{-3})^2 \cdot (2^4)}{(2^{-2})^{-2}} = \frac{2^{(-3)(2)} \cdot 2^4}{2^{(-2)(-2)}}$$
$$= \frac{2^{-6} \cdot 2^4}{2^4} = \frac{2^{-6+4}}{2^4}$$
$$= \frac{2^{-2}}{2^4} = 2^{-2-4}$$
$$= 2^{-6} = \frac{1}{2^6} = \frac{1}{64}$$

DO EXERCISES 39 THROUGH 44.

Note the following, which lead to the next rule.

1. $(x \cdot y)^2 = (x \cdot y)(x \cdot y) = x \cdot y \cdot x \cdot y$
 $= x \cdot x \cdot y \cdot y = x^2 \cdot y^2$
2. $(2x)^3 = (2x) \cdot (2x) \cdot (2x) = 2 \cdot x \cdot 2 \cdot x \cdot 2 \cdot x$
 $= 2 \cdot 2 \cdot 2 \cdot x \cdot x \cdot x = 2^3 \cdot x^3$

In general, to remove the parentheses when we have a product inside the parentheses and an exponent outside the parentheses, we apply the following rule.

RULE 2.15

$$(xy)^n = x^n y^n$$

for all integers n and $x \neq 0$, $y \neq 0$.

Simplify and/or calculate. Assume all variables are nonzero.

45. $(x^2y^3)^2$

46. $(3x^{-2})^{-2}$

Example 8 Calculate and simplify. Assume all variables are nonzero.

(a) $(xy)^3 = x^3 \cdot y^3$

(b) $(2x)^4 = 2^4 \cdot x^4 = 16x^4$

(c) $(2x^2y^{-1})^5 = 2^5(x^2)^5(y^{-1})^5 = 32x^{2 \cdot 5}y^{(-1)(5)}$
$$= 32x^{10}y^{-5} = \frac{32x^{10}}{y^5}$$

(d) $(3x^{-2})^{-2} = 3^{-2}(x^{-2})^{-2} = \frac{x^{(-2)(-2)}}{3^2} = \frac{x^4}{9}$

DO EXERCISES 45 AND 46.

Example 9 We cannot use Rule 2.15 if we have the operation of addition or subtraction inside the parentheses. That is,

$$(x + y)^2 \neq x^2 + y^2$$

For example,

$$25 = (2 + 3)^2 \neq 2^2 + 3^2 = 4 + 9 = 13$$

Later we will see that

$$(x + y)^2 = x^2 + 2xy + y^2$$

DO EXERCISE 47.

We have a similar rule for removing parentheses around a quotient.

RULE 2.16

$$\left(\frac{x}{y}\right)^n = \frac{x^n}{y^n}$$

for all integers n and $x \neq 0$, $y \neq 0$.

Example 10 Calculate and simplify. Assume all variables are nonzero.

(a) $\left(\frac{x}{y}\right)^3 = \frac{x^3}{y^3}$

(b) $\left(\frac{2x^2}{y}\right)^2 = \frac{(2x^2)^2}{y^2} = \frac{2^2(x^2)^2}{y^2} = \frac{4x^4}{y^2}$

(c) $\left(\frac{-2x^2y}{3^{-1}x^{-3}}\right)^{-2} = \frac{(-2x^2y)^{-2}}{(3^{-1}x^{-3})^{-2}} = \frac{(-2)^{-2}(x^2)^{-2}y^{-2}}{(3^{-1})^{-2}(x^{-3})^{-2}}$

$$= \frac{(-2)^{-2}x^{-4}y^{-2}}{3^2x^6} = \frac{x^{-4-6}y^{-2}}{(-2)^2 3^2}$$

$$= \frac{x^{-10}y^{-2}}{4 \cdot 9} = \frac{1}{36x^{10}y^2}$$

Note that we used many different exponential rules to simplify this example.

DO EXERCISES 48 THROUGH 52.

We still need to define what an exponent of zero is going to mean.

Note

1. In the rule

$$\frac{x^n}{x^m} = x^{n-m}$$

suppose that n and m are the same. Then we obtain

$$1 = \frac{x^n}{x^n} = x^{n-n} = x^0$$

47. Show that $(7 - 3)^2$ and $7^2 - 3^2$ are not the same number.

Simplify.

48. $\left(\frac{2^3}{2}\right)^2$

49. $\left(\frac{x^{-1}}{y^2}\right)^3$

50. $\frac{10^{-2} \cdot 10^6}{10^4 \cdot 10}$

51. $\frac{(x^3y)^{-2}}{(x^{-2}y^2)^3}$

52. $\left(\frac{xy^2}{x^{-2}y}\right)^{-2}$

Simplify and/or calculate. Assume all variables are nonzero.

53. $(23)^0$

54. $(-4x^2)^0$

55. $\dfrac{(2x^2y)^3}{(-3x^{-3}y^2)^{-2}}$

2. Recall that earlier we had $10^0 = 1$.

Thus, the rule (it is actually a definition) for an exponent of zero is as follows.

RULE 2.17

$$x^0 = 1$$

for all $x \neq 0$. The expression 0^0 is not defined.

Example 11 Calculate and simplify. Assume all variables are nonzero.

(a) $7^0 = 1$

(b) $(-3.4)^0 = 1$

(c) $(2x^4y^{-3})^0 = 1$

DO EXERCISES 53 THROUGH 55.

DO SECTION PROBLEMS 2.7.

ANSWERS TO MARGINAL EXERCISES

1. 3^8 **2.** $(-4)^{10}$ **3.** x^{10} **4.** y^6 **5.** -13^{27} **6.** x^2y^2 **7.** a^{11} **8.** $(xy)^6$ **9.** $\left(\frac{1}{2}\right)^5$ **10.** a^5b^3 **11.** $\frac{1}{7^3}$ **12.** $\frac{1}{81}$ **13.** $\frac{1}{16}$ **14.** 64 **15.** $\frac{1}{6}$ **16.** $\frac{81}{16}$ **17.** $\frac{1}{y^3}$ **18.** $\frac{1}{(4x)^3}$ **19.** $\frac{4}{x^3}$ **20.** $\frac{2}{x^2y^5}$ **21.** 4^2 **22.** x^5 **23.** $\frac{1}{3^3}$ **24.** $\frac{1}{7^3}$ **25.** 3^4 **26.** $\frac{y^4}{x^2}$ **27.** $\frac{9}{4}$ **28.** x^4 **29.** $\frac{1}{x^6}$ **30.** $\frac{3^2}{y^2}$ **31.** y^6 **32.** $\frac{1}{x^5}$ **33.** x^7 **34.** a^3 **35.** 27 **36.** 10^2 **37.** $\frac{1}{16}$ **38.** 4 **39.** 10^6 **40.** $\frac{1}{y^{15}}$ **41.** y^2 **42.** $\frac{1}{81}$ **43.** 32 **44.** $\frac{1}{4}$ **45.** x^4y^6 **46.** $\frac{x^4}{9}$ **47.** $(7-3)^2 = 4^2 = 16 \neq 7^2 - 3^2 = 40$ **48.** 16 **49.** $\frac{1}{x^3y^6}$ **50.** $\frac{1}{10}$ **51.** $\frac{1}{y^8}$ **52.** $\frac{1}{x^6y^2}$ **53.** 1 **54.** 1 **55.** $72y^7$

NAME COURSE/SECTION DATE

SECTION PROBLEMS 2.7

Calculate or simplify. Assume all variables are nonzero.

1. 9^{-2} **2.** 3^{-3} **3.** $(-2)^{-4}$ **4.** $(-3)^{-3}$

5. $\left(\frac{1}{5}\right)^{-2}$ **6.** $\left(\frac{2}{3}\right)^{-3}$ **7.** $(1.5)^{-2}$ **8.** $-(-2)^{-2}$

9. $4^3 \cdot 4^3$ **10.** $x^2 \cdot x^5$ **11.** $\left(\frac{1}{2}\right)^3\left(\frac{1}{2}\right)\left(\frac{1}{2}\right)^{-2}$ **12.** $x^2 \cdot x^{-4}$

13. $\frac{6^{-3}}{6^{-1}}$ **14.** $\frac{x^4}{x^{-3}}$ **15.** $\frac{y^{-2}}{y^4}$ **16.** $\frac{x^4}{x^6}$

17. $\frac{2^7}{2^4}$ **18.** $\frac{3^{-2}}{3^{-3}}$ **19.** $\frac{y^3 \cdot y^{-4}}{y^5}$ **20.** $\frac{2^{-3} \cdot 2^4}{2^{-1}}$

21. $(2x^2y)^0$ **22.** $a^3 \cdot a \cdot a^{-4}$ **23.** $\frac{y^8}{y^{16}}$ **24.** $(2^2)^{-3}$

25. $(x^2)^3$ **26.** $\frac{x^{-2}}{x^3}$ **27.** $\frac{y^4}{y^{-2}}$ **28.** $(2x^2)^3$

29. $(x^2)^{-3}$ **30.** $\frac{(2^{-3})^2}{2^{-4}}$ **31.** $\frac{(2x)^{-2}}{x^4}$ **32.** $\frac{(3y)^{-2}}{(2y)^{-1}}$

ANSWERS

1. ______
2. ______
3. ______
4. ______
5. ______
6. ______
7. ______
8. ______
9. ______
10. ______
11. ______
12. ______
13. ______
14. ______
15. ______
16. ______
17. ______
18. ______
19. ______
20. ______
21. ______
22. ______
23. ______
24. ______
25. ______
26. ______
27. ______
28. ______
29. ______
30. ______
31. ______
32. ______

33. ______
34. ______
35. ______
36. ______
37. ______
38. ______
39. ______
40. ______
41. ______
42. ______
43. ______
44. ______
45. ______
46. ______
47. ______
48. ______
49. ______
50. ______
51. ______
52. ______
53. ______
54. ______
55. ______
56. ______

33. $\frac{x^{-3} \cdot x^5}{x^4}$

34. $(-2x^{-2}y)^4$

35. $\left(\frac{3x}{y}\right)^3$

36. $(2x^{-1}y)^2 \cdot (x^2y)^{-2}$

37. $(9x^2y^{-5}z)^0$

38. $\left(\frac{2^{-1}x^2}{y}\right)^{-3}$

39. $(2xy^{-1})^2 \cdot (x^2y)^{-3}$

40. $\left(\frac{2^{-3} \cdot 2^4}{2^{-2}}\right)^2$

41. $\frac{(x^2y^{-1})^3}{(x^{-2}y)^{-3}}$

42. $7x^{-1}$

43. $(7x)^{-1}$

44. $(-3x^2y)^0$

45. $\frac{(2^3)^{-2} \cdot 2^{-4}}{(2^{-5})^2}$

46. $\left(\frac{2}{3}\right)^{-3} \cdot \left(\frac{3}{2}\right)^2$

47. $\frac{-(-5)^{-3}}{(-5)^2}$

48. $\frac{3^2 \cdot 3^{-5}}{(3^{-1})^4}$

49. $x^4 \cdot x^{-3} \cdot x^2$

50. $\frac{7^{-1} \cdot 7^3}{7^4 \cdot 7^{-2}}$

51. $\frac{2^3}{2^{-2} \cdot 2^4}$

52. $(-2x^4)(-4x^{-7})$

53. $(9x^2y)^0$

54. $(-3xy^2)^4$

55. $\left(\frac{x^2y^{-1}}{x^2y}\right)^{-2}$

56. $\frac{(-2)^3}{-(3)^{-2}}$

2.8

Laws for Rational Numbers

LEARNING OBJECTIVES

After finishing Section 2.8, you should be able to:

- Identify the commutative, associative, and distributive laws.
- Multiply expressions such as $2(x + 5)$.
- Factor expressions such as $3x - 6y + 9$.

We looked at the commutative, associative, and distributive laws for whole numbers, fractions, and decimals in Chapter 1. These laws also hold for rational numbers, and we have been using them throughout this chapter. However, we will state them again formally, look at some examples, and then single out the distributive law for special attention.

In what follows, a, b, and c are arbitrary rational numbers.

COMMUTATIVE LAW FOR ADDITION

$$a + b = b + a$$

COMMUTATIVE LAW FOR MULTIPLICATION

$$ab = ba$$

ASSOCIATIVE LAW FOR ADDITION

$$(a + b) + c = a + (b + c)$$

ASSOCIATIVE LAW FOR MULTIPLICATION

$$(ab)c = a(bc)$$

DISTRIBUTIVE LAW OF MULTIPLICATION OVER ADDITION

$$a(b + c) = ab + ac$$

Example 1 The following are illustrations of which law(s)?

(a) $2 + \frac{1}{2} = \frac{1}{2} + 2$ — *Commutative* (addition)

(b) $\left(\frac{3}{4} \cdot 2\right) \cdot (-3) = \frac{3}{4} \cdot [(2) \cdot (-3)]$ — *Associative* (multiplication)

(c) $-4\left(2.3 + \frac{2}{3}\right) = (-4)(2.3) + (-4)\left(\frac{2}{3}\right)$ — *Distributive*

(d) $\left(1\frac{3}{4}\right)\left(-2\frac{1}{3}\right) = \left(-2\frac{1}{3}\right)\left(1\frac{3}{4}\right)$ — *Commutative* (multiplication)

(e) $(2 + 0.4) + 1.5 = 2 + (1.5 + 0.4)$ — *Associative* and *Commutative* (addition)

(f) $3(2x) = (3 \cdot 2)x$ — *Associative* (multiplication)

Which law(s) are demonstrated by the following?

1. $(7.2)(3.1) = (3.1)(7.2)$

2. $\left(-\frac{1}{2} + 3\right) + \frac{2}{7} = -\frac{1}{2} + \left(3 + \frac{2}{7}\right)$

3. $1.4(-3 + 2.5)$
$= (1.4)(-3) + (1.4)(2.5)$

4. $(3 \cdot 4) \cdot x = (4 \cdot 3) \cdot x$

5. $-4(7x) = [(-4)(7)]x$

Identify each term in the following algebraic expressions.

6. $6x + 7xy$

7. $0.14r - 3.5rst + 7$

8. $\frac{7}{2}a^2 + 13b^2$

Identify the numerical coefficient of each term in the following algebraic expressions.

9. $6x + 7xy$

10. $0.14r - 3.5rst$

11. $\frac{7}{2}a^2 - 13b^2$

(g) $-4 + y = y + (-4)$ — *Commutative* (addition)

(h) $3 + (4 + x) = 3 + (x + 4)$ — *Commutative* (addition)

DO EXERCISES 1 THROUGH 5.

Before we look more closely at the distributive law, we need to establish some terminology that we will be using from now on.

DEFINITION 2.7 If we combine variables or numbers (or both) with other variables or numbers (or both) by adding, subtracting, multiplying, or dividing, then we have an *algebraic expression.*

Example 2 The following are examples of algebraic expressions.

(a) $3x + 4y$ (b) $0.5st + \frac{7s}{t}$ (c) $2x - 4m + 5$

(d) $abc - 2ab$ (e) $2\pi r^2 + 2\pi rh$ (f) t

DEFINITION 2.8 The *terms* of an algebraic expression are separated by addition signs. A term may consist of the product of a number and a variable(s), just a number, or just a variable(s).

Example 3

(a) The terms of the algebraic expression $3x + 4y$ are $3x$ and $4y$.

(b) The terms of $7st - 2s + 15 = 7st + (-2s) + 15$ are $7st$, $-2s$, and 15. The 15 is called a *constant term.*

(c) Note that the single term $7st$ consists of three factors, a 7, an s, and a t.

DO EXERCISES 6 THROUGH 8.

DEFINITION 2.9 The *numerical coefficient* of a term is the number in the term that is multiplied times the variable(s).

Example 4

(a) In the algebraic expression $3x - 4xy$, the numerical coefficient of the $3x$ term is 3 and the numerical coefficient of the $-4xy$ term is -4.

(b) In the expression $10t + x$, the numerical coefficient of the x term is 1. That is, $x = 1 \cdot x$.

DO EXERCISES 9 THROUGH 11.

Now we will look more closely at the distributive law, $a(b + c) = ab + ac$. First note that:

1. The distributive law can be used with numbers or variables or both.

   EXAMPLES: $3(2 + 7) = 3 \cdot 2 + 3 \cdot 7$
   $4(x + 5) = 4 \cdot x + 4 \cdot 5$
   $-3(x + y) = (-3)(x) + (-3)(y)$

2. There is a distributive law for multiplication over subtraction which says that $a(b - c) = ab - ac$. However, if we change subtraction to addition, we can do with just a distributive law for multiplication over addition. That is, $a(b - c) = a[b + (-c)] = ab + a(-c)$.

From the distributive law we notice that the expressions $2(x + 5)$ and $2x + 10$ are equal. It is often necessary to be able to change from one form of an expression to the other form. Before we do some examples of this we need the following definition.

DEFINITION 2.10

$$\underbrace{2(x + 5)}_{\text{Factored form}} = \underbrace{2x + 10}_{\text{Multiplied form}}$$

If we start with the form $2(x + 5)$ and rewrite this as $2x + 10$, we are *multiplying*. If we start with the form $2x + 10$ and rewrite it as $2(x + 5)$, we are *factoring*.

Multiplying

$$2(x + 5) = 2x + 10$$

Factoring

Note that multiplying in Definition 2.10 has the effect of removing the parentheses.

Example 5 Multiply the following expressions.

(a) $3(2x + 5) = 3(2x) + 3(5)$
$= 6x + 15$

Note that $3(2x) = (3 \cdot 2)x = 6x$ and demonstrates a use of the associative law.

(b) $-2(x - 4) = -2[x + (-4)] = (-2)(x) + (-2)(-4) = -2x + 8$

(c) $3(-2x - 8) = 3[-2x + (-8)] = 3(-2x) + 3(-8)$
$= -6x + (-24)$ (or $-6x - 24$)

(d) $\frac{2}{3}(3x - 6y + 9) = \frac{2}{3}[3x + (-6y) + 9] = \frac{2}{3}(3x) + \frac{2}{3}(-6y) + \frac{2}{3}(9)$
$= 2x + (-4y) + 6$ (or $2x - 4y + 6$)

DO EXERCISES 12 THROUGH 16.

Multiply.

12. $3(8x - 2)$

13. $-2(y + 4)$

14. $-5(2x - 4)$

15. $-3(-x - 4y + 1)$

16. $\frac{1}{2}(2x - 6y + 5)$

Factor.

17. $5x - 20$

18. $-8x + 12$

19. $9x + 3y - 15$

20. $-3x - 6$

21. $3bx + 6b$

22. Are $ba + ca$ and $(b + c)a$ equal?

When we factor an expression, we first look for a common factor (which may be a number or a variable) in each term of the algebraic expression.

Example 6 Factor the following expressions.

(a) $2x + 6$ We can factor a 2 from each term.

$$2x + 6 = \underline{2} \cdot x + \underline{2} \cdot 3 = 2(x + 3)$$

(b) $20x + 12y + 8$ We can factor a 4 from each term.

$$20x + 12y + 8 = \underline{4} \cdot 5x + \underline{4} \cdot 3y + \underline{4} \cdot 2 = 4(5x + 3y + 2)$$

(c) $-3x + 15$ We can factor a -3 from each term.

$$-3x + 15 = \underline{-3}x + (\underline{-3})(-5) = -3[x + (-5)] \quad [\text{or } -3(x - 5)]$$

or we can factor a 3 from each term.

$$-3x + 15 = \underline{3}(-x) + \underline{3}(5) = 3(-x + 5)$$

(d) $3xy + 5x$ We can factor an x from each term.

$$3xy + 5x = 3\underline{x}y + 5\underline{x} = x(3y + 5)$$

(e) $-4ab - 26ac$ We can factor a $-2a$ from each term.

$$-4ab - 26ac = -4ab + (-26ac) = \underline{(-2a)}(2b) + \underline{(-2a)}(13c) = -2a(2b + 13c)$$

Note that (1) $4x + 6 = 2(2x + 3)$ but also that (2) $4x + 6 = 4\left(x + \frac{3}{2}\right)$. The first factorization is "legal" but the second is not. When we factor we want all coefficients to be integers, that is, no rational numbers like $\frac{3}{2}$ are allowed.

In Chapter 5 we will go into more detail on factoring.

DO EXERCISES 17 THROUGH 22.

DO SECTION PROBLEMS 2.8.

ANSWERS TO MARGINAL EXERCISES

1. Commutative (multiplication) **2.** Associative (addition) **3.** Distributive **4.** Commutative (multiplication) **5.** Associative (multiplication) **6.** $6x$; $7xy$ **7.** $0.14r$; $-3.5rst$; 7 **8.** $\frac{7}{2}a^2$; $13b^2$ **9.** 6; 7 **10.** 0.14; -3.5 **11.** $\frac{7}{2}$; -13 **12.** $24x + (-6)$ or $24x - 6$ **13.** $-2y + (-8)$ or $-2y - 8$ **14.** $-10x + 20$ **15.** $3x + 12y + (-3)$ or $3x + 12y - 3$ **16.** $x + (-3y) + \frac{5}{2}$ or $x - 3y + \frac{5}{2}$ **17.** $5[x + (-4)]$ or $5(x - 4)$ **18.** $-4[2x + (-3)]$ or $-4(2x - 3)$ or $4(-2x + 3)$ **19.** $3[3x + y + (-5)]$ or $3(3x + y - 5)$ **20.** $-3(x + 2)$ or $3[-x + (-2)]$ **21.** $3b(x + 2)$ **22.** Yes

NAME ______ COURSE/SECTION ______ DATE ______

SECTION PROBLEMS 2.8

Which law(s) is demonstrated in the following?

1. $-3 + 7 = 7 + (-3)$

2. $4\left(-\frac{2}{3} + 5\right) = (4)\left(-\frac{2}{3}\right) + (4)(5)$

3. $\left(3 \cdot \frac{1}{5}\right) \cdot (-2) = 3 \cdot \left[\left(\frac{1}{5}\right)(-2)\right]$

4. $(-2)(-5) = (-5)(-2)$

5. $-2(x + 3) = (-2)(x) + (-2)(3)$

6. $7(3x) = (7 \cdot 3)x$

7. $(7 + 2) + x = (2 + 7) + x$

8. $2 + (3 + x) = (2 + 3) + x$

9. $5x + 5y = 5(x + y)$

10. $ab = ba$

Identify each term in the following algebraic expressions.

11. $3x + 4y$

12. $\frac{7}{2}x^2 - 3x + 4$

13. $0.5y - 1.4x$

14. $1.1x^2 - xy + y + \frac{3}{4}$

Identify the numerical coefficient of each term in the following algebraic expressions.

15. $\frac{7}{2}x^2 - 3x$

16. $1.1x^2 - xy + 1$

Multiply.

17. $3(x + 4)$

18. $-2(3x + 7)$

19. $7(x - y)$

20. $6(2a - 3b + 4)$

ANSWERS

1. ______
2. ______
3. ______
4. ______
5. ______
6. ______
7. ______
8. ______
9. ______
10. ______
11. ______
12. ______
13. ______
14. ______
15. ______
16. ______
17. ______
18. ______
19. ______
20. ______

21. ______
22. ______
23. ______
24. ______
25. ______
26. ______
27. ______
28. ______
29. ______
30. ______
31. ______
32. ______
33. ______
34. ______
35. ______
36. ______
37. ______
38. ______
39. ______
40. ______
41. ______
42. ______

21. $-4(7x - y - 5)$

22. $-1(-2 - x)$

23. $-\frac{3}{5}(10x - 3)$

24. $4(-3x - 2y + 1)$

25. $-\frac{1}{2}(-2x + 4y + 5)$

26. $-3(2x - y + 4)$

27. $-1.8(4x - 5)$

28. $2.3(-0.5x + 0.4y)$

Factor.

29. $3x + 12$

30. $4x - 10$

31. $-5 + 25a$

32. $12 - 3t + 6s$

33. $-6x + 3y - 18$

34. $6ax - 10ay$

35. $-4x + 6y - 8$

36. $10x - 25y$

37. $-6ab - 3ac + 3a$

38. $4x - 12y + 5$

39. $-3x + 12$

40. $4x + 10y - 8$

41. $7x^2y - 14xy^2$

42. $-15abc - 10ab$

2.9

Simplifying Algebraic Expressions

We begin with a definition.

DEFINITION 2.11 *Like* (or similar) *terms* of an algebraic expression are terms (1) that have the same variable (or variables) and (2) whose variable(s) must have the same exponent(s).

Recall that the terms of an algebraic expression are separated by addition signs and that the numerical coefficient of a term is the number that is multiplied times the variable(s).

Example 1 Consider $2x + y + 3x + 4x^2$.

(a) $2x$ and $3x$ are like terms.

(b) $2x$ and y are not like terms (different variables).

(c) $2x$ and $4x^2$ are not like terms (different exponents).

Example 2 Consider $6ab + 2b + (-ab) + 4b$.

(a) $6ab$ and $-ab$ are like terms.

(b) $2b$ and $4b$ are like terms.

Example 3 Consider $2x^2 + xy + 3x^2 + y^2 + 4xy$.

(a) $2x^2$ and $3x^2$ are like terms.

(b) xy and $4xy$ are like terms.

DO EXERCISES 1 THROUGH 6.

If we have like terms in an algebraic expression, then we may combine the like terms. The distributive law is our justification for this. For example,

$$\underbrace{2x + 3x = (2 + 3)x}_{\text{Distributive law}} = 5x$$

RULE 2.18 To combine like terms, we add their numerical coefficients.

Example 4 Simplify the following algebraic expressions by combining all like terms.

(a) $4x - 2y + 3x = 4x + (-2y) + 3x = 4x + 3x + (-2y)$
$= 7x + (-2y) \quad$ (or $7x - 2y$)

(b) $1 + x + 2y + (-2x) + 3 = 1 + 3 + x + (-2x) + 2y$
$= 4 + (-x) + 2y \quad$ (or $4 - x + 2y$)

LEARNING OBJECTIVES

After finishing Section 2.9, you should be able to:

- Identify and combine like terms.
- Simplify algebraic expressions by removing grouping symbols and combining like terms.

Identify the like terms in the following algebraic expressions.

1. $3x - 4y + x + 7$

2. $2a^2 + 7ab - 4a + a^2 - 2ab^2 - 3a^2 + 6a + 1$

3. $4 - x^2y^2 + 3x^2 + 6 + (2xy)^2$

4. Why aren't x^3y^2 and x^2y^3 like terms (or are they)?

5. Why aren't $2x$ and $3xy$ like terms (or are they)?

6. Why aren't $-3xy^4$ and $\frac{5}{2}xy^4$ like terms (or are they)?

Combine like terms.

7. $2x - 4x$

8. $x + 3y + 4x - y$

9. $a^2 + 3a^2b^2 - 2b^2 + 7a^2b^2 - 5b^2 + a$

10. $7s - 2st + 3t - 4st + s - 5$

11. $1.3x - 0.4x$

12. $\frac{3}{5}y - \frac{1}{2} - \frac{11}{15}y - 2$

Simplify.

13. $3(x - 4) - 4x + 1$

14. $7x - 2(-x + 7) - 10xy$

15. $2(a^2 - b) - 3(2a^2 + b)$

16. $-3s + 2(s - t) - 4t$

Remove the parentheses.

17. $-(2x + 3)$

18. $-4(x + 1)$

19. $-(-3s - t)$

20. $-(a - 3)$

21. $-(-x + y - 2)$

(c) $y^2 - 3y + 2y^2 + 4y + 5 = y^2 + (-3y) + 2y^2 + 4y + 5$
$= y^2 + 2y^2 + (-3y) + 4y + 5$
$= 3y^2 + y + 5$

(d) $\frac{2}{3}x + \frac{3}{4}x = \frac{8}{12}x + \frac{9}{12}x = \frac{17}{12}x \quad \left(\text{or } \frac{17x}{12}\right)$

Note that $\frac{17}{12}x = \frac{17}{12} \cdot \frac{x}{1} = \frac{17x}{12}$.

DO EXERCISES 7 THROUGH 12.

At times it is necessary to remove parentheses before we can simplify by combining like terms.

Example 5 Simplify.

(a) $x + 3(2x + y) - y = x + 6x + 3y + (-y) = 7x + 2y$

(b) $x - 4(x - y) + 3y = x + (-4)[x + (-y)] + 3y$
$= x + (-4x) + 4y + 3y$
$= -3x + 7y$

(c) $0.2(1.3x - 4) + 0.5x + 0.3 = 0.2[1.3x + (-4)] + 0.5x + 0.3$
$= (0.2)(1.3x) + (0.2)(-4) + 0.5x + 0.3$
$= 0.26x + (-0.8) + 0.5x + 0.3$
$= 0.76x + (-0.5) \quad (\text{or } 0.76x - 0.5)$

DO EXERCISES 13 THROUGH 16.

Note the following:

1. $-5 = -(5) = (-1)(5)$
2. $-12 = -(12) = (-1)(12)$

In general, we have the following rule.

RULE 2.19 We may replace a negative sign in front of a parenthesis with -1. That is, $-(x) = (-1)(x)$.

Example 6 Remove the parentheses.

(a) $-(x + 2) = (-1)(x + 2)$
$= (-1)(x) + (-1)(2) = -x + (-2) \quad (\text{or } -x - 2)$

(b) $-(-x - 2y + 3) = (-1)[-x + (-2y) + 3]$
$= (-1)(-x) + (-1)(-2y) + (-1)(3)$
$= x + 2y + (-3) \quad (\text{or } x + 2y - 3)$

DO EXERCISES 17 THROUGH 21.

In the following examples we show all the steps involved, including changing subtraction to addition. Eventually, as we gain more experience and confidence, we would expect that some shortcuts could be used, for example, not changing subtraction to addition. However, it is never wrong to put in all the steps, and until the ideas and patterns involved here are mastered, be very careful in taking shortcuts.

Example 7 Simplify.

(a) $3x - (x + 4) = 3x + -(x + 4) = 3x + (-1)(x + 4)$
$= 3x + (-1)(x) + (-1)(4)$
$= 3x + (-x) + (-4)$
$= 2x + (-4) \quad (\text{or } 2x - 4)$

(b) $(x + y) - (x - y) = (x + y) + -(x - y)$
$= (x + y) + (-1)[x + (-y)]$
$= x + y + (-1)(x) + (-1)(-y)$
$= x + y + (-x) + y$
$= 2y$

(c) $-2a - (-a + 4b) + b = -2a + -(-a + 4b) + b$
$= -2a + (-1)(-a + 4b) + b$
$= -2a + a + (-4b) + b$
$= -a + (-3b) \quad (\text{or } -a - 3b)$

(d) $2x^2 - x - (-x - 2y) - x^2 = 2x^2 + (-x) + -[-x + (-2y)] + (-x^2)$
$= 2x^2 + (-x) + (-1)[-x + (-2y)] + (-x^2)$
$= 2x^2 + (-x) + x + 2y + (-x^2)$
$= x^2 + 2y$

DO EXERCISES 22 THROUGH 25.

Recall Rule 2.8 on the Order of Operations.

1. Perform operations within parentheses. Recall that the grouping symbols ( ), { }, and [ ] all have the same meaning in an algebraic expression, namely multiplication.
2. Perform operations involving exponents.
3. Multiply or divide, proceeding in order from left to right.
4. Add or subtract, proceeding in order from left to right.

Quite often we will have grouping symbols inside grouping symbols. The best method for handling this situation is to start with the innermost set of grouping symbols and gradually work your way out.

Simplify.

22. $y - (2x + y)$

23. $-x + y - (-x - y)$

24. $4(s - 2t) - (s - t) + 2s$

25. $-2(x^2 + x - 1) - (-3x^2 - 2x + 4)$

Simplify.

26. $3(4+2) - \{7 - [4 - (5+6)]\}$

27. $-2 - \{-1 - [3 - 2(-1+4)]\}$

28. $2x - y - (x - y)$

29. $\dfrac{(4+2)3}{-2} \cdot \dfrac{[3+(-1)]}{2(5-3)}$

30. $2(x - 3y) + 3[4 - (-x - 2y) + x]$

Example 8 Simplify.

(a)
$$\begin{aligned} -\{7 - 3[4 + (2-3)]\} &= -\{7 + (-3)[4 + (2 + (-3))]\} \\ &= -\{7 + (-3)[4 + (-1)]\} \\ &= -\{7 + (-3)[3]\} \\ &= -\{7 + (-9)\} \\ &= -\{-2\} = 2 \end{aligned}$$

(b)
$$\begin{aligned} x - [2 - (3x+4) + x] &= x + -[2 + -(3x+4) + x] \\ &= x + (-1)[2 + (-1)(3x+4) + x] \\ &= x + (-1)[2 + (-3x) + (-4) + x] \\ &= x + (-1)[-2 + (-2x)] \\ &= x + 2 + 2x = 3x + 2 \end{aligned}$$

DO EXERCISES 26 THROUGH 30.

DO SECTION PROBLEMS 2.9.

ANSWERS TO MARGINAL EXERCISES

1. $3x$ and x **2.** $2a^2$, a^2, and $-3a^2$; $-4a$ and $6a$ **3.** 4 and 6; $-x^2y^2$ and $4x^2y^2$ **4.** Different exponents **5.** Different variables **6.** They are. **7.** $-2x$ **8.** $5x + 2y$ **9.** $a^2 + 10a^2b^2 + (-7b^2) + a$ **10.** $8s + (-6st) + 3t + (-5)$ **11.** $0.9x$ **12.** $\frac{-2}{15}y + \left(-\frac{5}{2}\right)$ **13.** $-x + (-11)$ **14.** $9x + (-14) + (-10xy)$ **15.** $-4a^2 + (-5b)$ **16.** $-s + (-6t)$ **17.** $-2x + (-3)$ **18.** $-4x + (-4)$ **19.** $3s + t$ **20.** $-a + 3$ **21.** $x + (-y) + 2$ **22.** $-2x$ **23.** $2y$ **24.** $5s + (-7t)$ **25.** $x^2 + (-2)$ **26.** 4 **27.** -4 **28.** x **29.** $-\frac{9}{2}$ **30.** $8x + 12$

NAME COURSE/SECTION DATE

SECTION PROBLEMS 2.9

Identify like terms in the following algebraic expressions.

1. $3x - y + 2x$

2. $7x^2 - 3x + x^2 - 4 + x$

3. $x^2y - xy + xy^2 - 2x^2y$

4. $a^2 - ab + 3a^2 - a + ab$

Remove the parentheses.

5. $2(3x - 4)$

6. $-5(-x + y)$

7. $-(x + 2)$

8. $-(-x + 3)$

9. $-(-x - 4y)$

10. $-(x - y - 3)$

Simplify.

11. $2x + 7x$

12. $x - y + 4x$

13. $2a - a^2 + 3a - b + 2a^2$

14. $1 + 3t - 4 + 7t$

15. $2m - 3mn + m - n + 4m^2$

16. $x^2 - 3x + 2x^2 + 5 + 4x$

17. $4(x + 2y) - 3x$

18. $-2(s + 2t) - (3s - t)$

19. $3R - 4(R - 3R)$

20. $a - 3(a - 4) + 1$

ANSWERS

1. ______
2. ______
3. ______
4. ______
5. ______
6. ______
7. ______
8. ______
9. ______
10. ______
11. ______
12. ______
13. ______
14. ______
15. ______
16. ______
17. ______
18. ______
19. ______
20. ______

21. $-4(2x - 1) - 3(-x + 5)$

22. $\frac{1}{2}(4x - 6) + 2x$

23. $0.12(x + 0.4y) - 1.2(0.3x - 2y)$

24. $-4(-2x^2 + 5) - (x^3 - x^2 + x + 1)$

25. $2x - 3\pi + 4 - 7\pi + 1$

26. $x - (-x - y) - y$

27. $-\{5 - 2[3 + (2 - 4)]\}$

28. $-(-6 - 4) - [-2 - 1(-3 - 5)]$

29. $x - [-x + 2(3x - 4) - x]$

30. $x - [3x - 2(x - 4) + x]$

31. $x - (-x + y)$

32. $-2x - (-x - y) - 3y$

33. $a - 2(a - b) - (a - b)$

34. $-2\{4 - 2[4 - 3(2 - 4) + 1]\}$

35. $-x - [x - 3(-x + 4) - 2x] + 5$

36. $-2a + b - 6b - 3a$

37. $-2x - 3(x + 4)$

38. $2s - (-s + t) - 3t$

39. $2(x^2 - x + 1) - (-3x^2 + 4)$

40. $-2\{3 - 2[6 - (3 - 4)]\}$

21. ______
22. ______
23. ______
24. ______
25. ______
26. ______
27. ______
28. ______
29. ______
30. ______
31. ______
32. ______
33. ______
34. ______
35. ______
36. ______
37. ______
38. ______
39. ______
40. ______

2.10

Evaluations

In this section we evaluate certain formulas and certain algebraic expressions by substituting a number for the variable. We start with evaluating formulas.

Example 1 Use the formula $A = h(B + b)/2$ to find the area A of a trapezoid having a height h of 6 inches, a large base B of 10 inches, and a small base b of 4 inches.

$$A = \frac{h(B + b)}{2} = \frac{6(10 + 4)}{2} = \frac{6(14)}{2} = \frac{84}{2} \text{ sq in.} = 42 \text{ sq in.}$$

Example 2 How much interest would you pay if you borrow \$1200 at an annual interest rate of 9% for 6 months? Let I = interest, P = principal, R = the annual interest rate, and T = time. Then use $I = PRT$ where R is expressed as a decimal and T is in years.

$$I = PRT = (\$1200)(0.09)(0.5) = \$54$$

Example 3 Determine the horsepower rating of a 4-cylinder automobile engine with a bore of 3.5 inches. Let N = the number of cylinders and D = the bore. Then use $HP = D^2N/2.5$.

$$HP = \frac{D^2N}{2.5} = \frac{(3.5)^2(4)}{2.5} = 19.6 \text{ horsepower}$$

DO EXERCISES 1 THROUGH 4.

In the algebraic expression $2x + 4$, we know that x is a variable, and we can substitute any number we wish for x. Once we substitute a number for x, the algebraic expression $2x + 4$ will become a number.

Example 4 Evaluate $2x + 4$ when (a) $x = 3$ and (b) $x = -5$.

(a) $x = 3$; $2(3) + 4 = 6 + 4 = 10$

(b) $x = -5$; $2(-5) + 4 = -10 + 4 = -6$

DO EXERCISE 5.

Example 5 Evaluate $\frac{x^2}{2} - 3x + 1$ when (a) $x = 2$, (b) $x = -4$, and (c) $x = 0$.

(a) $x = 2$; $\frac{2^2}{2} - 3(2) + 1 = \frac{4}{2} - 6 + 1 = 2 - 6 + 1 = -3$

(b) $x = -4$; $\frac{(-4)^2}{2} - 3(-4) + 1 = \frac{16}{2} + (-3)(-4) + 1 = 8 + 12 + 1 = 21$

(c) $x = 0$; $\frac{0^2}{2} - 3(0) + 1 = 0 - 0 + 1 = 1$

LEARNING OBJECTIVES

After finishing Section 2.10, you should be able to:

- Evaluate formulas when numerical values are given for the variables.
- Evaluate algebraic expressions when numerical values are given for the variables.

1. Find the area of a circle with a radius of 14 inches. Use $A = \pi r^2$ where $\pi \cong 3.14$.

2. Find the power P in watts when the current I is 15 amperes and the resistance R is 25 ohms. Use $P = I^2R$.

3. How far (d) will Marian travel in 3 hours 12 minutes (t) at a speed r of 40 miles per hour? Use $d = rt$. (*Hint:* $t = 3\frac{1}{5}$ hours.)

4. Find the volume V of a rectangular solid if it has a length L of 4 feet, a width W of 2.5 feet, and a height H of 1.3 feet. Use $V = LWH$.

5. Evaluate $2(x + 4)/3$ when (a) $x = -2$, (b) $x = 0$, and (c) $x = 8$.

Example 6 Evaluate $-x^3 + 2x^2 - x + 1$ when (a) $x = 3$ and when (b) $x = -2$.

(a) $x = 3$; $-(3)^3 + 2(3)^2 - 3 + 1 = -27 + 18 - 3 + 1$
$= -11$

(b) $x = -2$; $-(-2)^3 + 2(-2)^2 - (-2) + 1 = -(-8) + 2(4) - (-2) + 1$
$= 8 + 8 + -(-2) + 1$
$= 16 + 2 + 1$
$= 19$

Example 7 Evaluate $\frac{x - 3}{y}$ when $x = -5$ and $y = 2$.

$$\frac{-5 - 3}{2} = \frac{-5 + (-3)}{2} = \frac{-8}{2} = -4$$

DO EXERCISES 6 AND 7.

Now here are a few more algebraic expressions to evaluate.

Example 8 Evaluate $a^2b - 2abc$ when $a = -3$, $b = -2$, and $c = 4$.

$(-3)^2(-2) - 2(-3)(-2)(4) = 9(-2) + (-2)(-3)(-2)(4)$
$= -18 + (6)(-8)$
$= -18 + (-48)$
$= -66$

Example 9 Evaluate $B^2 - 4AC$ when $A = 2$, $B = -1$, and $C = -3$.

$(-1)^2 - 4(2)(-3) = 1 + (-4)(2)(-3)$
$= 1 + 24$
$= 25$

Example 10 Evaluate $2x - 3y$ when $x = 3$ and $y = -2$.

$2(3) - 3(-2) = 6 + (-3)(-2)$
$= 6 + 6$
$= 12$

DO EXERCISES 8 THROUGH 11.

DO SECTION PROBLEMS 2.10.

ANSWERS TO MARGINAL EXERCISES

1. 615.44 sq in. **2.** 5625 watts **3.** 128 miles **4.** 13 cu ft
5. (a) $\frac{4}{3}$ (b) $\frac{8}{3}$ (c) 8 **6.** (a) 1 (b) 3 (c) 13 **7.** (a) −3 (b) −3 (c) −21 **8.** 14 **9.** 3 **10.** −7 **11.** −14

---

6. Evaluate $x^2 - x + 1$ when (a) $x = 0$, (b) $x = 2$, and (c) $x = -3$.

7. Evaluate $x^3 - 2x^2 + x - 3$ when (a) $x = 0$, (b) $x = 1$, and (c) $x = -2$.

8. Evaluate $2A^2 + BC^2$ when $A = -1$, $B = 3$, and $C = -2$.

9. Evaluate $(2x - y) - (-x - 2y)$ when $x = 2$ and $y = -3$.

10. Evaluate $A^2 - 2B^2$ when $A = 1$ and $B = 2$.

11. Evaluate $2s - 3t + 4$ when $s = -3$ and $t = 4$.

NAME COURSE/SECTION DATE

ANSWERS

1. ____________
2. ____________
3. ____________
4. ____________
5. ____________
6. ____________
7. ____________
8. ____________
9. ____________
10. ____________
11. ____________

SECTION PROBLEMS 2.10

1. Evaluate $2x - 3$ when (a) $x = 3$, (b) $x = 0$, (c) $x = -5$, (d) $x = \frac{7}{2}$.

2. Evaluate $-2x^2 + 3x - 4$ when (a) $x = 0$, (b) $x = 4$, (c) $x = -5$, (d) $x = 1.2$.

Given $a = 3$, $b = -4$, $c = 5$, and $d = 1.3$, evaluate the following.

3. $b^2 - 4ac$

4. $2ab + c - d$

5. $a^2b^2 - abc$

6. $-2b + c - d$

7. $3a^2 - 2bc + b^2$

8. $a - bc + d^2$

9. $3a^2b - 2b^2c$

10. $abc - \frac{c}{a}$

11. Find the area A of a triangle with base b 4 ft and height h 5.2 ft. Use $A = bh/2$.

12. ______________

13. ______________

14. ______________

15. ______________

16. ______________

17. ______________

18. ______________

19. ______________

12. Find the taper per inch T of a tapered plug that has a diameter D of $2\frac{3}{8}$ inches at the large end, a diameter d of $1\frac{7}{8}$ inches at the small end, and is 3 inches long (L). Use

$$T = \frac{D - d}{L}$$

13. How much power P is lost in a transmission line having a resistance R of 1.5 ohms and a carrying current I of 50 amperes? Use $P = I^2R$.

14. What current I is needed to run a clothes dryer with a resistance R of 40 ohms and a voltage E of 220 volts? Use $I = E/R$.

15. Find the circumference of a circle that has a 14-inch diameter. Use $C = 2\pi r$ where r = radius and $\pi \cong 3.14$.

16. Find the volume V of a cube that has a side s of 4 cm. Use $V = s^3$.

Find the value of the missing letter in each formula.

17. $C = \frac{5}{9}(F - 32°)$ where $F = 98°$.

18. $SA = 2\pi rh + 2\pi r^2$ where $r = 5$ cm, $h = 4$ cm, and $\pi \cong 3.14$.

19. $A = \dfrac{a_1 + a_2 + a_3 + a_4}{4}$ where $a_1 = 76$ lb, $a_2 = 69$ lb, $a_3 = 94$ lb, and $a_4 = 105$ lb.

NAME | COURSE/SECTION | DATE | ANSWERS

20. $A = h(B + b)/2$ where $h = 2$ inches, $B = 5$ inches, and $b = 3$ inches.

21. $S = \frac{1}{2}gt^2$ where $g = 32$ ft/sec and $t = 20$ sec.

22. $I = PRT$ where $P = \$500$, $R = 10.5\%$, and $T = 2$ years.

23. Evaluate $x^3 - x^2 + 3x + 4$ when (a) $x = -2$ and (b) $x = 3$.

24. Evaluate $(x - y)^2$ when (a) $x = 2$, $y = -3$ and (b) $x = -1$, $y = -2$.

25. Evaluate $(x - y)(x + y)$ when $x = 4$ and $y = -3$.

26. Evaluate $(2x - y - z)(x^2 - y)$ when $x = -1$, $y = 2$, and $z = -3$.

27. Evaluate $2x - 3y$ when $x = 4$ and $y = -2$.

28. Evaluate $-3x + 5y$ when $x = -2$ and $y = -3$.

ANSWERS

20. ____________

21. ____________

22. ____________

23. (a) ____________

(b) ____________

24. (a) ____________

(b) ____________

25. ____________

26. ____________

27. ____________

28. ____________

29. ______________

30. ______________

31. ______________

32. (a) ______________

(b) ______________

33. (a) ______________

(b) ______________

34. ______________

29. Evaluate $b^2 - 4ac$ when $a = 2$, $b = 3$, and $c = -2$.

30. Evaluate $\frac{x}{x^2 - 3x}$ when $x = 4$.

31. Evaluate $3x - y + 2z$ when $x = -2$, $y = 3$, and $z = -2$.

32. Evaluate $2x^2 - x + 1$ when (a) $x = 2$ and (b) $x = -3$.

33. Evaluate $3x^3 - 2x^2 - x + 1$ when (a) $x = 2$ and (b) $x = -2$.

34. Evaluate ab^2 when $a = \frac{1}{3}$ and $b = -\frac{5}{2}$.

NAME ______ COURSE/SECTION ______ DATE ______

CHAPTER 2 REVIEW PROBLEMS

SECTION 2.2

Calculate.

1. $-2 + 7 + 4$

2. $-3 + (-4)$

3. $-53 + (-60) + 75$

4. $-310 + (-75)$

5. $453 + 92$

6. $4 + (-5) + (-10) + 17 + (-15)$

SECTION 2.3

Calculate.

7. $-3 - 5$

8. $15 - 20$

9. $4 - (-7)$

10. $-2 + 7 - 4$

11. $-(-3) + (-4)$

12. $-472 - 316$

13. $-72 + 63 - (-213)$

14. $|7 - 10|$

15. $|-8| + |-10|$

16. $512 - (-204)$

17. $-78 - (-43) - 27$

ANSWERS

1. ______
2. ______
3. ______
4. ______
5. ______
6. ______
7. ______
8. ______
9. ______
10. ______
11. ______
12. ______
13. ______
14. ______
15. ______
16. ______
17. ______

18. ______

19. ______

20. ______

21. ______

22. ______

23. ______

24. ______

25. ______

26. ______

27. ______

28. ______

29. ______

30. ______

31. ______

32. ______

33. ______

SECTION 2.4

Calculate.

18. $(-2)(-5)$

19. $\frac{-20}{-10}$

20. $-12 \div 1$

21. $(-2)(-5)(-4)$

22. $0 \div (-15)$

23. $7 - 4(-3)$

24. $\frac{(4)(-3)}{-2}$

25. $\frac{-12}{-2}$

26. $\frac{(-2 - 4)(-4 + 1)}{-6}$

27. $4 - 3[2 + (-6)]$

SECTION 2.5

Calculate.

28. $\left(-1\frac{4}{5}\right)\left(3\frac{1}{2}\right)$

29. $-2\frac{6}{7} \div -\frac{2}{5}$

30. $-2\frac{1}{4} + 3\frac{1}{2}$

31. $\frac{3}{8} + \left(-\frac{1}{16}\right)$

32. $3\frac{3}{8} - 2\frac{1}{4} + 1\frac{1}{16} + \frac{-1}{8}$

33. $7.1 + (-3.4)$

NAME ______ COURSE/SECTION ______ DATE ______

ANSWERS

34. $-24.51 + 16.04$

35. $-4.3 - (-5.2)$

36. $(-14.22) \div (-0.03)$

37. $(-5.126) \div 1$

SECTION 2.6

38. Expand $\left(\frac{1}{3}\right)^4$.

39. Expand $x^4 + y^2$.

Write using exponents.

40. $(-5)(-5)(-5)(-5)$

41. $xx + (2y)(2y)(2y) + z$

Calculate.

42. $(-8)^2$

43. $-(-2)^5$

44. $-\left(-\frac{1}{2}\right)^4$

45. $2 + 3 \cdot 2^3$

46. $(4 - 2)^3 \cdot 3 - 2$

47. $\frac{(3 - 1)^2(-7 + 2)}{5} - \frac{(-4)(-3)}{5}$

48. $\frac{[2 - (-4)](-4 + 1)^2}{-3} - \frac{(2)(5)}{5}$

34. ______
35. ______
36. ______
37. ______
38. ______
39. ______
40. ______
41. ______
42. ______
43. ______
44. ______
45. ______
46. ______
47. ______
48. ______

49. ______
50. ______
51. ______
52. ______
53. ______
54. ______
55. ______
56. ______
57. ______
58. ______
59. ______
60. ______
61. ______
62. ______
63. ______
64. ______

SECTION 2.7

Calculate or simplify.

49. $(-3)^{-2}$

50. $\left(\frac{1}{4}\right)^{-3}$

51. $x^5 \cdot x^{-7}$

52. $\frac{(2^{-3})^2}{2^4}$

53. $(9x^2y^{-5})^0$

54. $(2x^{-2}y)^2 \cdot (3^{-1}xy^{-1})^{-2}$

55. 7^{-2}

56. $3^{-4} \cdot 3^2$

57. $\frac{x^{-2}}{x^5}$

58. $\frac{3^2 \cdot 3^{-4}}{3^{-1} \cdot 3^{-2}}$

59. $(2x^2y^3)^2$

60. $\left(\frac{2x^{-2}y}{x^2y^{-3}}\right)^2$

SECTION 2.8

Which law is demonstrated by each of the following?

61. $2x + 5 = 5 + 2x$

62. $3 \cdot 4 + 3x = 3(4 + x)$

63. $2(7x) = (2 \cdot 7)x$

64. $3 + (4 + 5) = (3 + 4) + 5$

NAME COURSE/SECTION DATE

65. $xy = yx$

66. $-3(4 + y) = (-3)(4) + (-3)(y)$

Multiply.

67. $3(2x + 4)$

68. $4(-x - 2y + 5)$

Factor.

69. $-6x + 9$

70. $4x - 6y + 12$

71. $3st - 9s$

SECTION 2.9

Simplify.

72. $3x + y - 5x$

73. $x - (x - 3y)$

74. $3(x - 4y) - (-x + y)$

75. $2a - (-a + 3b) - 2a + 3b$

76. $-4\{-2 - 3[4 + 2(3 - 5) - 2]\}$

ANSWERS

65. ________

66. ________

67. ________

68. ________

69. ________

70. ________

71. ________

72. ________

73. ________

74. ________

75. ________

76. ________

77. ____________

78. ____________

79. ____________

80. ____________

81. ____________

SECTION 2.10

77. Find the area A of a triangle with a base b of 12 inches and a height h of 9 inches. Use $A = bh/2$.

78. Evaluate $b^2 - 4ac$ when $a = -2$, $b = -3$, and $c = 4$.

79. Evaluate $2ab + bc^2$ when $a = -2$, $b = -2$, and $c = 3$.

80. Evaluate $(x - y)(x^2 - 2y)$ when $x = 2$ and $y = -3$.

81. Given $C = \frac{5}{9}(F - 32°)$ and $F = 14°$. Find C.

NAME ______ COURSE/SECTION ______ DATE ______

ANSWERS

1. ______
2. ______
3. ______
4. ______
5. ______
6. ______
7. ______
8. ______
9. ______
10. ______
11. ______
12. ______
13. ______
14. ______
15. ______
16. ______

CHAPTER 2 TEST

Calculate.

1. $-5 + 6 - 3$

2. $-4 - 6$

3. $-2 - (2 - 6)$

4. $\dfrac{(-12)(5)}{-4}$

5. $|2 - 7| - |-2|$

6. $410 - (-150) - 260 + 80$

7. $\left(-2\frac{3}{4}\right)\left(1\frac{1}{3}\right)$

8. $-\frac{3}{5} - \left(-\frac{1}{2}\right)$

9. $1.56 - (-0.9)$

10. $\left(-\frac{3}{4}\right) \div \left(-1\frac{2}{3}\right)$

11. $-2\frac{1}{3} + 3\frac{1}{2} - \left(-\frac{1}{6}\right) - 4\frac{3}{4}$

12. $\dfrac{7 - 2 \cdot 3}{2}$

13. $\dfrac{(-1 - 5)(-7 + 4)}{-6} - \dfrac{(-2)(-3)}{5}$

14. $3\{5 - 3[1 - 4(3 - 2)] - 1\}$

15. Multiply $-3(-x + 2y - 5)$

16. Factor $-6x + 15y$

17. ____________

18. ____________

19. ____________

20. ____________

21. ____________

22. ____________

23. ____________

24. ____________

25. ____________

26. ____________

27. (a) ____________

(b) ____________

28. ____________

29. ____________

30. (a) ____________

(b) ____________

(c) ____________

Simplify.

17. $-2x - (3x - y)$

18. $2(a + 4) - (3 - a) + 4a$

19. $-3(t^2 - 2t - 3) - (2t^2 - 4)$

Calculate or simplify. Assume all variables are nonzero.

20. $\left(\frac{2}{3}\right)^4$

21. 7^{-2}

22. $(-2)^{-5}$

23. $x^3 \cdot x \cdot x^{-5}$

24. $\dfrac{3^3 \cdot 3^{-5}}{3^{-2} \cdot 3^{-1}}$

25. $(-3x^2y^{-4})^2$

26. $\left(\dfrac{2xy^{-1}}{x^{-2}y^2}\right)^3$

27. Evaluate $-x^2 + 4x - 3$ when (a) $x = 2$ and (b) $x = -2$.

28. Evaluate $(x - 2y)(-x + 3y^2)$ when $x = 3$ and $y = -2$.

29. Evaluate $b^2 - 4ac$ when $a = -3$, $b = 4$, and $c = -1$.

30. What law is demonstrated by each of the following?

(a) $7 + 0.53 = 0.53 + 7$

(b) $4(-2 + x) = (4)(-2) + (4)(x)$

(c) $2(3x) = (2 \cdot 3)x$

CHAPTER 3

Linear Equations in One Variable

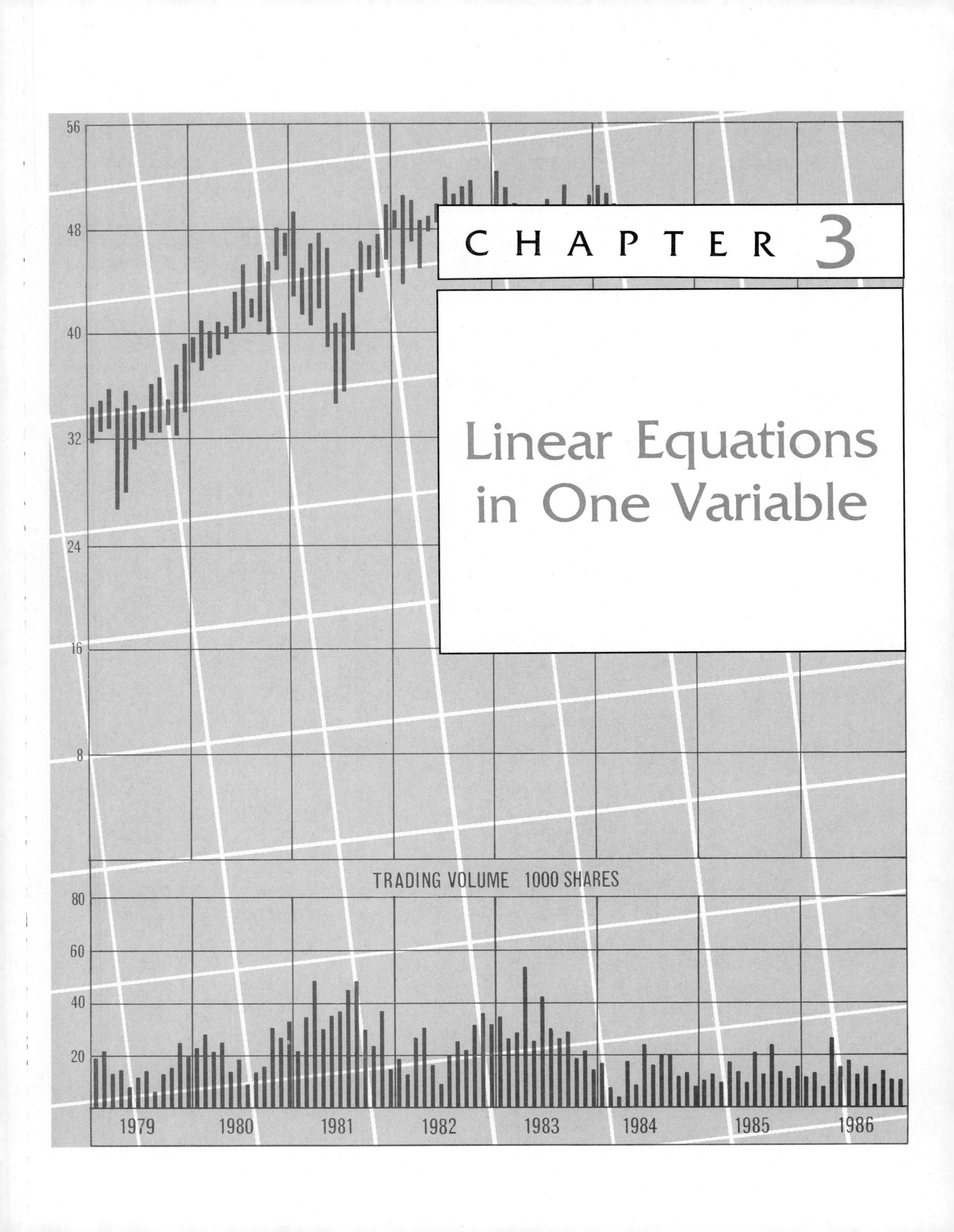

LEARNING OBJECTIVES

After finishing Section 3.1, you should be able to:

- Identify linear equations in one variable.
- Define the solution to an equation.
- Define equivalent equations.
- Use the Addition and Multiplication Principles to change one equation to an equivalent equation.

1. Which of the following are linear equations in one variable? If not, why not?
(a) $2x = 10$
(b) $-3a + 2t = 1$
(c) $y = x^2 - 2x + 4$
(d) $-3(x + 4) = 2(-3x - 5) + 1$
(e) $x - 3 = \frac{1}{x}$

3.1

Introduction

Now we want to use some of the algebra we developed in Chapter 2 to help us solve linear (or first-degree) equations in one variable. Such equations often arise when we try to find solutions to practical (word) problems. We will look at word problems in Chapter 4.

First we need to define an equation.

DEFINITION 3.1 An *equation* is a sentence or statement that contains numbers or variables or both and an equality sign. The equality sign, =, is read "is equal to."

An equation may be true, false, or neither true nor false.

Example 1

(a) The equation $7 + 2 = 9$ is a true statement.

(b) The equation $-3 + 4 = -5$ is a false statement.

(c) The equation $x + 2 = 8$ is neither true nor false since we do not know what the variable x stands for.

DEFINITION 3.2 A *linear equation* (or first-degree equation) *in one variable* is an equation that involves only one variable, and the exponent on the variable must be a positive 1.

Example 2 The following *are* linear equations in one variable.

(a) $2x - 3 = 4$

(b) $5(a - 7) = 2a - 3$

(c) $\frac{-2(t - 3)}{5} + t = 2 - 3t$

(d) $-2y + 3 = -1.05$

Example 3 The following *are not* linear equations in one variable.

(a) $2x - 3y = 2$	More than one variable
(b) $y = y^2 - 3$	An exponent other than +1
(c) $-2a - 3b + 4c = 1$	More than one variable
(d) $3 = \frac{1}{x}$	An exponent other than +1, that is, $\frac{1}{x} = x^{-1}$

■ **DO EXERCISE 1.**

DEFINITION 3.3 A *solution* (root) of a linear equation in one variable is any number that gives us a true statement when it is substituted for the variable in the equation.

Example 4 Check whether the following numbers are solutions for the equation $2x + 4 = 0$.
(a) 2 (b) -3 (c) -2

(a) Substitute 2 for x in the equation $2x + 4 = 0$.

$2x + 4$	0
$2(2) + 4$	0
$4 + 4$	0
8	0

Since the equation $8 = 0$ is not true, 2 is not a solution.

(b) Substitute -3 for x in the equation.

$2x + 4$	0
$2(-3) + 4$	0
$-6 + 4$	0
-2	0

Since the equation $-2 = 0$ is not true, -3 is not a solution.

(c) Substitute -2 for x in the equation.

$2x + 4$	0
$2(-2) + 4$	0
$-4 + 4$	0
0	0

Since the equation $0 = 0$ is true, -2 is a solution. In fact, as we will see later, it is the only solution.

Example 5 Check whether the following numbers are solutions for the equation $3y - 1 = 2y + 4$.
(a) 2 (b) 5

(a) Substitute 2 for y in the equation.

$3y - 1$	$2y + 4$
$3(2) - 1$	$2(2) + 4$
$6 - 1$	$4 + 4$
5	8

Since the equation $5 = 8$ is not true, 2 is not a solution.

(b) Substitute 5 for y in the equation.

$3y - 1$	$2y + 4$
$3(5) - 1$	$2(5) + 4$
$15 - 1$	$10 + 4$
14	14

Since the equation $14 = 14$ is true, 5 is a solution; in fact, it is the only solution.

DO EXERCISES 2 THROUGH 7.

If we have a linear equation in one variable, the type of solution falls into one of three general categories.

1. *Unique solution.* For example, $2x + 4 = 0$ has only one solution (namely,

2. Is 4 a solution of $x + 4 = 5$?

3. Is -2 a solution of $2a - 1 = -5$?

4. Is $\frac{1}{2}$ a solution of $3x - \frac{1}{2} = 1$?

5. Is -1.3 a solution of $m + 2.6 = 1.4$?

6. Is -2 a solution of $2y - 4 = 7y + 6$?

7. Is -1 a solution of $-2 - a = -1$?

8. Give an example of an equation that has no solution.

-2). Most of the equations we will look at will fall into this category. However, there are two other categories that we need to be aware of.

2. *No solution.* For example, $x + 1 = x$ has no solution, that is, we cannot take a number and add 1 and have the same number again.
3. *Infinitely many solutions.* For example, $2x + 10 = 2(x + 5)$ has every number as a solution.

Although categories 2 and 3 are not as common as category 1, the unique solution, we should know of their existence. We will solve equations in these two categories later in this chapter.

DO EXERCISE 8.

Before we give a strategy for solving linear equations in one variable, we note the following.

1. The equation $x + 3 = 0$ has solution -3.
2. The equation $2x = -6$ has solution -3.
3. The equation $4x - 2 = -14$ has solution -3.
4. The equation $x = -3$ has solution -3.

These four equations are not the same equation, but they all have the same solution.

DEFINITION 3.4 *Equivalent equations* are equations that have exactly the same solution(s).

9. The solution to the equation $2x + 4 = 10$ is 3. Is the equation $x - 5 = -2$ equivalent to the equation $2x + 4 = 10$? Why or why not?

Example 6 The equations $x + 3 = 0$, $2x = -6$, $4x - 2 = -14$, and $x = -3$ all have -3 for their solution, and thus are equivalent equations.

Example 7 The equation $x + 3 = 5$ has solution 2 while the equation $x^2 = 4$ has solutions 2 and -2 [note that $2^2 = 4$ and $(-2)^2 = 4$]. Thus, $x + 3 = 5$ and $x^2 = 4$ are *not* equivalent equations.

DO EXERCISE 9.

Our strategy for solving an equation for which the solution is not obvious is to change the original equation to an equivalent equation whose solution is obvious. For example, we change the original equation to an equivalent equation of the form $x =$ (some number). Then, since the two equations are equivalent (same solution), we have also found the solution to the original equation. Before proceeding, one question that we must answer is, how do we change from one equation to an equivalent equation? It turns out there are basically only two ways. The first way involves the operation of addition.

Note

1. The equation $x + 3 = 5$ has solution 2.
2. If we arbitrarily pick 7 to add to both sides of the equation $x + 3 = 5$, we obtain the following.

$$x + 3 = 5$$
$$x + 3 + 7 = 5 + 7$$
$$x + 10 = 12$$

But note that the solution to $x + 10 = 12$ is still 2.

3. If we arbitrarily pick -10 to add to both sides of the equation $x + 3 = 5$, we obtain the following.

$$x + 3 = 5$$
$$x + 3 + (-10) = 5 + (-10)$$
$$x + (-7) = -5$$ But the solution here is still 2.

4. If we arbitrarily pick $4x$ to add to both sides of the equation $x + 3 = 5$, we obtain the following.

$$x + 3 = 5$$
$$x + 3 + 4x = 5 + 4x$$
$$5x + 3 = 4x + 5$$ But this equation also has 2 for its solution.

The above motivates the following principle.

ADDITION PRINCIPLE If we add the same quantity (a number or a variable or both) to both sides of an equation, we obtain an equivalent equation.

Note

1. We may add a number or a variable, or both, to both sides of an equation to obtain an equivalent equation.
2. We could have a subtraction principle; that is, if we subtract the same quantity from both sides of an equation, we obtain an equivalent equation. But subtracting 2 is the same as adding negative 2, so we really do not need a separate rule for subtraction.

DO EXERCISE 10.

The second way to change from one equation to an equivalent equation involves the operation of multiplication.

Note

1. The equation $4x = 12$ has solution 3.
2. If we multiply both sides of the equation $4x = 12$ by -2, we obtain the following.

$$4x = 12$$
$$-2(4x) = -2(12)$$
$$-8x = -24$$ But this equation also has solution 3.

3. If we multiply both sides of the equation $4x = 12$ by $\frac{1}{2}$, we obtain the following.

$$4x = 12$$
$$\left(\frac{1}{2}\right)(4x) = \left(\frac{1}{2}\right)(12)$$
$$2x = 6$$ But the solution here is still 3.

The above motivates the following principle.

10. Use the Addition Principle to find two equations that are equivalent to the equation $x + 4 = -3$.

11. Use the Multiplication Principle to find two equations that are equivalent to the equation $-2x = 4$.

MULTIPLICATION PRINCIPLE If we multiply both sides of an equation by the same nonzero quantity, we obtain an equivalent equation.

We could have a division principle, that is, if we divide both sides of an equation by the same nonzero quantity, we obtain an equivalent equation. Such a rule is not really needed since dividing by 2, for example, is the same as multiplying by $\frac{1}{2}$.

DO EXERCISE 11.

In the next section we will use the strategy of solving an equation by changing to equivalent equations until we obtain an equation of the form $x =$ (some number). There are five steps that we will follow in solving linear equations in one variable. We will see that *not* every step is used in each problem. In fact, in the next section we will use just Step 4 and Step 5. Gradually, as the problems become more involved, we will begin to include the other steps.

STEP 1: CLEAR FRACTIONS

This step should be used only when it is more convenient to solve an equation by first removing the denominators of any fractions in the equation. The Multiplication Principle is used.

STEP 2: ALGEBRAIC SIMPLIFICATION

This step involves removing parentheses and/or combining like terms on the same side of the equality sign. Neither the Addition Principle nor the Multiplication Principle is used.

STEP 3: GATHER VARIABLES

All the terms that have a variable should be gathered on one side (either side will do) of the equality sign. The Addition Principle is used.

STEP 4: ISOLATE

We want to isolate the variable on one side of the equality sign; that is, we want to change the equation to the form $x =$ (some number). The Addition Principle and/or the Multiplication Principle are used.

STEP 5: CHECK

We can always check our solution by substituting it in the original equation to see if we obtain a true statement.

DO SECTION PROBLEMS 3.1.

ANSWERS TO MARGINAL EXERCISES

1. (a) Yes (b) No; two variables (c) No; exponent other than 1 and two variables (d) Yes (e) No; exponent other than one $\left(\frac{1}{x} = x^{-1}\right)$ **2.** No **3.** Yes **4.** Yes **5.** No **6.** Yes **7.** Yes **8.** Answers will vary. **9.** Yes **10.** Answers will vary. **11.** Answers will vary.

NAME ____________ COURSE/SECTION ____________ DATE ____________

SECTION PROBLEMS 3.1

Which of the following are linear equations in one variable? If not, why not?

1. $y = 3x + 1$

2. $\frac{2x - 4}{5} = -3y$

3. $\frac{1}{s} = 5$

4. $x^2 - 3x + 4 = 0$

5. $2(t - 5) - 3t = t + 1$

6. $4 = \frac{1}{t}$

7. $\frac{m}{2} + 4 = 2(m - 3)$

8. $3x + 2y + 4z = 0$

9. Is 4 a solution to $3x - 4 = 8$?

10. Is 2 a solution to $2x - 4 = x - 1$?

11. Is -2 a solution to $-2t + 5 = 4$?

12. Is -3 a solution to $\frac{x - 4}{7} = -1$?

13. Is 4 a solution to $2a - 5 = a - 1$?

14. Is 1 a solution to $1 - 3y = 2y + 6$?

15. Use the Addition Principle and find three equations that are equivalent to the equation $x = 5$.

16. Repeat Problem 15 for the equation $y - 1 = 4$.

ANSWERS

1. ____________
2. ____________
3. ____________
4. ____________
5. ____________
6. ____________
7. ____________
8. ____________
9. ____________
10. ____________
11. ____________
12. ____________
13. ____________
14. ____________
15. ____________
16. ____________

17. ______________

18. ______________

19. ______________

20. ______________

21. ______________

22. ______________

23. ______________

24. ______________

25. ______________

26. ______________

17. Use the Multiplication Principle and find three equations that are equivalent to the equation $x = -3$.

18. Repeat Problem 17 for the equation $2x - 1 = 3$.

19. Give an example of an equation that is false.

20. Give an example of an equation that is true.

21. Give an example of an equation that is neither true nor false.

22. Is the equation $x - 3 = 4$ true, false, or neither true nor false?

23. State the Addition Principle.

24. State the Multiplication Principle.

25. Define "equivalent equations."

26. Define "solution."

3.2

The Addition Principle and the Multiplication Principle

LEARNING OBJECTIVES

After finishing Section 3.2, you should be able to:

- Solve a linear equation in one variable by using the Addition Principle and the Multiplication Principle.

Recall from Section 3.1 the five steps for solving a linear equation in one variable: Step 1: *Clear fractions;* Step 2: *Algebraic simplification;* Step 3: *Gather variables;* Step 4: *Isolate;* and Step 5: *Check.* For most of the equations we solve in this section, it is only necessary to use Step 4 and Step 5. Gradually, as we encounter more involved equations, we will include the other steps.

Example 1 Solve $x + 4 = -5$.

STEP 4: ISOLATE

We want to isolate x (that is, "get rid of" the 4 on the left-hand side of the equation). This will give us an equivalent equation of the form $x =$ (some number).

(a) Add -4 (the additive inverse of 4) to both sides of the equation and simplify.

$$x + 4 = -5$$
$$x + 4 + (-4) = -5 + (-4)$$
$$x + 0 = -9$$
$$x = -9$$

(b) By the Addition Principle, $x + 4 = -5$ and $x = -9$ are equivalent equations and hence must have the same solution. Thus, the solution to $x + 4 = -5$ is -9.

STEP 5: CHECK

Substitute -9 for x in the equation $x + 4 = -5$.

$x + 4$	-5
$-9 + 4$	-5
-5	-5

DO EXERCISE 1.

Solve.

1. $x + 3 = 2$

Example 2 Solve $3 = x - 9$.

STEP 4: ISOLATE

(a) (Optional) Change subtraction to addition.

$$3 = x + (-9)$$

(b) To isolate x, add 9 (the additive inverse of -9) to both sides of the equation.

$$3 + 9 = x + (-9) + 9$$
$$12 = x + 0$$
$$12 = x$$

(c) By the Addition Principle $3 = x - 9$ and $12 = x$ (or $x = 12$) are equivalent equations. Hence, the solution to $3 = x - 9$ is 12.

Solve.

2. $9 = y - 5$

STEP 5: CHECK

Substitute 12 for x.

3	$x - 9$
3	$12 - 9$
3	3

■ **DO EXERCISE 2.**

Example 3 Solve $t + \frac{1}{2} = 3$.

STEP 4: ISOLATE

(a) Add $-\frac{1}{2}$ (the additive inverse of $\frac{1}{2}$) to both sides of the equation.

$$t + \frac{1}{2} = 3$$

$$t + \frac{1}{2} + \left(-\frac{1}{2}\right) = 3 + \left(-\frac{1}{2}\right)$$

$$t = \frac{6}{2} + \frac{-1}{2}$$

$$t = \frac{5}{2}$$

(b) Thus, $t + \frac{1}{2} = 3$ and $t = \frac{5}{2}$ are equivalent equations, and the solution to $t + \frac{1}{2} = 3$ is $\frac{5}{2}$.

STEP 5: CHECK

Substitute $\frac{5}{2}$ for t.

$t + \frac{1}{2}$	3
$\frac{5}{2} + \frac{1}{2}$	3
$\frac{6}{2}$	3
3	3

Solve.

3. $m - \frac{4}{5} = 2$

■ **DO EXERCISE 3.**

Example 4 Solve $y - 1.3 = -4.6$.

STEP 4: ISOLATE

(a) (Optional) Change subtraction to addition.

$$y + (-1.3) = -4.6$$

(b) Add 1.3 to both sides of the equation.

$$y + (-1.3) + (1.3) = -4.6 + 1.3$$

$$y + 0 = -3.3$$

$$y = -3.3$$

(c) Thus, the solution to $y - 1.3 = -4.6$ is -3.3.

STEP 5: CHECK

Substitute -3.3 for y.

$y - 1.3$	-4.6
$-3.3 - 1.3$	-4.6
$-3.3 + (-1.3)$	-4.6
-4.6	-4.6

DO EXERCISES 4 THROUGH 6.

Next we are going to solve some equations using the Multiplication Principle. The strategy here will be the same as before; namely, to change the original equation to an equivalent equation of the form $x =$ (some number). We want to isolate the variable in the equation.

Example 5 Solve $2x = 18$.

STEP 4: ISOLATE

To isolate the x, we need to get rid of the 2. The 2 is in the numerator ($2x = 2x/1$) and we have 2 times x. To get rid of a 2 in the numerator, we must have a 2 in the denominator. Thus we multiply both sides of the equation by $\frac{1}{2}$, the multiplicative inverse of 2.

(a) Multiply both sides of the equation by $\frac{1}{2}$ and simplify.

$$2x = 18$$
$$\left(\frac{1}{2}\right)\left(\frac{2x}{1}\right) = \left(\frac{1}{2}\right)\left(\frac{18}{1}\right)$$
$$\frac{2x}{2} = \frac{18}{2}$$
$$x = 9$$

(b) By the Multiplication Principle, $2x = 18$ and $x = 9$ are equivalent equations. Hence, the solution to $2x = 18$ is 9.

STEP 5: CHECK

Substitute 9 for x.

$2x$	18
2(9)	18
18	18

Note

1. We could also get rid of the 2 in $2x = 18$ by dividing each side by 2; that is,

$$2x = 18$$
$$\frac{2x}{2} = \frac{18}{2}$$
$$x = 9$$

2. Notice the difference between solving the equation $2x = 18$, where we *multiply* by $\frac{1}{2}$ to isolate x, and solving the equation $x + 2 = 18$, where we *add* -2 to isolate x.

DO EXERCISES 7 AND 8.

Solve.

4. $t - 4 = -\frac{5}{2}$

5. $x + 3.4 = -0.014$

6. $0 = n - 13$

Solve.

7. $3x = 15$

8. $y + 4 = -10$

Solve.

9. $-3 = \frac{x}{6}$

Example 6 Solve $-2 = \frac{n}{4}$

STEP 4: ISOLATE

To isolate n, we need to get rid of the 4 in the denominator. Hence we need a 4 in the numerator.

(a) Multiply both sides of the equation by 4, the multiplicative inverse of $\frac{1}{4}$.

$$-2 = \frac{n}{4}$$
$$(4)(-2) = \frac{4}{1}\left(\frac{n}{4}\right)$$
$$-8 = \frac{4n}{4}$$
$$-8 = n$$

(b) Thus, the solution to $-2 = \frac{n}{4}$ is -8.

STEP 5: CHECK

Substitute -8 for n.

-2	$\frac{n}{4}$
-2	$\frac{-8}{4}$
-2	-2

DO EXERCISE 9.

Example 7 Solve $-3x = -12$.

STEP 4: ISOLATE

(a) Multiply both sides of the equation by $-\frac{1}{3}$, the multiplicative inverse of -3.

$$-3x = -12$$
$$\left(-\frac{1}{3}\right)\left(\frac{-3x}{1}\right) = \left(-\frac{1}{3}\right)(-12)$$
$$\frac{-3x}{-3} = \frac{-12}{-3}$$
$$x = 4$$

(b) Thus, the solution to $-3x = -12$ is 4.

STEP 5: CHECK

Substitute 4 for x in the original equation.

$-3x$	-12
$-3(4)$	-12
-12	-12

Note

1. In Step 4(a) we multiply by $-\frac{1}{3}$ (instead of $+\frac{1}{3}$) since we want positive x on the left-hand side of the equality sign.
2. We could also solve this equation by dividing each side by -3. That is,

$$-3x = -12$$

$$\frac{-3x}{-3} = \frac{-12}{-3}$$

$$x = 4$$

DO EXERCISE 10.

Solve.

10. $-4y = 20$

Example 8 Solve $2.4t = 0.48$.

STEP 4: ISOLATE

(a) Multiply both sides of the equation by $\frac{1}{2.4}$, the multiplicative inverse of 2.4.

$$2.4t = 0.48$$

$$\left(\frac{1}{2.4}\right)\left(\frac{2.4t}{1}\right) = \left(\frac{1}{2.4}\right)\left(\frac{0.48}{1}\right)$$

$$\frac{2.4t}{2.4} = \frac{0.48}{2.4}$$

$$t = 0.2$$

(b) Thus, the solution to $2.4t = 0.48$ is 0.2.

STEP 5: CHECK

Substitute 0.2 for t in the original equation.

$2.4t$	0.48
$(2.4)(0.2)$	0.48
0.48	0.48

DO EXERCISES 11 THROUGH 13.

Solve.

11. $0.1a = 4.5$

12. $\frac{t}{-2} = 3$

13. $t - 2 = 3$

Example 9 Solve $\frac{2}{3}x = -4$.

STEP 4: ISOLATE

To isolate x, we need to get rid of a 2 in the numerator and a 3 in the denominator.

(a) Multiply both sides of the equation by $\frac{3}{2}$, the multiplicative inverse of $\frac{2}{3}$.

$$\frac{2x}{3} = -4$$

$$\left(\frac{3}{2}\right)\left(\frac{2x}{3}\right) = \left(\frac{3}{2}\right)(-4)$$

$$\frac{6x}{6} = \frac{-12}{2}$$

$$x = -6$$

(b) Thus, the solution to $\frac{2}{3}x = -4$ is -6.

Solve.

14. $\frac{2x}{3} = 8$

15. $-\frac{5}{4}y = \frac{10}{3}$

16. $\frac{1.1x}{1.4} = 2.53$

STEP 5: CHECK

Substitute -6 for x in the original equation.

$\frac{2}{3}x$	-4
$\left(\frac{2}{3}\right)\left(\frac{-6}{1}\right)$	-4
$\frac{-12}{3}$	-4
-4	-4

DO EXERCISES 14 THROUGH 16.

DO SECTION PROBLEMS 3.2.

ANSWERS TO MARGINAL EXERCISES

1. -1 **2.** 14 **3.** $2\frac{4}{5}$ **4.** $\frac{3}{2}$ **5.** -3.414 **6.** 13 **7.** 5
8. -14 **9.** -18 **10.** -5 **11.** 45 **12.** -6 **13.** 5 **14.** 12
15. $-\frac{8}{3}$ **16.** 3.22

NAME ______ COURSE/SECTION ______ DATE ______

SECTION PROBLEMS 3.2

Solve.

1. $x + 4 = 5$
2. $a + 10 = -3$
3. $y - 6 = 9$
4. $m - 2 = -5$
5. $x + \frac{1}{2} = \frac{3}{4}$
6. $y - \frac{1}{3} = -\frac{2}{3}$
7. $-\frac{3}{2} = t - 3$
8. $-10 = 5 + s$
9. $-7 + x = 0$
10. $x + 4 = 12$
11. $4x = 12$
12. $-3t = 9$
13. $-10 = 2t$
14. $4 = \frac{x}{5}$
15. $-3 + y = 4$
16. $x - 3.2 = 4.5$
17. $\frac{y}{3} = -1$
18. $\frac{3}{4}x = -2$

ANSWERS

1. ______
2. ______
3. ______
4. ______
5. ______
6. ______
7. ______
8. ______
9. ______
10. ______
11. ______
12. ______
13. ______
14. ______
15. ______
16. ______
17. ______
18. ______

19. ____________

20. ____________

21. ____________

22. ____________

23. ____________

24. ____________

25. ____________

26. ____________

27. ____________

28. ____________

29. ____________

30. ____________

31. ____________

32. ____________

33. ____________

34. ____________

35. ____________

19. $-\frac{5}{3}t = -10$

20. $0.2x = -1.4$

21. $-0.5x = -2.2$

22. $\frac{s}{-3} = -4$

23. $-3t = 6$

24. $x - 0.5 = 1.4$

25. $s - \frac{1}{3} = -\frac{1}{4}$

26. $\frac{x}{2} = -\frac{5}{3}$

27. $\frac{0.1x}{0.2} = -40$

28. $-\frac{0.5y}{1.1} = -0.3$

29. $-1.4 = y + 2.3$

30. $\frac{1}{2} = -\frac{x}{5}$

31. $\frac{2x}{3} = -\frac{4}{5}$

32. $-n = 4$

33. $3a = -\frac{1}{4}$

34. $\frac{x}{2} = -3$

35. $x - \frac{1}{2} = -\frac{3}{4}$

3.3

Using the Principles Together

LEARNING OBJECTIVES

After finishing Section 3.3, you should be able to:

- Solve linear equations in one variable by using the Addition and Multiplication Principles together.

Now we look at some examples in which we use both the Addition and Multiplication Principles to solve an equation.

Example 1 Solve $2x + 5 = 9$.

STEP 4: ISOLATE

To isolate the variable x, we must get rid of the 2 and the 5. This can be done in any order, that is, the 2 first and then the 5, or vice versa. However, it is usually easier to get rid of what is added to the variable first (the 5 in this case) and then what is multiplied times the variable (the 2 in this case).

(a) Add -5 (the additive inverse of 5) to both sides of the equation.

$$2x + 5 = 9$$
$$2x + 5 + (-5) = 9 + (-5)$$
$$2x = 4$$

(b) Multiply both sides of the equation by $\frac{1}{2}$, the multiplicative inverse of 2.

$$\frac{1}{2}\left(\frac{2x}{1}\right) = \frac{1}{2}(4)$$
$$\frac{2x}{2} = \frac{4}{2}$$
$$x = 2$$

(c) Thus, by the Addition and Multiplication Principles the equations $2x + 5 = 9$ and $x = 2$ are equivalent. Hence the solution to $2x + 5 = 9$ is 2.

STEP 5: CHECK

Substitute 2 for x in the original equation.

$2x + 5$	9
$2(2) + 5$	9
$4 + 5$	9
9	9

DO EXERCISE 1.

Solve.

1. $3x + 4 = 1$

Example 2 Solve $-3x - 4 = 3$.

STEP 4: ISOLATE

(a) (Optional) Change subtraction to addition.

$$-3x - 4 = 3$$
$$-3x + (-4) = 3$$

(b) Add 4 to both sides of the equation.

$$-3x + (-4) + 4 = 3 + 4$$
$$-3x = 7$$

(c) Multiply both sides of the equation by $-\frac{1}{3}$.

$$\left(-\frac{1}{3}\right)\left(\frac{-3x}{1}\right) = \left(-\frac{1}{3}\right)\left(\frac{7}{1}\right)$$

$$\frac{-3x}{-3} = -\frac{7}{3}$$

$$x = -\frac{7}{3}$$

(d) Thus, the solution to $-3x - 4 = 3$ is $-\frac{7}{3}$.

STEP 5: CHECK

Substitute $-\frac{7}{3}$ for x in the original equation.

$-3x - 4$	3
$-3\left(-\frac{7}{3}\right) - 4$	3
$\frac{21}{3} + (-4)$	3
$7 + (-4)$	3
3	3

■ **DO EXERCISES 2 AND 3.**

Solve.

2. $2y - 6 = 0$

3. $-6t - 10 = 8$

Example 3 Solve $-5 = 4 - t$

STEP 4: ISOLATE

(a) (Optional) Change subtraction to addition.

$$-5 = 4 + (-t)$$

(b) Add -4 to both sides of the equation.

$$-5 + (-4) = 4 + (-t) + (-4)$$
$$-9 = -t$$

(c) One way to get rid of the -1 in front of the t is to multiply both sides of the equation by -1.

$$(-1)(-9) = (-1)(-t)$$
$$9 = t$$

(d) Thus, the solution to $-5 = 4 - t$ is 9.

STEP 5: CHECK

Substitute 9 for t in the original equation.

-5	$4 - t$
-5	$4 - (9)$
-5	-5

■ **DO EXERCISES 4 AND 5.**

Solve.

4. $4 = -1 - y$

5. $3 - 2t = -5$

Example 4 Solve $\frac{y}{2} - 3 = -1$.

STEP 4: ISOLATE

(a) (Optional) Change subtraction to addition.

$$\frac{y}{2} + (-3) = -1$$

(b) Add 3 to both sides of the equation.

$$\frac{y}{2} + (-3) + 3 = -1 + 3$$

$$\frac{y}{2} = 2$$

(c) Multiply both sides of the equation by 2.

$$2\left(\frac{y}{2}\right) = 2(2)$$

$$\frac{2y}{2} = 4$$

$$y = 4$$

(d) Thus the solution is 4.

STEP 5: CHECK

Substitute 4 for y in the original equation.

$\frac{y}{2} - 3$	-1
$\frac{4}{2} - 3$	-1
$2 - 3$	-1
-1	-1

DO EXERCISES 6 AND 7.

Solve.

6. $\frac{x}{4} + 5 = -2$

7. $6 - \frac{y}{3} = 4$

Example 5 Solve $\frac{x + 1.3}{2.1} = 4.4$.

STEP 1: CLEAR FRACTIONS

Remove the denominator 2.1 by multiplying both sides of the equation by 2.1. Note that we are using the Multiplication Principle.

$$\frac{x + 1.3}{2.1} = 4.4$$

$$\frac{2.1}{1}\left(\frac{x + 1.3}{2.1}\right) = (4.4)(2.1)$$

$$\frac{(2.1)(x + 1.3)}{(2.1)} = 9.24$$

$$x + 1.3 = 9.24$$

STEP 4: ISOLATE

(a) Add -1.3 to both sides of the equation.

$$x + 1.3 + (-1.3) = 9.24 + (-1.3)$$

$$x = 7.94$$

(b) Thus, the solution to $\frac{x + 1.3}{2.1} = 4.4$ is 7.94.

Solve.

8. $\dfrac{x - 1.1}{0.5} = 2.4$

9. $-1.1b + 0.4 = 0.62$

STEP 5: CHECK

Substitute 7.94 for x in original equation.

$\dfrac{x + 1.3}{2.1}$	4.4
$\dfrac{7.94 + 1.3}{2.1}$	4.4
$\dfrac{9.24}{2.1}$	4.4
4.4	4.4

DO EXERCISES 8 AND 9.

DO SECTION PROBLEMS 3.3.

ANSWERS TO MARGINAL EXERCISES

1. -1 **2.** 3 **3.** -3 **4.** -5 **5.** 4 **6.** -28 **7.** 6
8. 2.3 **9.** -0.2

NAME COURSE/SECTION DATE

ANSWERS

SECTION PROBLEMS 3.3

Solve.

1. $x + 5 = 7$
2. $y - 3 = -2$
3. $0 = 5 + t$
4. $4 - t = 0$
5. $2a + 4 = 10$
6. $3Q - 1 = 5$
7. $\frac{x}{-5} = -3$
8. $-10 = 2m$
9. $\frac{x}{4} - 3 = 2$
10. $2 - \frac{t}{3} = 1$
11. $-a + 4 = -3$
12. $2 = 4 - y$
13. $-2 - x = 5$
14. $2r + 1 = -5$
15. $-3s - 2 = 7$
16. $2 - y = 5$
17. $6 = 5x - 4$
18. $-5 - 2s = 3$

1. ____
2. ____
3. ____
4. ____
5. ____
6. ____
7. ____
8. ____
9. ____
10. ____
11. ____
12. ____
13. ____
14. ____
15. ____
16. ____
17. ____
18. ____

19. ____________

20. ____________

21. ____________

22. ____________

23. ____________

24. ____________

25. ____________

26. ____________

27. ____________

28. ____________

29. ____________

30. ____________

31. ____________

32. ____________

33. ____________

34. ____________

35. ____________

19. $0 = 2x + 12$

20. $x - 2 = 3.4$

21. $2.8 = 0.07y$

22. $-6y = 1.8$

23. $6 = 6c$

24. $-3a + 4 = 24$

25. $\frac{t}{5} = 6$

26. $r + 7 = -25$

27. $\frac{M}{-3} + 4 = 5$

28. $\frac{2x}{3} = 4$

29. $3 = 6 - \frac{y}{4}$

30. $2x - 2.5 = -1.4$

31. $-2a - 4 = 1$

32. $\frac{-7x}{3} = \frac{5}{6}$

33. $\frac{x - 5}{1.3} = 42.5$

34. $\frac{P + 0.5}{4} = 1.2$

35. $\frac{2x - 15}{3} = -1$

3.4

Equations with Parentheses and/or Variables on Both Sides of the Equation

In this section we continue solving various forms of linear equations in one variable. The first form we look at involves equations in which the variable is on both sides of the equation.

Example 1 Solve $3x + 4 = 2x + 10$.

STEP 3: GATHER VARIABLES

We must gather all the terms that involve an x on one side (either left or right) of the equality sign. We will gather them on the left-hand side and, thus, we need to get rid of the $2x$ term on the right-hand side.

Add $-2x$ to both sides of the equation and simplify. Note that we are using the Addition Principle.

$$3x + 4 = 2x + 10$$
$$3x + 4 + (-2x) = 2x + 10 + (-2x)$$
$$x + 4 = 10$$

STEP 4: ISOLATE

(a) Add -4 to both sides of the equation.

$$x + 4 + (-4) = 10 + (-4)$$
$$x = 6$$

(b) Thus, the solution to $3x + 4 = 2x + 10$ is 6.

STEP 5: CHECK

Substitute 6 for x in the original equation.

$3x + 4$	$2x + 10$
$3(6) + 4$	$2(6) + 10$
$18 + 4$	$12 + 10$
22	22

DO EXERCISE 1.

Example 2 Solve $-2y - 3 = -5y + 6$.

STEP 3: GATHER VARIABLES

(a) (Optional) Change subtraction to addition.

$$-2y - 3 = -5y + 6$$
$$-2y + (-3) = -5y + 6$$

(b) Just to differ from Example 1 we will gather the variables in this equation on the right-hand side. Thus add $2y$ to both sides of the equation.

$$-2y + (-3) + 2y = -5y + 6 + 2y$$
$$-3 = -3y + 6$$

STEP 4: ISOLATE

(a) Add -6 to both sides of the equation.

$$-3 + (-6) = -3y + 6 + (-6)$$
$$-9 = -3y$$

LEARNING OBJECTIVES

After finishing Section 3.4, you should be able to:

- Solve a linear equation in one variable that involves parentheses and has variables on both sides of the equality sign.
- Solve a linear equation in one variable that has a unique solution, no solution, or infinitely many solutions.

Solve.

1. $4x - 5 = 2x + 7$

Solve.

2. $-3t + 4 = 4t - 10$

3. $5 = 4y - 3$

(b) Multiply both sides of the equation by $-\frac{1}{3}$.

$$\left(-\frac{1}{3}\right)(-9) = \left(-\frac{1}{3}\right)\left(-\frac{3y}{1}\right)$$
$$\frac{9}{3} = \frac{3y}{3}$$
$$3 = y$$

(c) Thus, the solution of $-2y - 3 = -5y + 6$ is 3.

STEP 5: CHECK

Substitute 3 for y in the original equation.

$-2y - 3$	$-5y + 6$
$-2(3) + (-3)$	$-5(3) + 6$
$-6 + (-3)$	$-15 + 6$
-9	-9

■ **DO EXERCISES 2 AND 3.**

The following examples involve equations that contain parentheses.

Example 3 Solve $3(x + 4) - 5 = 2x + 10$.

STEP 2: ALGEBRAIC SIMPLIFICATION

(a) (Optional) Change subtraction to addition.

$$3(x + 4) - 5 = 2x + 10$$
$$3(x + 4) + (-5) = 2x + 10$$

Algebraic simplification involves removing parentheses and/or combining any like terms that are on the same side of the equality sign. Note that this step does not involve the Addition and Multiplication principles.

(b) Remove parentheses and simplify.

$$3(x + 4) + (-5) = 2x + 10$$
$$3x + 12 + (-5) = 2x + 10$$
$$3x + 7 = 2x + 10$$

STEP 3: GATHER VARIABLES

Add $-2x$ to both sides of the equation.

$$3x + 7 + (-2x) = 2x + 10 + (-2x)$$
$$x + 7 = 10$$

Solve.

4. $3(x + 4) = 2x + 6$

STEP 4: ISOLATE

(a) Add -7 to both sides of the equation.

$$x + 7 + (-7) = 10 + (-7)$$
$$x = 3$$

(b) Thus, 3 is the solution to $3(x + 4) - 5 = 2x + 10$.

STEP 5: CHECK

Substitute 3 for x in the original equation.

$3(x + 4) - 5$	$2x + 10$
$3(3 + 4) + (-5)$	$2(3) + 10$
$3(7) + (-5)$	$6 + 10$
$21 + (-5)$	16
16	16

■ **DO EXERCISE 4.**

Example 4 Solve $-2(4t - 5) + 3t = 3(-t + 2) - 5$.

STEP 2: ALGEBRAIC SIMPLIFICATION

(a) (Optional) Change subtraction to addition.

$$-2[4t + (-5)] + 3t = 3(-t + 2) + (-5)$$

(b) Remove parentheses and simplify.

$$-8t + 10 + 3t = -3t + 6 + (-5)$$
$$-5t + 10 = -3t + 1$$

STEP 3: GATHER VARIABLES

Add $3t$ to both sides of the equation.

$$-5t + 10 + 3t = -3t + 1 + 3t$$
$$-2t + 10 = 1$$

STEP 4: ISOLATE

(a) Add -10 to both sides of the equation.

$$-2t + 10 + (-10) = 1 + (-10)$$
$$-2t = -9$$

(b) Multiply both sides of the equation by $-\frac{1}{2}$.

$$\left(-\frac{1}{2}\right)\left(\frac{-2t}{1}\right) = \left(-\frac{1}{2}\right)\left(-\frac{9}{1}\right)$$
$$\frac{-2t}{-2} = \frac{9}{2}$$
$$t = \frac{9}{2}$$

(c) Thus, the solution to the original equation is $\frac{9}{2}$.

STEP 5: CHECK

Substitute $\frac{9}{2}$ for t in original equation.

$-2(4t - 5) + 3t$	$3(-t + 2) - 5$
$-2\left[4\left(\frac{9}{2}\right) + (-5)\right] + 3\left(\frac{9}{2}\right)$	$3\left(-\frac{9}{2} + 2\right) + (-5)$
$-2[18 + (-5)] + \frac{27}{2}$	$3\left(-\frac{5}{2}\right) + (-5)$
$-2(13) + \frac{27}{2}$	$\frac{-15}{2} + (-5)$
$-26 + \frac{27}{2}$	$-\frac{25}{2}$
$\frac{-52}{2} + \frac{27}{2}$	$\frac{-25}{2}$
$\frac{-25}{2}$	$\frac{-25}{2}$

DO EXERCISES 5 AND 6.

Solve.

5. $2(3a - 1) - (a + 4) = -2(a - 4)$

6. $5(2y - 1) + 3y = 4$

We mentioned at the beginning of this chapter that the types of solutions to linear equations in one variable fall into three categories, namely

1. *Unique solution.* This is the type of equation we have looked at up to now.
2. *No solution.* An example would be $x + 1 = x$.
3. *Infinitely many solutions.* An example would be $3x + 12 = 3(x + 4)$.

Categories 2 and 3 are not as common as category 1, but we must be able to handle such equations when they occur. Thus, we will look at an example after we present Rule 3.1.

RULE 3.1 A. When we are solving an equation, if all the variables disappear and we are left with a false statement (such as $3 = -4$ or $0 = 5$), then there is *no solution* to the original equation since the false statement we are left with is equivalent to the original equation.

B. When we are solving an equation, if all the variables disappear and we are left with a true statement (such as $3 = 3$ or $-4 = -4$), then there are *infinitely many solutions* to the original equation since the true statement we are left with is equivalent to the original equation.

Example 5 Solve $2(x + 4) - 3x = -x + 4$.

STEP 2: ALGEBRAIC SIMPLIFICATION

(a) (Optional) Change subtraction to addition.

$$2(x + 4) + (-3x) = -x + 4$$

(b) Remove parentheses and simplify.

$$2x + 8 + (-3x) = -x + 4$$
$$-x + 8 = -x + 4$$

STEP 3: GATHER VARIABLES

(a) Add x to both sides of the equation.

$$-x + 8 + x = -x + 4 + x$$
$$8 = 4 \quad \text{False statement}$$

(b) Thus, from Rule 3.3, we see there are *no solutions* to the original equation.

Example 6 Solve $-3(t + 2) + 6t = 2(t - 3) + t$.

STEP 2: ALGEBRAIC SIMPLIFICATION

(a) (Optional) Change subtraction to addition.

$$-3(t + 2) + 6t = 2[t + (-3)] + t$$

(b) Remove parentheses and simplify.

$$-3t + (-6) + 6t = 2t + (-6) + t$$
$$3t + (-6) = 3t + (-6)$$

STEP 3: GATHER VARIABLES

(a) Add $-3t$ to both sides of the equation.

$$3t + (-6) + (-3t) = 3t + (-6) + (-3t)$$
$$-6 = -6 \quad \text{True statement}$$

(b) Thus, from Rule 3.3, we see there are *infinitely many solutions* to the original equation.

DO EXERCISES 7 THROUGH 11.

DO SECTION PROBLEMS 3.4.

Solve.

7. $4(x - 1) = -2(3x + 1)$

8. $5 - 2x = -2(-3 + x) + 1$

9. $2(t - 3) + t = 3(t - 2)$

10. $2s - 4 = 4(s + 1) + 6s$

11. $-3(1 - n) + 2n = -5n + 2$

ANSWERS TO MARGINAL EXERCISES

1. 6 **2.** 2 **3.** 2 **4.** -6 **5.** 2 **6.** $\frac{9}{13}$ **7.** $\frac{1}{5}$ **8.** No solution **9.** Infinitely many solutions **10.** -1 **11.** $\frac{1}{2}$

NAME ______ COURSE/SECTION ______ DATE ______

SECTION PROBLEMS 3.4

Solve.

1. $2x - 3 = x + 4$

2. $y - 6 = -2y + 3$

3. $4 - 3t = 2t - 1$

4. $-2s - 4 = -3s + 1$

5. $1 - x = 3x - 7$

6. $-2y - 5 = 4y - 7$

7. $6 - 2x = 3x - 4$

8. $6 - 7P = -3P + 5$

9. $-3t + 6 = 9$

10. $-10 = 2 - 4M$

11. $2(x - 1) = -6$

12. $4(3x + 2) = 68$

13. $3 - (y - 4) = 5$

14. $4(x - 2) - 3x = 3$

15. $7x - 3(x - 2) = 26$

16. $y - (2 - 3y) = -4$

17. $5 = 2(a + 1) - 3(1 - a)$

18. $-4 = -2(b + 3) + 3b$

ANSWERS

1. ______
2. ______
3. ______
4. ______
5. ______
6. ______
7. ______
8. ______
9. ______
10. ______
11. ______
12. ______
13. ______
14. ______
15. ______
16. ______
17. ______
18. ______

19. ____________
20. ____________
21. ____________
22. ____________
23. ____________
24. ____________
25. ____________
26. ____________
27. ____________
28. ____________
29. ____________
30. ____________
31. ____________
32. ____________
33. ____________
34. ____________

19. $3(2 - x) = -2(x + 1)$

20. $3(t + 4) = -t + 12$

21. $-3(y + 4) + 6 = 2(y + 1) - 5y$

22. $2(-3 - t) + 2(t + 1) = t$

23. $2t - (t + 1) = 3t + 4$

24. $4(t - 2) = -3(t + 2)$

25. $2x - 2 = 3(1 - x) + 5(x - 1)$

26. $2(a - 3) - (a - 4) = 5$

27. $Q - 3(Q + 1) = -2(1 - Q)$

28. $-2(-x + 3) = -x - (-3x + 1)$

29. $\frac{1}{2}(4x - 2) = -2x - 1$

30. $4(1 - y) + 2y = -3(y + 1) - 4$

31. $0.5t - 1.4 = 2.3$

32. $16 - 3s = 2s + 1$

33. $1 - y = 2(y + 1) - (3y + 1)$

34. $2(x + 5) - (3 - x) = -(1 - x) + 2(2x - 3)$

3.5

Equations Involving Fractions

In this section we look at equations involving fractions.

Example 1 Solve $x + \frac{1}{2} = \frac{2}{3}$.

STEP 1: CLEAR FRACTIONS

We have the option of clearing the denominators of the fractions or working the problem by leaving the fractions in it. Normally we do whichever is most convenient in a particular problem. In this example, we will not clear the fractions, but in the next example we will.

STEP 4: ISOLATE

(a) Add $-\frac{1}{2}$ to both sides of the equation.

$$x + \frac{1}{2} + \left(-\frac{1}{2}\right) = \frac{2}{3} + \left(-\frac{1}{2}\right)$$

$$x = \frac{4}{6} + \frac{-3}{6}$$

$$x = \frac{1}{6}$$

(b) Thus, the solution to $x + \frac{1}{2} = \frac{2}{3}$ is $\frac{1}{6}$.

STEP 5: CHECK

Substitute $\frac{1}{6}$ for x in original equation.

$x + \frac{1}{2}$	$\frac{2}{3}$
$\frac{1}{6} + \frac{1}{2}$	$\frac{2}{3}$
$\frac{1}{6} + \frac{3}{6}$	$\frac{2}{3}$
$\frac{4}{6}$	$\frac{2}{3}$
$\frac{2}{3}$	$\frac{2}{3}$

DO EXERCISE 1.

Example 2 Solve $\frac{3y}{4} - \frac{1}{2} = 4$.

STEP 1: CLEAR FRACTIONS

(a) (Optional) Change subtraction to addition.

$$\frac{3y}{4} + \left(-\frac{1}{2}\right) = 4$$

LEARNING OBJECTIVES

After finishing Section 3.5, you should be able to:

- Solve linear equations in one variable that involve fractions.

Solve.

1. $t - \frac{4}{7} = 5$

Solve.

2. $\dfrac{-2x}{3} + \dfrac{5}{2} = -\dfrac{5}{6}$

3. $\dfrac{3y}{10} = -\dfrac{3}{5}$

(b) Clear the denominators of the fractions by multiplying both sides of the equation by the LCD (see Definition 1.10) of 4 and 2. (Note that we are using the Multiplication Principle.)

$$4\left[\frac{3y}{4} + \left(-\frac{1}{2}\right)\right] = 4(4)$$
$$\frac{4}{1}\left(\frac{3y}{4}\right) + \frac{4}{1}\left(-\frac{1}{2}\right) = 16$$
$$3y + (-2) = 16$$

STEP 4: ISOLATE

(a) Add 2 to both sides of the equation.

$$3y + (-2) + 2 = 16 + 2$$
$$3y = 18$$

(b) Multiply both sides of the equation by $\frac{1}{3}$.

$$\left(\frac{1}{3}\right)\left(\frac{3y}{1}\right) = \left(\frac{1}{3}\right)\left(\frac{18}{1}\right)$$
$$y = 6$$

(c) Thus, the solution to $\dfrac{3y}{4} - \dfrac{1}{2} = 4$ is 6.

STEP 5: CHECK

Substitute 6 for y in the original equation.

$\frac{3y}{4} - \frac{1}{2}$	4
$\frac{3(6)}{4} - \frac{1}{2}$	4
$\frac{18}{4} - \frac{1}{2}$	4
$\frac{9}{2} - \frac{1}{2}$	4
$\frac{8}{2}$	4
4	4

Note

Once again, remember that in solving equations that involve fractions, we may either remove the fractions or leave them in. This decision normally depends on whether it is easier to solve the equation without the fractions in it or it is more convenient to keep the fractions.

DO EXERCISES 2 AND 3.

Example 3 Solve $\dfrac{2}{3} - \dfrac{x}{2} = -2$.

STEP 1: CLEAR FRACTIONS

(a) (Optional) Change subtraction to addition.

$$\frac{2}{3} + \left(-\frac{x}{2}\right) = -2$$

(b) Multiply both sides of the equation by 6.

$$6\left[\frac{2}{3} + \left(-\frac{x}{2}\right)\right] = 6(-2)$$

$$6\left(\frac{2}{3}\right) + 6\left(-\frac{x}{2}\right) = -12$$

$$4 + (-3x) = -12$$

STEP 4: ISOLATE

(a) Add -4 to both sides of the equation.

$$4 + (-3x) + (-4) = -12 + (-4)$$

$$-3x = -16$$

(b) Multiply both sides of the equation by $-\frac{1}{3}$.

$$\left(-\frac{1}{3}\right)\left(\frac{-3x}{1}\right) = \left(-\frac{1}{3}\right)(-16)$$

$$x = \frac{16}{3}$$

(c) Thus, the solution to $\frac{2}{3} - \frac{x}{2} = -2$ is $\frac{16}{3}$.

STEP 5: CHECK

Substitute $\frac{16}{3}$ for x in original equation.

$\frac{2}{3} + \left(-\frac{1}{2}x\right)$	-2
$\frac{2}{3} + \left(-\frac{1}{2}\right)\left(\frac{16}{3}\right)$	-2
$\frac{2}{3} + \left(-\frac{8}{3}\right)$	-2
$\frac{-6}{3}$	-2
-2	-2

DO EXERCISES 4 AND 5.

Solve.

4. $\frac{3}{5} - \frac{2t}{3} = \frac{1}{10}$

5. $-\frac{m}{3} - \frac{1}{4} = \frac{3}{5}$

Example 4 Solve $0.45x + \frac{1}{2} = 1.4(x - 5) - \frac{x}{5}$.

STEP 1: CLEAR FRACTIONS

Multiply both sides of the equation by 10.

$$10\left[0.45x + \frac{1}{2}\right] = 10\left[1.4(x + (-5)) + \left(-\frac{x}{5}\right)\right]$$

$$(10)(0.45x) + \left(\frac{10}{1}\right)\left(\frac{1}{2}\right) = 10(1.4)(x + (-5)) + \left(\frac{10}{1}\right)\left(-\frac{x}{5}\right)$$

STEP 2: ALGEBRAIC SIMPLIFICATION

Remove parentheses and simplify.

$$4.5x + 5 = 14(x + (-5)) + (-2x)$$

$$4.5x + 5 = 14x + (-70) + (-2x)$$

$$4.5x + 5 = 12x + (-70)$$

Solve.

6. $1.2x - \frac{1}{4} = \frac{x}{2} + 0.1$

7. $\frac{m}{1.3} - 0.4 = -1.5$

STEP 3: GATHER VARIABLES

Add $-4.5x$ to both sides of the equation.

$$4.5x + 5 + (-4.5x) = 12x + (-70) + (-4.5x)$$
$$5 = 7.5x + (-70)$$

STEP 4: ISOLATE

(a) Add 70 to both sides of the equation.

$$5 + 70 = 7.5x + (-70) + 70$$
$$75 = 7.5x$$

(b) Multiply both sides of the equation by $\frac{1}{7.5}$.

$$\left(\frac{1}{7.5}\right)\left(\frac{75}{1}\right) = \left(\frac{1}{7.5}\right)\left(\frac{7.5x}{1}\right)$$
$$\frac{75}{7.5} = \frac{7.5x}{7.5}$$
$$10 = x$$

(c) Thus the solution to the original equation is 10.

STEP 5: CHECK

Substitute 10 in the original equation. (Note that we changed 0.45 to $\frac{45}{100} = \frac{9}{20}$ and 1.4 to $\frac{14}{10} = \frac{7}{5}$.)

$0.45x + \frac{1}{2}$	$1.4(x - 5) - \frac{x}{5}$
$\left(\frac{9}{20}\right)\left(\frac{10}{1}\right) + \frac{1}{2}$	$\frac{7}{5}(10 - 5) + \left(-\frac{10}{5}\right)$
$\frac{90}{20} + \frac{1}{2}$	$\frac{7}{5}(5) + \left(-\frac{10}{5}\right)$
$\frac{9}{2} + \frac{1}{2}$	$\frac{35}{5} + \left(-\frac{10}{5}\right)$
$\frac{10}{2}$	$\frac{25}{5}$
5	5

DO EXERCISES 6 AND 7.

Example 5 Solve $\frac{2x + 3}{3} + \frac{x}{4} = 2 + \frac{x}{2}$.

STEP 1: CLEAR FRACTIONS

Multiply both sides of the equation by 12.

$$12\left(\frac{2x + 3}{3} + \frac{x}{4}\right) = 12\left(2 + \frac{x}{2}\right)$$

$$\left(\frac{12}{1}\right)\left(\frac{2x+3}{3}\right)+\left(\frac{12}{1}\right)\left(\frac{x}{4}\right)=12\cdot 2+\frac{12}{1}\left(\frac{x}{2}\right)$$

$$\frac{12(2x+3)}{3}+3x=24+6x$$

$$4(2x+3)+3x=24+6x$$

STEP 2: ALGEBRAIC SIMPLIFICATION

Remove parentheses and simplify.

$$8x+12+3x=24+6x$$
$$11x+12=24+6x$$

STEP 3: GATHER VARIABLES

Add $-6x$ to both sides of the equation.

$$11x+12+(-6x)=6x+24+(-6x)$$
$$5x+12=24$$

STEP 4: ISOLATE

(a) Add -12 to both sides of the equation.

$$5x+12+(-12)=24+(-12)$$
$$5x=12$$

(b) Multiply both sides of the equation by $\frac{1}{5}$.

$$\left(\frac{1}{5}\right)(5x)=\left(\frac{1}{5}\right)(12)$$
$$\frac{5x}{5}=\frac{12}{5}$$
$$x=\frac{12}{5}$$

(c) Thus, our solution to the original equation is $\frac{12}{5}$.

STEP 5: CHECK

Substitute $\frac{12}{5}$ for x in original equation.

$\frac{2x+3}{3}+\frac{x}{4}$	$2+\frac{x}{2}$
$\dfrac{2\left(\frac{12}{5}\right)+3}{3}+\left(\frac{1}{4}\right)\left(\frac{12}{5}\right)$	$2+\left(\frac{1}{2}\right)\left(\frac{12}{5}\right)$
$\dfrac{\frac{24}{5}+\frac{15}{5}}{3}+\frac{3}{5}$	$\frac{10}{5}+\frac{6}{5}$
$\frac{39}{15}+\frac{9}{15}$	$\frac{16}{5}$
$\frac{48}{15}$	$\frac{16}{5}$
$\frac{16}{5}$	$\frac{16}{5}$

DO EXERCISES 8 AND 9.

DO SECTION PROBLEMS 3.5.

Solve.

8. $\frac{x+1}{4}-\frac{2x+3}{3}=\frac{1}{6}$

9. $-3(2s+1)+4s=2(s+3)+7$

ANSWERS TO MARGINAL EXERCISES

1. $5\frac{4}{7}$ **2.** 5 **3.** -2 **4.** $\frac{3}{4}$ **5.** $-\frac{51}{20}$ **6.** $\frac{1}{2}$ **7.** -1.43
8. $-\frac{11}{5}$ **9.** -4

NAME ______ COURSE/SECTION ______ DATE ______

SECTION PROBLEMS 3.5

Solve.

1. $\frac{x}{2} + \frac{3}{4} = \frac{1}{4}$

2. $\frac{a}{2} + \frac{2}{3} = 4$

3. $x + 5 = -3(x + 1) + 2x$

4. $2.3y - 4.1 = 0.73$

5. $\frac{t}{3} = -\frac{4}{15}$

6. $\frac{x}{12} = \frac{9}{4}$

7. $-\frac{5y}{6} = \frac{3}{4}$

8. $\frac{4M}{3} = -\frac{4}{5}$

9. $2(s - 5) = 3s - (s + 10)$

10. $-2(m + 1) = 3m + 8$

11. $\frac{x - 1}{3} = 4$

12. $\frac{1 - 2t}{4} = -3$

13. $\frac{0.4t + 3}{1.2} = -3.6$

14. $\frac{2x - 2.4}{0.5} = 1.4$

15. $1.3x - 4 = 1.1 - 0.4x$

16. $0.5(1.2t - 3.4) = 0.6t + 3$

17. $\frac{2}{5} + \frac{x}{3} = \frac{2}{3}$

18. $-\frac{3}{8} + \frac{y}{4} = 2$

ANSWERS

1. ______
2. ______
3. ______
4. ______
5. ______
6. ______
7. ______
8. ______
9. ______
10. ______
11. ______
12. ______
13. ______
14. ______
15. ______
16. ______
17. ______
18. ______

19. ______________

20. ______________

21. ______________

22. ______________

23. ______________

24. ______________

25. ______________

26. ______________

27. ______________

28. ______________

29. ______________

30. ______________

31. ______________

32. ______________

33. ______________

34. ______________

19. $\frac{x}{6} + \frac{9}{2} = \frac{2}{3}$

20. $-\frac{1}{2} + \frac{2x}{3} = 3$

21. $\frac{3}{4} = \frac{1}{2} - \frac{b}{8}$

22. $\frac{2x}{-3} = \frac{-12}{7}$

23. $\frac{x}{3} + \frac{3}{4} = \frac{5}{6}$

24. $-\frac{y}{2} - \frac{5}{6} = 1$

25. $3(x + 4) = 2x - 1$

26. $4(3 - x) + 2(x + 4) = 0$

27. $\frac{3x}{2} - \frac{1}{4} = \frac{x}{3} + \frac{1}{6}$

28. $2P - \frac{1}{3} = -\frac{1}{4}$

29. $0.5t + \frac{1}{4} = -1.25$

30. $\frac{1}{2} - 0.2y = \frac{3}{5} + 1.4y$

31. $\frac{x + 3}{2} + \frac{x}{6} = 4$

32. $\frac{2b - 1}{3} - \frac{b + 4}{2} = \frac{3}{4}$

33. $\frac{1 - x}{3} - \frac{x}{2} = 1$

34. $\frac{x + 4}{3} - \frac{x - 1}{2} = 1 + \frac{x}{6}$

3.6

Literal Equations

Recall the formula (also called a literal equation) for finding the circumference of a circle, $C = 2\pi r$, where C = circumference, r = radius, and $\pi \cong 3.14$ ($\cong$ means "approximately equal to"). This formula gives the circumference of a circle in terms of the radius and π, and it would be used if we knew the radius of a circle and wanted to determine the circumference. However, suppose we had to work ten problems where, in each case, we knew the circumference of a circle and were looking for its radius. Then it might be convenient to have a formula that expressed the radius in terms of the circumference and π, that is, $r = ?$. We can obtain such a formula by taking the formula $C = 2\pi r$ and solving it for the variable r. This is the type of problem we are dealing with in this section: take a formula and solve it for one of its variables. The principles we apply to isolate the variable we are solving for are exactly the same rules we saw in Section 3.1 for solving equations, namely, the Addition and Multiplication Principles.

> *ADDITION PRINCIPLE* If we add the same quantity to both sides of an equation, we obtain an equivalent equation.

> *MULTIPLICATION PRINCIPLE* If we multiply both sides of an equation by the same nonzero quantity, we obtain an equivalent equation.

Example 1 Solve $C = 2\pi r$ for r.

STEP 4: ISOLATE

We need to get rid of the 2π. Since we have 2π times r, we multiply both sides of the equation by $1/2\pi$.

$$C = 2\pi r$$

$$\frac{1}{2\pi} \cdot \frac{C}{1} = \frac{1}{2\pi} \cdot \frac{2\pi r}{1}$$

$$\frac{C}{2\pi} = \frac{2\pi r}{2\pi}$$

$$\frac{C}{2\pi} = r$$

Thus, our new formula is $r = C/2\pi$.

DO EXERCISES 1 AND 2.

Example 2 Solve $S = P + PRT$ for T. This formula gives the amount S to which a given principal P will accumulate at an interest rate R for T years.

STEP 4: ISOLATE

We get rid of what is added to the variable first and then eliminate what is multiplied times the variable.

(a) Add $-P$ to both sides of the equation.

$$S + (-P) = P + PRT + (-P)$$

$$S - P = PRT$$

LEARNING OBJECTIVES

After finishing Section 3.6, you should be able to:

- Solve literal equations for a particular variable.

1. Solve $I = PRT$ for T.

2. Solve $C^2 = A^2 + B^2$ for A^2.

3. Solve $S = P + PRT$ for R.

4. Solve $A = 2\pi r(h + r)$ for h.

(b) Multiply both sides of the equation by $1/PR$.

$$\frac{1}{PR} \cdot \frac{(S - P)}{1} = \frac{1}{PR} \cdot \frac{PRT}{1}$$

$$\frac{S - P}{PR} = \frac{PRT}{PR}$$

$$\frac{S - P}{PR} = T$$

Thus, our new formula is $T = \frac{S - P}{PR}$.

Note that neither P nor R could be 0 in this formula since if they were, we would have 0 in the denominator.

DO EXERCISE 3.

Example 3 Solve $A = \frac{h}{2}(B + b)$ for B. This is the formula for the area of a trapezoid.

STEP 1: CLEAR FRACTIONS

Multiply both sides by 2.

$$2A = \frac{2}{1}\left(\frac{h}{2}\right)(B + b)$$

$$2A = h(B + b)$$

STEP 2: ALGEBRAIC SIMPLIFICATION

Remove parentheses.

$$2A = hB + hb$$

STEP 4: ISOLATE

(a) Add $-hb$ to both sides of the equation.

$$2A + (-hb) = hB + hb + (-hb)$$

$$2A - hb = hB$$

(b) Multiply both sides by $1/h$.

$$\frac{1}{h}(2A - hb) = \frac{1}{h} \cdot \frac{hB}{1}$$

$$\frac{2A}{h} - \frac{hb}{h} = \frac{hB}{h}$$

$$\frac{2A}{h} - b = B$$

Thus, our new formula is $B = \frac{2A}{h} - b$ $(h \neq 0)$.

DO EXERCISE 4.

Example 4 Solve $S = P + PRT$ for P.

STEP 4: ISOLATE

(a) Notice that we have P in two terms, P and PRT. To get P by itself, we must first factor P out of the terms P and PRT.

$$S = P + PRT$$

$$S = P \cdot 1 + PRT$$

$$S = P(1 + RT)$$

(b) Multiply both sides of the equation by $\frac{1}{1 + RT}$ to get rid of the factor $1 + RT$.

$$\frac{1}{1 + RT} \cdot \frac{S}{1} = \frac{1}{(1 + RT)} \cdot \frac{P(1 + RT)}{1}$$

$$\frac{S}{1 + RT} = \frac{P(1 + RT)}{(1 + RT)}$$

$$\frac{S}{1 + RT} = P$$

(c) Thus our new formula is $P = \frac{S}{1 + RT}$.

DO EXERCISE 5.

Example 5 Solve $B = \frac{BC + D}{A}$ for B.

STEP 1: CLEAR FRACTIONS

Multiply both sides of the equation by A.

$$A \cdot B = \frac{A}{1} \cdot \frac{(BC + D)}{A}$$

$$AB = BC + D$$

STEP 3: GATHER VARIABLES

(a) We must gather the terms (namely AB and BC) that contain B on the same side of the equality sign.

(b) Add $-BC$ to both sides of the equation.

$$AB = BC + D$$

$$AB + (-BC) = BC + D + (-BC)$$

$$AB + (-BC) = D$$

STEP 4: ISOLATE

(a) Factor B out of both terms on the left-hand side of the equation.

$$B[A + (-C)] = D$$

or

$$B(A - C) = D$$

(b) Multiply both sides of the equation by $\frac{1}{A - C}$.

$$\frac{1}{A - C} \cdot \frac{B(A - C)}{1} = \frac{1}{A - C} \cdot \frac{D}{1}$$

$$\frac{B(A - C)}{(A - C)} = \frac{D}{A - C}$$

$$B = \frac{D}{A - C}$$

Thus, our new formula is $B = \frac{D}{A - C}$ $(A - C \neq 0)$.

DO EXERCISE 6.

Example 6 Solve $2x - 3y = 4$ for x.

STEP 4: ISOLATE

(a) (Optional) Change subtraction to addition.

$$2x + (-3y) = 4$$

5. Solve $PQ - Q = P$ for Q.

6. Solve $B - C = \frac{DC}{A}$ for C.

(b) Add $3y$ to both sides of the equation.

$$2x + (-3y) + 3y = 3y + 4$$
$$2x = 3y + 4$$

(c) Multiply both sides of the equation by $\frac{1}{2}$.

$$\frac{1}{2}\left(\frac{2x}{1}\right) = \frac{1}{2}(3y + 4)$$
$$\frac{2x}{2} = \frac{3y}{2} + \frac{4}{2}$$
$$x = \frac{3y}{2} + 2$$

Thus, $x = \dfrac{3y}{2} + 2$.

DO EXERCISE 7.

Example 7 Solve $4x - 6y = 3$ for y.

STEP 4: ISOLATE

(a) (Optional) Change subtraction to addition.

$$4x + (-6y) = 3$$

(b) Add $-4x$ to both sides of the equation.

$$4x + (-6y) + (-4x) = -4x + 3$$
$$-6y = -4x + 3$$

(c) Multiply both sides of the equation by $-\frac{1}{6}$.

$$\left(-\frac{1}{6}\right)\left(\frac{-6y}{1}\right) = \left(-\frac{1}{6}\right)\left(\frac{-4x}{1}\right) + \left(-\frac{1}{6}\right)\left(\frac{3}{1}\right)$$
$$\frac{-6y}{-6} = \frac{-4x}{-6} + \left(-\frac{1}{2}\right)$$
$$y = \frac{2x}{3} + \left(-\frac{1}{2}\right)$$

Thus, $y = \dfrac{2x}{3} + \left(-\dfrac{1}{2}\right)$, or $y = \dfrac{2x}{3} - \dfrac{1}{2}$.

DO EXERCISES 8 AND 9.

Example 8 Solve $-2ax + 3by = 2cy + 3ax$ for y.

STEP 3: GATHER VARIABLES

We need to gather the terms that contain y on the same side of the equality sign.

Add $-2cy$ to both sides of the equation.

$$-2ax + 3by + (-2cy) = 2cy + 3ax + (-2cy)$$
$$-2ax + 3by + (-2cy) = 3ax$$

7. Solve $-3x + y = 4$ for x.

8. Solve $2t - 3s = 2s + a$ for s.

9. Solve $2x - 5y = 4$ for y.

STEP 4: ISOLATE

(a) Add $2ax$ to both sides of the equation.

$$-2ax + 3by + (-2cy) + 2ax = 3ax + 2ax$$
$$3by + (-2cy) = 5ax$$

(b) Factor the y out of the terms $3by$ and $-2cy$.

$$y[3b + (-2c)] = 5ax$$

(c) Divide both sides of the equation by $3b + (-2c)$, or multiply both sides by $\dfrac{1}{3b + (-2c)}$.

$$\frac{y[3b + (-2c)]}{[3b + (-2c)]} = \frac{5ax}{[3b + (-2c)]}$$
$$y = \frac{5ax}{[3b + (-2c)]}$$

Thus, $y = \dfrac{5ax}{3b + (-2c)}$, or $y = \dfrac{5ax}{3b - 2c}$.

DO EXERCISE 10.

DO SECTION PROBLEMS 3.6.

10. Solve $-2ax + 3by = 2cy + 3ax$ for x.

ANSWERS TO MARGINAL EXERCISES

1. $T = \frac{I}{PR}$ 2. $A^2 = C^2 - B^2$ 3. $R = \frac{S - P}{PT}$ 4. $h = \frac{A}{2\pi r} - r$ or $h = \frac{A - 2\pi r^2}{2\pi r}$ 5. $Q = \frac{P}{P - 1}$ 6. $C = \frac{AB}{A + D}$ 7. $x = \frac{y}{3} - \frac{4}{3}$ 8. $s = \frac{2t - a}{5}$ 9. $y = \frac{2}{5}x - \frac{4}{5}$ 10. $x = \frac{3by - 2cy}{5a}$

NAME ______ COURSE/SECTION ______ DATE ______

SECTION PROBLEMS 3.6

1. Solve $A = \frac{bh}{2}$ for h.

2. Solve $F = ma$ for a.

3. Solve $I = PRT$ for R.

4. Solve $E = IR$ for I.

5. Solve $A = \frac{h(B + b)}{2}$ for h.

6. Solve $C = \frac{5}{9}(F - 32°)$ for F.

7. Solve $t = \frac{d}{r}$ for d.

8. Solve $h = \frac{v}{2g}$ for v.

9. Solve $HP = \frac{D^2NL}{2g}$ for D^2.

10. Solve $F = \frac{kmM}{d^2}$ for m.

11. Solve $p = 2l + 2w$ for l.

12. Solve $v_0 = 2v - v_t$ for v.

13. $F = \frac{9}{5}C + 32°$ for C.

14. Solve $ax = bx + c$ for x.

15. Solve $3x - 5y = 10$ for x.

16. Solve $3x - 5y = 10$ for y.

ANSWERS

1. ______

2. ______

3. ______

4. ______

5. ______

6. ______

7. ______

8. ______

9. ______

10. ______

11. ______

12. ______

13. ______

14. ______

15. ______

16. ______

17. ______________

18. ______________

19. ______________

20. ______________

21. ______________

22. ______________

23. ______________

24. ______________

25. ______________

26. ______________

27. ______________

28. ______________

29. ______________

30. ______________

31. ______________

32. ______________

33. ______________

34. ______________

17. Solve $2x + y = -3$ for x.

18. Solve $2x + y = -3$ for y.

19. Solve $-2a + 5b = -3$ for a.

20. Solve $-2a + 5b = -3$ for b.

21. Solve $-2x + y = 3$ for x.

22. Solve $3s - 2t = 4$ for t.

23. Solve $-3(t + s) = 7t + s - 20$ for t.

24. Solve $2x - 3y + 5 = -x + 4y$ for x.

25. Solve $2t - (s + t) = 3 - 4s$ for s.

26. Solve $-2(x - y) = 3y + 5$ for y.

27. Solve $A = 2\pi rh + 2\pi r^2$ for h.

28. Solve $A = 2\pi rh + 2\pi r^2$ for π.

29. Solve $PQ = P + Q$ for P.

30. Solve $s - a_1r = a_2r + a_3r$ for r.

31. Solve $A - MX = NX$ for X.

32. Solve $B - D = \frac{B}{C}$ for B.

33. Solve $2ax - 3by = 4 + 7cy$ for y.

34. Solve $-2at + 5bv = -3cv + 8$ for v.

NAME COURSE/SECTION DATE

CHAPTER 3 REVIEW PROBLEMS

SECTION 3.1

Are the following linear equations in one variable? If not, why not?

1. $2s - 3t = 1$

2. $5x - 4 = \frac{3x}{2}$

3. $y = x^2 + 1$

4. Is 2 a solution to $3x - 5 = 3$?

5. Give an example of two equations that are equivalent.

SECTION 3.2

Solve.

6. $x + 7 = -5$

7. $t - 2 = -6$

8. $\frac{x}{4} = -3$

9. $4y = -5$

10. $0.5s = -1.4$

SECTION 3.3

Solve.

11. $3x - 4 = 5$

12. $-3t + 7 = -4$

13. $-1.1t + 4.2 = -1.3$

14. $0.4 - 0.3y = -1.4$

15. $-\frac{2t}{3} - 5 = 7$

ANSWERS

1. ______
2. ______
3. ______
4. ______
5. ______
6. ______
7. ______
8. ______
9. ______
10. ______
11. ______
12. ______
13. ______
14. ______
15. ______

16. ____________

17. ____________

18. ____________

19. ____________

20. ____________

21. ____________

22. ____________

23. ____________

24. ____________

25. ____________

26. ____________

27. ____________

28. ____________

29. ____________

30. ____________

31. ____________

32. ____________

33. ____________

34. ____________

SECTION 3.4

Solve.

16. $2y - 5 = 3y + 1$

17. $6t - 3 = 5 + 8t - 4$

18. $7(x + 4) - x = 2(3x - 2)$

19. $2(x - 3) + x = 3x - 6$

20. $3(y + 4) - 2y = -3y - (-4y - 12)$

21. $\frac{y}{-5} - 4 = -3$

22. $-x - 2.2 = 0$

SECTION 3.5

Solve.

23. $\frac{x}{2} - \frac{7}{10} = \frac{1}{5}$

24. $\frac{2y}{3} = -\frac{5}{6}$

25. $\frac{2(1 - x)}{3} + \frac{1}{4} = \frac{3x - 6}{2}$

26. $\frac{2s}{3} - 1.4 = 0.5$

27. $\frac{2x + 1}{3} - \frac{3 - x}{4} = \frac{x}{6}$

SECTION 3.6

28. Solve $A = PQ + N$ for P.

29. Solve $F = \frac{9}{5}C + 32°$ for C.

30. Solve $V = \frac{dc}{e}$ for d.

31. Solve $I = PRT$ for P.

32. Solve $A = bh$ for h.

33. Solve $2x - 5y = 10$ for x.

34. Solve $2x - 5y = 10$ for y.

NAME COURSE/SECTION DATE ANSWERS

CHAPTER 3 TEST

1. Are the following linear equations in one variable? If not, why not?

(a) $-2a = 3 + b$ (b) $y = 2x^2 + 3$ (c) $s - \frac{1}{t} = 4$

2. Is -2 a solution to the equation $1 - 2y = -4y + 5$?

3. Give an example of two equations that are equivalent.

Solve.

4. $x - 4 = -5$

5. $5y = -15$

6. $2t + 1 = -7$

7. $-5 = 2a - 3$

8. $\frac{2M - 1}{3} = -4$

9. $-2 - s = 6$

ANSWERS

1. (a) ________

(b) ________

(c) ________

2. ________

3. ________

4. ________

5. ________

6. ________

7. ________

8. ________

9. ________

10. ____________

11. ____________

12. ____________

13. ____________

14. ____________

15. ____________

16. ____________

17. ____________

18. ____________

19. ____________

20. ____________

10. $-\frac{2}{3}a = 4$

11. $-2y + \frac{1}{3} = -\frac{1}{2}$

12. $3x - 4 = -2x + 6$

13. $3(m - 1) - m = -7$

14. $-2(x + 3) + 4x = 3(x - 2) - x$

15. $0.5y - 4.3 = -1.1$

16. $\frac{t}{2} - \frac{2}{3} = -\frac{5}{6}$

17. $\frac{1 - 2x}{3} + \frac{x}{2} = 1$

18. Solve $Q = \frac{100M}{C}$ for M

19. Solve $P = 2l + 2w$ for l.

20. Solve $2x - 3y = -4$ for y.

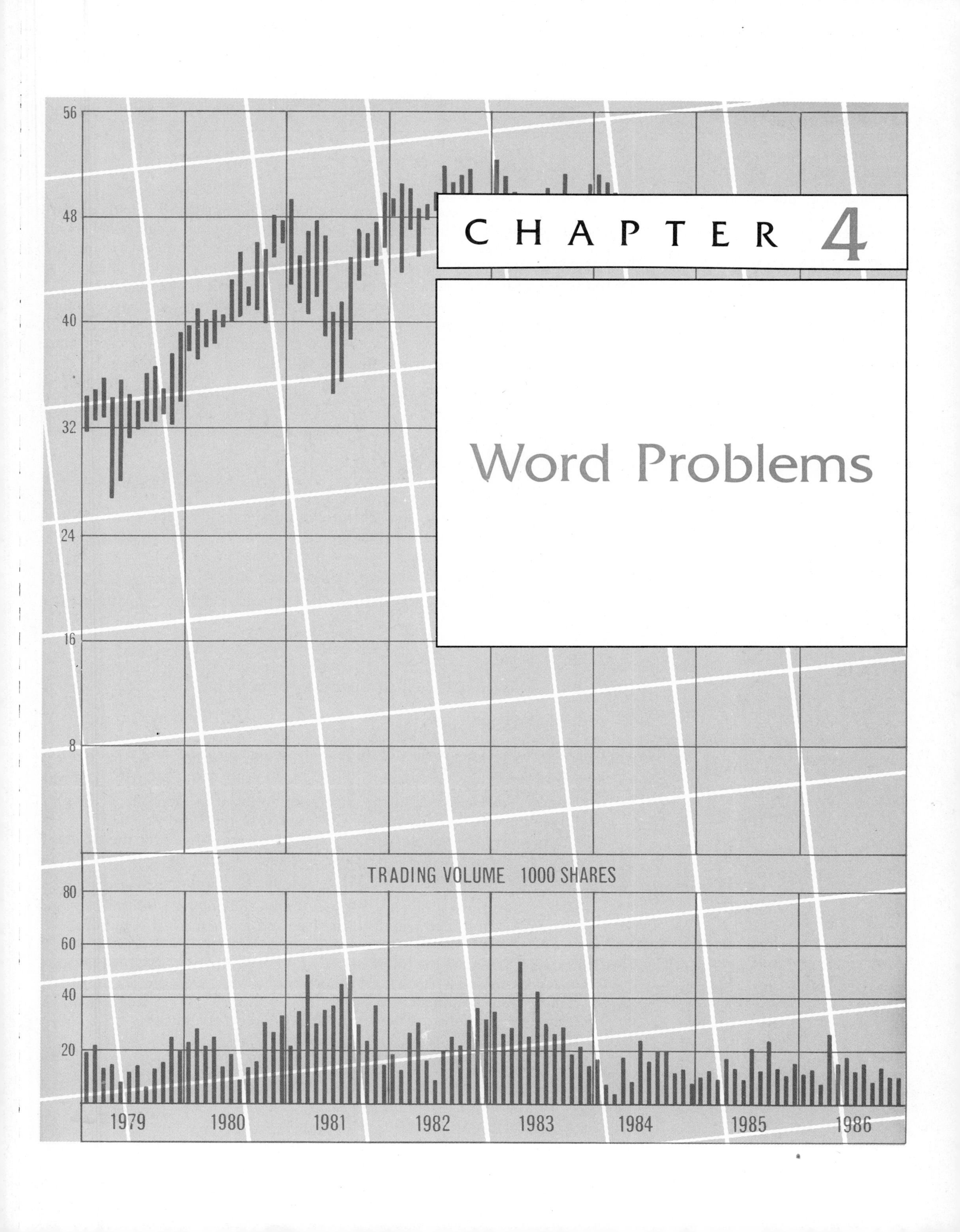
CHAPTER 4
Word Problems
56
48
40
32
24
16
8
TRADING VOLUME 1000 SHARES
80
60
40
20
1979
1980
1981
1982
1983
1984
1985
1986

LEARNING OBJECTIVES

After finishing Section 4.1, you should be able to:

- Change word phrases and equations to mathematical phrases and equations.

4.1

Introduction to Word Problems

As we discussed in Section 2.1, we can use algebra to solve problems that we cannot solve with just a knowledge of arithmetic. Now that we have a working knowledge of algebra, we can put it to use in solving word problems. Being able to solve word problems that we might come across in our work or in everyday life is certainly one of the benefits of having a knowledge of algebra.

We will look at a variety of word problems. First, we must develop a method for handling word problems. Unfortunately, word problems vary greatly in type and difficulty and we cannot give a set formula for solving every kind. However, there are some steps that we will follow in every problem.

STEP 1: PICTURE

Not every word problem lends itself to a picture, but when it does, a picture is very helpful.

STEP 2: VARIABLE

This is a very important step. Be sure you understand in the word problem exactly what your variable stands for.

STEP 3: EQUATION

Quite often this means translating a word equation into a mathematical equation.

STEP 4: SOLVE

In this step, solve the equation we developed in Step 3.

STEP 5: CHECK

This step can mean a variety of things depending on the particular word problem. For example:

1. Does our answer check in the equation developed in Step 4?
2. Did we answer the question that was asked in the word problem?
3. Is our answer reasonable? For example, if we are solving for the time and we arrive at a negative answer, then the answer is not reasonable, since time must be nonnegative.

Before we do word problems, we will practice changing word phrases and word equations into mathematical phrases and mathematical equations.

Example 1 Translate the following word phrases into mathematical phrases.

(a) 3 more than a number
Let x = the number. (The word "more" indicates addition.)

Word phrase: 3 more than a number
Mathematical phrase: $x + 3$

(b) Twice a number
Let y = the number. (The word "twice" indicates multiplication by 2.)

Word phrase: Twice a number
Mathematical phrase: $2y$

(c) A number decreased by 4
Let n = the number. (The word "decreased" indicates subtraction.)

Word phrase: A number decreased by 4
Mathematical phrase: $n - 4$

(d) A number divided by 6
Let x = the number.

Word phrase: A number divided by 6

Mathematical phrase: $x \div 6$ or $\frac{x}{6}$

DO EXERCISES 1 THROUGH 5.

Example 2 Translate the following word phrases into mathematical phrases.

(a) 3 times the sum of two numbers
Let one of the numbers be x. Let the other number be y. Then the sum of the two numbers is $x + y$.

Word phrase: 3 times the sum of two numbers
Mathematical phrase: $3(x + y)$

Note

The parentheses are definitely needed here since the mathematical phrase $3x + y$ would translate as "3 times a number plus another number," and that is not what we want.

(b) Decrease a number by 1 and then double the result.
Let x be the number.

Word phrase: Decrease a number by 1.
Mathematical phrase: $x - 1$
Word phrase: Decrease a number by 1 and then double the result.
Mathematical phrase: $2(x - 1)$ The parentheses are definitely needed here.

(c) Divide a number by 2 and then add 3.
Let n be the number.

Word phrase: Divide a number by 2.

Mathematical phrase: $\frac{n}{2}$

Word phrase: Divide a number by 2 and then add 3.

Mathematical phrase: $\frac{n}{2} + 3$

Change the word phrases into mathematical phrases.

1. A number less 10

2. The product of a number and -3

3. A number increased by 5

4. 4 times a number

5. 12 divided by a number

Translate the word phrases into mathematical phrases.

6. Increase a number by 5 and then double the result.

7. Twice a number plus 5

8. A number squared minus 4 times the number plus 7

9. The difference of two numbers divided by 3

10. Try to add at least one word to each of the lists of words which invoke addition, subtraction, multiplication, and division.

(d) Divide a number plus 3 by 2.
Let n be the number.

Word phrase: A number plus 3
Mathematical phrase: $n + 3$
Word phrase: Divide a number plus 3 by 2.
Mathematical phrase: $\frac{n + 3}{2}$

Note the difference between (c) and (d).

(e) A number squared plus 3 times the number
Let x be the number.

Word phrase: A number squared
Mathematical phrase: x^2
Word phrase: A number squared plus 3 times the number
Mathematical phrase: $x^2 + 3x$

DO EXERCISES 6 THROUGH 9.

We have noticed by now that certain key words in a phrase indicate whether we are going to add, subtract, multiply, or divide. What follows is a list, far from complete, of some common words and the operations (addition, subtraction, multiplication, or division) they invoke.

Addition

1. a number *plus* 2
2. 3 *more than* a number
3. 6 *added* to x
4. a number *increased* by 5
5. the result of *adding* x to y
6. the *sum* of x and 2

Subtraction

1. a number *decreased* by 3
2. a number *minus* 4
3. a number *less* 5
4. the *difference* between x and y
5. *subtract* x from y
6. x *take away* 4

Multiplication

1. 2 *times* a number
2. *twice* a number
3. the *product* of x and 3
4. the result of *multiplying* a by 4
5. *double* a number
6. 3 *times as much* as x
7. two-thirds *of* a number

Division

1. the *quotient* of x and y
2. a number *divided* by 3
3. *divide* x by y

DO EXERCISE 10.

Now we want to practice changing word equations into mathematical equations. Do not worry about solving these equations—we will do that in the next sections. In the following examples, we will be practicing Step 2 and Step 3 of the five steps that are used in solving a word problem.

Example 3 A number plus 2 is the same as -5.

STEP 2: VARIABLE

Choose your variable. Let x = the number.

STEP 3: EQUATION

Translate the word equation into a mathematical equation.

$$\underbrace{\text{a number}}_{x}\ \underbrace{\text{plus}}_{+}\ \underbrace{2}_{2}\ \underbrace{\text{is the same as}}_{=}\ \underbrace{-5}_{-5}$$

or $x + 2 = -5$.

Note

The following phrases always translate as equals, =.

1. is
2. is the same as
3. equals
4. will be
5. was
6. gives the same result as
7. is equivalent to
8. is equal to

DO EXERCISE 11.

Example 4 Twice a number minus 4 gives the same result as the number plus 7.

STEP 2: VARIABLE

Choose your variable. Let y = the number.

STEP 3: EQUATION

Translate the word equation into a mathematical equation.

$$\underbrace{\text{Twice a number}}_{2y}\ \underbrace{\text{minus}}_{-}\ \underbrace{4}_{4}\ \underbrace{\text{gives the same result as}}_{=}\ \underbrace{\text{the number}}_{y}\ \underbrace{\text{plus}}_{+}\ \underbrace{7}_{7}$$

or $2y - 4 = y + 7$.

DO EXERCISE 12.

Translate the word equation into a mathematical equation. (You do not have to solve the equation.)

11. A number less 4 is the same as -10.

Translate the word equation into a mathematical equation.

12. If you take twice a number and add 7 the result will be 15.

Translate the word equation into a mathematical equation.

13. -3 times a number equals double the result of adding 4 to the number.

Example 5 Four times a number plus 3 times the same number equals -3 times the sum of the number and 2.

STEP 2: VARIABLE

Let $t =$ the number.

STEP 3: EQUATION

Translate the word equation into a mathematical equation.

$$\underbrace{\text{4 times a number}}_{4t} \quad \underbrace{\text{plus}}_{+} \quad \underbrace{\text{3 times the same number}}_{3t} \quad \underbrace{\text{equals}}_{=} \quad \underbrace{-3\text{ times}}_{-3} \quad \underbrace{\text{the sum of the number and 2}}_{(t+2)}$$

or $4t + 3t = -3(t + 2)$.

DO EXERCISE 13.

DO SECTION PROBLEMS 4.1.

ANSWERS TO MARGINAL EXERCISES

1. $x - 10$ **2.** $(-3)(x)$ **3.** $x + 5$ **4.** $4x$ **5.** $\frac{12}{x}$ **6.** $2(x + 5)$ **7.** $2x + 5$ **8.** $x^2 - 4x + 7$ **9.** $\frac{x - y}{3}$ **10.** Answers will vary. **11.** $x - 4 = -10$ **12.** $2x + 7 = 15$ **13.** $-3x = 2(x + 4)$

NAME COURSE/SECTION DATE

SECTION PROBLEMS 4.1

Change the following word phrases (equations) into mathematical phrases (equations).

1. 7 less than a number

2. A number plus 3

3. The product of a number and 4

4. A number subtracted from 10

5. A number divided by -3

6. The sum of two numbers

7. The quotient of two numbers

8. 3 minus twice a number

9. Double the result of 10 added to n.

10. The difference between a number and -2

11. $\frac{3}{5}$ as much as a number

12. Divide 5 by twice a number.

13. 4 times a number plus 5

14. Triple the sum of x and 14.

15. A number divided by 2 minus 8

16. A number minus 8 divided by 2

17. 4 times the result of increasing a number by 2

ANSWERS

1. ________

2. ________

3. ________

4. ________

5. ________

6. ________

7. ________

8. ________

9. ________

10. ________

11. ________

12. ________

13. ________

14. ________

15. ________

16. ________

17. ________

18. ________

19. ________

20. ________

21. ________

22. ________

23. ________

24. ________

25. ________

26. ________

27. ________

28. ________

29. ________

30. ________

31. ________

32. ________

33. ________

34. ________

35. ________

18. Take 6 away from -2 times x.

19. $\frac{2}{3}$ a number plus 4

20. 15 more than twice a number

21. The sum of two numbers is the same as 17.

22. A number plus three equals the number divided by two.

23. 3 more than a number is the same as 4 times the number minus 7.

24. If you add 3 to a number and then divide by 2, the result will be 1.5.

25. $\frac{3}{4}$ of a number is the same as -3.

26. 24 divided by a number is the same as -3.

27. A number plus 1 minus twice the number is equal to -4 times the number plus 6.

28. A number is 4.

29. One-half a number plus $\frac{3}{4}$ is the same as $-\frac{5}{6}$ plus the number.

30. Twice a number minus 3 times another number is -14.

31. 7 minus twice a number and the result divided by 4

32. Increase a number by 6.

33. 6 minus twice a number is the same as four times the number plus 1.

34. Take a number and subtract 3; double the result; divide by 3; and the result is the number minus 5.

35. The product of a number and negative 2 is the same as the number divided by 2 plus $\frac{1}{3}$.

4.2

General Word Problems

LEARNING OBJECTIVES

After finishing Section 4.2, you should be able to:

- Solve word problems by using the five general procedural steps.

Recall the five steps we use to solve word problems: Step 1: *Picture;* Step 2: *Variable;* Step 3: *Equation;* Step 4: *Solve;* and Step 5: *Check.*

Example 1 Lynda has 120 feet of chicken wire that she plans to put around her garden to keep animals out. Lynda wants the dimensions of her garden to have nice propòrtions, so she decides to make the length of the garden twice the width. If she plans to use all 120 feet of chicken wire, what should the dimensions of the garden be?

STEP 1: PICTURE

A picture does help.

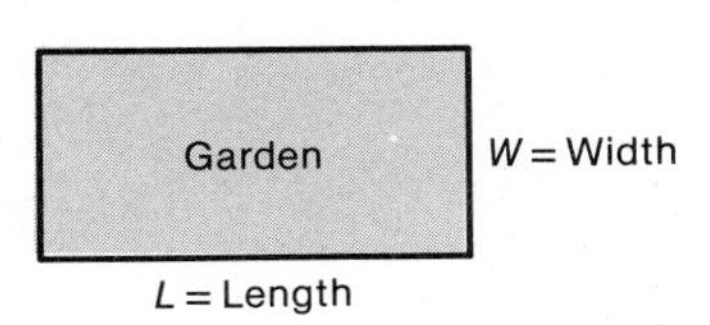

STEP 2: VARIABLE

Note that the length is twice the width.

Let x = width; then
$2x$ = length.

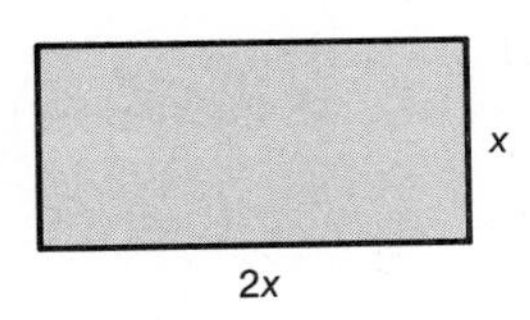

STEP 3: EQUATION

To write the mathematical equation, we use the geometric formula for the perimeter of a rectangle: $P = 2L + 2W$.

$$P = 2L + 2W$$
$$120 = 2(2x) + 2(x)$$
$$120 = 6x$$

STEP 4: SOLVE

(a) Solve the equation.

$$6x = 120$$
$$\frac{6x}{6} = \frac{120}{6}$$
$$x = 20$$

(b) Thus, the width should be 20 feet and the length $(2x)$ should be 40 feet.

STEP 5: CHECK

Our answers are reasonable, and a rectangle with a length of 40 feet and a width of 20 feet does have a perimeter of 120 feet.

DO EXERCISE 1.

Example 2 Marian has 18 cookies that she wants to divide among her husband and four daughters. If each daughter receives a single share and the husband receives a double share, how many cookies does each person receive?

1. Find the length and width of a rectangle if you know that the perimeter is 60 centimeters and the length is 5 centimeters more than the width.

2. Martha Young has 24 acres of land that she wants to divide among her 3 sons, who are 13, 12, and 5 years old. She feels the oldest should receive twice as much as the youngest and the 12-year-old son should get one less than the 13-year-old. How many acres does each son receive?

STEP 1: PICTURE

A picture really does not help in this problem.

STEP 2: VARIABLE

Let c = number of cookies that one daughter receives, that is, a single portion; then
$2c$ = double share or number of cookies Marian's husband receives.

STEP 3: EQUATION

We write a word equation first and then translate it into a mathematical equation.

$$\begin{pmatrix}\text{Total number}\\ \text{of cookies}\\ \text{husband}\\ \text{receives}\end{pmatrix} + \begin{pmatrix}\text{Total}\\ \text{number}\\ \text{daughters}\\ \text{receive}\end{pmatrix} = \begin{pmatrix}\text{Total number of}\\ \text{cookies}\\ \text{available}\end{pmatrix}$$

$$2c + 4c = 18$$

Note

Each daughter receives c cookies, so 4 daughters receive a total of $4c$ cookies.

STEP 4: SOLVE

$$\begin{aligned} 2c + 4c &= 18 \\ 6c &= 18 \\ c &= 3 \end{aligned}$$

Thus, each daughter receives 3 cookies and Marian's husband receives 6 cookies.

STEP 5: CHECK

Our answers are certainly reasonable; also, $(2 \cdot 3) + (4 \cdot 3)$ is 18.

DO EXERCISE 2.

Example 3 David took his wife and four daughters to see *The Revenge of the Pink Panther.* The total cost for all of them to see the show (not including popcorn, candy, and soda) was \$13.50. An adult's ticket costs $2\frac{1}{2}$ times a child's ticket. What would have been the cost if David had left the four children at home?

STEP 1: PICTURE

A picture is possibly of some help.

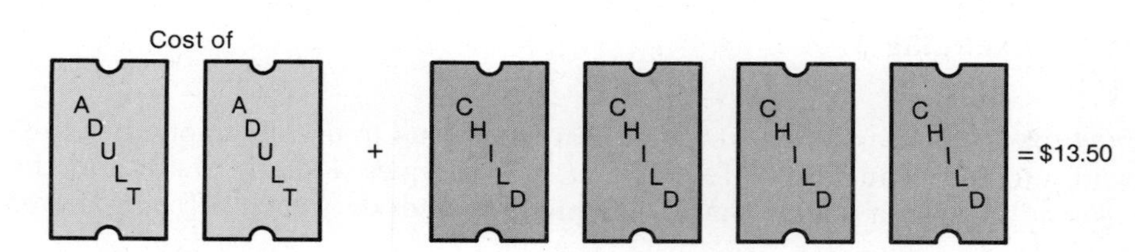

STEP 2: VARIABLE

Note that the adult's ticket is $2\frac{1}{2}$ times as expensive as the child's ticket.

Let x = cost of child's ticket; then $\left(2\frac{1}{2}\right)(x) = \frac{5}{2}x$ = cost of adult's ticket.

STEP 3: EQUATION

Start with a word equation and change to a mathematical equation.

$$\begin{pmatrix}\text{Cost of}\\ \text{2 adult's}\\ \text{tickets}\end{pmatrix} + \begin{pmatrix}\text{Cost of 4}\\ \text{child's}\\ \text{tickets}\end{pmatrix} = \$13.50$$

$$2\left(\frac{5}{2}x\right) + 4(x) = \$13.50$$

STEP 4: SOLVE

$$\frac{10}{2}x + 4x = \$13.50$$

$$5x + 4x = \$13.50$$

$$9x = \$13.50$$

$$x = \$1.50$$

Thus, a child's ticket costs \$1.50 and an adult's ticket costs $2\frac{1}{2}x$ or \$3.75.

STEP 5: CHECK

Our answers are certainly reasonable, and 2(\$3.75) + 4(\$1.50) is \$13.50. However, we have not yet actually answered the question that was asked in the problem. The cost if David had left the four children at home would have been 2(\$3.75) = \$7.50.

DO EXERCISE 3.

3. Natalie takes her family to Taco Heaven for dinner. They have 8 tacos and 5 sodas, and the total bill (without tax) is \$9.30. If a taco costs \$.35 more than a soda, how much does a taco cost?

Example 4 Luis has 80 feet of very nice old Victorian molding. Luis is remodeling his living room and he wants the molding to go all the way around the room. The width of the living room is fixed at 15 feet. However, Luis has some flexibility in how long to make the living room. What is the longest the living room can be and still leave Luis with enough of the Victorian molding to go all the way around?

STEP 1: PICTURE

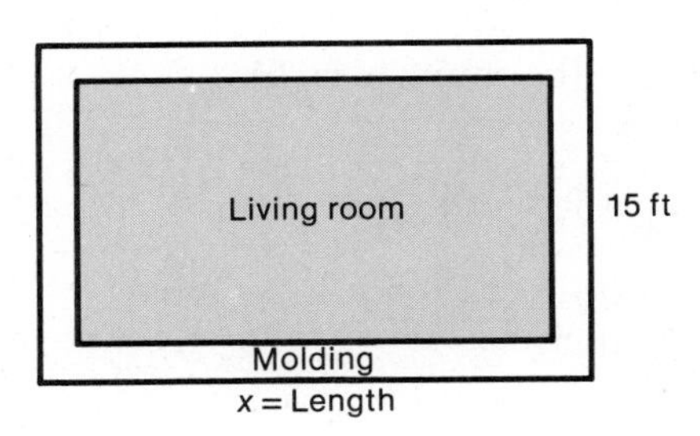

STEP 2: VARIABLE

Let x = length.

4. The sum of three consecutive integers is 99. Find the three integers. (*Hint:* If you consider the integer to be 8, then the next consecutive integer is $8 + 1 = 9$; or if n is the integer, the next consecutive integer is $n + 1$.)

5. A number increased by seven is the same as one-half the number minus three. Find the number.

STEP 3: EQUATION

We will use the formula for the perimeter of a rectangle.

$$P = 2L + 2W$$
$$80 = 2x + 2(15)$$
$$80 = 2x + 30$$

STEP 4: SOLVE

$$2x + 30 = 80$$
$$2x = 50$$
$$x = 25$$

Thus, the length of the living room can be 25 feet at the most.

STEP 5: CHECK

The answer is reasonable, and $2(25) + 2(15)$ does give a perimeter of 80.

DO EXERCISE 4.

Example 5 Twice the sum of a number and five is the same as three times the number minus four. Find the number.

STEP 2: VARIABLE

Let $n =$ the number.

STEP 3: EQUATION

Translate the word equation into a mathematical equation.

Twice the sum of a number and 5 equals 3 times the number minus 4.

$$2(n + 5) = 3n - 4$$

STEP 4: SOLVE

$$2n + 10 = 3n + (-4)$$
$$2n + 10 + (-3n) = 3n + (-4) + (-3n)$$
$$-n + 10 = -4$$
$$-n = -14$$
$$n = 14$$

Thus, the number is 14.

STEP 5: CHECK

Substitute 14 for n in original equation.

$2n + 10$	$3n - 4$
$2(14) + 10$	$3(14) + (-4)$
$28 + 10$	$42 + (-4)$
38	38

DO EXERCISE 5.

Example 6 Dennis is in Reno. He starts the evening by playing Blackjack, where he first doubles his money and then loses \$400. He then goes to the roulette table, where he doubles his remaining money, but then loses \$300. Then he goes to the craps table, where he triples his remaining money before losing \$200. He finally returns to Blackjack where he again doubles his remaining money, loses \$150, and then retires for the evening with \$50 left. How much money did he start with?

STEP 1: PICTURE

A diagram may be helpful in tracing Dennis's steps.

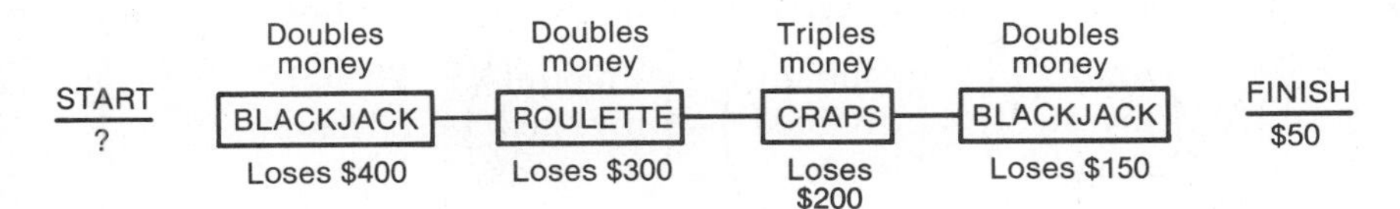

STEP 2: VARIABLE

Let x = amount Dennis started with.

STEP 3: EQUATION

We will write the mathematical equation in stages.

Dennis starts the evening by playing Blackjack, where he doubles his money and then loses \$400.

$$2x - 400$$

At the roulette table he doubles his money and then loses \$300.

$$2(2x - 400) - 300$$

At the craps table he triples his money and then loses \$200.

$$3[2(2x - 400) - 300] - 200$$

Back at the Blackjack table, he doubles his money, loses \$150, and has \$50 left.

$$2\{3[2(2x - 400) - 300] - 200\} - 150 = 50$$

STEP 4: SOLVE

$$\begin{aligned}
2\{3[2(2x - 400) - 300] - 200\} - 150 &= 50 \\
2\{3[4x - 800 - 300] - 200\} - 150 &= 50 \\
2\{12x - 2400 - 900 - 200\} - 150 &= 50 \\
24x - 4800 - 1800 - 400 - 150 &= 50 \\
24x - 7150 &= 50 \\
24x &= 7200 \\
x &= 300
\end{aligned}$$

Thus, Dennis started with \$300.

STEP 5: CHECK

We could check this problem by starting with \$300 and going through the various steps in the problem. Details are left to you.

DO EXERCISE 6.

6. Kay Start puts her savings in a stock. First she doubles her money and then loses \$4000. She then transfers all of her money into some speculative real estate where she triples her money and then loses \$12,000. She then puts her money in a money market account where she immediately loses \$4000, but rebounds by doubling her money soon after. At this point she has had enough. She takes all her money (\$16,000 by this time) and goes to Europe. How much did Kay have in her savings when she started?

7. One angle of a triangle is 30° more than another angle. The third angle of the triangle is 3 times the smallest angle minus 40°. What are the three angles of this triangle? (*Hint:* The sum of the interior angles of a triangle is always 180°.)

Example 7 Denise Darling has $45,000 to invest in gold, stocks, and bonds. She wants to invest twice as much in stocks as in gold and $5000 less in bonds than in stocks. How much should she invest in each?

STEP 2: VARIABLE

Let x = amount (in dollars) to invest in gold; then
$2x$ = amount (in dollars) to invest in stocks, and
$2x - 5000$ = amount (in dollars) to invest in bonds.

STEP 3: EQUATION

The total amount to invest is $45,000. Thus we have the following equation.

$$\begin{pmatrix}\text{Amount}\\ \text{invested}\\ \text{in gold}\end{pmatrix} + \begin{pmatrix}\text{Amount}\\ \text{invested}\\ \text{in stocks}\end{pmatrix} + \begin{pmatrix}\text{Amount}\\ \text{invested}\\ \text{in bonds}\end{pmatrix} = \begin{pmatrix}\text{Total}\\ \text{amount}\\ \text{invested}\end{pmatrix}$$

$$x + 2x + 2x - 5000 = 45000$$

STEP 4: SOLVE

$$x + 2x + 2x - 5000 = 45000$$
$$5x - 5000 = 45000$$
$$5x = 50000$$
$$x = 10000$$

Thus Denise should invest $10,000 in gold, $20,000 in stocks, and $15,000 in bonds.

STEP 5: CHECK

The answers are reasonable, and $10,000 + $20,000 + $15,000 is $45,000.

■ **DO EXERCISE 7.**

Example 8 Picken's Hardware sells $\frac{3}{4}$ of its available stock of wheelbarrows during the month of May. If Picken's has 6 wheelbarrows left, how many did it have at the beginning of the month?

STEP 1: PICTURE

A picture is only useful if you can draw wheelbarrows.

STEP 2: VARIABLE

Let w = number of wheelbarrows in stock at the beginning of the month.

STEP 3: EQUATION

$$\begin{pmatrix}\text{Number in}\\ \text{stock at}\\ \text{beginning of}\\ \text{month}\end{pmatrix} - \begin{pmatrix}\text{Number sold}\\ \text{during the}\\ \text{month}\end{pmatrix} = \begin{pmatrix}\text{Number left}\\ \text{in stock at}\\ \text{the end of}\\ \text{the month}\end{pmatrix}$$

$$w - \left(\frac{3}{4} \text{ of } w\right) = 6$$

$$w - \frac{3}{4}w = 6$$

STEP 4: SOLVE

$$1w - \frac{3}{4}w = 6$$

$$\frac{1}{4}w = 6$$

$$4\left(\frac{1}{4}w\right) = 4(6)$$

$$w = 24$$

Thus Picken's had 24 wheelbarrows in stock at the beginning of the month.

DO EXERCISE 8.

As you have probably noticed in the word problems we have worked thus far, in STEP 3: EQUATION we often write our equation first in "English" words and then translate the word equation into an equation that uses mathematical symbols and numbers. The reason for first writing the equation in words is because we first form our thoughts in words. We think in words and not in mathematical symbols and numbers.

DO SECTION PROBLEMS 4.2.

8. Justrite takes $\frac{3}{5}$ of his allotted vacation time. He uses four days of this to work in the yard and eight days to go fishing. How many vacation days did Justrite have coming?

ANSWERS TO MARGINAL EXERCISES

1. $12\frac{1}{2}$ cm, $17\frac{1}{2}$ cm **2.** 5, 9, and 10 **3.** \$.85 **4.** 32, 33, and 34

5. -20 **6.** \$6000 **7.** 38°, 68°, and 74° **8.** 20

NAME COURSE/SECTION DATE

ANSWERS

1. ______
2. ______
3. ______
4. ______
5. ______
6. ______

SECTION PROBLEMS 4.2

1. The length of the perimeter of a rectangle is $2\frac{1}{2}$ times the width. If the perimeter is 35 inches, find the length and width.

2. The perimeter of a triangle is 28 inches. If the longest side is twice the length of the shortest side and the third side is four inches more than the shortest side, find the length of each side of the triangle.

3. If 9 is subtracted from five times a certain number, the result is 21. Find the number.

4. Twice a number plus 4 is the same as three times the number minus 2. Find the number.

5. An integer plus 5 times the next consecutive integer is 53. What is the value of the smaller integer?

6. The sum of three consecutive even integers is 36. What are the integers?

7. ____________

8. ____________

9. ____________

10. ____________

11. ____________

12. ____________

7. In a triangle, one angle is twice as large as another angle and the third angle is 40° larger than the smallest angle. Find all three angles. Recall that the sum of the interior angles of a triangle is 180°.

8. The sum of the interior angles of a trapezoid is 360°. If one angle is twice the smallest angle, another angle is three times the smallest angle, and the fourth angle is the sum of the other three angles minus 60°, what are the measurements of the four angles?

9. Three-fourths of the students in Math 200 received a grade of C on the last test and one-fifth received an A. If twenty-two more students received C's than received A's, how many students are in the class?

10. One-fourth of the Largent's monthly budget goes for rent and one-sixth goes for food. If the Largents spend $200 more on rent than on food, what is their monthly budget?

11. Mr. Frick, a purchasing agent for Delwar Industries, buys 8 cars (all of the same type) and 6 pickups (all of the same type) for his company. The total bill (without tax) is $114,000. If a car costs $600 more than a pickup, what did Mr. Frick pay for each car and for each pickup?

12. At Barney's take-out service, the price of an order of french fries is one-half the price of a cheeseburger. If an order of five cheeseburgers and three french fries comes to $10.40 (without tax), what is the cost of a cheeseburger?

NAME COURSE/SECTION DATE

ANSWERS

13. ____

14. ____

15. ____

16. ____

17. ____

18. ____

19. ____

20. ____

13. Stark plumbing is installing a new furnace for the Harrison family. Mr. Stark divides the job into three phases—the preparation phase, installation phase, and clean-up phase. Mr. Stark knows from past experience that the installation phase takes twice as long as the preparation phase, but the clean-up phase only takes $\frac{1}{3}$ the time of the preparation phase. If the whole job must be completed in 10 hours, how long should Mr. Stark allow for each phase?

14. A rope 180 feet long is cut into three pieces such that the second piece is twice the length of the first piece and the third piece is three times the length of the second piece. How long is each piece?

15. Murphy and LeBlanc Associates are preparing tax returns. As this is the busy time of the year, Murphy takes $\frac{3}{5}$ of the unfinished tax returns home for the weekend. Murphy gives 5 of these returns to LeBlanc to finish over the weekend and keeps 7 for himself. At the beginning of the weekend, how many unfinished tax returns did Murphy and LeBlanc Associates have?

16. Horrible Huey takes $\frac{2}{3}$ of the cookies out of the cookie jar and gives 7 of them to his sister, Sweet Lil. If Huey has 11 cookies left, how many cookies were originally in the cookie jar?

17. The width of a rectangle is $\frac{1}{4}$ the length and the perimeter is seven times the width plus 12 feet. Find the dimensions of the rectangle.

18. Hugo Stump is building two rectangular cow pens with wire fencing (see figure below). Hugo has 180 feet of fencing and wants the width of each pen to be $\frac{1}{2}$ the length. Find the dimensions of each pen.

l

$\vdash w \dashv$

19. Dena Tabb buys 14 pencils and 5 pens. If the total bill (without tax) is \$2.62, and the cost of a pen is twice the cost of a pencil plus 2¢, what does a pencil cost?

20. The Stitch and Thread Store buys 8 bolts of a blended wool material and 15 bolts of a cotton material for a total of \$1925. A bolt of the blended wool material costs \$25 more than a bolt of the cotton material. What is the cost of a bolt of the blended wool?

21. ____________

22. ____________

23. ____________

24. ____________

21. Dan, Donald, and Doug drink 22 sodas. If Dan drinks one-half as much as Donald, and Doug drinks 2 less than Dan, how many sodas does each person drink?

22. Mr. Kaplan left his three sons an inheritance of $50,000. John and Henry received the same amount, but Mike only received one-half of what John received. How much did each son receive?

23. Helen wants to cook three turkeys for her church's Thanksgiving dinner. She plans to use three different types of stuffing, so she wants three different size turkeys—a medium-size turkey to be 6 pounds heavier that the smallest turkey and the heaviest turkey to be 4 pounds heavier than the medium-size turkey. If Helen also calculates that she needs $\frac{1}{2}$ pound of turkey per person and there are going to be 92 persons present, how much should each turkey weigh?

24. Jerry Graza wants to divide a 110-foot rope into 3 pieces. He wants one piece to be twice the length of another piece and the third piece to be 30 feet longer than the shortest piece. How long should each piece of rope be?

4.3

Percent Problems

In this section we will work some word problems that involve percents.

Example 1 In July, Chris buys 28% of the A–Z Company stock. In August he has a chance to buy another 10% of the stock. If after this second purchase he owns 5700 shares of stock, what is the total number of shares of stock in the A–Z Company?

STEP 2: VARIABLE

Let x = total number of shares in the A–Z Company.

STEP 3: EQUATION

$$\begin{pmatrix}\text{Number of shares}\\ \text{bought in July}\end{pmatrix} + \begin{pmatrix}\text{Number of shares}\\ \text{bought in August}\end{pmatrix} = \begin{pmatrix}\text{Total number of}\\ \text{shares Chris owns}\end{pmatrix}$$

$$\begin{pmatrix}\text{28\% of total}\\ \text{shares in}\\ \text{the company}\end{pmatrix} + \begin{pmatrix}\text{10\% of total}\\ \text{shares in}\\ \text{the company}\end{pmatrix} = \begin{pmatrix}\text{Total number}\\ \text{of shares}\\ \text{Chris owns}\end{pmatrix}$$

$$28\% \text{ of } x + 10\% \text{ of } x = 5700$$

$$0.28x + 0.10x = 5700$$

STEP 4: SOLVE

$$0.38x = 5700$$

$$\frac{0.38x}{0.38} = \frac{5700}{0.38}$$

$$x = 15{,}000$$

Thus, there are 15,000 shares of stock in the A–Z Company.

STEP 5: CHECK

Note that 38% of 15,000 is 5700.

■ **DO EXERCISE 1.**

Example 2 The price of hamburger has gone up 60% in the last six months. If the price today is $1.92 a pound, what was the price six months ago?

STEP 2: VARIABLE

Let p = price of hamburger six months ago (old price).

STEP 3: EQUATION

$$\begin{pmatrix}\text{Price of hamburger}\\ \text{6 months ago}\end{pmatrix} + \begin{pmatrix}\text{Increase}\\ \text{in price}\end{pmatrix} = \text{Today's price}$$

$$p + 60\% \text{ of old price} = 1.92$$

$$p + 60\% \text{ of } p = 1.92$$

$$p + 0.60p = 1.92$$

LEARNING OBJECTIVES

After finishing Section 4.3, you should be able to:

- Solve word problems involving percents.

1. Carmen is presently earning $8 an hour. If she receives a 12% raise, what is her new hourly wage?

2. A grocer sells a quart of berries for $1.28. The grocer makes a profit of 60% of the cost of the berries. What did the quart of berries cost the grocer?

3. A radio sold for $60 after a discount of 20% off the list price. What was the list price of the radio?

STEP 4: SOLVE

$$1.6p = 1.92$$

$$p = \frac{1.92}{1.6}$$

$$p = 1.20$$

Thus, the price six months ago was $1.20 per pound.

STEP 5: CHECK

The answer is reasonable. Also, note that 60% of $1.20 is $.72 and $1.20 + $.72 = $1.92.

DO EXERCISE 2.

Example 3 Profits for the Hammer Company fell 14% from the First Quarter to the Second Quarter of the year. If the Second Quarter profits were $129,000, what were the First Quarter profits?

STEP 2: VARIABLE

Let P = First Quarter profits.

STEP 3: EQUATION

$$\begin{pmatrix}\text{First Quarter}\\ \text{profits}\end{pmatrix} - \begin{pmatrix}\text{Decrease in}\\ \text{profits}\end{pmatrix} = \begin{pmatrix}\text{Second Quarter}\\ \text{profits}\end{pmatrix}$$

$$P - \begin{pmatrix}\text{14\% of First}\\ \text{Quarter profits}\end{pmatrix} = 129{,}000$$

$$P - 14\% \text{ of } P = 129{,}000$$

$$P - 0.14P = 129{,}000$$

STEP 4: SOLVE

$$P - 0.14P = 129{,}000$$

$$0.86P = 129{,}000$$

$$P = \frac{129{,}000}{0.86}$$

$$P = 150{,}000$$

Thus the First Quarter profits were $150,000.

STEP 5: CHECK

The answer is reasonable. Also note that 14% of $150,000 is $21,000 and $150,000 − $21,000 = $129,000.

DO EXERCISE 3.

Example 4 Candidates A, B, and C were in an election in which 1000 votes were cast. Candidate A received 35% of the votes. Candidate B received 52 more votes than Candidate C. Who won the election?

STEP 2: VARIABLE

Let x = number of votes received by C; then
$x + 52$ = number of votes received by B, and
35% of 1000 = 350 = number of votes received by A.

STEP 3: EQUATION

$$\begin{pmatrix}\text{Number of votes}\\ \text{received by A}\end{pmatrix} + \begin{pmatrix}\text{Number of votes}\\ \text{received by B}\end{pmatrix} + \begin{pmatrix}\text{Number of votes}\\ \text{received by C}\end{pmatrix} = 1000$$

$$350 + x + 52 + x = 1000$$

STEP 4: SOLVE

$$2x + 402 = 1000$$
$$2x = 598$$
$$x = 299$$

Thus, Candidate C received 299 votes, Candidate B received 351 votes, and Candidate A received 350 votes. Candidate B won the election.

STEP 5: CHECK

The answer is reasonable (except perhaps to Candidate A).

DO EXERCISE 4.

DO SECTION PROBLEMS 4.3.

4. Twenty percent of Mary Dangelo's monthly salary is spent on food, twice the amount spent on food is spent on a house payment, and one-half the amount spent on food is spent for utilities. If the total of these three amounts is $1050, what is Mary's monthly salary?

ANSWERS TO MARGINAL EXERCISES

1. \$8.96 **2.** \$.80 **3.** \$75 **4.** \$1500

NAME COURSE/SECTION DATE

ANSWERS

1. ________
2. ________
3. ________
4. ________
5. ________
6. ________
7. ________
8. ________

SECTION PROBLEMS 4.3

1. A dress sells for \$200 after a 20% discount. What was the original price of the dress?

2. Janis Mulen buys a stereo on sale for \$450. If the sale is a 25% off sale, how much did Janis save by buying the stereo on sale?

3. In 1985 Century Oak Farms made 30% of its profit from its apple orchard and 25% of its profit from timber. If the profit from apples and timber amounted to \$44,000, what was Century Oak Farm's profit in 1985?

4. Five hundred persons are in a cola taste test for three different colas. Cola Bola received 36% of the votes as best cola and Bitsy Cola received 60 more votes than Mora Cola. Which cola won the taste test?

5. If May car sales are down 20% from April sales, and in May 240 cars were sold, how many cars were sold in April at Delmar's Dodge Distributor?

6. A number decreased by 15% of itself is 68. What is the number?

7. Adult tickets for a show cost \$3.50 and children's tickets cost \$1.50. If the total receipts for the show were \$510 and there were twice as many adults as children at the show, how many adults were at the show?

8. A man has four daughters. The oldest is one year older than the second daughter. The second daughter is twice as old as the youngest daughter, and the third daughter's age is the average of the ages of the oldest and youngest daughters. How old is each daughter if the sum of their ages is 34?

9. ____________

10. ____________

11. ____________

12. ____________

13. ____________

14. ____________

15. ____________

16. ____________

17. ____________

9. Nathan received a 12% increase in his monthly salary. If he now makes $2016 per month, what was his salary before his raise?

10. John spends 25% more time than Jim in driving to work. If it takes John 50 minutes to drive to work, how long does it take Jim?

11. Beth Idol has a rectangular-shaped garden. If the width is 80% of the length and the perimeter is 26 feet, what is the area of the garden?

12. The size of the smallest angle of a triangle is 30% of the size of the largest angle, and the size of the third angle is 20° more than the smallest angle. Find the size of each angle. (Recall that the sum of the interior angles of a triangle is 180°.)

13. The Dow Jones Average was down 3% on Tuesday from what it was on Monday. If the Tuesday Dow Jones Average was 1358, what was Monday's Dow Jones Average?

14. The number of incoming freshmen at Western College in 1982 was up 6% over the number in 1981. If this increase amounted to 24 students, how many freshmen came to Western in 1982?

15. Over a year's period, the dairy herd at Wholesome Dairy increased from 200 to 240. What percent increase is this? Hint: Note the amount of increase is 40. Thus another way to ask the above question is, 40 is what percent of 200?

16. Wilson's salary increased from $1100 per month to $1250 per month. What percent increase did Wilson receive?

17. College tuition at EUI in 1984 is up 12% over 1983. As large as this increase may be, it is only three-fifths of the increase from 1982 to 1983. If tuition in 1984 is $5040, what was the tuition in 1982?

4.4

Age Problems

In this section we will look at a type of word problem that generally comes under the heading of age problems. Age problems are interesting and beneficial to do as another example of word problems; however, there really are not many practical applications of age problems.

Example 1 Jeannie has a book her father gave her that is 8 times as old as she is. However, the book's age in 3 years will be the same as 6 times Jeannie's age in 3 years plus 9. How old is Jeannie now?

STEP 2: VARIABLE

Note that the book is 8 times as old as Jeannie.

Let x = Jeannie's present age; then $8x$ = book's present age.

STEP 3: EQUATION

Consider the word equation: the book's age *in* 3 *years* equals 6 times Jeannie's age *in* 3 *years* plus 9. This equation compares the book's age in 3 years to Jeannie's age in 3 years. Thus we write:

$$x + 3 = \text{Jeannie's age in 3 years}$$
$$8x + 3 = \text{book's age in 3 years}$$

And the word equation translates as:

$$8x + 3 = 6(x + 3) + 9$$

STEP 4: SOLVE

$$\begin{aligned} 8x + 3 &= 6x + 18 + 9 \\ 8x + 3 &= 6x + 27 \\ 2x + 3 &= 27 \\ 2x &= 24 \\ x &= 12 \end{aligned}$$

Thus, Jeannie is presently 12 years old and the book is $8(12) = 96$ years old.

STEP 5: CHECK

The answers are reasonable. In 3 years Jeannie will be 15 and the book will be 99, and $6 \cdot 15 + 9 = 99$.

DO EXERCISE 1.

LEARNING OBJECTIVES

After finishing Section 4.4, you should be able to:

- Solve word problems known as age problems.

1. Dan is 8 times as old as his daughter Lucy; however, Dan's age in 3 years will be only 5 times as old as Lucy's age in 3 years. How old is each?

2. Lynda is two years older than Beth. Their father is 37 years old, and twice the sum of Lynda's and Beth's ages in two years is the same as their father's age one year ago. How old are Beth and Lynda now?

Example 2 Henry is 8 years older than his brother, Joseph. Twice Henry's age in two years will be the same as three times Joseph's age three years ago minus 1. How old is each now?

STEP 2: VARIABLE

Note that Henry is 8 years older than Joseph.

Let x = Joseph's age now; then
$x + 8$ = Henry's age now.

STEP 3: EQUATION

The comparison in ages is between Henry's age in 2 years and Joseph's age 3 years ago. Thus

$$x + 10 = \text{Henry's age in 2 years}$$
$$x - 3 = \text{Joseph's age 3 years ago}$$

Twice Henry's age in 2 years = 3 times Joseph's age 3 years ago minus 1.

$$2(x + 10) = 3(x - 3) - 1$$

STEP 4: SOLVE

$$2x + 20 = 3x - 9 - 1$$
$$2x + 20 = 3x - 10$$
$$-x + 20 = -10$$
$$-x = -30$$
$$x = 30$$

Thus, Joseph is 30 now and Henry is 38.

STEP 5: CHECK

Three years ago Joseph was 27, and two years from now Henry will be 40. We also see that 2(40) is the same as 3(27) − 1.

DO EXERCISE 2.

Example 3 Mr. Nicherson has two daughters, Sharon (the older) and Ruth, whose ages are two years apart. The father's age is three times Sharon's age plus 2. The sum of Sharon's and Ruth's ages is 60% of their father's age. How old is each daughter?

STEP 2: VARIABLE

Let x = Ruth's age now. Then
$x + 2$ = Sharon's age now, and
$3(x + 2) + 2$ = their father's age now.

STEP 3: EQUATION

$$\text{Sharon's age} + \text{Ruth's age} = 60\% \text{ of father's age}$$

$$x + 2 + x = 0.6[3(x + 2) + 2]$$

STEP 4: SOLVE

$$2x + 2 = 0.6(3x + 8)$$
$$2x + 2 = 1.8x + 4.8$$
$$0.2x + 2 = 4.8$$
$$0.2x = 2.8$$
$$x = \frac{2.8}{0.2} = 14$$

Thus Ruth is 14 and Sharon is 16. (Their father is 50.)

STEP 5: CHECK

The answers are reasonable, and 60% of 50 is the same as $14 + 16 = 30$.

DO EXERCISE 3.

DO SECTION PROBLEMS 4.4.

3. Jake's age is 60% of his mother's age. However, Jake's age in 6 years will be 75% of his mother's age 10 years ago. How old is each now?

ANSWERS TO MARGINAL EXERCISES

1. 4 and 32 **2.** 6 and 8 **3.** 54 and 90

NAME COURSE/SECTION DATE

ANSWERS

1. ______
2. ______
3. ______
4. ______
5. ______
6. ______
7. ______
8. ______

SECTION PROBLEMS 4.4

1. Henry McGill is 6 years older than his wife, Louise. Twice Henry's age in 4 years plus 1 will be the same as 3 times Louise's age 3 years ago. How old is each now?

2. Joseph is 7 years older than his brother Tony. Twice Joseph's age in 4 years minus 2 will equal three times Tony's age in 4 years. How old is Joseph now?

3. Jean is 4 years older than Domenic. Three times Jean's age in 3 years will equal five times Domenic's age in three years minus 2. How old is each now?

4. Bill Nichols is 10 years older than Sam Twitchell. Twice Bill's age in 8 years will be the same as 3 times Sam's age in 8 years. How old will each be in two years?

5. The Lafer family had to pay $5400 in federal taxes in 1984. In 1985 they paid $6000. What is the percent of increase from 1984 to 1985?

6. It costs the Candy Nook $1.80 a pound to buy Yum Yum Chocolates. If the Candy Nook wants a profit of 60% of the selling price, for what price should they sell a pound of Yum Yum Chocolates?

7. The sum of a daughter's age and father's age is 41 and the difference is 31. How old is each?

8. Don's age is 70% of Mark's age. The difference between their ages is 12. How old is Don?

9. ______________

10. ______________

11. ______________

12. ______________

13. ______________

14. ______________

15. ______________

16. ______________

9. Jennifer Brake is 1 year older than Lauren Colby. Three times Lauren's age in 2 years is the same as twice Jennifer's age in 3 years minus 1. How old will each be in 4 years?

10. My age is one-half my father's age plus 3. In 2 years my age plus 20 will be the same as my mother's age 6 years ago. If my mother is 4 years younger than my father, how old will I be next year?

11. Deborah, Nick, and John are practicing for their piano lessons. Deborah practices two-thirds as long as Nick and John practices one-half as long as Nick. If their combined practice time is 65 minutes, how long does Deborah practice?

12. A rectangle has a width that is 2 inches less than its length. If the perimeter of the rectangle is 24 inches, find the length and the width.

13. A television sells for $510 after a 15% discount. How much was the set before the discount?

14. The sum of Karen and Tina's ages is 19 and the difference is 3. If Tina is older than Karen, how old is each?

15. Jim Hart is 5 years younger than Joan Chaput. Five years ago Jim was two-thirds as old as Joan. How old will they be in 2 years?

16. Larry Cook is 10 years older than Sue Bake, and Shirley Stew's age is the average of Larry and Sue's ages. The sum of Sue's age in 5 years, of twice Larry's age 2 years ago, and of Shirley's age now is 146. How old is each now?

4.5

Mixture Problems

LEARNING OBJECTIVES

After finishing Section 4.5, you should be able to:

- Solve word problems called mixture problems.

In this section we will look at a variety of word problems generally called mixture problems.

Example 1 Marian is mixing peppermint tea and comfrey tea together to form a mixture of $1\frac{1}{2}$ pounds (24 ounces) of her famous Peppfrey tea. The value of peppermint tea is \$.48 an ounce and the value of comfrey tea is \$.36 an ounce. If she sells her Peppfrey tea for \$.44 an ounce, how many ounces of each did she use to make her mixture?

STEP 1: PICTURE

A picture is somewhat helpful.

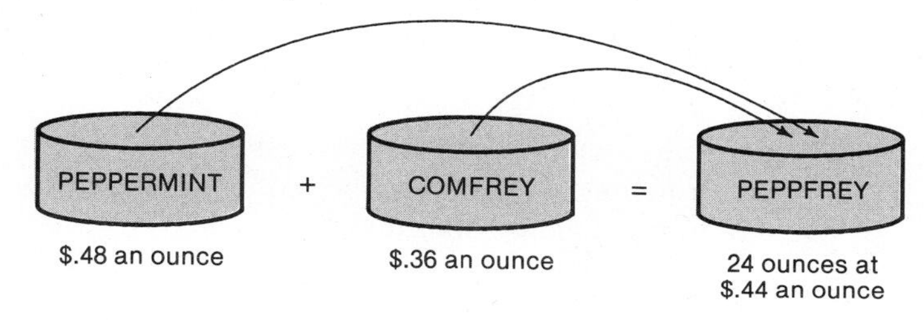

STEP 2: VARIABLE

Let x = number of ounces of peppermint tea in the mixture; then $24 - x$ = number of ounces of comfrey tea in the mixture (since we want a total of 24 ounces in the final mixture).

STEP 3: EQUATION

Note

1. $\begin{pmatrix}\text{Value of peppermint} \\ \text{tea in the final} \\ \text{mixture}\end{pmatrix} + \begin{pmatrix}\text{Value of comfrey} \\ \text{tea in the} \\ \text{final mixture}\end{pmatrix} = \begin{pmatrix}\text{Total value} \\ \text{of final mixture}\end{pmatrix}$

2. If peppermint tea is worth \$.48 for one ounce and if we have x ounces of peppermint tea, the total value must be (\$.48)(x). Similarly, the total value of comfrey tea is (\$.36)($24 - x$).

Thus, word equation 1 becomes

$$(\$.48)(x) + (\$.36)(24 - x) = (\$.44)(24)$$

or, if we do everything in cents,

$$48x + 36(24 - x) = (44)(24)$$

STEP 4: SOLVE

$$\begin{aligned} 48x + 864 - 36x &= 1056 \\ 12x + 864 &= 1056 \\ 12x &= 192 \\ x &= 16 \end{aligned}$$

1. An order of candy costs \$14. It contains one kind of candy worth \$0.50 a pound and another kind of candy worth \$0.60 a pound. If there are 5 more pounds of the \$0.60 candy, how many pounds of each kind were in the order?

Thus, the mixture has 16 ounces of peppermint tea and $24 - 16 = 8$ ounces of comfrey tea.

STEP 5: CHECK

The answers are reasonable. Note that the final mixture is worth 44¢ an ounce and since 44¢ is closer to 48¢ (the value of peppermint tea) than to 36¢ (the value of comfrey), we would expect to have more peppermint tea than comfrey in the final mixture.

DO EXERCISE 1.

Example 2 Jamie Troy has money invested in two different mutual funds. The Lomax-1 is a fairly conservative fund and gave Jamie a 12% return on her money last year. The Lomax-3 is a more aggressive fund and gave Jamie a 17% return on her money last year. Jamie has \$8000 invested in the two funds and earned \$1200 on her two investments last year. How much does Jamie have invested in each fund?

STEP 2: VARIABLE

Let x = amount of money (in dollars) invested in Lomax-1; then
$8000 - x$ = amount of money (in dollars) invested in Lomax-3 (since 8000 is the sum of the amount invested).

STEP 3: EQUATION

$$\begin{pmatrix}\text{Amount earned}\\ \text{from Lomax-1}\end{pmatrix} + \begin{pmatrix}\text{Amount earned}\\ \text{from Lomax-3}\end{pmatrix} = \begin{pmatrix}\text{Total amount}\\ \text{earned}\end{pmatrix}$$

$$12\% \text{ of } x + 17\% \text{ of } (8000 - x) = 1200$$

(Recall that Lomax-1 earned 12% and Lomax-3 earned 17% last year.)

STEP 4: SOLVE

$$\begin{aligned} 0.12x + 0.17(8000 - x) &= 1200 \\ 0.12x + (0.17)(8000) - 0.17x &= 1200 \\ -0.05x + 1360 &= 1200 \\ -0.05x &= -160 \\ x &= \frac{-160}{-0.05} \\ x &= 3200 \end{aligned}$$

Thus Jamie has \$3200 invested in Lomax-1 and \$4800 ($8000 - 3200 = 4800$) invested in Lomax-3.

STEP 5: CHECK

The answers are reasonable. Also 12% of 3200 is 384, 17% of 4800 is 816, and $384 + 816$ is 1200.

DO EXERCISE 2.

Example 3 Cheryl has $3.30 in quarters, dimes, and nickels. If she has four more quarters than nickels and one-half as many dimes as quarters, how many of each type of coin does she have?

STEP 2: VARIABLE

Let x = number of nickels; then
$x + 4$ = number of quarters, and
$\frac{x+4}{2}$ = number of dimes.

STEP 3: EQUATION

$$\begin{pmatrix}\text{Value of all}\\ \text{the nickels}\\ \text{(in cents)}\end{pmatrix} + \begin{pmatrix}\text{Value of all}\\ \text{the dimes}\\ \text{(in cents)}\end{pmatrix} + \begin{pmatrix}\text{Value of all}\\ \text{the quarters}\\ \text{(in cents)}\end{pmatrix} = \begin{pmatrix}\text{Total}\\ \text{value}\\ \text{(in cents)}\end{pmatrix}$$

$$5x + 10\left(\frac{x+4}{2}\right) + 25(x+4) = 330$$

(Each nickel is worth 5¢, so if I have x number of them, then the total value of all the nickels is 5 times x or $5x$.)

STEP 4: SOLVE

$$\begin{aligned} 5x + \frac{10(x+4)}{2} + 25(x+4) &= 330 \\ 5x + 5(x+4) + 25x + 100 &= 330 \\ 5x + 5x + 20 + 25x + 100 &= 330 \\ 35x + 120 &= 330 \\ 35x &= 210 \\ x &= 6 \end{aligned}$$

Thus, Cheryl has 6 nickels, 10 quarters, and 5 dimes.

STEP 5: CHECK

The answers are reasonable and 10 quarters ($2.50), 5 dimes ($.50), and 6 nickels ($.30) add up to $3.30.

DO EXERCISE 3.

DO SECTION PROBLEMS 4.5.

2. Harry Drwyer has $6000 invested in stocks and bonds. The stocks yielded a 15% return on Harry's money last year and the bonds returned 8%. Harry's $6000 earned him $725 last year. How much of the $6000 is invested in stocks?

3. Lynda mixed three kinds of candy (worth $1.50 a pound, $1.80 a pound, and $2.00 a pound, respectively) and ended up with a mixture worth $26. If she used three pounds more of the $1.50 kind than of the $1.80 kind and one-half as much of the $2.00 kind as of the $1.80 kind, how many pounds of each kind did she use?

ANSWERS TO MARGINAL EXERCISES

1. 10 pounds of the 50¢ kind and 15 pounds of the 60¢ kind **2.** \$3500
3. 8 pounds of the \$1.50 kind, 5 pounds of the \$1.80 kind, and $2\frac{1}{2}$ pounds of the \$2.00 kind

NAME COURSE/SECTION DATE

ANSWERS

1. ____________
2. ____________
3. ____________
4. ____________
5. ____________
6. ____________
7. ____________
8. ____________
9. ____________

SECTION PROBLEMS 4.5

1. Rose Bellini has $1.85 in nickels and dimes. If there are 25 coins altogether, how many nickels and dimes does she have?

2. If we mix a tea worth $1.00 an ounce with another tea worth $.60 an ounce, how many ounces of each do we need to use if we want $2\frac{1}{2}$ pounds of a mixture worth $.75 an ounce?

3. Cathy Proed has a 70% alcohol solution and a 50% alcohol solution. How much of each should she use if she wants to obtain 25 pints of a mixture that is 55% alcohol?

4. Sid Moore uses regular gas and gasohol in his car. He knows he gets two more miles to the gallon with gasohol than with regular. On a recent trip of 1160 miles he used 20 gallons of regular gas and 30 gallons of gasohol. How many miles per gallon did he get for regular and how many for gasohol?

5. I have two sums of money invested, one at 5% per year and one at 6% per year. My interest income last year was $700. How much do I have invested at each interest rate if the total amount invested is $12,500?

6. Charlie Crafts invested $4000 in two different mutual funds. One fund returned a yield of 15% last year, but the other fund only returned 5% last year. Charlie made $440 on his two investments last year. How much was invested in each mutual fund?

7. John is 6 years older than Marilyn. Three times Marilyn's age in two years is the same as twice John's age in five years plus 10. How old is each now?

8. A class of 90 students was asked to pick their favorite color among blue, red, and green. Thirty percent of the class picked blue, and twice as many picked red as picked green. How many students picked red?

9. If you have a 40% alcohol solution and a 70% alcohol solution, how much of each do you need to make 20 gallons of a solution that is 50% alcohol?

10. ______________

11. ______________

12. ______________

13. ______________

14. (a) ____________

(b) ____________

15. ______________

16. ______________

10. Marty Furnchild has $5.85 in quarters and dimes. If he has 4 times as many dimes as quarters, how many dimes and how many quarters does he have?

11. By combining water with pure alcohol, Louis wants to obtain 2 liters of a solution that is 20% alcohol. On his first attempt at arbitrarily mixing the two together, Louis obtains 2 liters of a solution that tests out at 36% alcohol. On his second attempt, he obtains 2 liters of a solution that is 16% alcohol. Louis now has two solutions (neither of which is correct), no more pure alcohol, and no idea of what he is going to do. His sister Dena tells him that by using algebra and combining certain amounts of the 36% solution with the 16% solution, she can solve his problem. How much of each solution does Dena combine to get Louis his 2 liters of a 20% alcohol solution?

12. Beth Muloney is going to combine a solution that is 40% salt with another solution that is 75% salt to obtain 14 pints of a 60% salt solution. How many pints of the 40% solution and how many pints of the 75% solution should Beth use?

13. Jean Apple wants to mix a 20% antifreeze solution with a 45% antifreeze solution to obtain 15 pints of a solution that is 30% antifreeze. How much of each solution should she use in her mixture?

14. I have 20 gallons of a solution that is 40% alcohol. (a) If I want to increase the level of alcohol in my solution to 60%, how much pure alcohol would I have to add to the 20 gallons of solution? (b) If I want to end up with 20 gallons of a 60% alcohol solution, then how much pure alcohol would I have to add? (Hint for (b): If you add x number of gallons of pure alcohol, you first must take away x number of gallons of the 40% solution.)

15. Martin's is having a 20% off sale. If the sale price of a coat is $128, how much is saved by buying the coat on sale instead of at the regular price?

16. Stella Schmidt has 3 more quarters than dimes, and the total value of her coins is $2.15. How many quarters and dimes does she have?

4.6

Miscellaneous Word Problems

LEARNING OBJECTIVES

After finishing Section 4.6, you should be able to:

- Solve lever and fulcrum problems, depreciation problems, and averaging problems.

The first type of problem we look at in this section involves a lever and a fulcrum.

Example 1 Charles, who weighs 90 pounds, and his younger brother, Joe, who weighs 60 pounds, are going to play on a 12-foot seesaw. If the bar on which the seesaw sits is in the middle of the 12-foot seesaw and if Joe sits on one end, where should Charles sit in order to balance the seesaw?

Before we do this problem, we need to become familiar with the idea of a lever and a fulcrum. A picture will help.

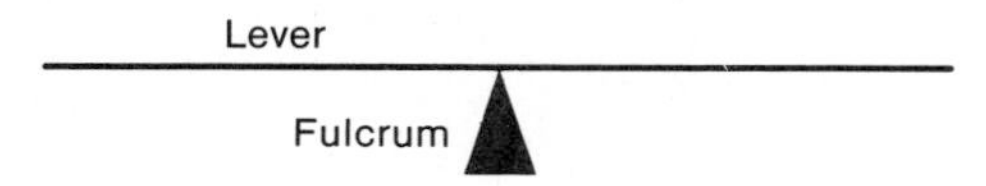

We are interested in what happens if we put a weight on one or both ends of the lever. (The weight can actually be placed anywhere along the lever and not just on the ends.)

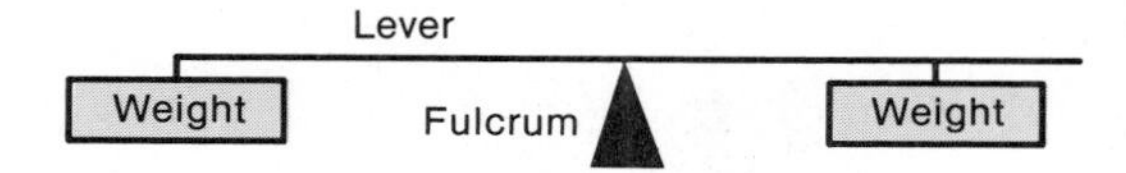

The weight will create a torque about the fulcrum, and the amount of torque is found by multiplying the weight times the distance the weight is from the fulcrum.

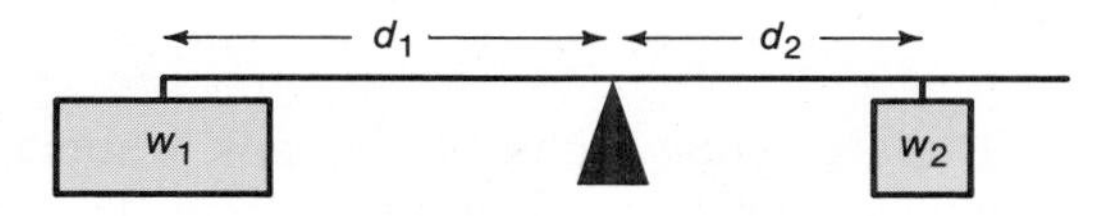

In our picture, the torque exerted by the weight on the left is $w_1 \cdot d_1$ and the torque exerted by the weight on the right is $w_2 \cdot d_2$. If these two torques are equal, then the lever is in equilibrium, that is, the lever will be balanced. Thus the lever is in equilibrium if

$$w_1 \cdot d_1 = w_2 \cdot d_2$$

Note the use of the subscripts on the variables

d_1 and d_2
↖ ↗
Subscripts

The subscripts are used to indicate that the variable d_1 is different from the variable d_2.

Now we can return to our seesaw problem. (A seesaw is an example of a lever and a fulcrum.)

1. Where should a 60-pound weight be placed to balance the lever in the following picture?

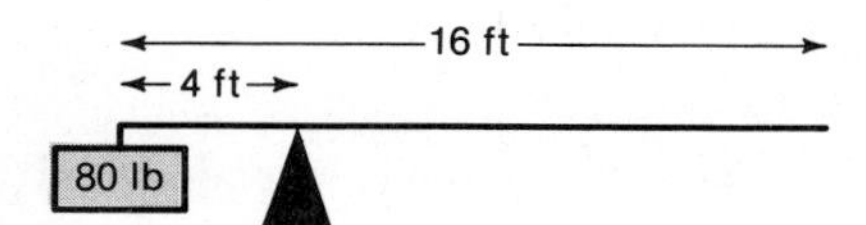

STEP 1: PICTURE

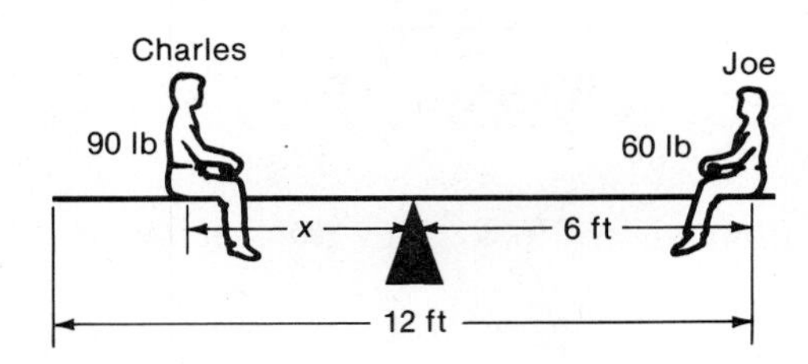

STEP 2: VARIABLE

Let x = distance Charles should sit from the fulcrum.

STEP 3: EQUATION

For the seesaw to balance, we need $w_1 \cdot d_1 = w_2 \cdot d_2$, or in our case $90x = 60 \cdot 6$.

STEP 4: SOLVE

$$90x = 60 \cdot 6$$
$$90x = 360$$
$$x = 4$$

Thus, Charles should sit 4 feet from the fulcrum.

STEP 5: CHECK

The answer is reasonable since Charles is heavier than Joe and, hence, should sit closer to the fulcrum than Joe. Also, we see that $90 \cdot 4$ does equal $6 \cdot 60$.

DO EXERCISE 1.

Example 2 Lionel DuBerg, who weighs 140 pounds, wants to dislodge a rock that he calculates weighs approximately 1000 pounds. To accomplish this he places a 5-foot steel bar under the rock, places a small rock 5 inches from the end of the bar to act as a fulcrum (see picture below), and then applies all his weight to the other end of the bar. Will Lionel be able to dislodge the rock?

STEP 1: PICTURE

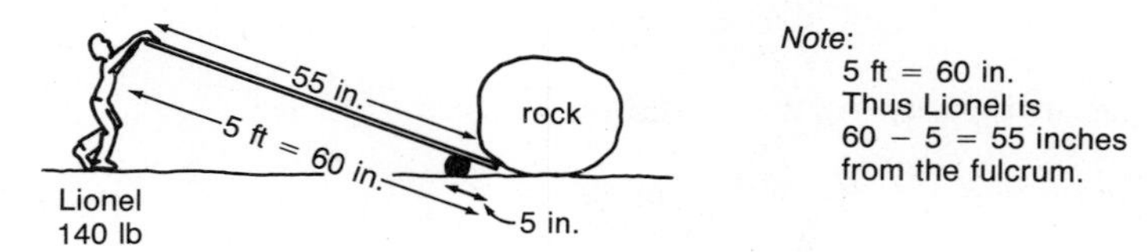

Note:
5 ft = 60 in.
Thus Lionel is
60 − 5 = 55 inches
from the fulcrum.

STEP 2: VARIABLE

Let w = the actual weight of the rock.

STEP 3: EQUATION

If the system is put in equilibrium (that is, the steel bar becomes parallel to the ground), then the rock will be dislodged. Our equilibrium equation, $w_1 \cdot d_1 = w_2 \cdot d_2$, becomes

$$(140)(55) = w(5)$$

STEP 4: SOLVE

$$7700 = 5w$$

$$\frac{7700}{5} = w$$

$$1540 = w$$

Thus Lionel should be able to dislodge a rock that weighs as much as 1540 pounds. Hence, if Lionel's estimate of the weight of this rock is correct, he should be able to dislodge it.

DO EXERCISE 2.

The following problem is an example of straight-line depreciation.

Example 3 Ray has a piece of machinery that is depreciating at the rate of \$2500 a year. If the book value of the machinery now is \$20,000, how long will it take the book value to depreciate to \$8750?

STEP 2: VARIABLE

Let t = number of years required to depreciate to \$8750.

STEP 3: EQUATION

We can write two word equations, as follows.

1. $$\begin{pmatrix}\text{Present}\\ \text{book}\\ \text{value}\end{pmatrix} - \begin{pmatrix}\text{Total amount of}\\ \text{depreciation}\end{pmatrix} = \begin{pmatrix}\text{New}\\ \text{book}\\ \text{value}\end{pmatrix}$$

2. $$\begin{pmatrix}\text{Total amount of}\\ \text{depreciation}\end{pmatrix} = \begin{pmatrix}\text{Amount of}\\ \text{depreciation}\\ \text{per year}\end{pmatrix} \cdot \begin{pmatrix}\text{Number}\\ \text{of years}\end{pmatrix}$$

$$\text{Total amount of depreciation} = (2500)(t)$$

Thus, equation 1 becomes $20{,}000 - 2500t = 8750$.

STEP 4: SOLVE

$$-2500t = -11{,}250$$

$$t = \frac{-11{,}250}{-2500}$$

$$t = 4.5$$

Thus, it will take $4\frac{1}{2}$ years to depreciate to \$8750.

STEP 5: CHECK

The answer is reasonable, and we see that $20{,}000 - 4.5(2500)$ is 8750.

DO EXERCISE 3.

2. To balance the lever in the following picture, where should we place the fulcrum?

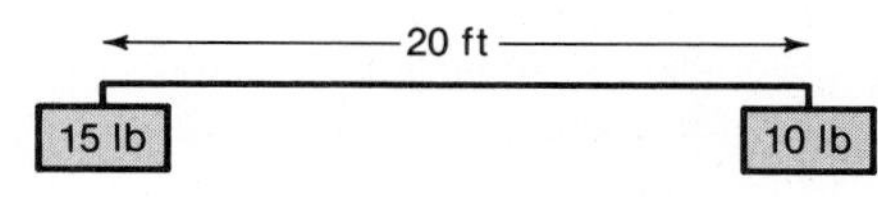

Hint: Let x = the distance the 15-pound weight is from the fulcrum. Thus $20 - x$ = the distance the 10-pound weight is from the fulcrum.

3. Vincent buys a typewriter for \$450. If it depreciates at a constant rate of \$60 a year, how long will it be before it depreciates to \$240?

4. Jerry needs an average of 92 on 4 tests to receive an A in English Literature. Thus far he has received grades of 90, 95, and 86. What grade must he receive on his fourth test if he wants to receive an A for the course?

Example 4 Rob is running to get ready for the Boston Marathon. He runs 5 times a week and likes to average 12 miles per run. So far this week he has run 4 times, recording distances of 10 miles, 12 miles, 9 miles, and 14 miles. How many miles must he run on his fifth outing to average 12 miles per run?

STEP 2: VARIABLE

Let d = distance he must run on the fifth day.

STEP 3: EQUATION

To find the average of a group of numbers, we add the numbers together and divide by how many numbers are in the group.

$$\frac{10 + 12 + 9 + 14 + d}{5} = 12$$

STEP 4: SOLVE

$$\frac{45 + d}{5} = 12$$

$$45 + d = (5)(12)$$

$$45 + d = 60$$

$$d = 15$$

Thus, Rob must run 15 miles on the fifth day.

STEP 5: CHECK

The answer is reasonable and we see that $\frac{10 + 12 + 9 + 14 + 15}{5}$ is 12.

DO EXERCISE 4.

Word problems are an important part of algebra. As we proceed through the text, you will be happy to know that we will learn how to do many more kinds of word problems besides those we have learned in this chapter.

DO SECTION PROBLEMS 4.6.

ANSWERS TO MARGINAL EXERCISES

1. $5\frac{1}{3}$ feet from fulcrum **2.** 8 feet from 15-lb weight **3.** $3\frac{1}{2}$ years
4. 97

NAME COURSE/SECTION DATE

ANSWERS

SECTION PROBLEMS 4.6

1. If I have a 20-foot board that has a 100-pound weight on one end and a 60-pound weight two feet from the other end, where would I place the fulcrum in order to balance the board?

2. If I have an 8-foot board with a 150-pound weight on one end and an 80-pound weight on the other end, where do I place the fulcrum to balance the board?

3. Martha Garven places a 6-foot steel bar under a large rock that she wants to move. She uses a smaller rock for a fulcrum and places it 8 inches from the end of the steel bar. If the rock weighs 1200 pounds, how much weight must Martha put on her end of the bar?.

4. Sean, who weighs 80 pounds, and Maureen, who weighs 60 pounds, want to balance a 14-foot seesaw. If the fulcrum is in the middle of the seesaw and if Maureen sits on one end, where should Sean sit to balance the seesaw?

5. Denise has a car that is worth $6000 but is depreciating at the rate of $1500 the first year and then $750 every succeeding year. How long will it take for the car to be worth $2700?

6. A piece of machinery that cost $50,000 depreciates $8000 the first year and then at a constant rate of $6000 per year. How long will it be before the piece of machinery has depreciated to $20,000?

7. The average age of five senior citizens, who gather for a poker game every Friday night, is 80. If four have ages of 73, 92, 65, and 80, what is the age of the fifth senior citizen?

8. Jenny wants to average 80 on four tests. On the first two tests she has scores of 72 and 90. What scores does she need on the third and fourth tests if she feels she can score five points more on the fourth test than on the third test?

1. ______
2. ______
3. ______
4. ______
5. ______
6. ______
7. ______
8. ______

9. ______________

10. ______________

11. ______________

12. ______________

13. ______________

14. ______________

15. ______________

16. ______________

9. Lois Shephard's salary went from \$16,500 to \$18,480. What percent increase is this?

10. Joanne is going to mix candy worth \$1.50 a pound with candy worth \$2.25 a pound to obtain 30 pounds of a mixture worth \$2.00 a pound. How much of each kind should she use in her mixture?

11. Susan likes to average 8 hours of sleep a night. For four consecutive nights, she slept 7 hours, 8 hours, 9.5 hours, and 6 hours. How many hours should she sleep on the fifth night in order to maintain her average of 8 hours a night?

12. Sol Lenowitz buys a computer for \$3000. He depreciates the computer over 5 years at a rate of \$600 a year. How long will it take for the computer to depreciate to \$1000?

13. The sum of three consecutive integers is the same as the average of the smallest and largest of the three consecutive integers plus 16. Find the integers.

14. One angle of a triangle is twice another angle. The third angle is 20° smaller than the smaller of the other two. Find all three angles.

15. To put the lever in the following picture in equilibrium, where should a 120-pound weight be placed?

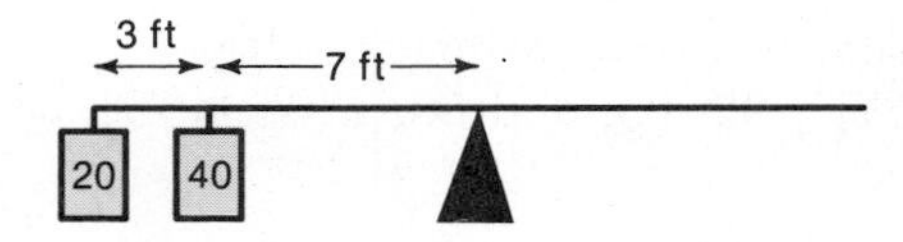

16. In the following picture, where should a 50-pound weight be placed to balance the board?

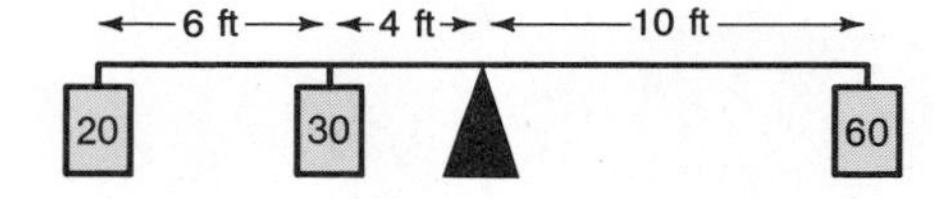

NAME COURSE/SECTION DATE

ANSWERS

1. ______
2. ______
3. ______
4. ______
5. ______
6. ______
7. ______
8. ______
9. ______

CHAPTER 4 REVIEW PROBLEMS

SECTION 4.1

Translate into mathematical phrases or equations.

1. Three times a number plus 7

2. Three times the result of a number increased by 7

3. Twice a number minus four times another number

4. Divide a number by another number and then add 6.

5. Twice a number plus 5 and the result divided by 3 is the same as 6.

SECTION 4.2

6. The length of a rectangle is twice the width plus 3. If the perimeter is 30 inches, find the dimensions of the rectangle.

7. The sum of the interior angles of a triangle is 180°. In a triangle, angle A is twice as large as angle B and angle C is 20° larger than angle A. Find the three angles.

8. Joe, Sue, and Helen are picking strawberries. For each box they pick, the farmer will pay them 50¢. Helen picks 50% more than Sue and Joe picks 3 boxes less than Helen. How much money does each person make if they pick a total of 29 boxes of strawberries?

9. Three hundred fifty people attended the Essex Historical Society's slide show. Tickets cost $2.50 for adults and $1.50 for children. The total income from the show was $681.00. How many children attended the show?

10. ____________

11. ____________

12. ____________

13. ____________

14. ____________

15. ____________

16. ____________

17. ____________

18. ____________

SECTION 4.3

10. A number increased by 15% of itself is 207. Find the number.

11. Honest Joe Fingers reduced the price of a new car from \$6550 to \$5900. What percent decrease is Joe giving his customers?

12. In History 102, 20% of the students receive a grade of B and 5% receive a grade of A. If 9 more students receive Bs than As, how many students are in History 102?

13. At a 30% off sale, a dress sells for \$112. What was the original price of the dress?

SECTION 4.4

14. A man has three daughters. The sum of the ages of the oldest and youngest is 34. The second daughter is 3 years older than the youngest, and the sum of all their ages is 51. How old is each daughter?

15. Jan is 2 years older than Buzz. Twice Jan's age in 3 years will be the same as 3 times Buzz's age 10 years ago. How old is each now?

16. Kathy's dog Louis has an age that is 30% Kathy's age. If we divide Kathy's age in half and then subtract 6, we also have Louis's age. How old is Louis?

SECTION 4.5

17. Lisa has 18 coins consisting of nickels, dimes, and quarters. If the total value is \$2.25 and there are two more dimes than quarters, how many of each coin does she have?

18. Five times a number minus $\frac{1}{2}$ is the same as 75% of the number plus 2. Find the number.

NAME COURSE/SECTION DATE

ANSWERS

19. ______
20. ______
21. ______
22. ______
23. ______
24. ______
25. ______
26. ______

19. How many pints of a solution that is 20% salt and how many pints of a solution that is 60% salt have to be mixed together to obtain 10 pints of a solution that is 30% salt?

20. Joe has a solution of 12 gallons that is 40% antifreeze. How many gallons of a solution that is 90% antifreeze does Joe have to add to his 12 gallons to obtain a solution that is 60% antifreeze?

SECTION 4.6

21. In the following picture, where should the fulcrum be placed in order to balance the lever?

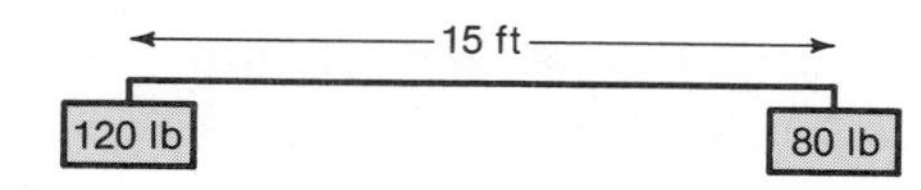

22. Beth, who weighs 60 pounds, and Lynda, who weighs 75 pounds, are playing on a seesaw that is 12 feet long. If Beth sits on one end, where should Lynda sit to balance the seesaw? Assume that the fulcrum is in the middle of the board.

23. The average price of 6 houses is $75,400. Two of the houses are priced at $83,000 each, two at $74,000 each, and one at $69,500. What is the price of the sixth house?

24. Sal buys a piece of office equipment for $6800. It depreciates at a constant rate of $1200 a year, and Sal wants to trade it in when its value reaches $1000. How long will Sal have to wait before he trades it in?

25. The sum of two consecutive even integers is 74. Find the integers.

26. The sum of three consecutive integers is 87. Find the integers.

27. ______________

28. ______________

29. ______________

30. ______________

31. ______________

32. ______________

33. ______________

34. ______________

27. The perimeter of a rectangle is 52 m. The length is 4 m greater than the width. Find the area of the rectangle.

28. Henry is 8 years older than Marsha. Twice Marsha's age in 3 years will be the same as Henry's age in 5 years plus 15. How old is each now?

29. The average of four numbers is 17.5. If three of the numbers are 23, 17, and 12, find the fourth number.

30. In a collection of quarters and dimes, there are 28 coins. If there are four more quarters than dimes, how many of each type of coin are in the collection?

31. Where should you place a 40-pound weight to balance the board in the following picture?

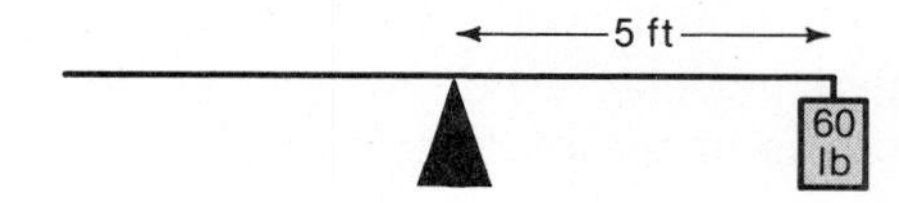

32. Where should you place a 60-pound weight to balance the board in the following picture?

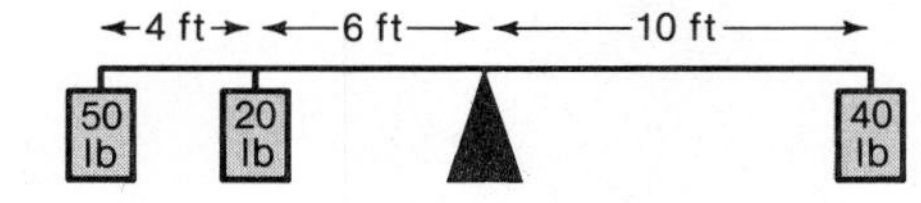

33. A wood-chipper that originally cost $1600 depreciates at a constant rate of $300 a year. How long will it take for it to depreciate to half its original value?

34. Mr. Meier has $14.40 worth of 60-cent American and 80-cent Cuban cigars. If he has two-thirds as many American cigars as Cuban cigars, how many of each does Mr. Meier have?

NAME COURSE/SECTION DATE

ANSWERS

1. ____________

2. ____________

3. ____________

4. ____________

5. ____________

CHAPTER 4 TEST

1. The sum of two consecutive integers is the same as 3 times the smaller integer minus 11. Find the integers.

2. Marlene Chateau buys a fall coat at a 25% off sale for \$135. How much did Marlene save by buying the coat on sale?

3. Vixen is 2 years older than Blixen. In 2 years twice Vixen's age will be the same as 3 times Blixen's age in 3 years minus 6. How old is each now?

4. Dorothy Small is going to mix tea worth \$1.30 an ounce with tea worth \$1.70 an ounce in order to obtain 2 pounds of a tea mixture that sells for \$1.45 an ounce. How many ounces of each kind should Dorothy use in her mixture?

5. Buford Paine is building a pen for his hogs, and he is going to use the barn for one side of the pen. For the 3 remaining sides Buford has 100 feet of fencing available. He wants the 2 sides that are perpendicular to the side of the barn to each be $\frac{3}{4}$ the length of the side that is parallel to the barn. Find the dimensions of the pen.

6. ______
7. ______
8. ______
9. ______
10. ______

6. Where should the fulcrum be placed to balance the lever in the following picture?

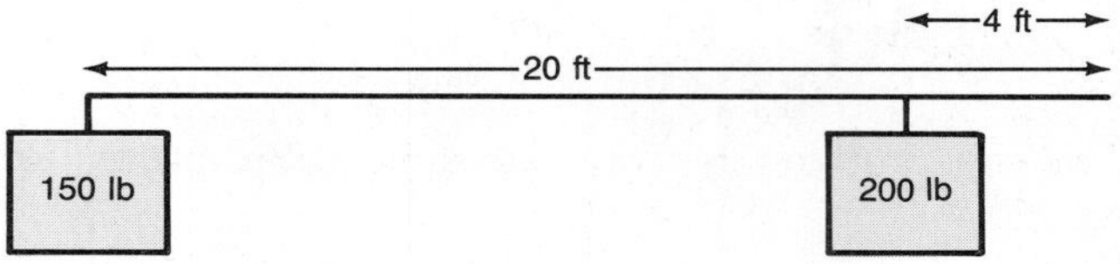

← 20 ft →	← 4 ft →
150 lb	200 lb

7. A number divided by 3 and then increased by $\frac{1}{2}$ is the same as the number decreased by 25% of itself. Find the number.

8. AV Construction buys a pickup for $12,500. They depreciate the pickup at a constant rate of $2500 yearly. How long will it take to depreciate the pickup to $7000?

9. Lucy Cook has $12,000 invested, some at a return of 10% and the rest at a return of 14%. The yearly return from the 10% investment is exactly the same as the yearly return from the 14% investment. How much does Lucy have invested at 14%?

10. Tad Bunting buys 50 bolts of material for the Sew-n-Save Shop. He buys three kinds of material: wool, blended wool, and cotton. There are 5 more bolts of blended wool than wool and $2\frac{1}{2}$ times as many cotton bolts as wool bolts. How many blended wool bolts did Tad buy?

CHAPTER 5

Polynomials

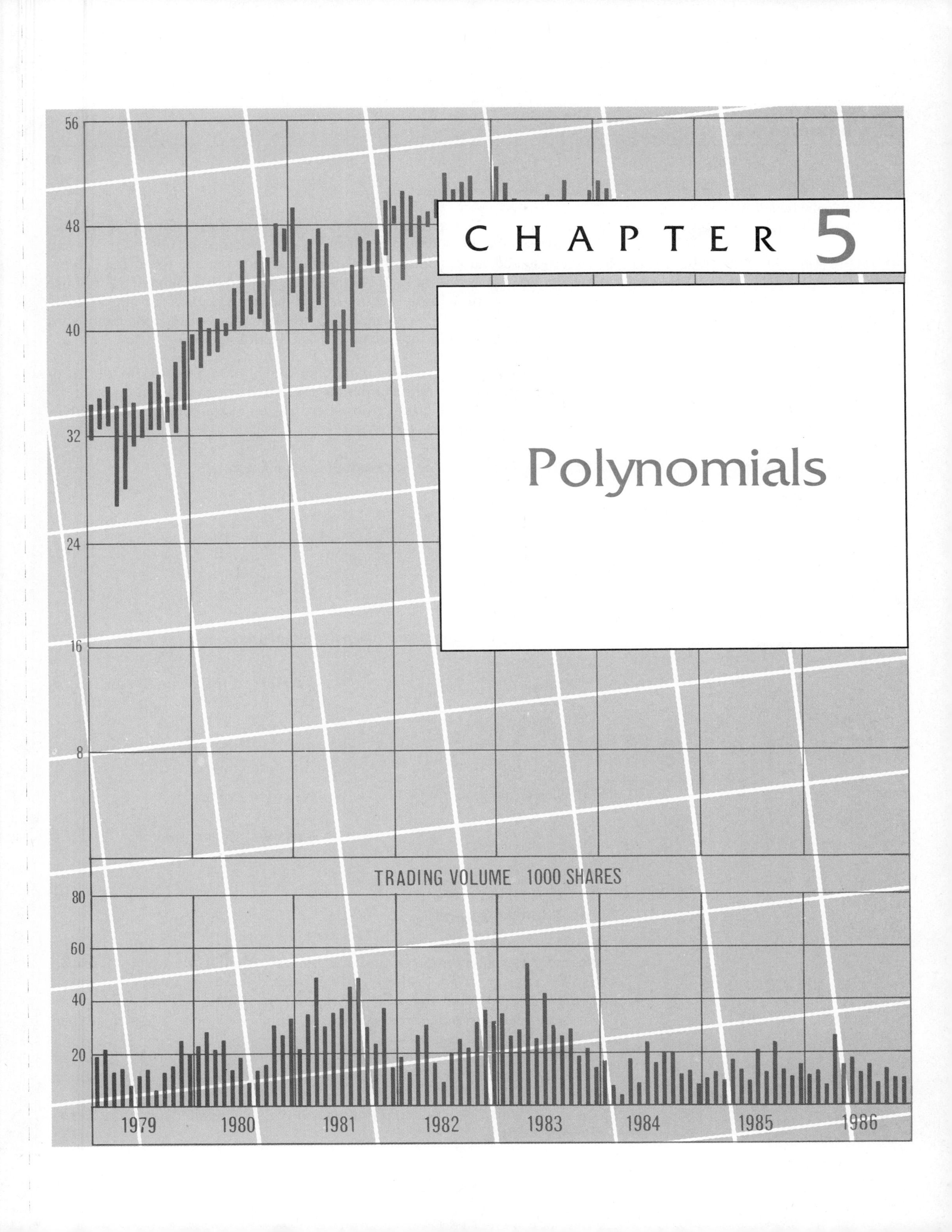

LEARNING OBJECTIVES

After finishing Section 5.1, you should be able to:

- Identify polynomials.
- Identify terms and numerical coefficients.
- Find the degree of terms and polynomials.

1. Which of the following are polynomials? If not, why not?

(a) $\frac{x^2}{y}$ (b) $x^3 - \frac{x^2}{3} + x - 4$

(c) $|x| - 3y$ (d) $a^2b + 3b^{-4}$

(e) -5

5.1

Introduction

In this chapter we are going to study one of the more important types of algebraic expressions, the polynomial. After learning what a polynomial is, we will learn how to add, subtract, multiply, and factor polynomials. Factoring polynomials is a necessary skill in many areas of algebra. In this chapter we will use factoring to help in solving certain second-degree equations in one variable (such as $x^2 - x - 6 = 0$). Such second-degree equations often arise from word problems, which we will look at in the last section.

DEFINITION 5.1 A *polynomial* is an algebraic expression made up of numbers and variables combined by the operations of addition, subtraction, and multiplication, and where all the exponents on the variables are whole numbers.

Example 1 The following are examples of polynomials.

(a) $x^2 - x + 4$ (b) $3ab - b^2$

(c) $-2x + \frac{3}{2}$ (d) $x^2y^2 - \frac{xyz}{5}$

(e) $\frac{s^4}{2} - \frac{t^4}{3}$ (f) 7

Example 2 The following are not polynomials:

(a) $\frac{1}{x} + y$ — Variable in the denominator (or exponent not a whole number, that is, $\frac{1}{x} = x^{-1}$)

(b) $|x| - 3$ — Absolute value

(c) $x^{-4} + x^{-2}$ — Negative exponents

(d) $\frac{x - y}{x^2}$ — Variable in denominator (or $\frac{1}{x^2} = x^{-2}$)

DO EXERCISE 1.

We may have polynomials in one variable (for example, $x^2 - x + 4$), in two variables ($x^2 - xy + y^2$), in three variables ($x^2y^2 - xyz$), or in as many variables as we like. We will focus on polynomials in one and two variables.

We need a few more definitions, some of which we saw in Chapter 2 when we studied algebraic expressions.

DEFINITION 5.2 A. The *terms* of a polynomial are separated by addition signs.

B. A term of a polynomial that has no variable is called a *constant term*.

DEFINITION 5.3 The *numerical coefficient* of a term is the number in the term that is multiplied times the variable(s) in the term. If a variable has no number in front of it, the coefficient is understood to be the number one ($x = 1x$) or the number negative one ($-x = -1x$).

Example 3 In the polynomial $x^2 - 3x + 1$ [or $x^2 + (-3x) + 1$], the terms are x^2, $-3x$, and 1 (constant term). The numerical coefficient of the x^2 term is 1, of the $-3x$ term is -3, and of 1 is 1.

Example 4 In the polynomial

$$3x^2y - 2xy - y^2 + \frac{7x}{2}$$

the terms are $3x^2y$, $-2xy$, $-y^2$, and $7x/2$. The numerical coefficients of the four terms are, respectively, 3, -2, -1, and $\frac{7}{2}$.

DO EXERCISE 2.

DEFINITION 5.4 A. A polynomial consisting of exactly one term is called a *monomial.*

EXAMPLES: x^2, $3s$, x^2y^2, and -4

B. A polynomial consisting of two terms is called a *binomial.*

EXAMPLES: $x - 4$, $4y^2 + y$, and $ts - 14$

C. A polynomial consisting of three terms is called a *trinomial.*

EXAMPLES: $x^2 - x + 5$ and $x^3y^3 - x + 1$

DO EXERCISE 3.

DEFINITION 5.5 A. The *degree of a term* that has only one variable is the exponent on that variable.

Example 5 Identify the degree of each term in the polynomial $-3x^4 + x^2 - 7x + 5$.

(a) $-3x^4$ has degree 4.

(b) x^2 has degree 2.

(c) $-7x$ has degree 1. (Note that $x = x^1$.)

(d) 5 has degree 0. (Note that $5 = 5 \cdot x^0$. A constant term of a polynomial will always have degree zero.)

DEFINITION 5.5 (continued) B. The *degree of a term* that has more than one variable is the sum of the exponents on the variables.

Example 6 Identify the degree of each term of $x^2y - x^3y^2 + y^4$.

(a) x^2y^1 has degree 3 since $2 + 1 = 3$.

(b) $-x^3y^2$ has degree 5 since $3 + 2 = 5$.

(c) y^4 has degree 4.

DO EXERCISE 4.

2. In the following polynomials, list each term and the numerical coefficient of each term.
(a) $-x^3 + 2x^2 - x + 4$
(b) $\frac{a^2}{2} - ab + 7b^2$

3. Classify the following polynomials as monomial, binomial, trinomial, or none of these.
(a) $x^2 - 7x$ (b) $a^4 - 3a^2 + 4$
(c) $x^2y^2 - xy + y^4$ (d) $7x^2$
(e) $x^5 - x^3 + x - 4$

4. Find the degree of each term of the following polynomials.
(a) $-x^3 + 2x^2 - x + 3$
(b) $7a^2b - a^3b^3 + b^4$

5. What is the degree of each of the following polynomials?
(a) $x^2 - 7x$ (b) $a^4 - 3a^2 + 4$
(c) $x^2y^2 - xy + y^3$ (d) $x - 3$
(e) $x^5 - x^3 + x - 4$

6. Arrange the following polynomials in descending and ascending order.
(a) $-2x^2 + x^3 - 4$
(b) $7 - x^2 - 3x^5 + x^4 - x$

7. Give the terms of the following polynomials, the coefficient of each term, the degree of each term, and the degree of the polynomial.
(a) $-2x^2 - x + 4$
(b) $x^2y^2 - xy^4 + 2y^5$
(c) $x - 3$ (d) 5

DEFINITION 5.6 The *degree of a polynomial* is the greatest degree of any of its individual terms.

Example 7 The polynomial $-3x^4 + x^2 - 7x + 5$ has degree 4 since the term with the largest degree is $-3x^4$.

Example 8 The polynomial $x^2y - x^3y^2 + y^4$ has degree 5 since the term with the largest degree is $-x^3y^2$.

DO EXERCISE 5.

The polynomial $x^2 - 3x + 5$ may be written in other forms by rearranging the terms. The following are different ways of writing the polynomial $x^2 - 3x + 5$:

1. $-3x + x^2 + 5$
2. $x^2 + 5 - 3x$
3. $5 + x^2 - 3x$

DEFINITION 5.7 A. A polynomial in one variable is in *descending order* if the terms are arranged so that the term with the largest degree is first, the term with the next largest degree is second, and so on.
B. A polynomial in one variable is in *ascending order* if the term with the smallest degree is first, the term with the next smallest degree is second, and so on.

Polynomials are most commonly written in descending order.

Example 9 Arrange $3x^2 + x^3 - x + 4x^4$ in descending order and then in ascending order.

Descending order: $4x^4 + x^3 + 3x^2 - x$

Ascending order: $-x + 3x^2 + x^3 + 4x^4$

DO EXERCISES 6 AND 7.

Before we look at addition and subtraction of polynomials in the next section, we must define like (similar) terms for polynomials. (Compare with Definition 2.11.)

DEFINITION 5.8 *Like* (similar) *terms* of a polynomial are terms (1) that have the same variable(s) and (2) whose corresponding variables have identical exponents.

Example 10 Pick out the like terms in the polynomial

$$y - x - 3x^2 + 4x - 5 - y^2 + 2y - x^2$$

(a) $-3x^2$ and $-x^2$ are like terms.

(b) $-x$ and $4x$ are like terms.

(c) y and $2y$ are like terms.

Recall from Chapter 2 that like terms may be combined by adding their numerical coefficients.

Example 11 Simplify the following polynomials by combining like terms.

(a) $2x^2 - y + 3x - x^2 + 2y - 4 + 7x^2 = 2x^2 + (-x^2) + 7x^2 + 3x$
$+ (-y) + 2y + (-4)$
$= 8x^2 + 3x + y + (-4)$
$= 8x^2 + 3x + y - 4$

(b) $x^2y^2 - 3xy + 2x^2y^2 - xy + 2x^2y = x^2y^2 + 2x^2y^2 + (-3xy)$
$+ (-xy) + 2x^2y$
$= 3x^2y^2 + (-4xy) + 2x^2y$
$= 3x^2y^2 - 4xy + 2x^2y$

DO EXERCISES 8 AND 9.

DO SECTION PROBLEMS 5.1.

Simplify the following polynomials by combining like terms.

8. $y - x - 3x^2 + 4x - 5 - y^2 + 2y - x^2$

9. $x^2y^2 - 3xy + 2x^2y^2 + xy^2 - xy + 2x^2y$

ANSWERS TO MARGINAL EXERCISES

1. (a) No; variable in denominator (b) Yes (c) No; absolute value (d) No; negative exponent (e) Yes **2.** (a) Terms are $-x^3$, $2x^2$, $-x$, and 4; coefficients are (respectively) -1, 2, -1, and 4. (b) Terms are $\frac{a^2}{2}$, $-ab$, and $7b^2$; coefficients are (respectively) $\frac{1}{2}$, -1, and 7. **3.** (a) Binomial (b) Trinomial (c) Trinomial (d) Monomial (e) None of these **4.** (a) 3, 2, 1, 0 (respectively) (b) 3, 6, 4 (respectively) **5.** (a) 2 (b) 4 (c) 4 (d) 1 (e) 5 **6.** (a) $x^3 - 2x^2 - 4$ (descending); $-4 - 2x^2 + x^3$ (ascending) (b) $-3x^5 + x^4 - x^2 - x + 7$ (descending); $7 - x - x^2 + x^4 - 3x^5$ (ascending) **7.** (a) $-2x^2$, $-x$, and 4 are the terms; -2, -1, 4 (respectively) are the coefficients; 2, 1, 0 (respectively) are the degrees of the terms; 2 is the degree of the polynomial. (b) x^2y^2, $-xy^4$, $2y^5$ are the terms; 1, -1, 2 (respectively) are the coefficients; 4, 5, 5 (respectively) are the degrees of the terms; 5 is the degree of the polynomial. (c) x, -3 are the terms; 1, -3 (respectively) are the coefficients; 1, 0 (respectively) are the degrees of the terms; 1 is the degree of the polynomial. (d) 5 is the term and the coefficient of that term; 0 is the degree of the term and of the polynomial. **8.** $-4x^2 + 3x + 3y - y^2 - 5$ **9.** $3x^2y^2 - 4xy + xy^2 + 2x^2y$

NAME COURSE/SECTION DATE

SECTION PROBLEMS 5.1

Are the following polynomials? If not, why not?

1. $x^3 - xy^2$ **2.** $x^2 - x$ **3.** $x^2y^{-2} + 1$

4. $x - 3 + y$ **5.** 4 **6.** $\dfrac{a-1}{a-2}$

7. $x^2 - x^{-1}$ **8.** $|t^2| - 2|t|$

Pick out the terms of each polynomial.

9. $-x^2 - 3x + 4$ **10.** $2a^2b^2 - 3ab^2 + 4ab - b^2$

11. $\dfrac{t}{2} - 5$ **12.** $0.1x^2 - 1.4x + 0.5$

13. $0.2y^2 - \dfrac{y}{4} + y$ **14.** $-\dfrac{7}{2}x^2yzt^3$

Give the numerical coefficient of each term.

15. $-x^3 - 3x + 4$ **16.** $2a^2b^2 - 3ab^2 + 4ab - b^2$

17. $\dfrac{t}{2} - 5$ **18.** $0.1x^2 - 1.4x + 0.5$

19. $-x^2y^2 - 3xy^2 - 1.4y$ **20.** $6 - \dfrac{3}{2}a^2 - a$

Give the degree of each term and the degree of the polynomial.

21. $3x^4 - 2x^2 + 5$ **22.** $-y^3 - 3y^2 + 3y$ **23.** $a^3b - a^2b^2 - ab^4$

24. $-2s^2t^2 - 3t^3 + t^4$ **25.** $x^7 - 3x^2 - x + 7$ **26.** $\dfrac{x}{5} + 4$

ANSWERS

1. ______
2. ______
3. ______
4. ______
5. ______
6. ______
7. ______
8. ______
9. ______
10. ______
11. ______
12. ______
13. ______
14. ______
15. ______
16. ______
17. ______
18. ______
19. ______
20. ______
21. ______
22. ______
23. ______
24. ______
25. ______
26. ______

27. ______

28. ______

29. ______

30. ______

31. ______

32. ______

33. ______

34. ______

35. ______

36. ______

37. ______

38. ______

39. ______

40. ______

41. ______

42. ______

43. ______

44. ______

27. Arrange the polynomial $3x^3 - 2x^5 + x^4 - x + 1$ in descending order.

28. Arrange the polynomial in Problem 27 in ascending order.

29. Arrange the polynomial $-4 + 3t^2 - t^3 + t - t^4$ in descending order.

30. Arrange the polynomial in Problem 29 in ascending order.

31. Give an example of a monomial of degree 3.

32. Give an example of a binomial in two variables.

33. Give an example of a trinomial of degree 3.

34. Give an example of a monomial of degree 0.

35. Give an example of a binomial in two variables of degree 6.

36. Give an example of a trinomial in three variables of degree 5.

37. Give an example of a binomial of degree 1.

38. Give an example of a trinomial in two variables of degree 1.

Combine like terms.

39. $x - 4x + 3$

40. $x^2 - 3x + 2x^2 + 5 - 7x$

41. $2xy - y^2 + xy - xy^2 + 2y^2$

42. $a^2b - ab^2 + 2a^2b - 3 + a + 4$

43. $y^2 - 3y + 4y - 5 - y^2 - 7$

44. $3 - t^2 + 4t - 6t - 5$

5.2

Addition and Subtraction of Polynomials

Addition and subtraction of polynomials is primarily an exercise in combining like terms.

Example 1 Add $(2x^3 - x^2 + 3x - 1)$ to $(-4x^3 - x + 6)$.

$$\begin{aligned}(2x^3 - x^2 + 3x - 1) + (-4x^3 - x + 6) &= 2x^3 + (-x^2) + 3x + (-1) \\ &\quad + (-4x^3) + (-x) + 6 \\ &= 2x^3 + (-4x^3) + (-x^2) + 3x \\ &\quad + (-x) + (-1) + 6 \\ &= -2x^3 + (-x^2) + 2x + 5 \\ &= -2x^3 - x^2 + 2x + 5\end{aligned}$$

DO EXERCISE 1.

Example 2 Add $(2x^3 + x + 5) + (-x^2 + 3x - 4) + (x^3 + 7x^2 + 1)$.

We will add these by using a column method, although we could use a horizontal method as in Example 1. Note that we keep like terms in the same column.

$$\begin{array}{rrrr} 2x^3 & & +\ x & +\ 5 \\ & -\ x^2 & +\ 3x & -\ 4 \\ x^3 & +\ 7x^2 & & +\ 1 \\ \hline 3x^3 & +\ 6x^2 & +\ 4x & +\ 2 \end{array}$$

Example 3 Add $x^2y^2 - 3xy + xy^2$, $-2x^2y^2 - 2xy^2 + x^2y$, and $7xy - 3xy^2 + 2x^2y$.

Notice that we keep like terms in the same column.

$$\begin{array}{rrrr} x^2y^2 & -\ 3xy & +\ xy^2 & \\ -2x^2y^2 & & -\ 2xy^2 & +\ x^2y \\ & 7xy & -\ 3xy^2 & +\ 2x^2y \\ \hline -x^2y^2 & +\ 4xy & -\ 4xy^2 & +\ 3x^2y \end{array}$$

DO EXERCISES 2 AND 3.

LEARNING OBJECTIVES

After finishing Section 5.2, you should be able to:

- Add polynomials.
- Subtract polynomials.

1. Add $(x^2 - 3x + 4)$ to $(-7x^2 - 5)$.

2. Add $(2y^2 - 3xy)$, $(x^2y^2 - y^2)$ and $-2x^2y^2 - 2xy + 3y^2$.

3. Add $(4a^3 - 3a^2 + 7)$ to $(-2a^2 - a + 5)$.

4. $(x^3 - 2x^2 + 4) - (-x^2 - x + 3)$

Subtraction of polynomials is similar to addition, except that we must be careful when we remove a minus sign from in front of parentheses.

Example 4 Subtract: $(3x^2 + 5x - 4) - (x^2 - 2x + 1)$

$$\begin{aligned}(3x^3 + 5x - 4) - (x^2 - 2x + 1) &= 3x^3 + 5x + (-4) + (-1)[x^2 + (-2x) + 1]\\ &= 3x^3 + 5x + (-4) + (-x^2) + 2x + (-1)\\ &= 3x^3 + (-x^2) + 7x + (-5)\\ &= 3x^3 - x^2 + 7x - 5\end{aligned}$$

DO EXERCISE 4.

Example 5 Subtract $-x^3 + x - 5$ from $2x - 1$.

5. Subtract $(a^2 - 3ab + b^2)$ from $(-2a^2 + ab + 6b^2)$.

$$\begin{aligned}(2x - 1) - (-x^3 + x - 5) &= 2x + (-1) + (-1)[-x^3 + x + (-5)]\\ &= 2x + (-1) + x^3 + (-x) + 5\\ &= x^3 + x + 4\end{aligned}$$

DO EXERCISE 5.

Example 6 $(s^3 - 2s^2 + 3) - (2s - 4) + (s^2 - s + 1)$

$$\begin{aligned}&[s^3 + (-2s^2) + 3] + (-1)[2s + (-4)] + [s^2 + (-s) + 1]\\ &\quad = s^3 + (-2s^2) + 3 + (-2s) + 4 + s^2 + (-s) + 1\\ &\quad = s^3 + (-s^2) + (-3s) + 8\\ &\quad = s^3 - s^2 - 3s + 8\end{aligned}$$

DO EXERCISE 6.

DO SECTION PROBLEMS 5.2.

6. $(s^2 - 3st + st^2 - t^2) +$ $(3st^2 + 4s^2 - t^2) +$ $(4s^2t^2 - st^2 + 3)$

ANSWERS TO MARGINAL EXERCISES

1. $-6x^2 - 3x - 1$ **2.** $4y^2 - 5xy - x^2y^2$ **3.** $4a^3 - 5a^2 - a + 12$
4. $x^3 - x^2 + x + 1$ **5.** $-3a^2 + 4ab + 5b^2$
6. $5s^2 - 3st + 3st^2 - 2t^2 + 4s^2t^2 + 3$

NAME ____________ COURSE/SECTION ____________ DATE ____________

SECTION PROBLEMS 5.2

Perform the indicated operations and simplify.

1. $(x^2 - 3x + 4) + (-2x^2 + 2x - 5)$

2. $(a^2b^2 - 2ab + b^2) + (3a^2b^2 + 6ab^2 - ab + b)$

3. $(t^3 - 2t^2 + 3) + (t^2 - t + 4) + (2t^3 - t^2 + t - 1)$

4. $(3x^2 - x + 1) - (2x^2 + x - 1)$

5. Subtract $-3x^2 - 6x + 4$ from $4x^3 - x^2 + 6$.

6. $(t^2 - 1) - (-t^3 + t^2 - t - 1)$

7. $(x^3 - x^2 + 3x) + (-2x^2 + x - 1) - (-x^2 - x + 1)$

8. $2(x^2 - x) + 3(x + 4)$

9. Subtract $x^4 - x^3$ from $2x - 5$.

10. $(-t^2 - t - 1) + (-3 + t^2 - 4) - (t^2 - 1) - (-t - t^2 + 1)$

ANSWERS

1. ____________

2. ____________

3. ____________

4. ____________

5. ____________

6. ____________

7. ____________

8. ____________

9. ____________

10. ____________

11. ____________

12. ____________

13. ____________

14. ____________

15. ____________

16. ____________

17. ____________

18. ____________

19. ____________

20. ____________

11. $(3x^2 + x - 5) + (x^3 - x + 4) + (-x^3 + 2x + 1)$

12. $(y^2 - y) - (3y + 1) - (y^2 + 5)$

13. Subtract $x^3 - x$ from $x^2 + 3x - 4$.

14. $(x^2y^2 - 3xy + 4) + (2x^2y^2 + xy)$

15. $(3a^2b - ab + b^2) - (2ab - b^2) - (a^2b + 4ab + 1)$

16. $(-t - 3) - (t^2 - 4t - 1)$

17. $2(x^2 - x + 3) - 4(3x + 2)$

18. $(x^3 + x) - (-x^2 - x + 1) + 3(-x^3 - 2x + 3)$

19. $2(s - 5) + (4s^2 + 10)$

20. $-2(x^2 + x - 3) - (x - x^2) + 4(1 - x)$

5.3

Multiplication of Polynomials

When we multiply polynomials, we will use exponential Rule 2.11 ($x^n \cdot x^m = x^{n+m}$), the distributive law [$a(b + c) = ab + ac$], and the commutative and associative laws.

We will first look at an example of a monomial times a monomial.

Example 1 Multiply the following.

(a) $(-2x^3)(3x^4) = (-2)(x^3)(3)(x^4) = (-2)(3)(x^3)(x^4) = -6x^7$

(b) $\left(\frac{x^4}{2}\right)(x)(-3x^3) = \left(\frac{1}{2}\right)(x^4)(x)(-3)(x^3) = \left(\frac{1}{2}\right)(-3)(x^4)(x)(x^3) = -\frac{3}{2}x^8$

(c) $(-3x^2)(4y^3) = (-3)(4)(x^2)(y^3) = -12x^2y^3$

DO EXERCISES 1 THROUGH 3.

We next multiply a monomial times a binomial or trinomial. (Note the use of the distributive law.)

Example 2 Multiply and, if possible, simplify.

(a) $2x(x^2 + 7x) = (2x)(x^2) + (2x)(7x) = 2x^3 + 14x^2$

(b) $$\begin{aligned} -3a^2(a - 2b + 4) &= -3a^2[a + (-2b) + 4] \\ &= (-3a^2)(a) + (-3a^2)(-2b) + (-3a^2)(4) \\ &= -3a^3 + 6a^2b + (-12a^2) \\ &= -3a^3 + 6a^2b - 12a^2 \end{aligned}$$

(c) $$\begin{aligned} -2x(x^2 - x + 1) - 3x^2(x + 4) &= -2x[x^2 + (-x) + 1] + (-3x^2)(x + 4) \\ &= (-2x)(x^2) + (-2x)(-x) + (-2x)(1) \\ &\quad + (-3x^2)(x) + (-3x^2)(4) \\ &= -2x^3 + 2x^2 + (-2x) + (-3x^3) \\ &\quad + (-12x^2) \\ &= -5x^3 + (-10x^2) + (-2x) \\ &= -5x^3 - 10x^2 - 2x \end{aligned}$$

(d) $$\begin{aligned} 2st(s^2 - 3st + t^2) &= 2st[s^2 + (-3st) + t^2] \\ &= (2st)(s^2) + (2st)(-3st) + (2st)(t^2) \\ &= 2s^3t + (-6s^2t^2) + 2st^3 \\ &= 2s^3t - 6s^2t^2 + 2st^3 \end{aligned}$$

DO EXERCISES 4 THROUGH 6.

When we multiply two binomials, we make use of the distributive law three times.

LEARNING OBJECTIVES

After finishing Section 5.3, you should be able to:

- Multiply polynomials.

Multiply and, if possible, simplify.

1. $(-3x)(2x^2)$

2. $\left(\frac{x}{4}\right)(6x^2)$

3. $(-2a^2b)(-5ab^2)$

Multiply and, if possible, simplify.

4. $-2t^2(t^2 - 3t + 1)$

5. $-x(x^2 - x - 4) + 2x(x + 5)$

6. Work the following four problems and notice the difference (if any) in the answers.
(a) $x^2 + x^2$ (b) $x^2 \cdot x^2$
(c) $x^2 + x^3$ (d) $x^2 \cdot x^3$

Multiply by using the horizontal method.

7. $(2x + 3)(x^2 + 1)$

8. $(3s - t)(b + c)$

Multiply by using the column method.

9. $(x^2 + 4x)(x - 3)$

10. $(a - b)(a + b)$

Example 3 Multiply $(x + 4)$ by $(x^2 + 3x)$.

We will demonstrate three methods.

(a) *Horizontal Method*

$$\begin{aligned}(x + 4)(x^2 + 3x) &= x(x^2 + 3x) + 4(x^2 + 3x) && \text{First use of distributive law}\\ &= x(x^2) + x(3x) + 4(x^2) + 4(3x) && \text{Second and third uses}\\ &= x^3 + 3x^2 + 4x^2 + 12x && \text{of distributive law}\\ &= x^3 + 7x^2 + 12x\end{aligned}$$

Notice that after the third use of the distributive law we have

$$\underbrace{x(x^2) + x(3x)}_{\text{These two terms represent } x \text{ times each term of the polynomial } (x^2+3x).} + \overbrace{4(x^2) + 4(3x)}^{\text{These two terms represent 4 times each term of the polynomial } (x^2+3x).}$$

Thus, we have multiplied each term of the polynomial $(x + 4)$ times each term of the polynomial $(x^2 + 3x)$. Using this fact, we look at another version of the horizontal method.

(b) *Short Horizontal Method*

$$\begin{aligned}(x + 4)(x^2 + 3x) &= \overset{1}{x(x^2)} + \overset{2}{x(3x)} + \overset{3}{4(x^2)} + \overset{4}{4(3x)}\\ &= x^3 + 3x^2 + 4x^2 + 12x\\ &= x^3 + 7x^2 + 12x\end{aligned}$$

(c) *Column Method* The steps in this method are identical to the steps in method (b); however, we arrange the polynomials in a column for convenience in combining like terms.

$$\begin{array}{rl} x^2 + 3x & \\ x + 4 & \\ \hline x^3 + 3x^2 & \leftarrow x \text{ times the polynomial } x^2 + 3x\\ + \quad 4x^2 + 12x & \leftarrow 4 \text{ times the polynomial } x^2 + 3x\\ \hline x^3 + 7x^2 + 12x & \end{array}$$

The short version of the Horizontal Method and the Column Method are the two methods we will use from now on.

Example 4 Multiply $(2a - b)$ by $(c + 3d)$.

We will use the short version of the Horizontal Method. This means we must multiply each term of $2a - b$ times each term of $c + 3d$.

$$\begin{aligned}(2a - b)(c + 3d) &= [2a + (-b)](c + 3d)\\ &= \overset{1}{2a(c)} + \overset{2}{2a(3d)} + \overset{3}{(-b)(c)} + \overset{4}{(-b)(3d)}\\ &= 2ac + 6ad + (-bc) + (-3bd)\\ &= 2ac + 6ad - bc - 3bd\end{aligned}$$

DO EXERCISES 7 THROUGH 10.

We now look at the problem of multiplying two polynomials no matter how many terms they may have.

RULE 5.1 When we multiply polynomials, we must multiply every term of one polynomial by every term of the other polynomial.

Example 5 Multiply $2x^2 - 3x + 4$ by $-x^2 + 4x - 1$.

For convenience in combining like terms, we will use the column method.

$$\begin{array}{lllll} -x^2 + & 4x & + (-1) & & \\ 2x^2 + (-3x) + & 4 & & & \\ \hline -2x^4 + & 8x^3 & + (-2x^2) & & \leftarrow 2x^2 \text{ times the polynomial } -x^2+4x+(-1) \\ & 3x^3 & + (-12x^2) & + 3x & \leftarrow -3x \text{ times the polynomial } -x^2+4x+(-1) \\ + & & -4x^2 & + 16x + (-4) & \leftarrow 4 \text{ times the polynomial } -x^2+4x+(-1) \\ \hline -2x^4 + & 11x^3 & + (-18x^2) & + 19x + (-4) & \end{array}$$

Example 6 Multiply $(x^2y - xy^2)$ by $(2x + y)$.

$$\begin{aligned}[x^2y + (-xy^2)](2x + y) &= (x^2y)(2x) + (x^2y)(y) + (-xy^2)(2x) + (-xy^2)(y) \\ &= 2x^3y + x^2y^2 + (-2x^2y^2) + (-xy^3) \\ &= 2x^3y + (-x^2y^2) + (-xy^3) \\ &= 2x^3y - x^2y^2 - xy^3\end{aligned}$$

DO EXERCISES 11 THROUGH 13.

Before we look at some special products of polynomials in the next section, we want to concentrate on the problem of multiplying two first-degree binomials. Since multiplication of this type comes up frequently, we want to become proficient at it.

Example 7 Multiply the following binomials.

(a) $$\begin{aligned}(x + 3)(x + 5) &= x(x) + x(5) + 3(x) + 3(5) \\ &= x^2 + 5x + 3x + 15 \\ &= x^2 + 8x + 15\end{aligned}$$

(b) $$\begin{aligned}(x - 3)(x + 4) &= x(x) + x(4) + (-3)(x) + (-3)(4) \\ &= x^2 + 4x + (-3x) + (-12) \\ &= x^2 + x - 12\end{aligned}$$

(c) $$\begin{aligned}(x - 5)(x - 2) &= x(x) + x(-2) + (-5)(x) + (-5)(-2) \\ &= x^2 + (-2x) + (-5x) + 10 \\ &= x^2 - 7x + 10\end{aligned}$$

(d) $$\begin{aligned}(2x - 3)(x + 7) &= 2x(x) + 2x(7) + (-3)(x) + (-3)(7) \\ &= 2x^2 + 14x + (-3x) + (-21) \\ &= 2x^2 + 11x - 21\end{aligned}$$

Multiply.

11. $(x^2 - 2x + 3)(-2x^2 + x - 2)$

12. $(xy - 2x)(x^2 - 4y)$

13. $(a - 2)(2a + 1)$

Notice that in Example 7 we did not always explicitly put in the step of changing subtraction to addition; however, in working these or any other problems, we may always put all the steps in if we so desire.

DO EXERCISES 14 THROUGH 16.

Multiply.

14. $(x + 5)(x + 2)$

15. $(2x + 1)(x - 3)$

16. $(x - 4)(x - 5)$

In Example 7(a), we had

$$(x + 3)(x + 5) = \overset{1}{x^2} + \overset{2}{5x} + \overset{3}{3x} + \overset{4}{15}$$

Some new terminology is appropriate here:

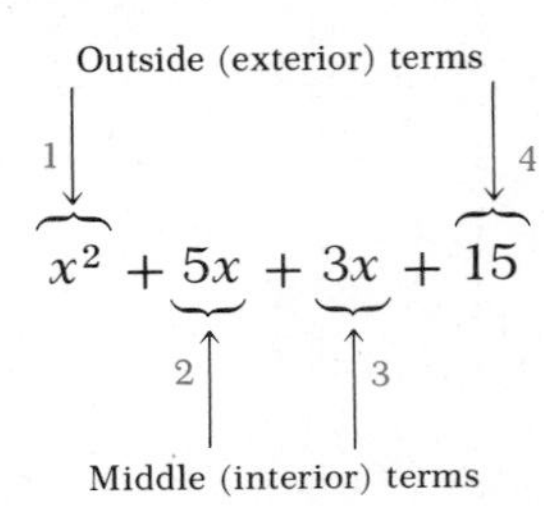

Note

(a) The (1) and (4) arrows give us the outside terms.

$$(x + 3)(x + 5)$$

(b) The (2) and (3) arrows give us the middle terms.

$$(x + 3)(x + 5)$$

(c) The middle terms can be combined.
(d) The middle terms are also called the first-degree terms.

Example 8 Calculate just the middle terms in the following products.

(a) $(x + 3)(x + 4)$ $\quad 4x + 3x = 7x$

(b) $(t - 5)(t + 7)$ $\quad 7t - 5t = 2t$

(c) $(2y + 3)(y - 5)$ $\quad -10y + 3y = -7y$

(d) $(x - 4)(x - 6)$ $\quad -6x - 4x = -10x$

DO EXERCISES 17 THROUGH 19.

Calculate *just* the middle terms.

17. $(x - 8)(x + 2)$

18. $(y + 3)(2y - 4)$

19. $(2x - 5)(3x - 1)$

Example 9 Multiply.

(a) $(x - 2)(x + 4) = x^2 + 4x - 2x - 8$
$= x^2 + 2x - 8$

(b) $(2x - 1)(3x - 2) = 6x^2 - 4x - 3x + 2$
$= 6x^2 - 7x + 2$

(c) $(t + 3)(t + 2) = t^2 + 2t + 3t + 6$
$= t^2 + 5t + 6$

Notice that in Example 9 we have somewhat shortened the multiplication process from previous examples. This ability normally comes after practice with multiplying two binomials. Remember, however, that it is never wrong to put all the steps in.

DO EXERCISES 20 THROUGH 23.

DO SECTION PROBLEMS 5.3.

Multiply.

20. $(y + 2)(y + 1)$

21. $(x + 2)(3x - 4)$

22. $(2x - 3)(x - 4)$

23. $(t - 1)(t - 5)$

ANSWERS TO MARGINAL EXERCISES

1. $-6x^3$ **2.** $\frac{3}{2}x^3$ **3.** $10a^3b^3$ **4.** $-2t^4 + 6t^3 - 2t^2$
5. $-x^3 + 3x^2 + 14x$ **6.** (a) $2x^2$ (b) x^4 (c) $x^2 + x^3$ (d) x^5
7. $2x^3 + 3x^2 + 2x + 3$ **8.** $3sb + 3sc - tb - tc$ **9.** $x^3 + x^2 - 12x$
10. $a^2 - b^2$ **11.** $-2x^4 + 5x^3 - 10x^2 + 7x - 6$
12. $x^3y - 4xy^2 - 2x^3 + 8xy$ **13.** $2a^2 - 3a - 2$ **14.** $x^2 + 7x + 10$
15. $2x^2 - 5x - 3$ **16.** $x^2 - 9x + 20$ **17.** $-6x$ **18.** $2y$
19. $-17x$ **20.** $y^2 + 3y + 2$ **21.** $3x^2 + 2x - 8$
22. $2x^2 - 11x + 12$ **23.** $t^2 - 6t + 5$

NAME ______ COURSE/SECTION ______ DATE ______

SECTION PROBLEMS 5.3

Multiply and simplify if possible.

1. $3x^2(-7x)$

2. $-ab(a^2b)$

3. $(-4t^2)(3t)$

4. $(6x^2)(7x^2y)$

5. $\left(\frac{x^2y}{2}\right)(-3xy^3)$

6. $(4x)(-3xy)(-2x^2y)$

7. $-4t(2t^2 - 3t + 1)$

8. $2xy(x^2 - xy + xy^2)$

9. $3x(-x^2 + 2x - 5)$

10. $-2ab^2(a^2 - ab + b^2)$

11. $\frac{x^2y}{2}(-3x + 4xy - 6)$

12. $-s^3(1 - 4s - 2s^2)$

13. $(2a - b)(c + d)$

14. $(2t + s)(t^2 - 1)$

15. $(2x + y)(x^2 - 3y)$

16. $(x^2 - 1)(2x^2 + x - 3)$

17. $(y + 1)(y^2 - 3y + 2)$

18. $(2y^2 - y + 3)(y^2 + y - 1)$

19. $(3x - 1)(x^3 - 2x^2 + x - 4)$

20. $(x^2 + 2x - 3)(x^4 - x^2 + 3x + 4)$

21. $(x - 4)(x + 5)$

22. $(t + 2)(t + 7)$

23. $(y - 3)(y - 4)$

24. $(x + 6)(x - 3)$

25. $(x + 5)(x - 5)$

26. $(x + 3)^2$

27. $(2t - 1)(t + 4)$

ANSWERS

1. ______
2. ______
3. ______
4. ______
5. ______
6. ______
7. ______
8. ______
9. ______
10. ______
11. ______
12. ______
13. ______
14. ______
15. ______
16. ______
17. ______
18. ______
19. ______
20. ______
21. ______
22. ______
23. ______
24. ______
25. ______
26. ______
27. ______

28. $(s - 3)(4s + 1)$

29. $(2t - 5)(3t + 1)$

30. $(2y - 3)(5y + 1)$

31. $(7x - 1)(6x - 3)$

32. $(a + 4)(a - 5)$

33. $(7x - 1)(6x + 3)$

34. $(2y + 5)(3y + 1)$

35. $(t - 4)(2t + 5)$

36. $(3x + 4)(2x + 1)$

37. $(t - 6)(3t - 7)$

38. $(2M + 3)(4M + 5)$

39. $(x - 3)(x + 3)(x + 4)$

40. $(y - 1)(y + 2)(y - 3)$

41. $(x + 1)^3$

42. $(P - 2)^3$

43. $(2s - t)(s^2 + t^2)$

44. $(3x^2 - x + 7)(-x^2 - 2x + 1)$

45. $-2y(3y - 1)(2y + 5)$

46. $(m + 4)(m + 7)^2$

47. $(1 - 3b)(2b + 1)$

48. $x(1 - x)(2 - 3x)$

49. $(1.4t + 2)(0.1t - 2.8)$

50. $(0.4x - 3)(0.2x + 1.5)$

28. ____________
29. ____________
30. ____________
31. ____________
32. ____________
33. ____________
34. ____________
35. ____________
36. ____________
37. ____________
38. ____________
39. ____________
40. ____________
41. ____________
42. ____________
43. ____________
44. ____________
45. ____________
46. ____________
47. ____________
48. ____________
49. ____________
50. ____________

5.4

Special Products of Binomials

There are some special products of binomials that come up often enough that we should take note of them.

Example 1 Multiply the following binomials.

(a) $(x - 5)(x + 5) = x^2 + 5x - 5x - (5)^2$
$= x^2 - 25$

(b) $(t - 2)(t + 2) = t^2 + 2t - 2t - (2)^2$
$= t^2 - 4$

(c) $(2x + 3)(2x - 3) = (2x)^2 - 6x + 6x - (3)^2$
$= 4x^2 - 9$

(d) $(3x - y)(3x + y) = (3x)^2 - 3xy + 3xy - y^2$
$= 9x^2 - y^2$

In each of these examples the terms of the two binomials we are multiplying are exactly the same, but the operation between the terms is addition for one of the binomials and subtraction for the other. Also, notice that in these cases the middle terms sum to 0 in the final answer. This is always the case when we multiply two binomials of this type, as Rule 5.2 shows.

RULE 5.2 $(A - B)(A + B) = A^2 - B^2$

Note

$A^2 - B^2$ is called the *difference of two squares.*

DO EXERCISES 1 THROUGH 3.

In Examples 2 and 3 we will look at the process of squaring a binomial.

Example 2 Multiply the following binomials.

(a) $(t + 4)^2 = (t + 4)(t + 4) = t^2 + 4t + 4t + (4)^2$
$= t^2 + 2(4t) + (4)^2$
$= t^2 + 8t + 16$

(b) $(y + 9)^2 = (y + 9)(y + 9) = y^2 + 9y + 9y + 9^2$
$= y^2 + 2(9y) + 9^2$
$= y^2 + 18y + 81$

(c) $(2x + 3)^2 = (2x + 3)(2x + 3) = (2x)^2 + (2x)(3) + (2x)(3) + (3)^2$
$= (2x)^2 + 2(2x)(3) + (3)^2$
$= 4x^2 + 12x + 9$

LEARNING OBJECTIVES

After finishing Section 5.4, you should be able to:

- Multiply binomials of the forms $(x + y)(x - y)$, $(x + y)^2$, and $(x - y)^2$.

Multiply.

1. $(x - 7)(x + 7)$

2. $(2t + 5)(2t - 5)$

3. $(y + 3)(y - 4)$

Multiply.

4. $(2t + 1)^2$

5. $(3x + 4)^2$

6. $(x - 6)(x + 6)$

Multiply.

7. $(2x - 7)^2$

8. $(y - 3)(y + 3)$

9. $(y - 3)^2$

10. $(y + 3)^2$

In general, then, we have the following rule.

RULE 5.3 $\quad (A + B)^2 = A^2 + 2AB + B^2$

Note that

$$(3 + 2)^2 = (5)^2 = 25$$

but

$$3^2 + 2^2 = 9 + 4 = 13$$

Thus, in general we have

$$(A + B)^2 \neq A^2 + B^2$$

DO EXERCISES 4 THROUGH 6.

Example 3 Multiply the following binomials.

(a) $(x - 3)^2 = (x - 3)(x - 3) = x^2 - 3x - 3x + (-3)^2$
$= x^2 - 2(3x) + (-3)^2$
$= x^2 - 6x + 9$

(b) $(y - 5)^2 = (y - 5)(y - 5) = y^2 - 5y - 5y + (-5)^2$
$= y^2 + 2(-5y) + (-5)^2$
$= y^2 - 10y + 25$

(c) $(3x - 4)^2 = (3x - 4)(3x - 4) = (3x)^2 - 12x - 12x + (-4)^2$
$= 9x^2 + 2(-12x) + (-4)^2$
$= 9x^2 - 24x + 16$

In general, then, we have the following rule.

RULE 5.4 $\quad (A - B)^2 = A^2 - 2AB + B^2$

Example 4 Note the difference in the following.

(a) $(x - 6)(x + 6) = x^2 + 6x - 6x - 36$
$= x^2 - 36$

(b) $(x - 6)^2 = (x - 6)(x - 6)$
$= x^2 - 6x - 6x + 36$
$= x^2 - 12x + 36$

(c) $(x + 6)^2 = (x + 6)(x + 6)$
$= x^2 + 6x + 6x + 36$
$= x^2 + 12x + 36$

DO EXERCISES 7 THROUGH 10.

DO SECTION PROBLEMS 5.4.

ANSWERS TO MARGINAL EXERCISES

1. $x^2 - 49$ **2.** $4t^2 - 25$ **3.** $y^2 - y - 12$ **4.** $4t^2 + 4t + 1$
5. $9x^2 + 24x + 16$ **6.** $x^2 - 36$ **7.** $4x^2 - 28x + 49$ **8.** $y^2 - 9$
9. $y^2 - 6y + 9$ **10.** $y^2 + 6y + 9$

NAME ____________ COURSE/SECTION ________ DATE ________

SECTION PROBLEMS 5.4

Multiply.

1. $(x - 4)(x + 4)$
2. $(t + 6)(t - 6)$
3. $(2y - 1)(2y + 1)$
4. $(3x - 2)(3x + 2)$
5. $(s - 3)(s + 3)$
6. $(x + 4)^2$
7. $(2y + 1)^2$
8. $(3x + 5)^2$
9. $(y - 1)^2$
10. $(y - 5)^2$
11. $(2t - 3)^2$
12. $(y - 4)(y + 4)$
13. $(6x - 3)^2$
14. $(2t + 5)^2$
15. $(t - 1.3)^2$
16. $(x - 5)(2x + 4)$
17. $(y - 3)^2$
18. $(2x + 9)^2$

ANSWERS

1. ________
2. ________
3. ________
4. ________
5. ________
6. ________
7. ________
8. ________
9. ________
10. ________
11. ________
12. ________
13. ________
14. ________
15. ________
16. ________
17. ________
18. ________

19. ______________

20. ______________

21. ______________

22. ______________

23. ______________

24. ______________

25. ______________

26. ______________

27. ______________

28. ______________

29. ______________

30. ______________

19. $(3x - 5)(x - 2)$

20. $(x - 3)^3$

21. $(2x + 1)^3$

22. $(y - 3)(y + 3)(y - 7)$

23. $(s + 2)^2(s - 1)$

24. $(2t - 1)^2(3t)$

25. $(x - 2)^2(x + 2)^2$

26. $(y - 2)^2$

27. $(2s - 3)(2s + 3)$

28. $(5y + 4)^2$

29. $(y - 2)(y + 2)(y - 1)$

30. $(2t - 1)^3$

5.5

Introduction to Factoring

Recall that we did some factoring (with the help of the distributive law) in Chapter 2, where we factored as much as we could from each term of a polynomial [for example, $3x + 6 = 3 \cdot x + 3 \cdot 2 = 3(x + 2)$]. In this section we will do more of this type of factoring, in addition to a new kind of factoring. The first thing we always look for when factoring is the greatest common factor in each term of a polynomial. If there is one, then we factor it out.

Example 1 Factor.

(a) $6x - 9 = (3)(2x) - (3)(3) = 3(2x - 3)$

(b) $x^3y^2 - x^2y^2 + xy = (xy)(x^2y) - (xy)(xy) + (xy)(1) = xy(x^2y - xy + 1)$

(c) $-4x^3 + 8x = (-4x)(x^2) + (-4x)(-2) = -4x(x^2 - 2)$

(d) $x(x - 2) - 3(x - 2) = (x - 2)(x - 3)$

Note that $(x - 2)$ is a common factor in $x(x - 2) - 3(x - 2)$. Thus we can factor it out.

(e) $10x - 9y$ (This polynomial cannot be factored.)

> *RULE 5.5* Whenever we factor we require the resulting polynomials to have only integers for coefficients and constant terms, that is, no nonintegral values are allowed.

As an example of Rule 5.5, note these two factorizations of $4x + 10$:

(a) $4x + 10 = 2(2x + 5)$ (b) $4x + 10 = 4\left(x + \frac{5}{2}\right)$

The factorization in (a) is a valid factorization, but the factorization in (b) is not valid because of the nonintegral value, $\frac{5}{2}$.

DO EXERCISES 1 THROUGH 7.

In the next section we will concentrate on factoring second-degree trinomials in one variable. However, in the rest of this section we will look at a special type of second-degree binomial in one variable, namely, the difference of two squares.

Recall Rule 5.2, $(A - B)(A + B) = A^2 - B^2$. In the previous section we used this rule to multiply certain binomials. If we think of the rule as $A^2 - B^2 = (A - B)(A + B)$, then we can use the rule to factor the difference of two squares.

Example 2 Factor $x^2 - 25$.

Note the lack of a middle (first-degree) term. Because this polynomial is the difference of two squares we can use Rule 5.2.

$$x^2 - 25 = x^2 - 5^2 = (x - 5)(x + 5)$$

LEARNING OBJECTIVES

After finishing Section 5.5, you should be able to:

- Factor certain polynomials.

Factor.

1. $3x - 6y + 9$

2. $-2ax^2 + 4a$

3. $x^3 - 2x^2 + 2x$

4. $3x^2y^2 - xy^2 + 3xy$

5. $2a^2b - 4ab + 3b^2$

6. $3x - 4y$

7. $y(x - 2) - 4(x - 2)$

Factor.

8. $a^2 - 9$

9. $4s^2 - 25$

10. $25x^2 - 49$

Factor.

11. $2x^2 - 8$

12. $4y^2 - 81$

13. $a(y^2 - 4) + b(y^2 - 4)$

A restatement of Definition 2.10 is appropriate.

$$\underbrace{x^2 - 25}_{\text{Multiplied form}} = \underbrace{(x - 5)(x + 5)}_{\text{Factored form}}$$

If we start with the second-degree polynomial $x^2 - 25$ and change it to the product, $(x - 5)(x + 5)$, of two first-degree polynomials, we are factoring. If we start with $(x - 5)(x + 5)$ and change it to $x^2 - 25$, we are multiplying. That is,

$$x^2 - 25 \underset{\text{Multiplying}}{\overset{\text{Factoring}}{\rightleftarrows}} (x - 5)(x + 5)$$

Multiplying has the effect of removing the parentheses and, of course, that is what we concentrated on in the previous two sections.

Example 3 Factor (a) $x^2 - 1$ and (b) $9y^2 - 16$.

(a) $x^2 - 1 = x^2 - 1^2 = (x - 1)(x + 1)$

(b) $9y^2 - 16 = (3y)^2 - 4^2 = (3y - 4)(3y + 4)$

■ **DO EXERCISES 8 THROUGH 10.**

Example 4 Factor (a) $3x^2 - 12$ and (b) $x(x^2 - 36) - 2(x^2 - 36)$.

Note: Anytime we are factoring we first look for a common factor—a 3 in problem (a)—in each term of the polynomial.

(a) $3x^2 - 12 = 3(x^2 - 4) = 3(x - 2)(x + 2)$

(b) $x(x^2 - 36) - 2(x^2 - 36) = (x^2 - 36)(x - 2)$
$= (x - 6)(x + 6)(x - 2)$

■ **DO EXERCISES 11 THROUGH 13.**

■ **DO SECTION PROBLEMS 5.5.**

ANSWERS TO MARGINAL EXERCISES

1. $3(x - 2y + 3)$ **2.** $-2a(x^2 - 2)$ **3.** $x(x^2 - 2x + 2)$
4. $xy(3xy - y + 3)$ **5.** $b(2a^2 - 4a + 3b)$ **6.** $3x - 4y$
7. $(x - 2)(y - 4)$ **8.** $(a - 3)(a + 3)$ **9.** $(2s + 5)(2s - 5)$
10. $(5x - 7)(5x + 7)$ **11.** $2(x + 2)(x - 2)$ **12.** $(2y - 9)(2y + 9)$
13. $(a + b)(y + 2)(y - 2)$

NAME ____ COURSE/SECTION ____ DATE ____

SECTION PROBLEMS 5.5

Factor.

1. $4x - 8y + 12$
2. $-6x + 9y$
3. $3a^2b^2 - 4ab^2$
4. $2t^2 - 6t$
5. $a(b - 4) + 7(b - 4)$
6. $3x(y + 2) - 2z(y + 2)$
7. $x^3 - 3x^2 + 9x$
8. $15xy - 12y + 10$
9. $y^2 - 36$
10. $x^2 - 144$
11. $t^2 - 16$
12. $9x^2 - 25$
13. $49y^2 - 36$
14. $x^3 - x$
15. $3y^2 - 27$

ANSWERS

1. ____
2. ____
3. ____
4. ____
5. ____
6. ____
7. ____
8. ____
9. ____
10. ____
11. ____
12. ____
13. ____
14. ____
15. ____

16. ____________

17. ____________

18. ____________

19. ____________

20. ____________

21. ____________

22. ____________

23. ____________

24. ____________

25. ____________

26. ____________

27. ____________

28. ____________

29. ____________

30. ____________

16. $3x - 9$

17. $2t^3 - 8t$

18. $s^2 - 4st^3$

19. $4 - x^2$

20. $25 - 9y^2$

21. $x^3y - xy^3$

22. $2y^2 - 8$

23. $y(x^2 - 1) + 3(x^2 - 1)$

24. $x(a^2 - 25) - 2(a^2 - 25)$

25. $2t^2 - 50$

26. $3a^2b - 6ab$

27. $3y^2 - 48$

28. $6 - 24t^2$

29. $-4ab^2 + 9x^2y$

30. $7x - 14y + 3$

5.6

Factoring Trinomials

LEARNING OBJECTIVES

After finishing Section 5.6, you should be able to:

- Factor trinomials.

Now we return to the problem of factoring trinomials of the second degree. If a second-degree trinomial can be factored (beyond factoring a constant out), the factors must be two first-degree polynomials. Before we start this type of factoring, we need a definition.

DEFINITION 5.9 Two polynomials are equal if all the terms of the same degree are exactly the same for each polynomial.

EXAMPLES: 1. $3x^2 - 4x + 1$ and $3x^2 - 4x + 1$ are equal.
2. $3x^2 - 4x + 1$ and $3x^2 + 6x + 1$ are not equal.

Further, two terms of the same degree are equal if their coefficients are the same.

EXAMPLE: If $6x = mx$, then m must be 6.

Example 1 Factor $x^2 - x - 6$.

(a) First, note that there is nothing we can factor from each term.

(b) If $x^2 - x - 6$ can be factored, the factors must be two first-degree binomials. The second-degree term in $x^2 - x - 6$ is x^2, and the only way to write x^2 as a product of two first-degree monomials is $x^2 = (x)(x)$. Thus, if $x^2 - x - 6$ can be factored, we must have

$$x^2 - x - 6 = (x + a)(x + b)$$

where, by Rule 5.5, a and b are integers yet to be determined.

(c) If we multiply $(x + a)(x + b)$, we obtain

$$(x + a)(x + b) = x^2 + bx + ax + ab = x^2 + (b + a)x + ab$$

and this must be equal to $x^2 - x - 6$. That is,

$$x^2 - x - 6 = x^2 - x + \overbrace{(-6)} = x^2 + (b + a)x + \overbrace{ab}$$

Using Definition 5.9, we have that the constant terms of the polynomial $x^2 - x + (-6)$ and the polynomial $x^2 + (a + b)x + ab$ must be the same. Thus we have the condition on the two integers a and b that their product must be -6; that is, $-6 = ab$.

(d) Remembering that a and b must be integers, we can make a table that shows all the possible integers whose product is -6.

a	b
1	−6
2	−3
3	−2
6	−1
−1	6
−2	3
−3	2
−6	1

(e) Finally, we can add another column to our table to see if any of the choices for a and b will give us the correct middle term in the polynomial $x^2 - x - 6$ when we multiply out $(x + a)(x + b)$.

a	b	$(x + a)(x + b)$
1	−6	$(x + 1)(x - 6) = x^2 - 5x - 6$
2	−3	$(x + 2)(x - 3) = x^2 - x - 6$ ⟵ This will work.
3	−2	$(x + 3)(x - 2) = x^2 + x - 6$
6	−1	$(x + 6)(x - 1) = x^2 + 5x - 6$
−1	6	$(x - 1)(x + 6) = x^2 + 5x - 6$
−2	3	$(x - 2)(x + 3) = x^2 + x - 6$
−3	2	$(x - 3)(x + 2) = x^2 - x - 6$ ⟵ This will work.
−6	1	$(x - 6)(x + 1) = x^2 - 5x - 6$

(f) We see from the table that there are two choices for a and b ($a = 2$ and $b = -3$ or $a = -3$ and $b = 2$). We choose $a = 2$ and $b = -3$. Thus, we have the following factorization:

$$x^2 - x - 6 = (x + 2)(x - 3)$$

If we had chosen $a = -3$ and $b = 2$, then we would have had the same factorization, namely $x^2 - x - 6 = (x - 3)(x + 2)$.

We use the following steps in every factorization problem.

STEP 1: COMMON FACTOR

Always look for a factor that is common to each term in the polynomial.

STEP 2: COEFFICIENT OF x^2 TERM

The coefficient of the x^2 term will determine how many different ways we can write this term as a product of two first-degree monomials. For example, $2x^2 = (2x)(x)$ but $12x^2 = (12x)(x)$ or $12x^2 = (6x)(2x)$ or $12x^2 = (4x)(3x)$.

STEP 3: CONDITION ON a AND b

The condition on a and b will always be that their product ab must equal the constant term of the polynomial we are trying to factor.

STEP 4: TABLE

All the possible choices for a and b can be listed in a table.

STEP 5: SOLUTION

If the polynomial can be factored, the a and b that work will be in the table.

Example 2 Factor $x^2 - 4x + 3$.

STEP 1: COMMON FACTOR

There is no common factor.

STEP 2: COEFFICIENT OF x^2 TERM

The only possibility is $x \cdot x$. Thus, we have

$$x^2 - 4x + 3 = (x + a)(x + b)$$

STEP 3: CONDITION ON a AND b

$$\text{Let } x^2 - 4x + 3 = (x + a)(x + b)$$
$$= x^2 + bx + ax + ab$$
$$= x^2 + (b + a)x + ab$$

Thus, we have

$$ab = 3$$

STEP 4: TABLE

a	b	$(x + a)(x + b)$
1	3	$(x + 1)(x + 3) = x^2 + 4x + 3$
3	1	$(x + 3)(x + 1) = x^2 + 4x + 3$
−1	−3	$(x - 1)(x - 3) = x^2 - 4x + 3$ ⟵ This will work.
−3	−1	$(x - 3)(x - 1) = x^2 - 4x + 3$ ⟵ This will work.

STEP 5: SOLUTION

Our polynomial can be factored, and we can let $a = -1$ and $b = -3$ (or $a = -3$ and $b = -1$). Thus, we have

$$x^2 - 4x + 3 = (x - 1)(x - 3)$$

DO EXERCISE 1.

Example 3 Factor $2x^2 + 16x + 30$.

STEP 1: COMMON FACTOR

We can factor a 2 from each term:

$$2x^2 + 16x + 30 = 2(x^2 + 8x + 15)$$

Now our problem reduces to factoring $x^2 + 8x + 15$.

STEP 2: COEFFICIENT OF x^2 TERM

The coefficient is 1. Thus we have

$$x^2 + 8x + 15 = (x + a)(x + b)$$

STEP 3: CONDITION ON a AND b

$$\text{Let } x^2 + 8x + 15 = (x + a)(x + b) = x^2 + (b + a)x + ab$$

Thus, we have

$$ab = 15$$

STEP 4: TABLE

a	b	$(x + a)(x + b)$
1	15	$(x + 1)(x + 15) = x^2 + 16x + 15$
3	5	$(x + 3)(x + 5) = x^2 + 8x + 15$
5	3	The above will work, and thus we do not need to fill in the rest of the table.
15	1	
−1	−15	
−3	−5	
−5	−3	
−15	−1	

Factor.

1. $x^2 + 3x + 2$

Factor.

2. $x^2 - 7x + 10$

3. $x^2 - 9$

4. $3x^2 + 3x - 6$

5. $2x^2 + 9x + 4$

STEP 5: SOLUTION

We have $2x^2 + 16x + 30 = 2(x + 3)(x + 5)$. (Do not forget to include the 2 from Step 1 in your final factorization.)

DO EXERCISES 2 THROUGH 4.

Example 4 Factor $2x^2 - 5x + 3$.

STEP 1: COMMON FACTOR

There is no common factor.

STEP 2: COEFFICIENT OF x^2 TERM

We can write $2x^2$ as $(2x)(x)$. Thus, we have

$$2x^2 - 5x + 3 = (2x + a)(x + b)$$

STEP 3: CONDITION ON a AND b

$$\text{Let } 2x^2 - 5x + 3 = (2x + a)(x + b)$$
$$= 2x^2 + 2bx + ax + ab$$
$$= 2x^2 + (2b + a)x + ab$$

Thus, we have

$$ab = 3$$

STEP 4: TABLE

a	b	$(2x + a)(x + b)$	
1	3	$(2x + 1)(x + 3) = 2x^2 + 7x + 3$	
3	1	$(2x + 3)(x + 1) = 2x^2 + 5x + 3$	
−1	−3	$(2x - 1)(x - 3) = 2x^2 - 7x + 3$	
−3	−1	$(2x - 3)(x - 1) = 2x^2 - 5x + 3$	⟵ This will work.

STEP 5: SOLUTION

We have $2x^2 - 5x + 3 = (2x - 3)(x - 1)$.

DO EXERCISE 5.

Example 5 Factor $t^2 - 3t + 3$.

STEP 1: COMMON FACTOR

There is no common factor.

STEP 2: COEFFICIENT OF x^2 (ACTUALLY t^2) TERM

We must have $t^2 - 3t + 3 = (t + a)(t + b)$.

STEP 3: CONDITION ON a AND b

$$\text{Let } t^2 - 3t + 3 = (t + a)(t + b)$$

Thus, we have

$$ab = 3$$

STEP 4: TABLE

a	b	$(t + a)(t + b)$
1	3	$(t + 1)(t + 3) = t^2 + 4t + 3$
3	1	$(t + 3)(t + 1) = t^2 + 4t + 3$
-1	-3	$(t - 1)(t - 3) = t^2 - 4t + 3$
-3	-1	$(t - 3)(t - 1) = t^2 - 4t + 3$

STEP 5: SOLUTION

We can see from the table that nothing works. Thus, $t^2 - 3t + 3$ cannot be factored in the sense described in Rule 5.5.

Example 6 Factor $x^2 - 36$.

We should be able to recognize $x^2 - 36$ as the difference of two squares. Thus, it may be factored as

$$x^2 - 36 = (x - 6)(x + 6)$$

If we did not recognize this as the difference of two squares and used Steps 1 through 5, we would arrive at the same factorization.

Example 7 Factor $3x^3 - 12x$.

STEP 1: COMMON FACTOR

We can factor a $3x$ from each term.

$$3x^3 - 12x = 3x(x^2 - 4)$$

This reduces our problem to factoring $x^2 - 4$. We recognize $x^2 - 4$ as the difference of two squares, and thus we factor it as

$$x^2 - 4 = (x - 2)(x + 2)$$

Hence, we have

$$3x^3 - 12x = 3x(x^2 - 4) = 3x(x - 2)(x + 2)$$

DO EXERCISES 6–8.

Example 8 Factor $4x^2 + 7x - 2$.

STEP 1: COMMON FACTOR

There is no common factor.

STEP 2: COEFFICIENT OF x^2 TERM

Since $4x^2 = (2x)(2x)$ or $4x^2 = (4x)(x)$, we have two possibilities:

Factor.

6. $x^2 - 9x + 14$

7. $x^2 + 3x + 1$

8. $4y^3 - y$

Factor.

9. $6x^2 + 7x - 10$

$$(1) \quad 4x^2 + 7x - 2 = (2x + a)(2x + b)$$

or

$$(2) \quad 4x^2 + 7x - 2 = (4x + a)(x + b)$$

We will try (1) first.

STEP 3: CONDITION ON a AND b

$$\text{Let} \quad 4x^2 + 7x - 2 = (2x + a)(2x + b)$$

Thus, we have

$$ab = -2$$

STEP 4: TABLE

a	b	$(2x + a)(2x + b)$
1	−2	$(2x + 1)(2x - 2) = 4x^2 - 2x - 2$
2	−1	$(2x + 2)(2x - 1) = 4x^2 + 2x - 2$
−1	2	$(2x - 1)(2x + 2) = 4x^2 + 2x - 2$
−2	1	$(2x - 2)(2x + 1) = 4x^2 - 2x - 2$

STEP 5: SOLUTION

Nothing works. However, this only means that $4x^2 + 7x - 2$ may not be factored as $(2x + a)(2x + b)$. We still have to go back and try our second possibility:

$$4x^2 + 7x - 2 = (4x + a)(x + b)$$

STEP 3: CONDITION ON a AND b

$$\text{Let} \quad 4x^2 + 7x - 2 = (4x + a)(x + b)$$

Thus, we have

$$ab = -2$$

STEP 4: TABLE

a	b	$(4x + a)(x + b)$
1	−2	$(4x + 1)(x - 2) = 4x^2 - 7x - 2$
2	−1	$(4x + 2)(x - 1) = 4x^2 - 2x - 2$
−1	2	$(4x - 1)(x + 2) = 4x^2 + 7x - 2$ ⟵ This will work.
−2	1	

STEP 5: SOLUTION

$$4x^2 + 7x - 2 = (4x - 1)(x + 2)$$

Obviously, we could have saved ourselves a lot of work if we had tried $(4x + a)(x + b)$ first.

■ **DO EXERCISE 9.**

Example 9 Factor $4x^2 + 4x + 1$.

STEP 1: COMMON FACTORS

There is no common factor.

STEP 2: COEFFICIENT OF x^2 TERM

There are two possibilities:

$$(1) \quad 4x^2 + 4x + 1 = (2x + a)(2x + b)$$

or

$$(2) \quad 4x^2 + 4x + 1 = (4x + a)(x + b)$$

We will try (1) first.

STEP 3: CONDITION ON a AND b

$$\text{Let} \quad 4x^2 + 4x + 1 = (2x + a)(2x + b)$$

Thus, we have

$$ab = 1$$

STEP 4: TABLE

a	b	$(2x + a)(2x + b)$
1	1	$(2x + 1)(2x + 1) = 4x^2 + 4x + 1$ ⟵ This will work.
−1	−1	

STEP 5: SOLUTION

$$4x^2 + 4x + 1 = (2x + 1)(2x + 1) = (2x + 1)^2$$

Note

We might have recognized the above as an example of Rule 5.3.

DO EXERCISES 10 AND 11.

As you practice factoring and gain experience with it, you can often "see" which a and b will work without using the table every time. However, the method used in this section will always work and if your trinomial factors you will be able to find the factors.

Factoring is an important skill in algebra and is used in numerous situations. But for now, we will use factoring in the next section to help solve certain second-degree equations in one variable.

DO SECTION PROBLEMS 5.6.

Factor.

10. $x^2 + 6x + 9$

11. $2x^2 - 5x - 12$

ANSWERS TO MARGINAL EXERCISES

1. $(x + 2)(x + 1)$ **2.** $(x - 2)(x - 5)$ **3.** $(x - 3)(x + 3)$
4. $3(x + 2)(x - 1)$ **5.** $(2x + 1)(x + 4)$ **6.** $(x - 2)(x - 7)$
7. $x^2 + 3x + 1$ **8.** $(y)(2y - 1)(2y + 1)$ **9.** $(6x - 5)(x + 2)$
10. $(x + 3)^2$ **11.** $(2x + 3)(x - 4)$

NAME COURSE/SECTION DATE

ANSWERS

SECTION PROBLEMS 5.6

Factor.

1. $x^2 - 9x + 20$

2. $t^2 + 5t - 6$

3. $x^2 + 5x - 14$

4. $y^2 - 2y - 35$

5. $s^2 + 8s + 12$

6. $3t^2 - 48$

7. $x^2 + x + 1$

8. $y^2 - 2y + 1$

9. $x^2 - 7x + 10$

10. $2t^2 - 2t - 24$

11. $x^3 - x$

12. $y^2 + 10y + 25$

13. $y^2 + 3y + 2$

14. $t^2 - 6t + 9$

15. $x^2 - 9$

16. $x^2 + 9$

17. $2y^2 + 11y + 12$

18. $6x^2 - 14x + 4$

19. $2s^2 + s - 6$

20. $3t^2 - 13t + 4$

21. $s^2t - 9t$

1. ______
2. ______
3. ______
4. ______
5. ______
6. ______
7. ______
8. ______
9. ______
10. ______
11. ______
12. ______
13. ______
14. ______
15. ______
16. ______
17. ______
18. ______
19. ______
20. ______
21. ______

22. ____________
23. ____________
24. ____________
25. ____________
26. ____________
27. ____________
28. ____________
29. ____________
30. ____________
31. ____________
32. ____________
33. ____________
34. ____________
35. ____________
36. ____________
37. ____________
38. ____________
39. ____________
40. ____________
41. ____________
42. ____________
43. ____________
44. ____________
45. ____________

22. $9m^2 + 6m + 1$

23. $6x^2 + 21x - 12$

24. $a^2 - 3a - 18$

25. $9y^2 - 12y + 4$

26. $16t^2 - 25$

27. $15x^2 - 2x - 1$

28. $x^3 - 12x^2 + 20x$

29. $2t^2 - t - 3$

30. $4x^2 + 12x + 9$

31. $2a^3b - 8ab$

32. $2y^2 - 3y + 4$

33. $12t^2 - 32t + 5$

34. $4x^2 - 400$

35. $9x^2 + 31x - 20$

36. $t^2 + 10t + 25$

37. $x^2 + 6x + 4$

38. $x^2 - 8x + 16$

39. $3x^2 - 75$

40. $12y^2 + 51y - 45$

41. $6x^2 - 14x + 10$

42. $9x^2 + 24x + 16$

43. $6p^2 - 7p + 2$

44. $12x^2 + 19x - 18$

45. $12t^2 - 11t - 5$

5.7

Solving Equations by Factoring

In this section we are going to use factoring to solve second-degree equations in one variable. First we need to define second-degree equations in one variable.

DEFINITION 5.10 A *second-degree equation in one variable* (or quadratic equation) has the standard form

$$ax^2 + bx + c = 0$$

where $a \neq 0$.

EXAMPLES: $2x^2 + 3x - 4 = 0$

$x^2 - 16 = 0$

$-t^2 - 2t + 1 = 0$

Not all second-degree (quadratic) equations start out in standard form, but they may be put in standard form.

Example 1 Put $x^2 = 3x - 4$ in standard form.

(a) Subtract $3x$ from each side of the equation.

$$x^2 = 3x + (-4)$$
$$x^2 - 3x = -4$$

(b) Add 4 to each side of the equation.

$$x^2 - 3x + 4 = 0 \quad \textit{Standard form}$$

DO EXERCISE 1.

Before we start to solve certain quadratic equations, we must look at the following law.

LAW OF ZERO PRODUCTS $PQ = 0$ if and only if $P = 0$, $Q = 0$, or both P and Q equal zero.

From this law we obtain the following useful fact:

If we have the product of two factors equal to zero, then the only way we can make this product zero is by setting one or both of the factors equal to zero. We will use this fact to help solve certain quadratic equations.

DEFINITION 5.11 A *solution* to the quadratic equation $ax^2 + bx + c = 0$ is any number that substituted for x gives us a true statement.

Example 2 Are the following values of x solutions to the quadratic equation $x^2 - x - 6 = 0$? (a) $x = 1$ (b) $x = 2$ (c) $x = -2$

(a) $x = 1$. Substitute 1 for x.

Thus, 1 is not a solution.

$x^2 - x - 6$	0
$1^2 - 1 - 6$	0
-6	0

LEARNING OBJECTIVES

After finishing Section 5.7, you should be able to:

- Solve certain second-degree equations in one variable.

1. Put $2x^2 = -3x + 4$ in standard form.

2. Are the following solutions to the quadratic equation $x^2 - 4x + 3 = 0$?
(a) 0 (b) 1
(c) 3 (d) -2

(b) $x = 2$. Substitute 2 for x.

Thus, 2 is not a solution.

$x^2 - x - 6$	0
$2^2 - 2 - 6$	0
$4 - 2 - 6$	0
-4	0

(c) $x = -2$. Substitute -2 for x.

Thus, -2 is a solution.

$x^2 - x - 6$	0
$(-2)^2 - (-2) - 6$	0
$4 + 2 - 6$	0
0	0

DO EXERCISE 2.

Guessing a solution for $x^2 - x - 6 = 0$ is not a very efficient method for solving this equation. With the help of the Law of Zero Products and factoring, however, we can solve the equation as follows:

1. Factor $x^2 - x - 6$ and obtain the equivalent equation

$$(x - 3)(x + 2) = 0$$

$$x^2 - x - 6 = 0$$
$$(x - 3)(x + 2) = 0$$

2. We can apply the Law of Zero Products since we now have the product of two factors equal to zero. That is,

$$(x - 3)(x + 2) = 0$$

There are two ways to make the product $(x - 3)(x + 2)$ equal zero.

(1) $x - 3 = 0$ or
(2) $x + 2 = 0$

3. Solve the first-degree equations in one variable for x.

(1) $x - 3 = 0$ gives us $x = 3$.
(2) $x + 2 = 0$ gives us $x = -2$.

4. Thus, our two solutions for $x^2 - x - 6 = 0$ are 3 and -2.

We will follow basically the same steps in solving other quadratic equations. However, we cannot solve every quadratic equation by use of the Law of Zero Products because we cannot factor every second-degree polynomial in one variable (in the sense described in Rule 5.5). The general procedural steps are as follows.

STEP 1: STANDARD FORM

The quadratic equation must be in standard form before we can apply the Law of Zero Products.

STEP 2: FACTOR

Factor the polynomial.

STEP 3: LAW OF ZERO PRODUCTS

Apply the Law of Zero Products.

STEP 4: SOLVE

Solve the resulting linear equations in one variable.

STEP 5: CHECK

We can always check our solutions by substituting in the original equation.

Example 3 Solve $x^2 - 16 = 0$.

STEP 1: STANDARD FORM

The equation is already in standard form.

STEP 2: FACTOR

Notice that we have the difference of two squares.

$$x^2 - 16 = 0$$
$$(x - 4)(x + 4) = 0$$

STEP 3: LAW OF ZERO PRODUCTS

There are two solutions to $(x - 4)(x + 4) = 0$; that is, there are two ways to make this product equal zero.

(1) $x - 4 = 0$ or
(2) $x + 4 = 0$

STEP 4: SOLVE

(1) $x - 4 = 0$ gives us $x = 4$.
(2) $x + 4 = 0$ gives us $x = -4$.

Thus, the two solutions are 4 and -4.

STEP 5: CHECK

Substitute in original equation.

(1)

$x^2 - 16$	0
$4^2 - 16$	0
$16 - 16$	0
0	0

(2)

$x^2 - 16$	0
$(-4)^2 - 16$	0
$16 - 16$	0
0	0

DO EXERCISES 3 THROUGH 5.

There are other quadratic equations we want to look at.

Example 4 Solve $x^2 = 7x - 10$.

STEP 1: STANDARD FORM

Subtract $7x$ from each side.

$$x^2 = 7x + (-10)$$
$$x^2 - 7x = -10$$

Add 10 to each side.

$$x^2 - 7x + 10 = 0$$

STEP 2: FACTOR

$$x^2 - 7x + 10 = 0$$
$$(x - 5)(x - 2) = 0$$

Solve.

3. $x^2 - 4x + 3 = 0$

4. $t^2 - 25 = 0$

5. $x^2 - x - 12 = 0$

Solve.

6. $x^2 - 10 = -3x$

7. $2(x^2 - x) = 3 - x$

STEP 3: LAW OF ZERO PRODUCTS

There are two ways to make this product zero.

$$(1) \quad x - 5 = 0 \quad \text{or}$$
$$(2) \quad x - 2 = 0$$

STEP 4: SOLVE

$$(1) \quad x - 5 = 0 \text{ gives us } x = 5.$$
$$(2) \quad x - 2 = 0 \text{ gives us } x = 2.$$

Thus, the two solutions are 5 and 2.

STEP 5: CHECK

Substitute in original equation.

(1)

x^2	$7x - 10$
$(5)^2$	$7(5) - 10$
25	$35 - 10$
25	25

(2)

x^2	$7x - 10$
$(2)^2$	$7(2) - 10$
4	$14 - 10$
4	4

DO EXERCISES 6 AND 7.

Example 5 Solve $2x^2 + 5x - 12 = 0$.

STEP 1: STANDARD FORM

The equation is already in standard form.

STEP 2: FACTOR

$$2x^2 + 5x - 12 = 0$$
$$(2x - 3)(x + 4) = 0$$

STEP 3: LAW OF ZERO PRODUCTS

$$(1) \quad 2x - 3 = 0 \quad \text{or}$$
$$(2) \quad x + 4 = 0$$

STEP 4: SOLVE

$$(1) \quad 2x - 3 = 0 \text{ gives us } 2x = 3 \text{ and } x = \tfrac{3}{2}.$$
$$(2) \quad x + 4 = 0 \text{ gives us } x = -4.$$

Thus, the two solutions are $\frac{3}{2}$ and (-4).

STEP 5: CHECK

Substitute in original equation.

(1)

$2x^2 + 5x - 12$	0
$2\left(\frac{3}{2}\right)^2 + 5\left(\frac{3}{2}\right) - 12$	0
$2\left(\frac{9}{4}\right) + \frac{15}{2} - 12$	0
$\frac{9}{2} + \frac{15}{2} - 12$	0
$\frac{24}{2} - 12$	0
0	0

(2)

$2x^2 + 5x - 12$	0
$2(-4)^2 + 5(-4) - 12$	0
$2(16) + (-20) + (-12)$	0
$32 + (-32)$	0
0	0

Example 6 Solve $(x - 1)(x + 4) = 6$.

STEP 1: STANDARD FORM

This equation is not in standard form since we do not have zero on one side of the equation.

$$\begin{aligned} (x - 1)(x + 4) &= 6 \\ x^2 + 4x - x - 4 &= 6 \\ x^2 + 3x - 4 &= 6 \\ x^2 + 3x - 4 - 6 &= 0 \\ x^2 + 3x - 10 &= 0 \end{aligned}$$

STEP 2: FACTOR

$$\begin{aligned} x^2 + 3x - 10 &= 0 \\ (x + 5)(x - 2) &= 0 \end{aligned}$$

STEP 3: LAW OF ZERO PRODUCTS

(1) $x + 5 = 0$ or

(2) $x - 2 = 0$

STEP 4: SOLVE

(1) $x + 5 = 0$ gives us $x = -5$.

(2) $x - 2 = 0$ gives us $x = 2$.

Thus, the two solutions are (-5) and 2.

Solve.

8. $(x - 2)(x + 3) = -4$

9. $3x^2 + 13x - 10 = 0$

STEP 5: CHECK

Substitute in original equation.

(1)

$(x - 1)(x + 4)$	6
$[(-5) - 1][(-5) + 4]$	6
$(-6)(-1)$	6
6	6

(2)

$(x - 1)(x + 4)$	6
$(2 - 1)(2 + 4)$	6
$(1)(6)$	6
6	6

DO EXERCISES 8 AND 9.

Once more, we cannot solve every quadratic equation in this manner (for example, $x^2 - 3x + 1 = 0$) since not every second-degree polynomial (for example, $x^2 - 3x + 1$) factors in the sense of Rule 5.5. In Chapter 10 we will develop other methods that will allow us to solve additional quadratic equations.

DO SECTION PROBLEMS 5.7.

ANSWERS TO MARGINAL EXERCISES

1. $2x^2 + 3x - 4 = 0$ **2.** (a) No (b) Yes (c) Yes (d) No **3.** 1, 3 **4.** $-5, 5$ **5.** $4, -3$ **6.** $-5, 2$ **7.** $\frac{3}{2}, -1$ **8.** $-2, 1$ **9.** $\frac{2}{3}, -5$

NAME COURSE/SECTION DATE

SECTION PROBLEMS 5.7

Solve.

1. $x^2 - 7x + 10 = 0$

2. $m^2 + 2m - 15 = 0$

3. $3s^2 - 12 = 0$

4. $x^2 = 3x + 4$

5. $c^2 + 6 = -7c$

6. $(y - 3)(2y + 1) = 0$

7. $t^2 - 8t + 15 = 0$

8. $s^2 = 4s + 5$

9. $2(y^2 - 3y) = y^2 - 9$

10. $3x^2 = 75$

11. $6x^2 + x - 15 = 0$

12. $3b^2 - 4b = 0$

13. $3M^2 = -5M + 2$

14. $2(x^2 - 5x) + 10 = x - 5$

ANSWERS

1. ____________
2. ____________
3. ____________
4. ____________
5. ____________
6. ____________
7. ____________
8. ____________
9. ____________
10. ____________
11. ____________
12. ____________
13. ____________
14. ____________

15. ______
16. ______
17. ______
18. ______
19. ______
20. ______
21. ______
22. ______
23. ______
24. ______
25. ______
26. ______
27. ______
28. ______
29. ______
30. ______

15. $(x - 3)(x + 2) = 6$

16. $6y^2 - 29y - 5 = 0$

17. $t(2t + 7) = -3$

18. $x(2x - 5) = 2(x^2 - 4x + 1)$

19. $4b^2 + 9 = 12b$

20. $(t + 5)(t + 1) = 21$

21. $y^2 - 6 = 5y$

22. $2y^2 + 11y + 12 = 0$

23. $3x^2 - 7x + 2 = 0$

24. $2(p^2 - p) = p^2 - 2p + 9$

25. $x^2 = 3 - 3x$

26. $(a + 1)(a - 5) = 7$

27. $2(y^2 - y - 3) = y^2 + 2$

28. $2t^2 = 15t + 50$

29. $x^2 - 7x = 0$

30. $t^2 = -9t - 14$

5.8

Word Problems

Now we turn to our favorite problems—word problems.

Example 1 Ken wants a rectangular garden plot that has a length 5 feet more than the width and has a total area of 750 square feet. What are the dimensions that Ken should use for his garden?

STEP 1: PICTURE

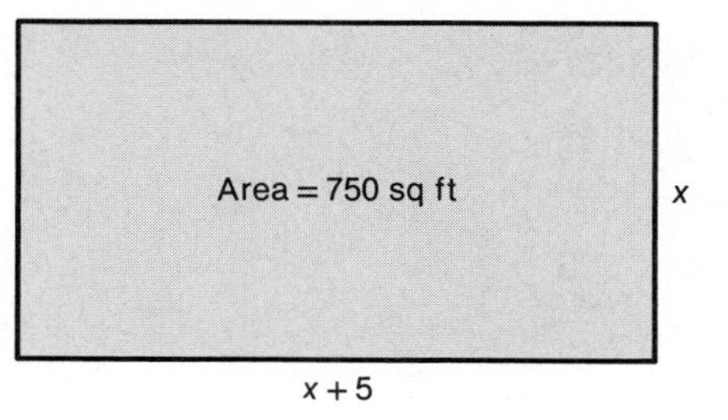

STEP 2: VARIABLE

Let x = width; then
$x + 5$ = length.

STEP 3: EQUATION

The area of a rectangle is given by $A = L \cdot W$ (L = length and W = width). This gives us

$$750 = (x + 5)(x)$$

STEP 4: SOLVE

$$x^2 + 5x = 750$$

Standard form $\quad x^2 + 5x - 750 = 0$

Factor $\quad (x + 30)(x - 25) = 0$

Law of Zero Products $\quad$ (1) $x + 30 = 0$ or (2) $x - 25 = 0$

Solve $\quad$ (1) $x = -30$ or (2) $x = 25$

Since width (x) is a distance and has to be positive, the only solution we need to consider is $x = 25$ feet (width) and $x + 5 = 30$ feet (length).

DO EXERCISE 1.

Example 2 A number squared is the same as three times the number plus four. Find the number.

STEP 2: VARIABLE

Let x = the number.

LEARNING OBJECTIVES

After finishing Section 5.8, you should be able to:

- Solve word problems that involve quadratic equations.

1. If the area of a triangle is 20 square inches and the height is 3 inches less than the base, what are the dimensions of the triangle? (Area $= A = bh/2$.)

2. Twice the square of a number minus four is the same as negative seven times the number. What is the number?

STEP 3: EQUATION

Translate the word equation into a mathematical equation.

$$\underbrace{\text{A number squared}}_{x^2} \quad \underbrace{\text{is the same as}}_{=} \quad \underbrace{\text{three times the number}}_{3x} \quad \underbrace{\text{plus 4}}_{+\,4}$$

STEP 4: SOLVE

$$x^2 = 3x + 4$$

Standard form $\quad x^2 - 3x - 4 = 0$

Factor $\quad (x - 4)(x + 1) = 0$

Law of Zero Products $\quad$ (1) $x - 4 = 0$ or (2) $x + 1 = 0$

Solve $\quad$ (1) $x = 4$ or (2) $x = -1$

Thus, there are two solutions to the problem, 4 and -1.

DO EXERCISE 2.

Example 3 The product of two consecutive integers is the same as their sum plus 5. Find the pair of integers.

3. The product of two consecutive even integers is the same as three times the sum of the two integers plus 6. Find the pair of integers.

STEP 2: VARIABLE

Let x = one integer; then
$x + 1$ = the next consecutive integer.

STEP 3: EQUATION

Translate the word equation into a mathematical equation and obtain

$$x(x + 1) = x + (x + 1) + 5$$

STEP 4: SOLVE

Standard form

$$x^2 + x = 2x + 6$$
$$x^2 - x = 6$$
$$x^2 - x - 6 = 0$$

Factor $\quad (x - 3)(x + 2) = 0$

Law of Zero Products $\quad$ (1) $x - 3 = 0$ or (2) $x + 2 = 0$

Solve $\quad$ (1) $x = 3$ or (2) $x = -2$

Thus, if $x = 3$, then $x + 1 = 4$, and the pair 3 and 4 would be one solution. If $x = -2$, then $x + 1 = -1$, and the pair -2 and -1 would be another solution.

DO EXERCISE 3.

Example 4 If the area of a trapezoid $[A = h(B + b)/2]$ is 72 square inches, the large base is twice the small base, and the height is two inches more than the small base, find the dimensions of the trapezoid.

STEP 1: PICTURE

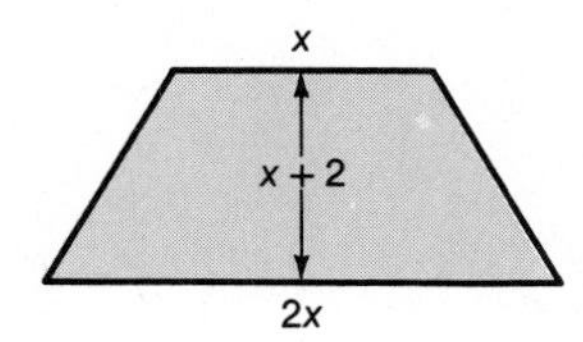

STEP 2: VARIABLE

Let $x =$ the small base; then
$2x =$ the large base, and
$x + 2 =$ the height.

STEP 3: EQUATION

We will use $A = h(B + b)/2$. Making the appropriate substitutions, we have

$$72 = \frac{(x + 2)(2x + x)}{2}$$

STEP 4: SOLVE

Standard form	$2(72) = (x + 2)(3x)$
	$144 = 3x^2 + 6x$
	$0 = 3x^2 + 6x - 144$
Factor	$0 = 3(x^2 + 2x - 48)$
	$0 = 3(x + 8)(x - 6)$
Law of Zero Products	(1) $x + 8 = 0$ or
	(2) $x - 6 = 0$
Solve	(1) $x = -8$
	(2) $x = 6$

Since the small base x is a distance and has to be positive, the only solution we need to consider is $x = 6$. Thus, we have

$$b = 6 \text{ inches}$$
$$B = 2x = 12 \text{ inches}$$
$$h = x + 2 = 8 \text{ inches}$$

DO EXERCISE 4.

DO SECTION PROBLEMS 5.8.

4. If the total surface area of a cylindrical vat is 42π square meters and the height is 2 meters less than the diameter of the vat, find the volume of the vat. The formula for surface area is $SA = 2\pi rh + 2\pi r^2$; the formula for the volume of a cylinder is $V = \pi r^2 h$.

ANSWERS TO MARGINAL EXERCISES

1. 5 in., 8 in. **2.** $\frac{1}{2}$ or -4 **3.** 6, 8 or -2, 0
4. $V = (3^2)(4)\pi$ cu m or 36π cu m

NAME COURSE/SECTION DATE

ANSWERS

1. ______
2. ______
3. ______
4. ______
5. ______
6. ______
7. ______

SECTION PROBLEMS 5.8

1. A number squared minus 3 times the number is equal to the number plus 12. Find the number.

2. A number squared minus four times the number is zero. Find the number.

3. The area of a trapezoid is 45 square inches. If the height is 4 inches less than the large base and the small base is half the large base, find the dimensions of the trapezoid. (Recall that $A = h(B + b)/2$ for a trapezoid.)

4. The area of a triangle ($A = bh/2$) is 150 square centimeters. If the base is 5 centimeters less than the height, find the dimensions of the triangle.

5. The product of 2 consecutive integers is the same as 5 times the smaller integer plus 4 times the larger integer minus 4. Find the integers.

6. The product of 2 consecutive integers is the same as 4 times the next consecutive integer plus 2. Find the 3 consecutive integers.

7. Darcy Bloom wants to have a garden with a width that is 5 feet less than the length and with an area of 500 square feet. How many feet of fencing will Darcy need to go around her garden?

8. ____________

9. ____________

10. ____________

11. ____________

12. ____________

13. ____________

8. The area of a rectangle is 54 square centimeters. If the width is 3 centimeters less than the length, find the dimensions of the rectangle.

9. A right circular cylinder has a surface area SA of 140π square inches. If the height h is twice the radius r minus 1, find the height of the cylinder. Use the formula $SA = 2\pi rh + 2\pi r^2$.

10. The area of a parallelogram ($A = bh$) is 96 square inches. If the height is three-fourths the length of the base minus 1, find the base and height of the parallelogram.

11. The sum of the squares of two consecutive integers is 61. Find the integers.

12. The sum of the squares of 2 numbers is 29. If one number is 3 less than the other, find the 2 numbers.

13. Faye Fitzgerald has a picture frame that is 10 inches by 12 inches. She wants to cut a mat to fit in the frame such that the border of the mat is the same width all the way around and the area left open in the middle for the picture is 63 square inches (see picture below). How wide should Faye make the border of the mat?

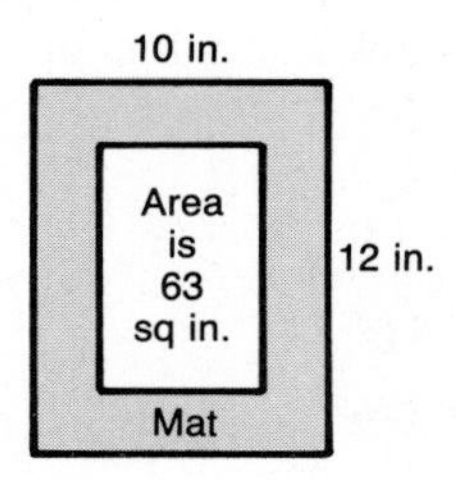

NAME ______ COURSE/SECTION ______ DATE ______

CHAPTER 5 REVIEW PROBLEMS

SECTION 5.1

Are the following expressions polynomials? If not, why not?

1. $2\sqrt{x} - \frac{3}{2}$ **2.** $xy - z$ **3.** $\frac{y - 3}{y^2 + y + 1}$

Give the terms of each polynomial.

4. $x^4 - x^3 + 3x^2 + 7x - 5$ **5.** $-x^2y^2 - 3xy^2 + 1.4y$

Give the numerical coefficient of each term.

6. $x^4 - x^3 + 3x^2 + 7x - 5$ **7.** $0.2y^2 - \frac{y}{4} + 7$

Give the degree of each term and the degree of the polynomial.

8. $2x^5 - 3x^3 + x^2 - 6$ **9.** $3xy^2 - 4x^2yz^3$

10. Arrange $3x + 6x^4 - x^5 + 2$ in descending order.

11. Arrange $t^2 - 7t^7 + t + 3t^4 - 1$ in ascending order.

12. Give an example of (a) a monomial, (b) a binomial, and (c) a trinomial each of degree 4.

ANSWERS

1. ______
2. ______
3. ______
4. ______
5. ______
6. ______
7. ______
8. ______
9. ______
10. ______
11. ______
12. (a) ______
(b) ______
(c) ______

13. ______

14. ______

15. ______

16. ______

17. ______

18. ______

19. ______

20. ______

21. ______

22. ______

23. ______

24. ______

25. ______

26. ______

SECTION 5.2

Perform the indicated operations and simplify.

13. $(3x^3 - x^2 + 7x) + (-2x^2 + x - 5)$

14. $(-t^4 + 3t^2 - 1) + (7t^3 + 6t^2 - t + 3)$

15. $(a^2b^2 - 3ab + b^3) + (4a^2b^2 + 7ab - a^3)$

16. $(y^3 - y^2 - y - 1) - (2y^3 - 7y^2 + y - 1)$

SECTION 5.3

Multiply.

17. $3x^2(-4x)$

18. $(-2x)(3x^2y)(-4y^2)$

19. $4ab^2(-7a^2 + ab - 3b)$

20. $(x + 4)(x^2 - 3x + 2)$

21. $(a + b)(c - d)$

22. $(y - 5)(y + 2)$

23. $(t - 6)(t - 5)$

24. $(2x - 5)(x + 4)$

25. $(6y - 5)(8y + 3)$

26. $(6x + 1)(2x + 5)$

NAME ______ COURSE/SECTION ______ DATE ______

ANSWERS

27. ______
28. ______
29. ______
30. ______
31. ______
32. ______
33. ______
34. ______
35. ______
36. ______

SECTION 5.4

Multiply.

27. $(x - 3)(x + 3)$

28. $(s + 5)^2$

29. $(2t - 3)^2$

30. $(10x - 7)(10x + 7)$

31. $(x + 1)^3$

SECTION 5.5

Factor.

32. $x^2 - 49$

33. $-4t^2 - 10t$

34. $a^2b - ab^2$

35. $y(x + 4) - 3(x + 4)$

36. $x^2(x^2 - 9) - 16(x^2 - 9)$

37. ____________

38. ____________

39. ____________

40. ____________

41. ____________

42. ____________

43. ____________

44. ____________

45. ____________

46. ____________

47. ____________

48. ____________

49. ____________

SECTION 5.6

Factor.

37. $y^2 - 4y + 3$

38. $x^2 + 7x + 10$

39. $3x^2 + 6x - 8$

40. $6x^2 + 13x + 6$

41. $12x^2 + 14x - 40$

42. $18a^2 + 35a + 12$

SECTION 5.7

Solve.

43. $x^2 + x - 12 = 0$

44. $x^2 - 6x + 5 = 0$

45. $10x^2 + 11x = -3$

46. $42x^2 + 19x - 35 = 0$

SECTION 5.8

47. A number squared is the same as three times the number plus 10. What is the number?

48. The sum of the squares of two numbers is 34. If the difference of the numbers is 2, find the numbers.

49. The product of two consecutive even integers is the same as 4 times the smaller integer plus 5 times the larger integer minus 2. Find the integers.

NAME | COURSE/SECTION | DATE | ANSWERS

CHAPTER 5 TEST

Perform the indicated operation and simplify.

1. $(3a^2 - 6a - 1) + (4 - a^2)$

2. $(2x^2 - x - 7) - (x^3 - 2x^2 + x)$

3. $(2m + n)(m^2 + n)$

4. $(t^2 + 1)(2t^2 - 4t - 5)$

5. $(2x - 5)(3x + 1)$

6. $(3a + 2)^2$

7. Given the polynomial $-x^3 + \frac{7}{2}x^2 + x - 1.5$

(a) What is the degree of the polynomial?
(b) List the terms of the polynomial.
(c) What is the coefficient of each term?

8. Arrange the polynomial $2t^2 - t^5 + 6t^3 - t + 4$ in descending order.

ANSWERS

1. ____________
2. ____________
3. ____________
4. ____________
5. ____________
6. ____________
7. (a) ____________
   (b) ____________
   (c) ____________
8. ____________

9. ____________

10. ____________

11. ____________

12. ____________

13. ____________

14. ____________

15. ____________

16. ____________

17. ____________

18. ____________

19. ____________

20. ____________

Factor.

9. $-6ax + 9bx$

10. $t^2 - 2t - 15$

11. $2y^2 + 5y - 12$

12. $3x^2 - 12$

13. $3x^2 + 7x + 4$

14. $12b^2 + 4b - 5$

Solve.

15. $t^2 + 3t - 10 = 0$

16. $2x^2 + 5x = -2$

17. $2y^2 = 8$

18. $(3a + 4)(a - 1) = -2$

19. The product of two consecutive even integers is the same as seven times their sum plus 14. Find the two integers.

20. Sharon Peters plants a rectangular flower bed that has a width that is 5 times the length plus 1. If the area is 48 square feet, find the dimensions of the flower bed.

CHAPTER 6

Graphing and Straight Lines

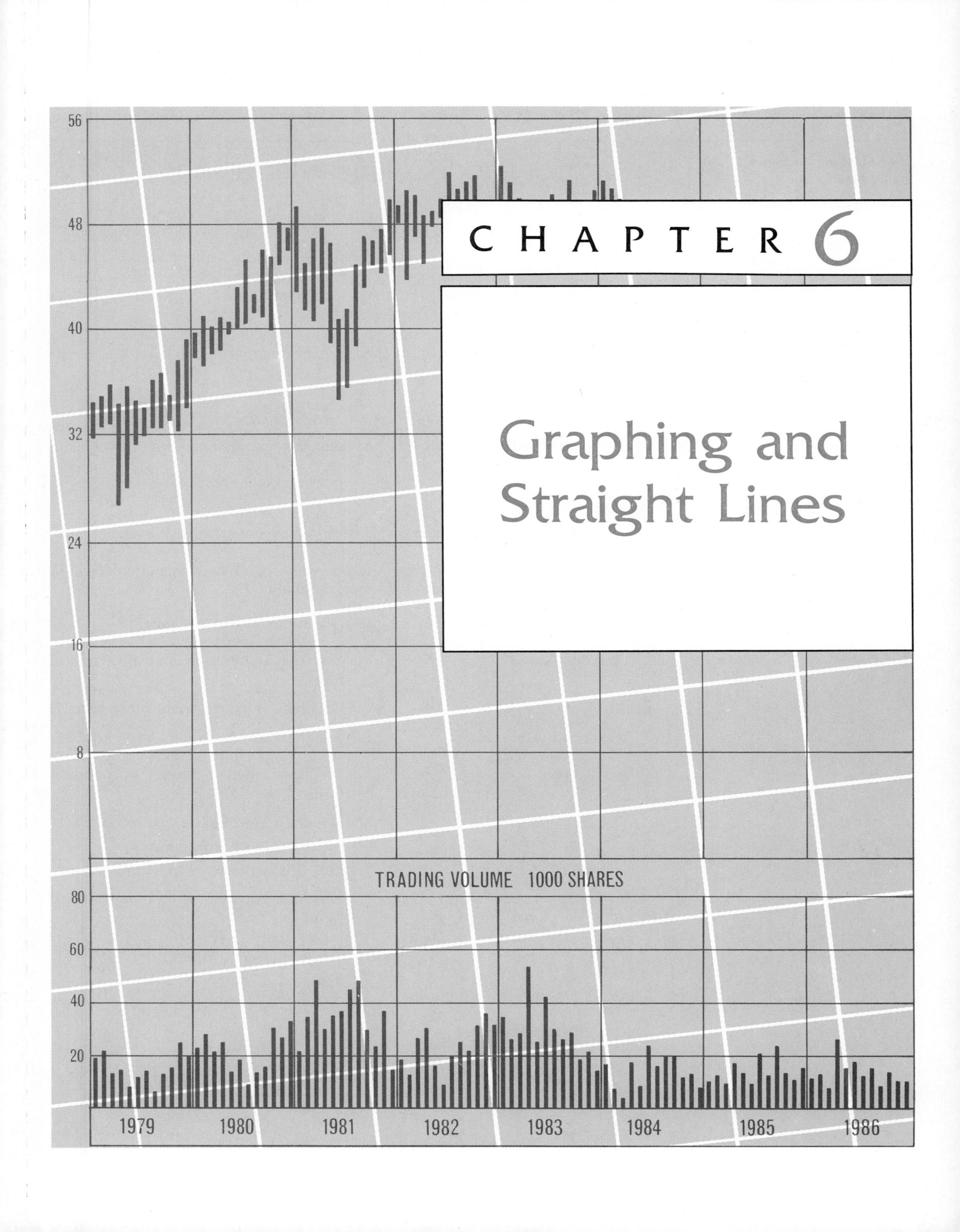

LEARNING OBJECTIVES

After finishing Section 6.1, you should be able to:

- Identify linear equations in two variables.
- Identify solutions to linear equations in two variables.

1. Which of the following are linear equations in two variables? If not, why not?
(a) $y = x^2 + x - 5$
(b) $3s - st = 4$
(c) $2a + b = -\frac{5}{2}$
(d) $x - y + z = 4$
(e) $x^2 + y^2 = 1$
(f) $mn = 1$

6.1

Introduction to Linear Equations in Two Variables

In Chapter 3 we studied linear (first-degree) equations in one variable, and in Chapter 5 we studied second-degree equations in one variable. In particular, we learned how to find solutions to such equations. Now we want to look at linear equations in two variables.

DEFINITION 6.1 A *linear* (first-degree) *equation in two variables* has the standard form

$$ax + by = c$$

where a, b, and c are arbitrary constants, except that a and b are not both zero.

Note

1. In a linear equation in two variables, the exponent on each variable must be 1.
2. In a linear equation in two variables, we cannot have a term that involves the product or quotient of the two variables.

Example 1 Are the following linear equations in two variables?

(a) $2x - 3y = 6$ — Yes; this equation is also in standard form.

(b) $y = x^2 + 1$ — No; there is an exponent other than 1.

(c) $2 - t + y = s$ — No; there are more than two variables.

(d) $\frac{7}{2}t = 4 - 0.3s$ — Yes; this equation is not in standard form.

(e) $3x - xy = 4$ — No; one term involves the product of x and y.

(f) $y = 7$ — Yes; this equation is in standard form.

(g) $3(a - 4) = 2a - (b + 7)$ — Yes; this equation is not in standard form.

(h) $a = \frac{1}{b} + 3$ — No; there is an exponent other than 1, that is, $\frac{1}{b} = b^{-1}$.

DO EXERCISE 1.

Any linear equation in two variables can be put in standard form.

Example 2 Put in standard form.

(a) $2y = -3x + 4$
Add $3x$ to each side.

$$2y + 3x = -3x + 4 + 3x \quad \text{or}$$
$$3x + 2y = 4$$

(b) $3(a - 2b) = -(b + 7) + 2a$

1. Remove parentheses. $\quad 3a - 6b = -b - 7 + 2a$

2. Add $-2a$ to each side. $\quad 3a - 6b + (-2a) = -b - 7 + 2a + (-2a)$
$$a - 6b = -b - 7$$

3. Add b to each side. $\quad a - 6b + b = -b - 7 + b$
$$a - 5b = -7$$

DO EXERCISE 2.

When we try to find solutions to a linear equation in one variable, we substitute a single number for the variable in the equation and see whether we have a true statement. However, if we have a linear equation in two variables, then to find a solution it will be necessary to substitute a number for each variable in the equation; that is, we must substitute a pair of numbers.

DEFINITION 6.2 A *solution* to $ax + by = c$ is any pair of numbers that, when substituted in the equation for x and y, gives us a true statement. Such a pair of numbers will be called an *ordered pair* and will be denoted by (x, y) where, by convention, the first element of the ordered pair is the number substituted for x and the second element of the ordered pair is the number substituted for y.

Example 3 Consider the equation $2x + y = 5$.

(a) Is (2, 1) a solution?
Substitute 2 for x and 1 for y in the equation.

$2x + y$	5
$2(2) + 1$	5
5	5

Thus (2, 1) is a solution.

(b) Is (3, -1) a solution?
Substitute 3 for x and -1 for y.

$2x + y$	5
$2(3) + (-1)$	5
5	5

Thus (3, -1) is a solution.

(c) Is (2, 2) a solution?
Substitute 2 for x and 2 for y.

$2x + y$	5
$2(2) + 2$	5
6	5

Thus (2, 2) is not a solution.

DO EXERCISES 3 AND 4.

Recall that (except for the special cases of no solution or infinitely many solutions) there is exactly one solution to a linear equation in one variable. However in Example 3 above we already have two solutions. Actually, there are infinitely many solutions to the equation $2x + y = 5$. In fact, given an arbitrary value of x, we can find the value for y that will make (x, y) a solution.

2. Put the following in standard form.
(a) $2x + 3y - 4 = 0$
(b) $\frac{7}{2}t = 4 - 0.3s$
(c) $2(s - t) = 3s + 1$
(d) $\frac{x}{3} - \frac{y}{4} = 1$

3. Which of the following are solutions to $-x + y = -5$?
(a) (3, -2) (b) (-1, 4) (c) (10, 5)

4. Find two solutions to $2x - 3y = 4$ and one ordered pair that is not a solution.

5. Find the solution to the equation $2y = 3x - 1$ such that $y = 4$.

6. Find the solution to the equation $2y = 3x - 1$ such that $x = -5$.

Example 4 Find a solution to the equation $2x + y = 5$ such that $x = 6$.

(a) Substitute 6 for x in the equation.

$$2(6) + y = 5$$
$$12 + y = 5$$

(b) Solve for y.

$$y = -7 \quad \text{We added } -12 \text{ to each side.}$$

Thus, the solution is $(6, -7)$.

We could also pick an arbitrary value for y and find the value for x that makes (x, y) a solution.

Example 5 Find a solution to the equation $2x + y = 5$ such that $y = 4$.

(a) Substitute 4 for y in the equation.

$$2x + 4 = 5$$

(b) Solve for x.

$$2x = 1 \quad \text{We added } -4 \text{ to each side.}$$
$$x = \frac{1}{2} \quad \text{We multiplied each side by } \frac{1}{2}.$$

Thus, the solution is $(\frac{1}{2}, 4)$.

In general, the equation $ax + by = c$ has infinitely many solutions. Since there are infinitely many solutions we obviously cannot list them all. However, in Sections 6.2 and 6.3 we will do the next best thing—we will obtain a geometric picture (a graph) of all the solutions.

DO EXERCISES 5 AND 6.

DO SECTION PROBLEMS 6.1.

ANSWERS TO MARGINAL EXERCISES

1. (a) No; exponent other than 1 (b) No; an st term (c) Yes (d) No; three variables (e) No; exponent other than 1 (f) No; mn term
2. (a) $2x + 3y = 4$ (b) $0.3s + \frac{7}{2}t = 4$ (c) $-s - 2t = 1$ (d) Already in standard form **3.** (a) and (c) **4.** Answers may vary. **5.** $(3, 4)$
6. $(-5, -8)$

NAME COURSE/SECTION DATE

SECTION PROBLEMS 6.1

Are the following linear equations in two variables? If not, why not?

1. $2y = 3x + 4$ **2.** $s + t = 3s - r$ **3.** $x^2 = x + 4$

4. $y = 2xy + 1$ **5.** $\frac{m - 4}{3} = 2(n + 5) - 3m$ **6.** $y = \frac{1}{x}$

7. $2(a - 1) = 4b + 5$ **8.** $x^2 - 3x + 1 = 0$

Put in standard form.

9. $x = 2y - 1$ **10.** $2s - 5 + 3t = 0$ **11.** $4(a - b) = 2a + 7$

12. $x - 3y = 4$ **13.** $\frac{x}{2} - \frac{y}{3} + \frac{1}{4} = 0$ **14.** $3(s - t) = 2t - 4$

15. $2y - 5 = 0$ **16.** $-3a = 2b - 5$

17. Which of the following are solutions to $2x - y = 7$?
(a) $(3, 1)$ (b) $(0, -7)$ (c) $(4, -3)$ (d) $(-5, -12)$

18. Which of the following are solutions to $2(x - y) = x + 3$?
(a) $(1, -1)$ (b) $(-3, 0)$ (c) $(4, \frac{1}{2})$ (d) $(2,2)$

19. Which of the following are solutions to $4 - 2y = 3(y - x)$?
(a) $(0, 0)$ (b) $(2, 2)$ (c) $(1, 2)$ (d) $(7, -5)$

20. Which of the following are solutions to $y = -2x + 5$?
(a) $(-1, 7)$ (b) $(0, 5)$ (c) $(2, 9)$ (d) $(-3, 11)$

ANSWERS

1. ______
2. ______
3. ______
4. ______
5. ______
6. ______
7. ______
8. ______
9. ______
10. ______
11. ______
12. ______
13. ______
14. ______
15. ______
16. ______
17. ______
18. ______
19. ______
20. ______

21. ____________

22. ____________

23. ____________

24. ____________

25. ____________

26. ____________

27. ____________

28. ____________

29. ____________

30. ____________

31. ____________

32. ____________

33. ____________

34. ____________

21. If $(a, 2)$ is a solution to the equation $2x - y = 6$, what is the value of a?

22. If $(-1, m)$ is a solution to the equation $y = -3x + 2$, what is the value of m?

23. If $(c, -5)$ is a solution to the equation $x - 3y = -2$, what is the value of c?

24. If $(\frac{3}{2}, d)$ is a solution to the equation $y = 4x - 6$, what is the value of d?

25. If $(-2, n)$ is a solution to the equation $2(x - y) = x + 3$, what is the value of n?

26. If $(m, 3)$ is a solution to the equation $-x - y = 4$, what is the value of m?

27. If $(c, \frac{5}{3})$ is a solution to the equation $x - 2y = -2$, what is the value of c?

28. If $\left(-\frac{1}{2}, n\right)$ is a solution to the equation $y = -\frac{x}{4} - \frac{1}{3}$, what is the value of n?

Find two ordered pairs that are solutions to the following equations and one ordered pair that is not.

29. $x + y = 4$

30. $3x - 2y = 6$

31. $y = -2x + 1$

32. $2(x - y) = x + 1$

33. $y = -4x + 1$

34. $\frac{x}{2} + 3y = 4$

6.2

The Rectangular Coordinate System

LEARNING OBJECTIVES

After finishing Section 6.2, you should be able to:

- Plot points on a rectangular coordinate system.

As we saw in the last section, there are infinitely many solutions (ordered pairs) to the equation $ax + by = c$. Hence, it is obviously impossible to list all of them. However, it is possible to obtain a geometric picture (a graph) of what all the solutions look like. To draw such a graph it is necessary to plot ordered pairs (or points). To do this we first form what is called a *rectangular coordinate system* by taking two number lines and putting them together so they are perpendicular and intersect at their origins.

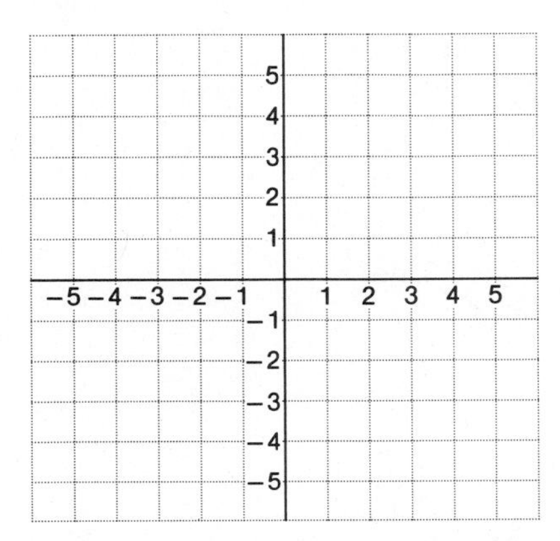

Note the following terminology used for our rectangular coordinate system:

1. The *origin* is the point where the two number lines intersect. It has coordinates (0, 0).
2. The horizontal number line is called the *x-axis* and the first element of an ordered pair is called the *x-coordinate* (or first-coordinate). The x-axis is positive to the right of the origin and negative to the left.
3. The vertical number line is called the *y-axis*, and the second element of an ordered pair is called the *y-coordinate* (or second-coordinate).* The y-axis is positive above the origin and negative below.
4. The rectangular coordinate system divides the plane into four *quadrants* named as follows:

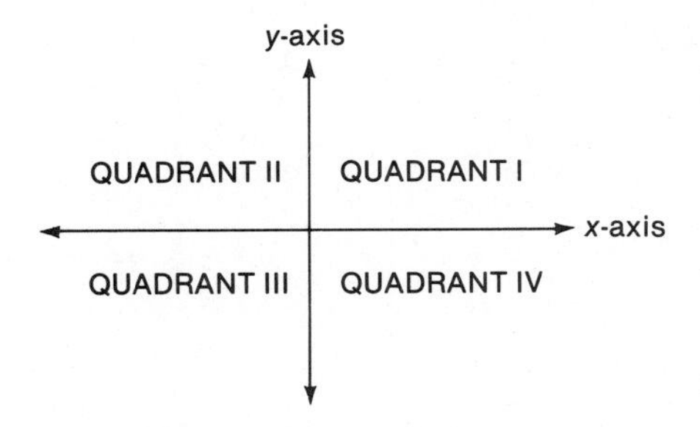

Now we want to plot (graph) some ordered pairs (points) on a rectangular coordinate system.

* The first coordinate of a point is sometimes called its *abscissa*. The second coordinate is sometimes called its *ordinate*.

Example 1 Plot the point (2, 3).

(a) Locate 2 on the x-axis and draw a line through 2 parallel to the y-axis.

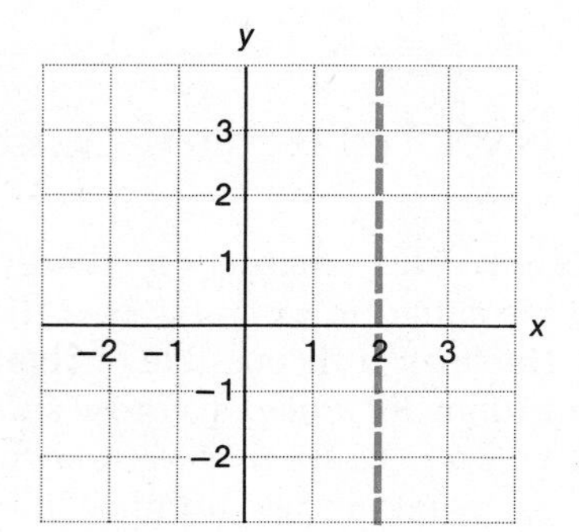

Note

All the points on this dotted line will have an x-coordinate of 2.

(b) Locate 3 on the y-axis and draw a line through 3 parallel to the x-axis.

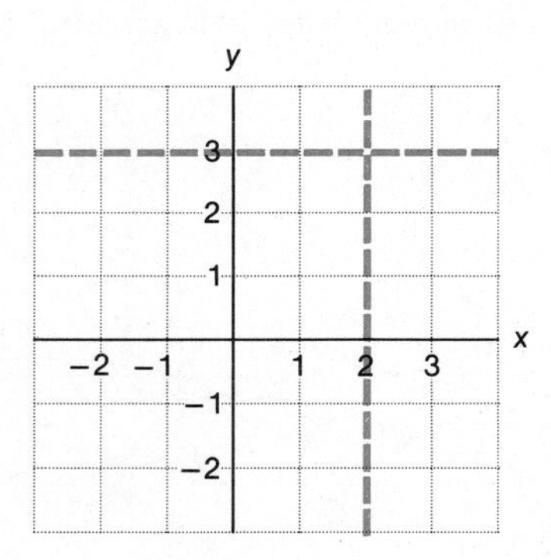

Note

All the points on this dotted line will have a y-coordinate of 3.

(c) The point where the two dotted lines intersect has coordinates (2, 3).

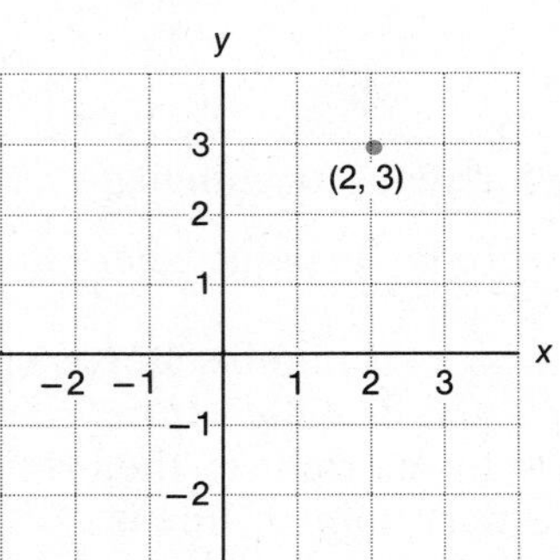

Example 2 Plot the ordered pair (−3, 1).

(a) Locate −3 on the x-axis and draw a line through −3 parallel to the y-axis.

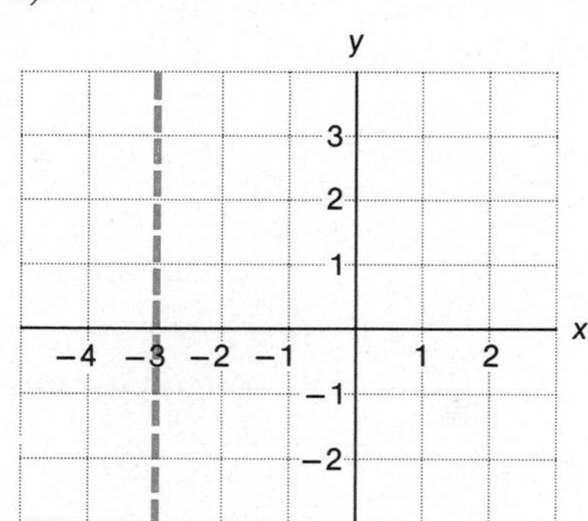

(b) Locate 1 on the y-axis and draw a line through 1 parallel to the x-axis.

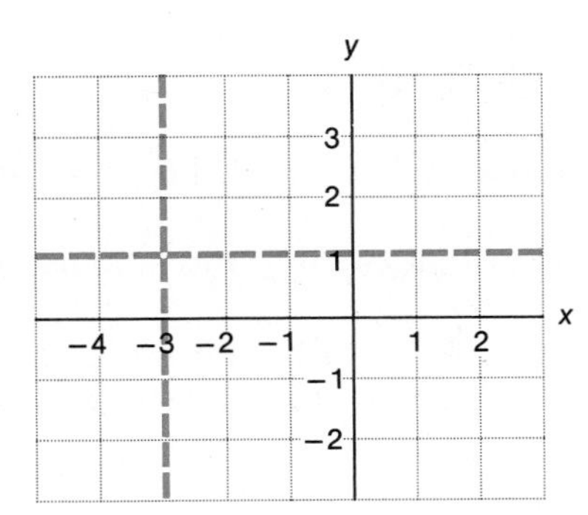

(c) The point where the two lines intersect has coordinates $(-3, 1)$.

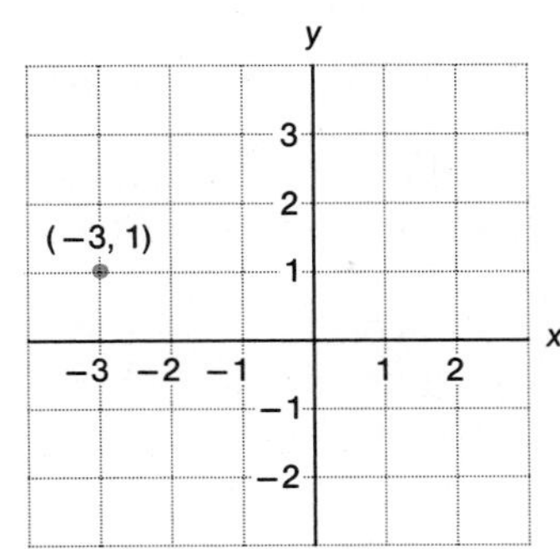

Example 3 Plot the point $(-2, -3)$.

(a) Locate -2 on the x-axis and draw a line through -2 parallel to the y-axis.

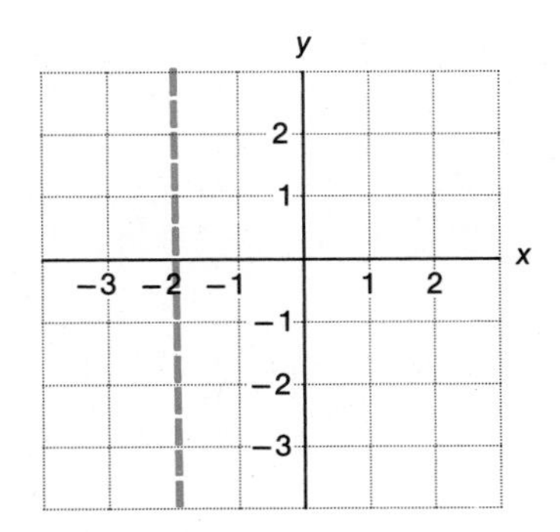

(b) Locate -3 on the y-axis and draw a line through -3 parallel to the x-axis.

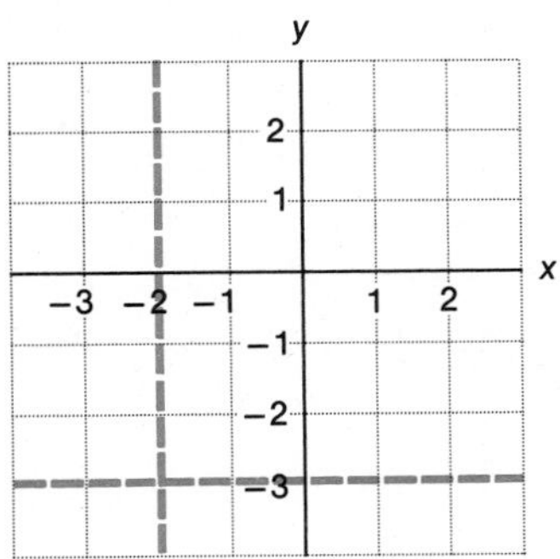

(c) The lines intersect at the point with coordinates $(-2, -3)$.

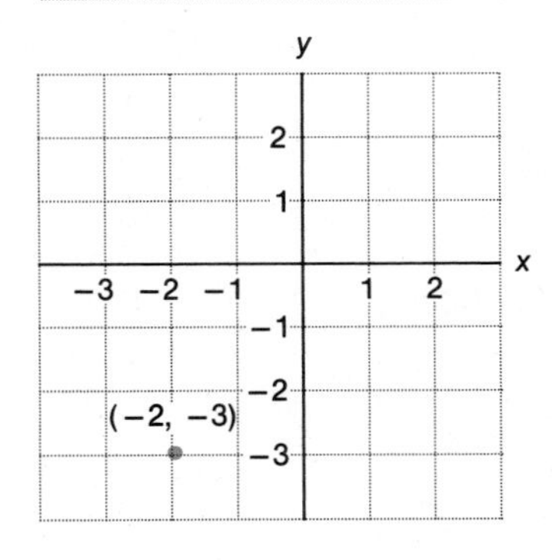

1. Plot the following points.
(a) (3, 4)
(b) (−2, 2)
(c) (−3, −1)
(d) (2, −4)

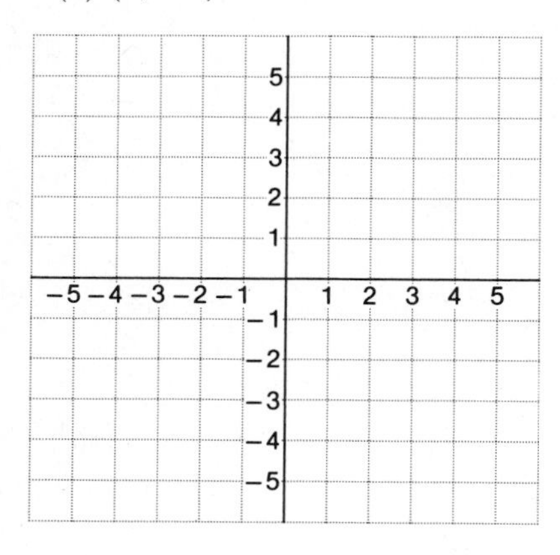

2. What do all the points in quadrant II have in common?

3. Plot the following points.
(a) $(-\frac{1}{2}, 4)$
(b) $(2\frac{1}{3}, -1\frac{1}{2})$
(c) (−3, 0)
(d) $(\frac{4}{3}, 0)$

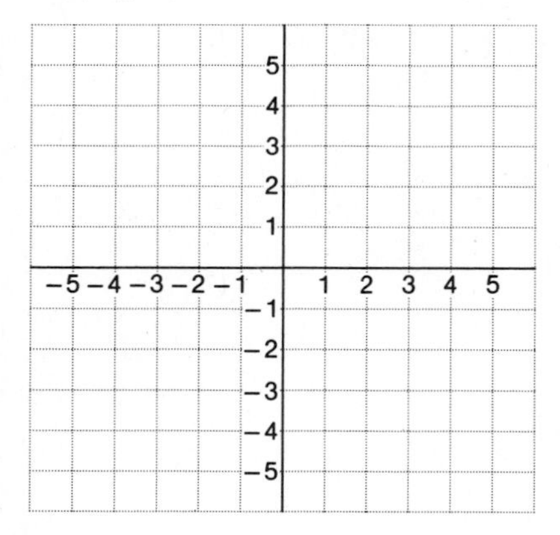

DO EXERCISE 1.

Example 4 What do all the points in quadrant IV have in common?

The x-coordinates are positive and the y-coordinates are negative.

Eventually, we will be able to plot points without first drawing in the dotted lines, but for now we will still put in the dotted lines.

DO EXERCISE 2.

Example 5 Plot the point $(1\frac{1}{2}, -2\frac{1}{3})$.

When we have rational numbers as coordinates, we estimate their location on the appropriate axis as best we can.

(a) Locate $1\frac{1}{2}$ on the x-axis and draw a line through $1\frac{1}{2}$ parallel to the y-axis.

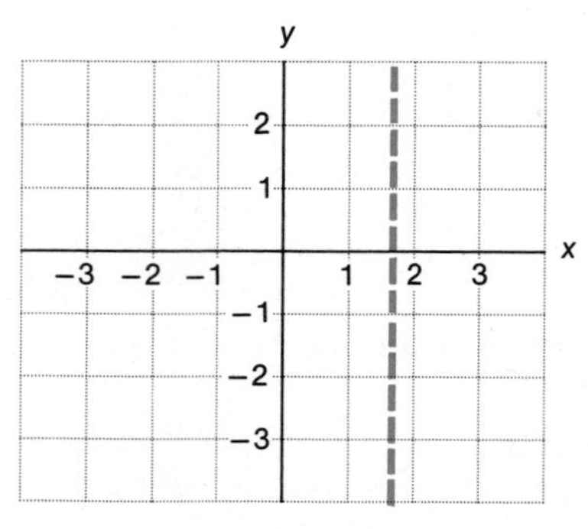

(b) Locate $-2\frac{1}{3}$ on the y-axis and draw a line through $-2\frac{1}{3}$ parallel to the x-axis.

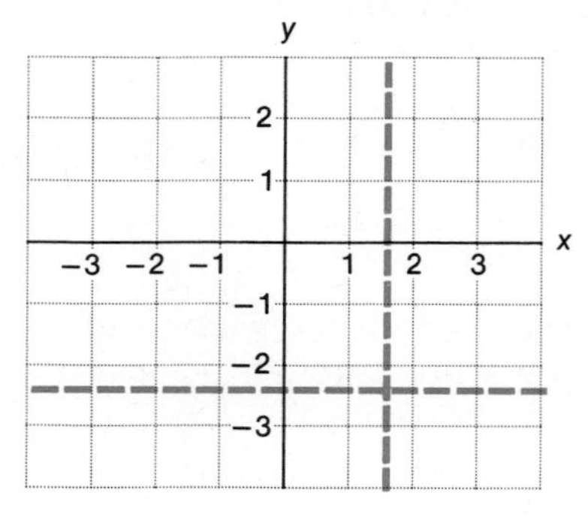

(c) The point where the lines intersect is $(1\frac{1}{2}, -2\frac{1}{3})$.

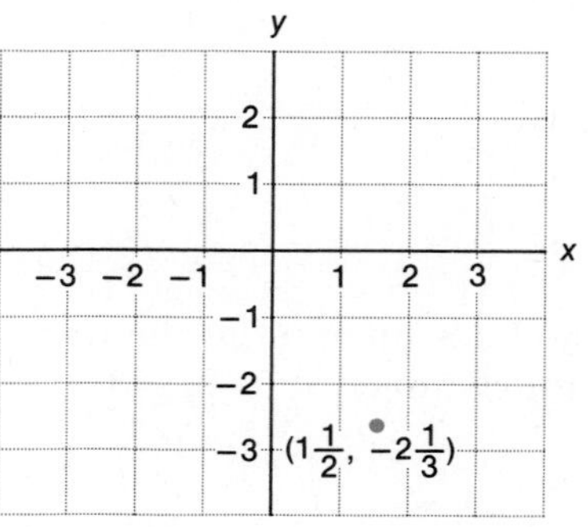

DO EXERCISE 3.

Sometimes we are given a point on a graph, and we want to find its approximate coordinates.

Example 6 Find the coordinates of point P.

(a) Draw a line through P parallel to the y-axis. Where this line intersects the x-axis is the x-coordinate of P.

(b) Draw a line through P parallel to the x-axis. Where this line intersects the y-axis is the y-coordinate of P.

(c) Thus, P has coordinates (2, 3).

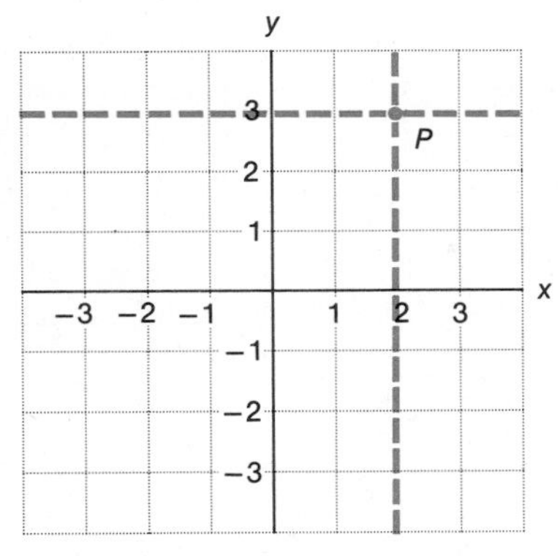

Example 7 Find the coordinates of point Q.

(a) Draw lines through Q parallel to the x-axis and y-axis.

(b) Thus, Q has (approximate) coordinates $(-1\frac{1}{3}, 2)$.

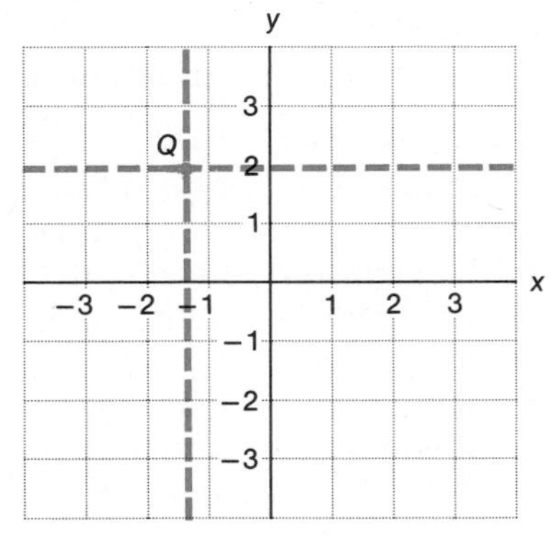

DO EXERCISE 4.

Example 8 Plot the points (0, 3), (0, −2), and $(0, 1\frac{1}{2})$.

All these points have x-coordinate zero. Since only points on the y-axis have x-coordinate zero, the required points are all on the y-axis.

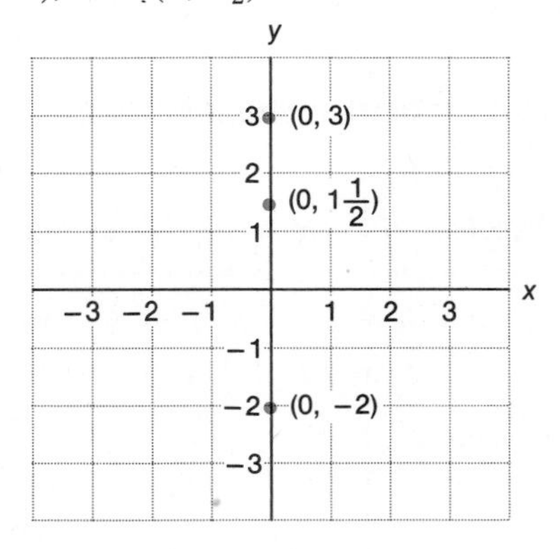

DO EXERCISES 5 AND 6.

Example 9 Find three solutions to $t = 2s - 1$ and plot them.

(a) Since our equation does not involve x and y, we need to decide whether the value for s or the value for t will be the first-coordinate of our ordered pair. We may choose either; however, if we follow convention, whichever variable comes first in the alphabet will be the first-coordinate. Thus, the value for s will be our first-coordinate, and the s-axis will be the horizontal axis.

4. Find the approximate coordinates of points P, Q, R, and S.

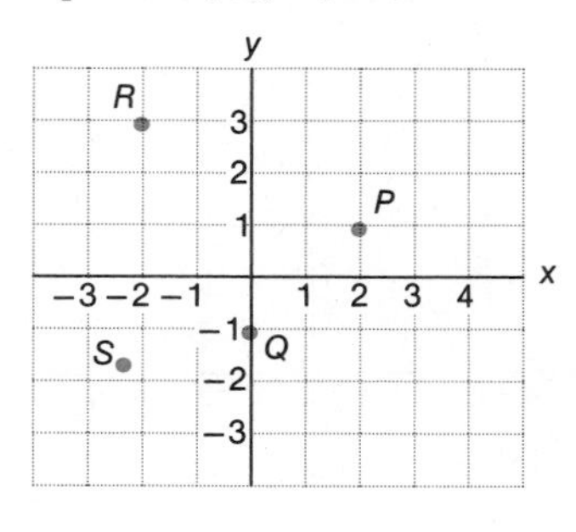

5. In which quadrant would you find a point with negative x-coordinate and positive y-coordinate?

6. What do all the points on the x-axis have in common?

7. Find three solutions to $a - 3b = 1$ and plot them.

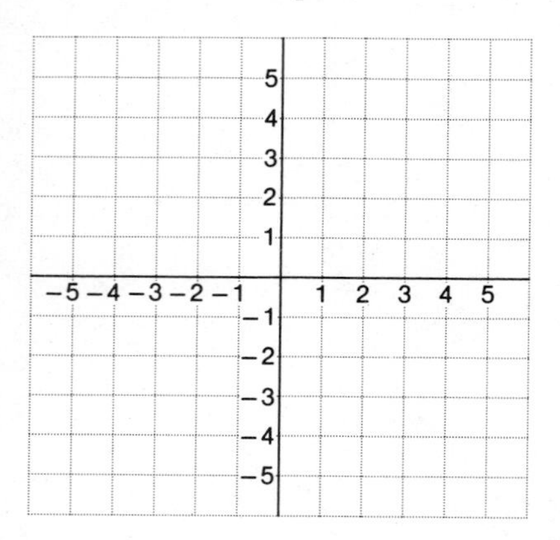

8. In Problem 7, how do we decide which axis is the a-axis and which is the b-axis?

(b) We find three arbitrary solutions: (0, −1), (2, 3), and (−2, −5).

(c) The three points are plotted as follows:

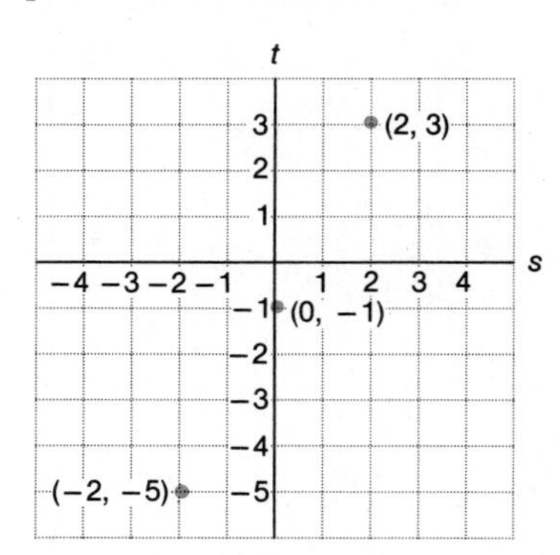

We have one final note. If we are given the point (x, y), then each coordinate tells us something very important about the point.

1. The absolute value of the x-coordinate tells us how many units from the y-axis the point is (to the right of the y-axis if the x-coordinate is positive and to the left if negative).
2. The absolute value of the y-coordinate tells us how many units above (if a positive y-coordinate) or below (if a negative y-coordinate) the x-axis the point is.

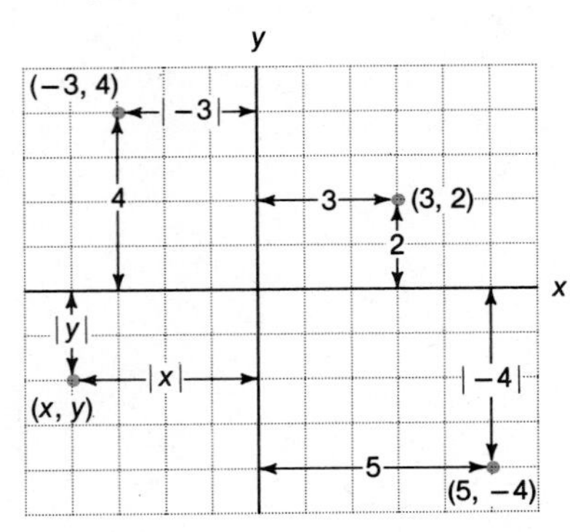

DO EXERCISES 7 AND 8.

DO SECTION PROBLEMS 6.2.

ANSWERS TO MARGINAL EXERCISES

1.

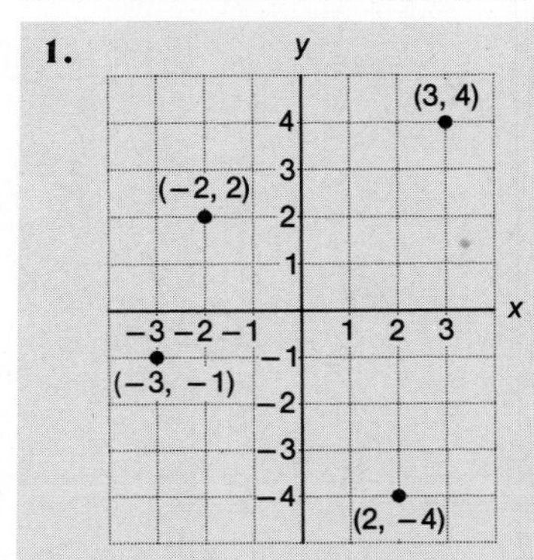

2. The x-coordinates are negative and y-coordinates are positive.

3.

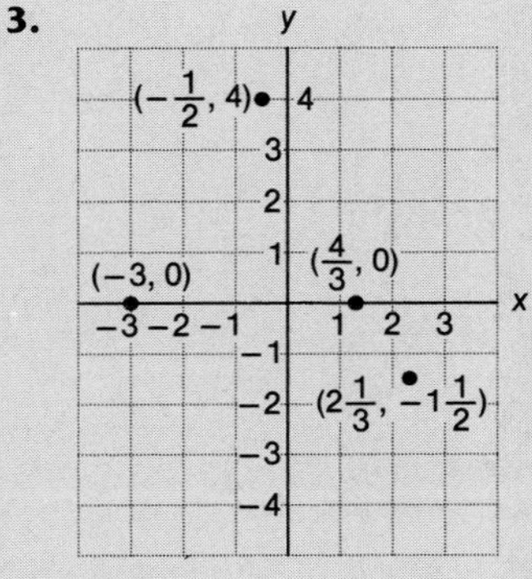

4. $P = (2, 1)$; $R = (-2, 3)$; $Q = (0, -1)$; $S = (-2\frac{1}{3}, -1\frac{2}{3})$ **5.** Quadrant II **6.** The y-coordinates are 0. **7.** Answers will vary. **8.** Since a comes before b in the alphabet, by convention the a-axis will be the horizontal axis.

NAME ____________ COURSE/SECTION ____________ DATE ____________

SECTION PROBLEMS 6.2

Without plotting, tell which quadrant the following points are in.

1. (−2, 3) **2.** (4, 7) **3.** (−2, −5) **4.** (3, −1)

5. (−1, −1) **6.** (−5, 6) **7.** (2, −2) **8.** $(1, \frac{1}{2})$

Plot the following points.

9. (−2, 1) **10.** $(1, 4\frac{1}{2})$ **11.** (−3, −2)

12. (−1, 0) **13.** (2, −5) **14.** $(0, \frac{1}{2})$

15. $(-3, -2\frac{1}{3})$ **16.** (4, −2) **17.** (−4, 1)

18. (0, −3)

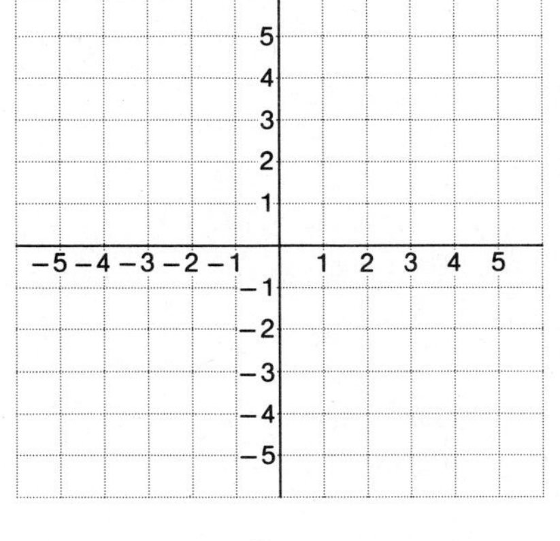

Find the approximate coordinates of the following points.

19. *P* **20.** *Q* **21.** *R*

22. *S* **23.** *T* **24.** *U*

25. *V* **26.** *W* **27.** *Y*

28. *Z* **29.** *A* **30.** *B*

31. *C* **32.** *D* **33.** *E*

34. *F*

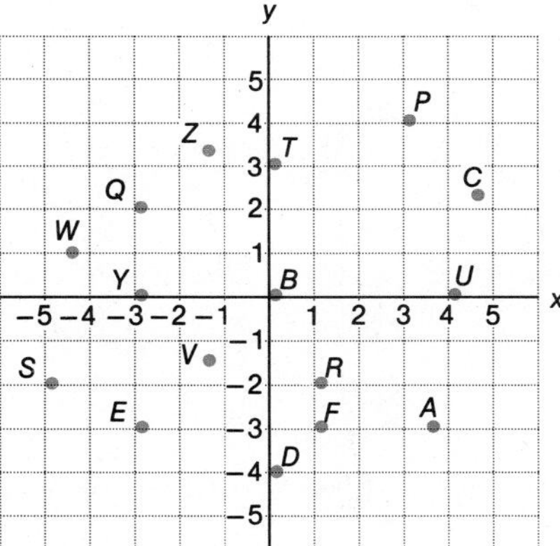

ANSWERS

1. ____________
2. ____________
3. ____________
4. ____________
5. ____________
6. ____________
7. ____________
8. ____________
9. See graph.
10. See graph.
11. See graph.
12. See graph.
13. See graph.
14. See graph.
15. See graph.
16. See graph.
17. See graph.
18. See graph.
19. ____________
20. ____________
21. ____________
22. ____________
23. ____________
24. ____________
25. ____________
26. ____________
27. ____________
28. ____________
29. ____________
30. ____________
31. ____________
32. ____________
33. ____________
34. ____________

35. See graph.

36. See graph.

37. See graph.

38. See graph.

39. ______________

35. Find three solutions to $2t + s = 3$ and plot them on a graph. If we stick to convention, which axis would be the s-axis and which the t-axis?

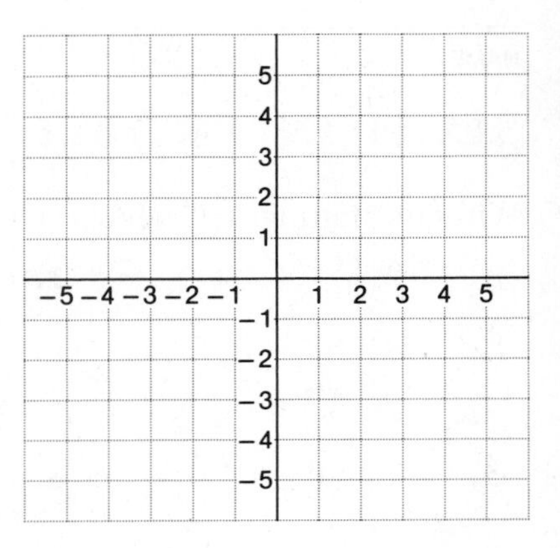

36. Find three solutions to $n = 5m - 4$ and plot them on a graph.

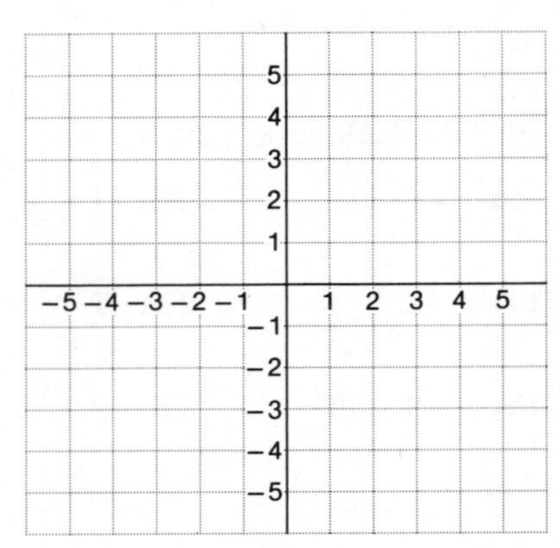

37. Find three solutions to $-3a + 2b = 8$ and plot them on a graph.

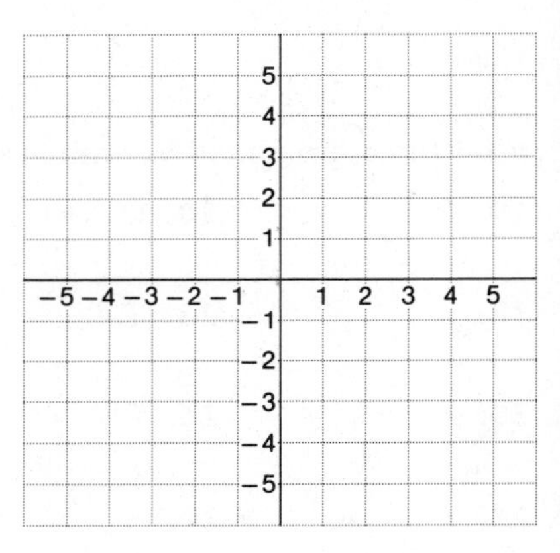

38. Find three solutions to $\frac{Q}{3} - \frac{P}{4} = \frac{1}{2}$ and plot them on a graph.

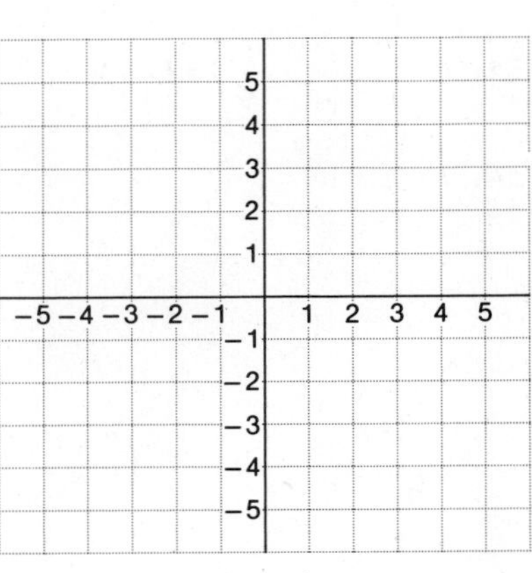

39. What do all the points on the y-axis have in common?

6.3

Graphing Equations

In the last section we mentioned that a linear equation in two variables, such as $y = 2x - 3$, has infinitely many solutions and, hence, it is impossible to list them all. However, we can obtain a geometric picture (graph) of what all the solutions to $y = 2x - 3$ look like. We will often refer to this process as "graphing the equation $y = 2x - 3$," but actually we are graphing all the solutions to the equation $y = 2x - 3$. In this section we will graph some linear equations in two variables and some equations in two variables that are not linear. When we graph an equation in two variables (that is, all the solutions to such an equation), if we have a sense for what the general shape of the graph should be, then it is often sufficient to find ordered pairs (points) that are solutions to the equation, plot these points, and then sketch the whole graph. In this section we will learn to recognize the general shape of the graph for some equations in two variables by working various examples.

LEARNING OBJECTIVES

After finishing Section 6.3, you should be able to:

- Graph certain equations in two variables such as $2x + 3y = 6$, $y = |x|$, and $y = 2^x$.

Example 1 Graph $y = 2x - 3$.

(a) Find some ordered pairs that are solutions to the equation. It is often convenient to use a table to record these ordered pairs.

x	y
0	-3
1	-1
-2	-7
3	3
-1	-5
2	1

If $x = 0$, then substituting in the equation $y = 2x - 3$ we obtain $y = 2(0) - 3 = -3$. Thus, $(0, -3)$ is a solution.

If $x = 1$, then $y = 2(1) - 3 = -1$. Thus, $(1, -1)$ is a solution.

If $x = -2$, then $y = -7$. Thus, $(-2, -7)$ is a solution.

$(3, 3)$ is a solution.

$(-1, -5)$ is a solution.

$(2, 1)$ is a solution.

(b) Plot the points (ordered pairs) we have found that satisfy the equation.

y
x
(3, 3)
(2, 1)
(1, −1)
(0, −3)
(−1, −5)
(−2, −7)

(c) It now is fairly obvious that the graph is going to be a straight line.

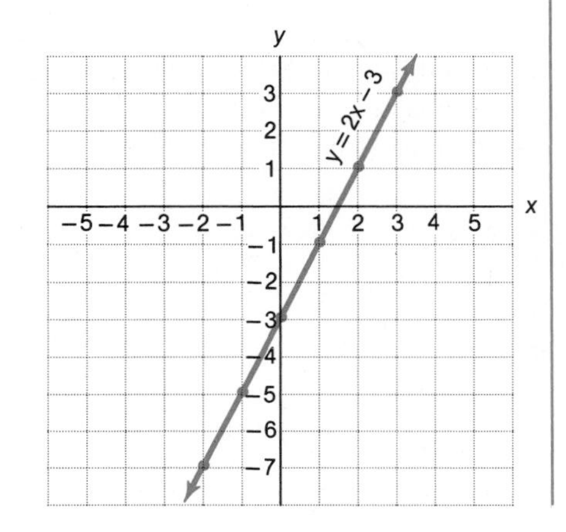

Graph the following equation.

1. $y = 2x - 2$

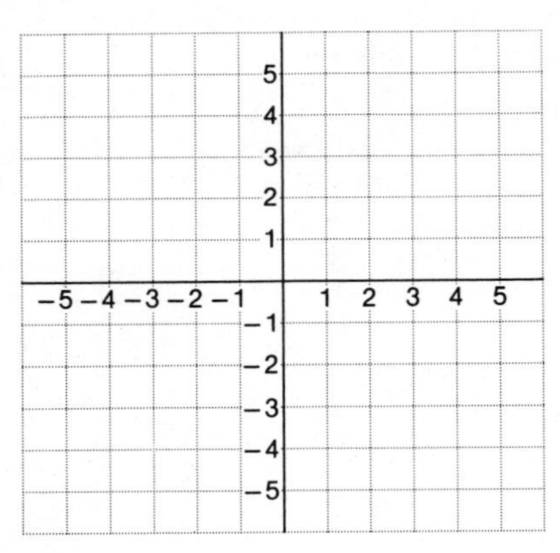

(d) The fact that the graph of all the solutions to the equation $y = 2x - 3$ is a straight line means that if we find a solution to the equation $y = 2x - 3$, it will be a point somewhere on that line. Conversely, if we consider any point on the line and find its x- and y-coordinates, then this ordered pair will satisfy the equation.

DO EXERCISE 1.

Example 2 Graph $2s - 3t = 6$.

(a) Find some ordered pairs that are solutions to the equation. The s-coordinate will be the first-coordinate.

If $s = 0$, then $t = -2$ and $(0, -2)$ is a solution.

If $t = 0$, then $s = 3$ and $(3, 0)$ is a solution.

$(-3, -4)$ is a solution.

$(2, -\frac{2}{3})$ is a solution.

$(-2, -\frac{10}{3})$ is a solution.

s	t
0	-2
3	0
-3	-4
2	$-\frac{2}{3}$
-2	$-\frac{10}{3}$

(b) Plot the points.

(c) Once again the graph is a straight line.

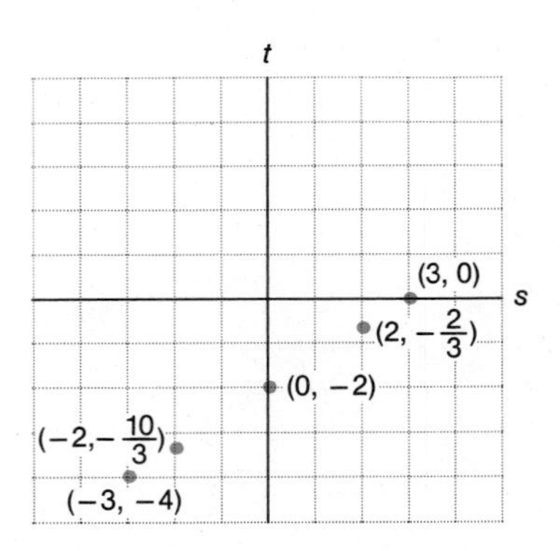

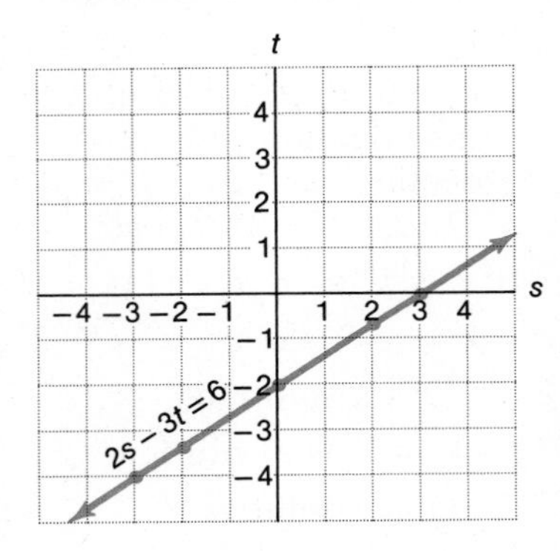

Graph the following equation.

2. $s - 2t = 4$

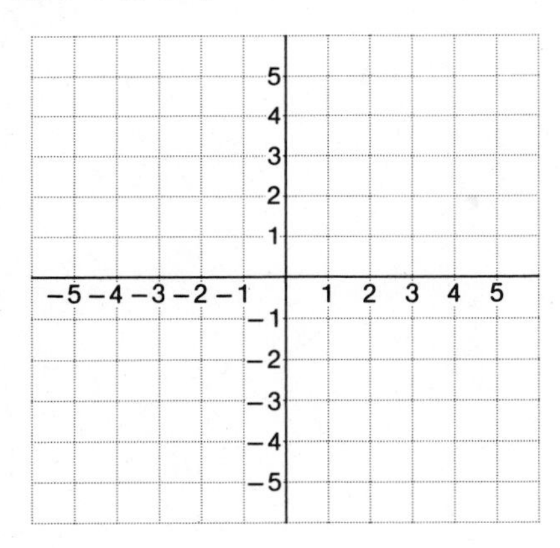

DO EXERCISE 2.

Note that in Examples 1 and 2 we were graphing linear equations in two variables and the graph was a straight line. This is no coincidence: the graph of a linear equation in two variables will always be a straight line. We will study straight lines in more detail in Sections 6.4 to 6.7.

Example 3 Graph $y = |x|$.

This is not a linear equation and its graph will not be a straight line.

(a) Find ordered pairs that satisfy the equation.
If $x = 0$, then $y = 0$.
If $x = 1$, then $y = 1$.
If $x = 2$, then $y = 2$.
If $x = 3$, then $y = 3$.
If $x = -1$, then $y = 1$.
If $x = -2$, then $y = 2$.
If $x = -3$, then $y = 3$. This may be enough points to tell what the shape of the graph will be.

x	y
0	0
1	1
2	2
3	3
−1	1
−2	2
−3	3

(b) Plot the ordered pairs.

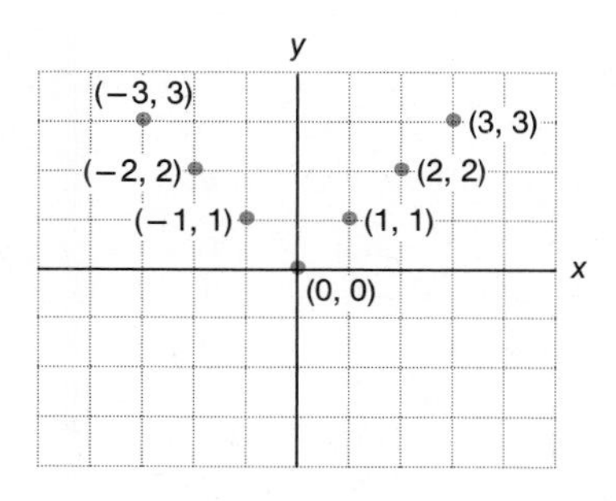

(c) It appears reasonable (and in fact it is true) that the graph will be shaped like a V.

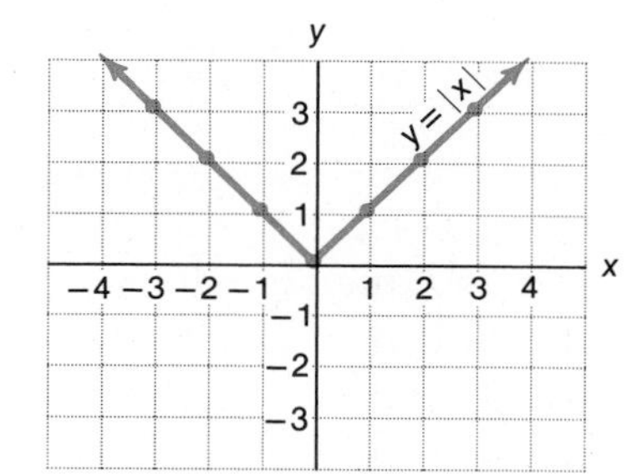

DO EXERCISE 3.

Example 4 Graph $b = 2|a - 1|$.

The first-coordinate will be the a-coordinate.

(a) Find ordered pairs that satisfy the equation.
If $a = 0$, then
$b = 2|0 - 1| = 2|-1| = 2$.
If $a = 1$, then $b = 2|1 - 1| = 0$.
If $a = 2$, then
$b = 2|2 - 1| = 2$.
If $a = 3$, then $b = 2|3 - 1| = 4$.
If $a = 4$, then $b = 2|4 - 1| = 6$.
If $a = -1$, then
$b = 2|-1 - 1| = 2|-2| = (2)(2) = 4$.
If $a = -2$, then
$b = 2|-2 - 1| = (2)(3) = 6$.

a	b
0	2
1	0
2	2
3	4
4	6
−1	4
−2	6

Graph the following equation.

3. $y = |x| - 1$

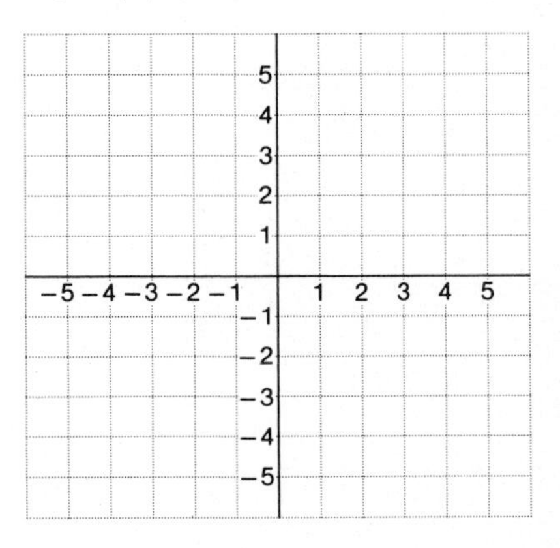

Graph the following equation.

4. $t = |s + 2|$

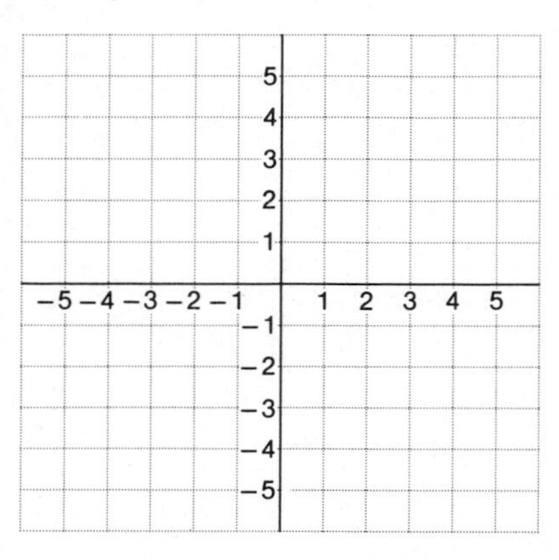

Graph the following equation.

5. $y = x^2 - 3$

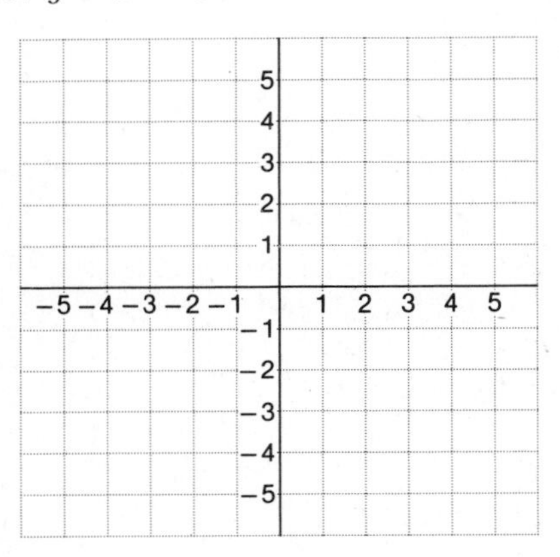

(b) Plot the ordered pairs.

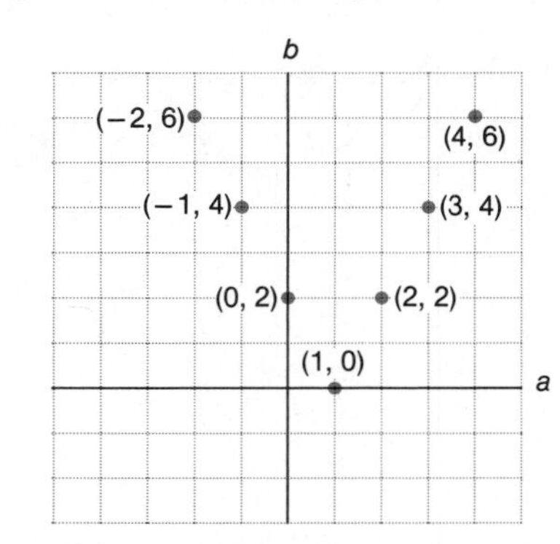

(c) The graph will be shaped like a V.

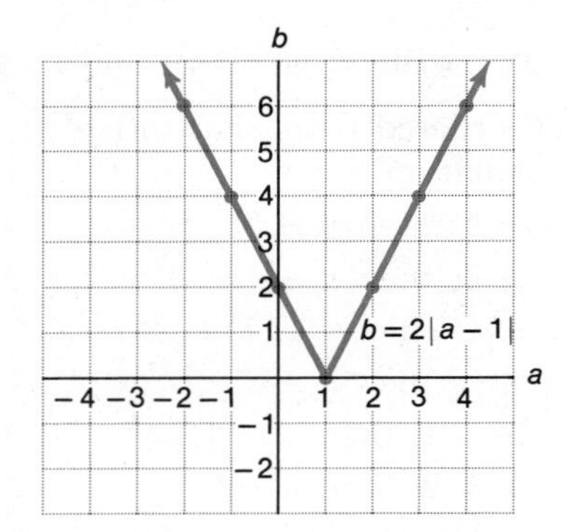

An equation in two variables involving an absolute value will often be shaped like a V.

DO EXERCISE 4.

Example 5 Graph $y = x^2$.

(a) Find ordered pairs that satisfy the equation.
If $x = 0$, then $y = 0$.
If $x = 1$, then $y = 1$.
If $x = 2$, then $y = 4$.
If $x = 3$, then $y = 9$.
If $x = -1$, then $y = 1$.
If $x = -2$, then $y = 4$.
If $x = -3$, then $y = 9$.

x	y
0	0
1	1
2	4
3	9
−1	1
−2	4
−3	9

(b) Plot the ordered pairs.

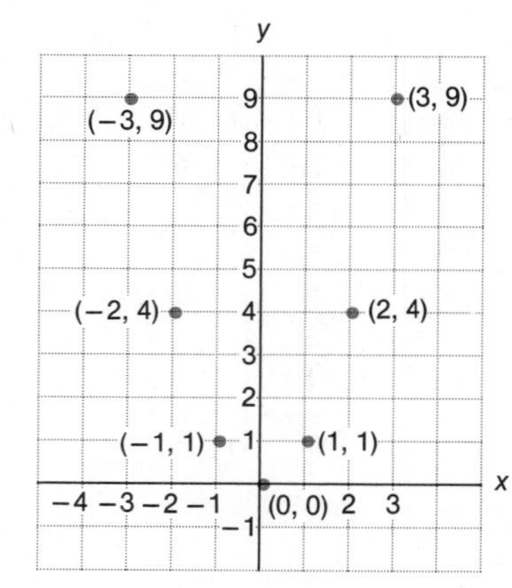

(c) The graph will be

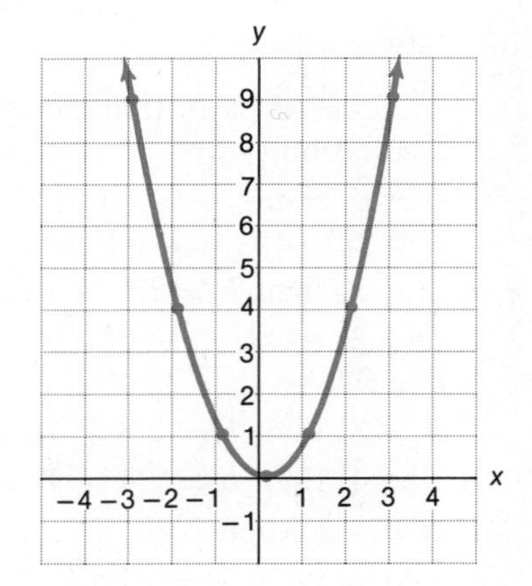

The graph in part (c) of Example 5 is called a *parabola*.

DO EXERCISE 5.

Example 6 Graph $t = (s + 1)^2$.

The first-coordinate will be the s-coordinate.

(a) Find ordered pairs that satisfy the equation.
If $s = 0$, then $t = 1$.
If $s = 1$, then
$t = (1 + 1)^2 = 4$.
If $s = 2$, then
$t = (2 + 1)^2 = 9$.
If $s = -1$, then
$t = (-1 + 1)^2 = 0$.
If $s = -2$, then
$t = (-2 + 1)^2 = (-1)^2 = 1$.
If $s = -3$, then
$t = (-3 + 1)^2 = (-2)^2 = 4$.
If $s = -4$, then
$t = (-4 + 1)^2 = 9$.

s	t
0	1
1	4
2	9
−1	0
−2	1
−3	4
−4	9

(b) Plot the ordered pairs.

(c) The graph will be

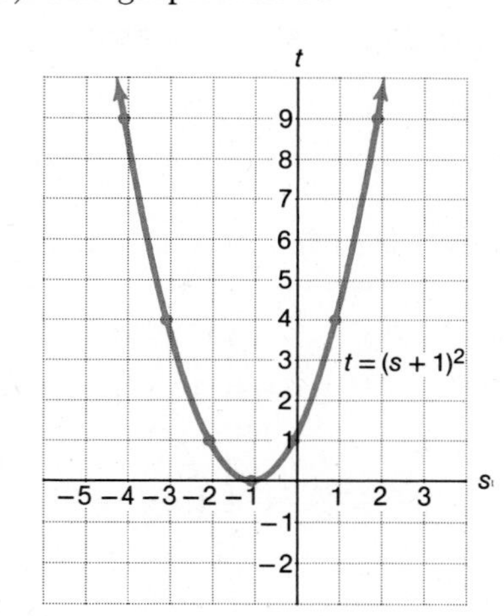

DO EXERCISE 6.

Example 7 Graph $y = 2^x$.

(a) Find ordered pairs that satisfy the equation.
If $x = 0$, then $y = 2^0 = 1$.
If $x = 1$, then $y = 2^1 = 2$.
If $x = 2$, then $y = 2^2 = 4$.
If $x = 3$, then $y = 2^3 = 8$.
If $x = -1$, then $y = 2^{-1} = \frac{1}{2}$.
If $x = -2$, then $y = 2^{-2} = \frac{1}{4}$.
If $x = -3$, then $y = 2^{-3} = \frac{1}{8}$.

x	y
0	1
1	2
2	4
3	8
−1	$\frac{1}{2}$
−2	$\frac{1}{4}$
−3	$\frac{1}{8}$

Graph the following equation.

6. $b = (a - 1)^2$

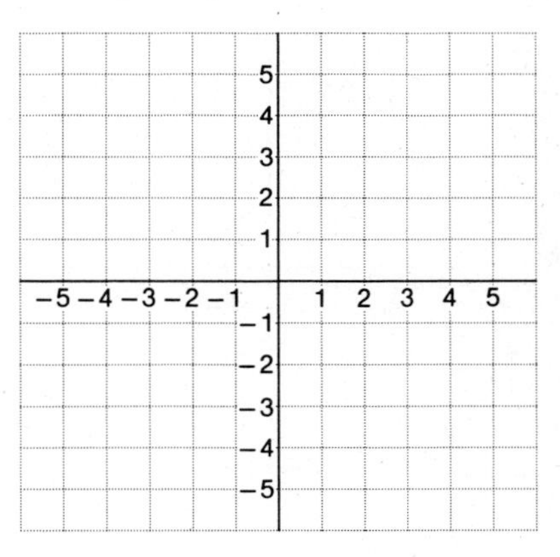

Graph the following equation.

7. $y = 3^x$

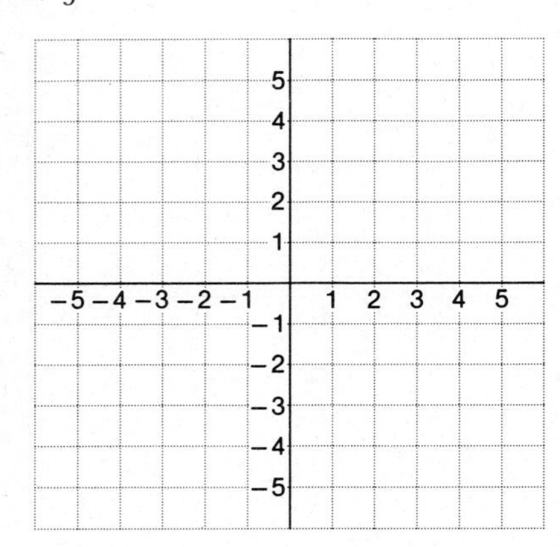

(b) Plot the ordered pairs.

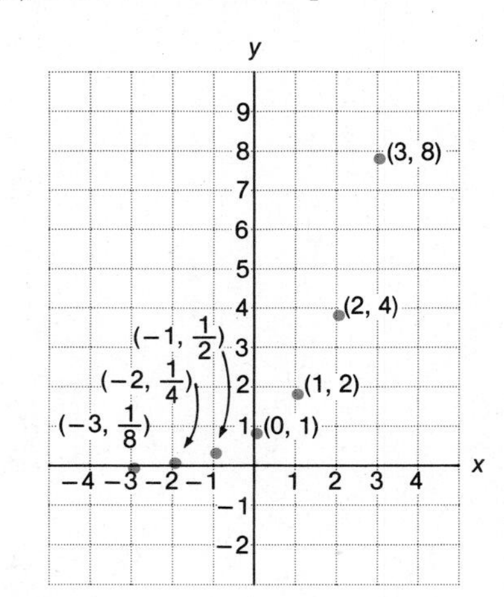

(c) The graph will be

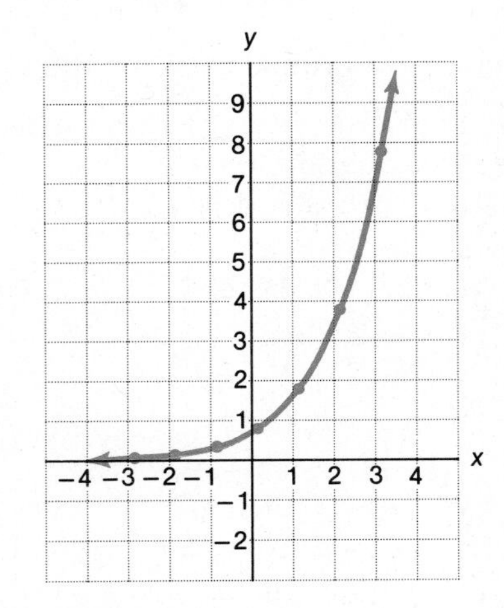

The equation $y = 2^x$ is called an *exponential equation.*

DO EXERCISE 7.

DO SECTION PROBLEMS 6.3.

ANSWERS TO MARGINAL EXERCISES

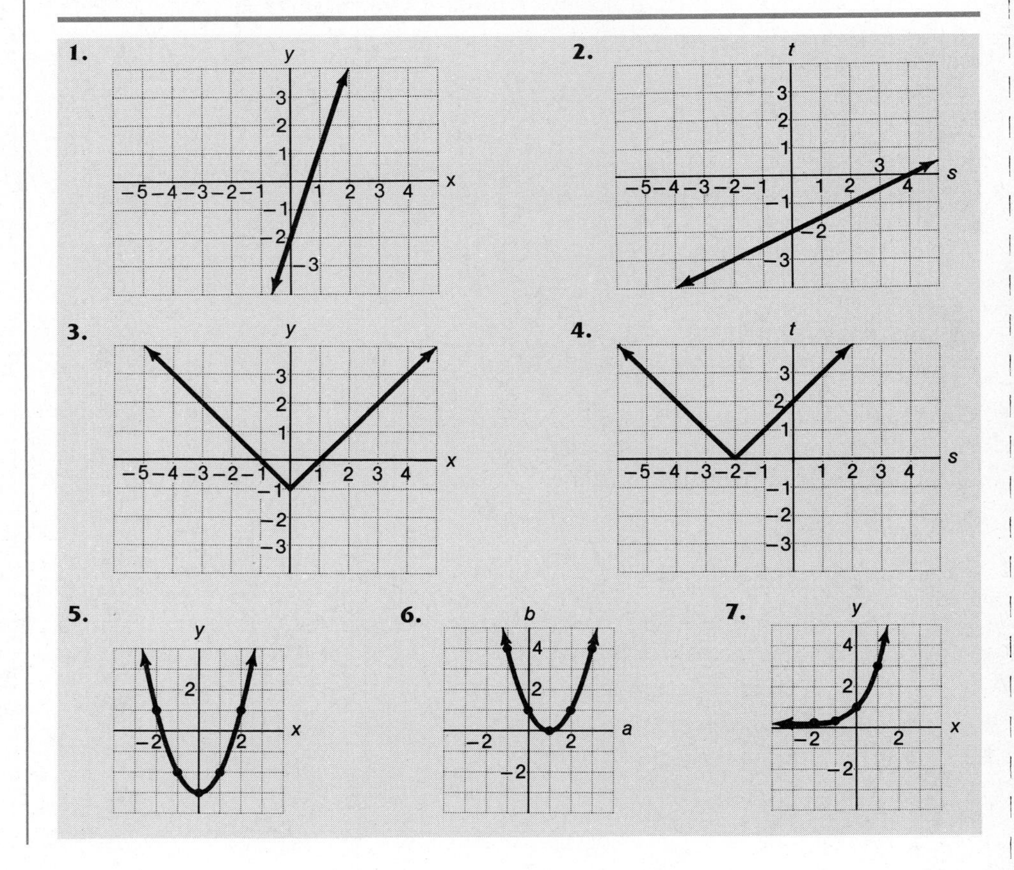

NAME COURSE/SECTION DATE

SECTION PROBLEMS 6.3

Graph the following equations.

1. $y = 3x - 2$

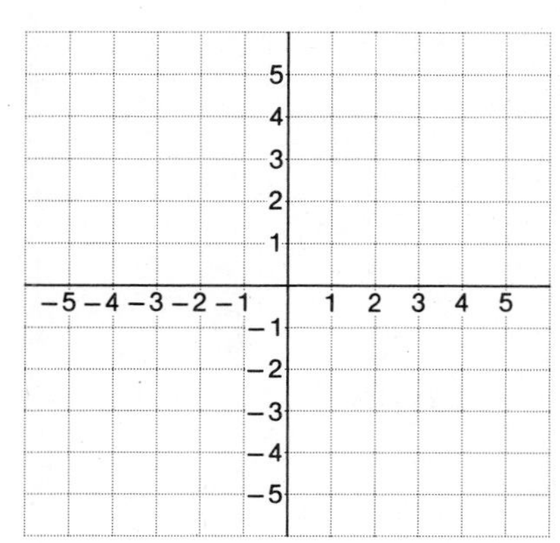

2. $a + 2b = 4$

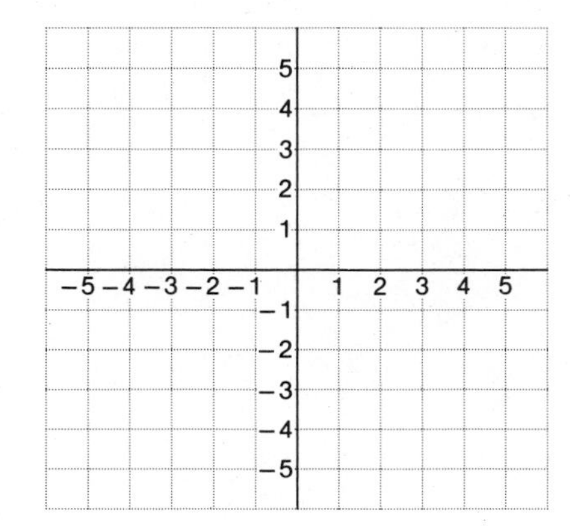

3. $2s - 3t = 6$

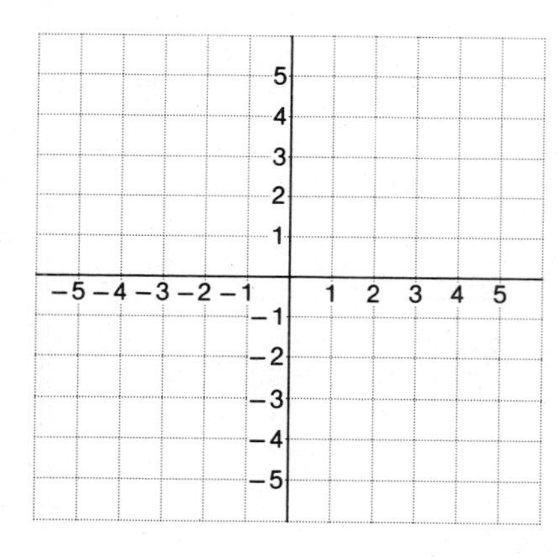

4. $-2x + 3y - 6 = 0$

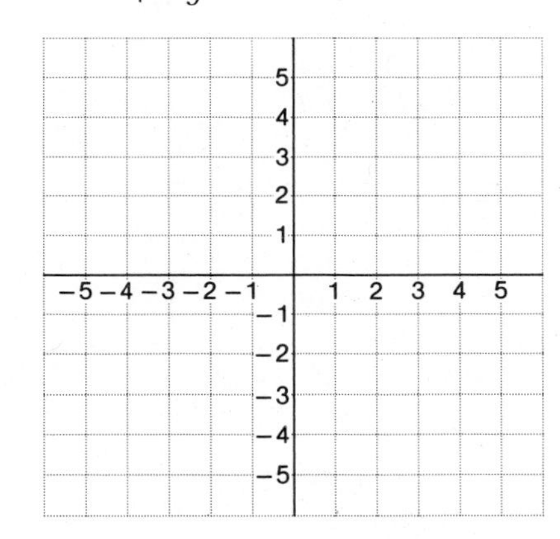

5. $y = |x| - 2$

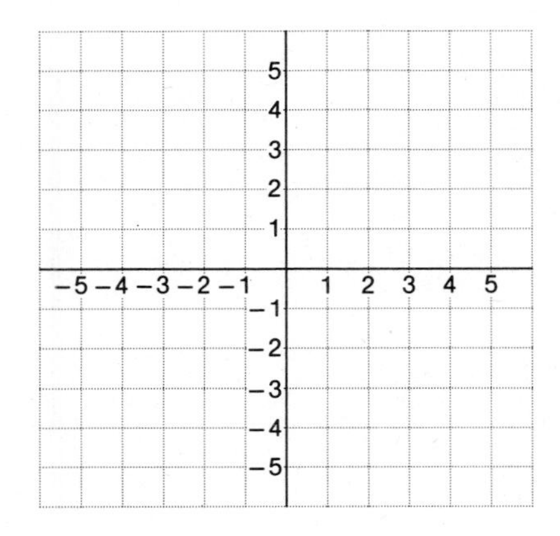

6. $n = |m - 3|$

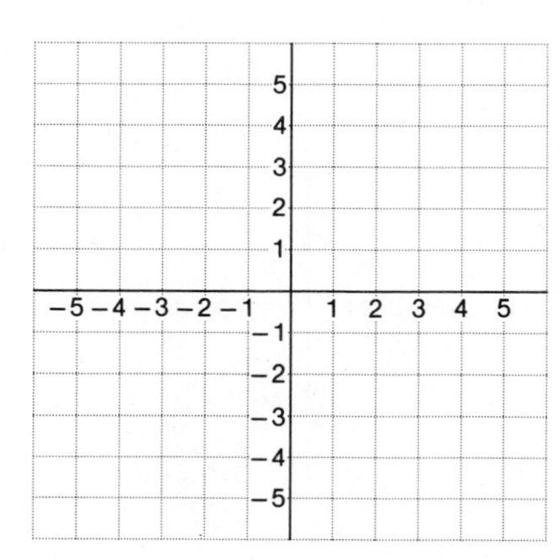

ANSWERS

1. See graph.
2. See graph.
3. See graph.
4. See graph.
5. See graph.
6. See graph.

7. See graph.

8. See graph.

9. See graph.

10. See graph.

11. See graph.

12. See graph.

7. $b = -2|a|$

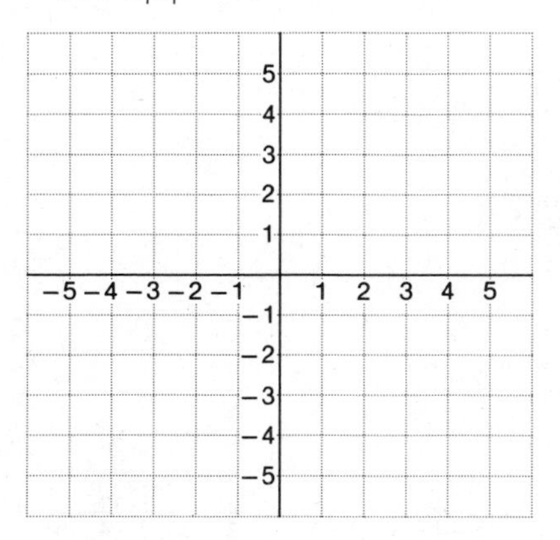

8. $n = |m| + 1$

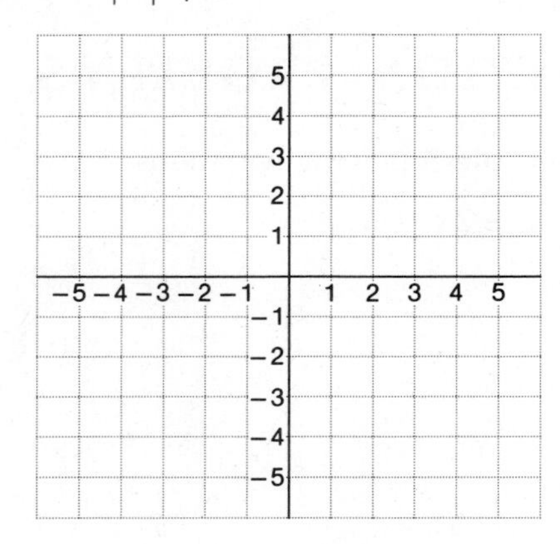

9. $y = x^2 + 1$

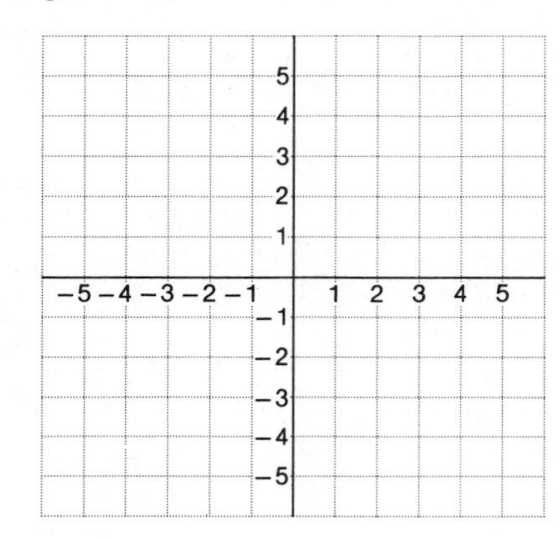

10. $y = (x - 1)^2$

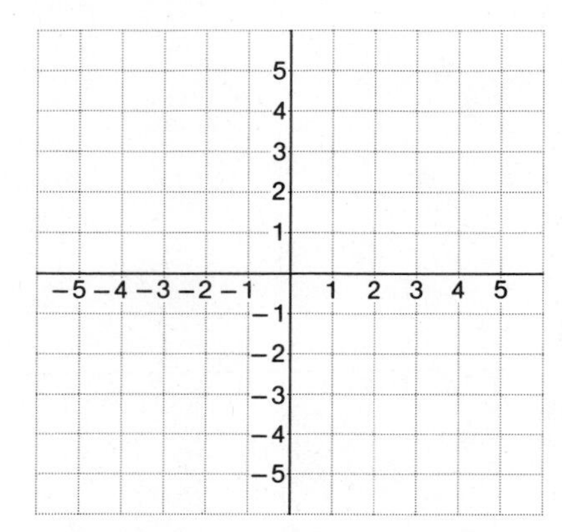

11. $b = a^2 - 2$

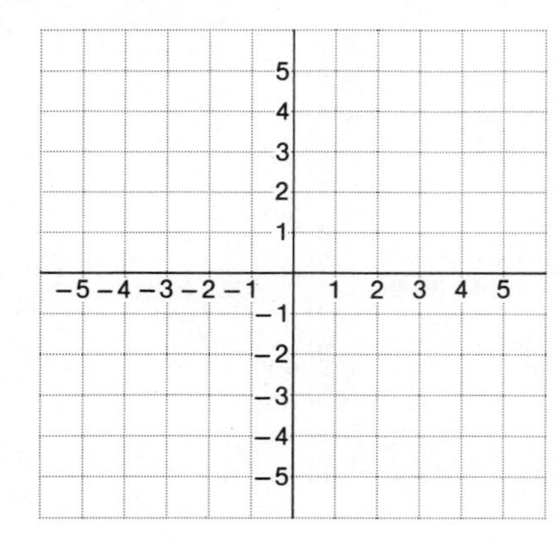

12. $t = -s^2$

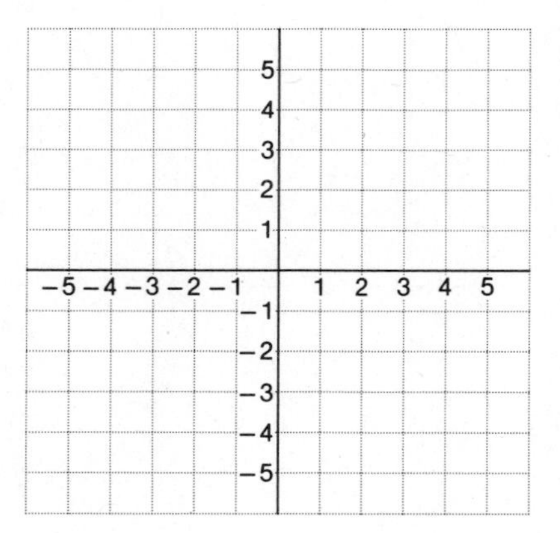

NAME COURSE/SECTION DATE

ANSWERS

13. See graph.

14. See graph.

15. See graph.

16. See graph.

17. See graph.

18. See graph.

13. $y = 2^x$

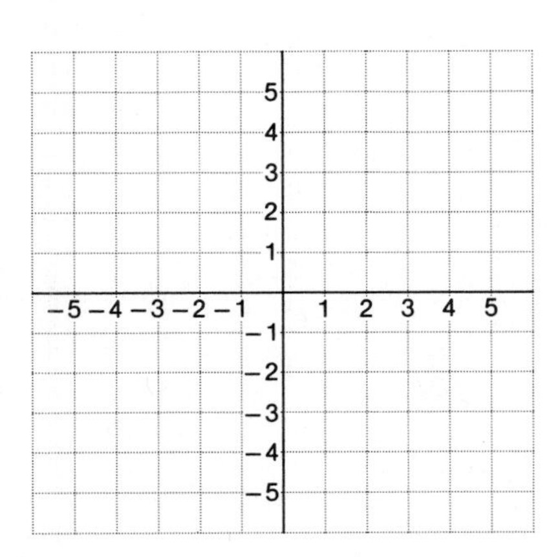

14. $y = 4^x$

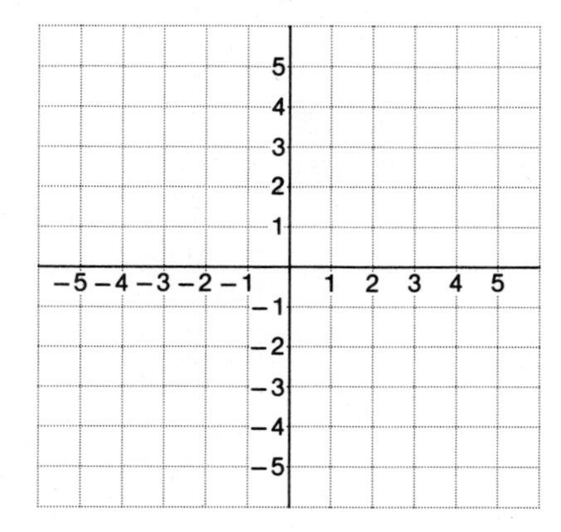

15. $t = \left(\frac{3}{2}\right)^s$

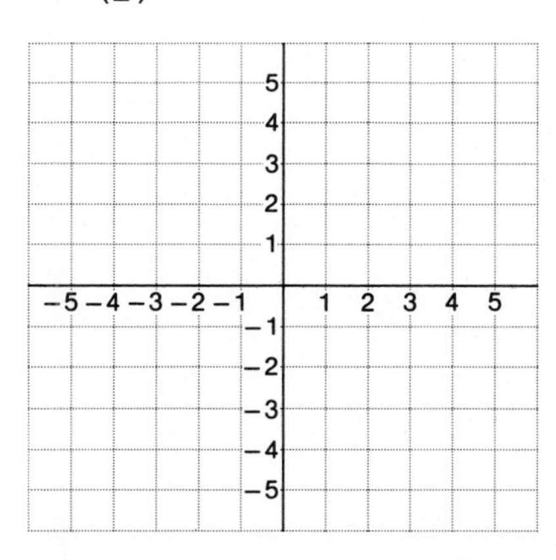

16. $b = \left(\frac{1}{2}\right)^a$

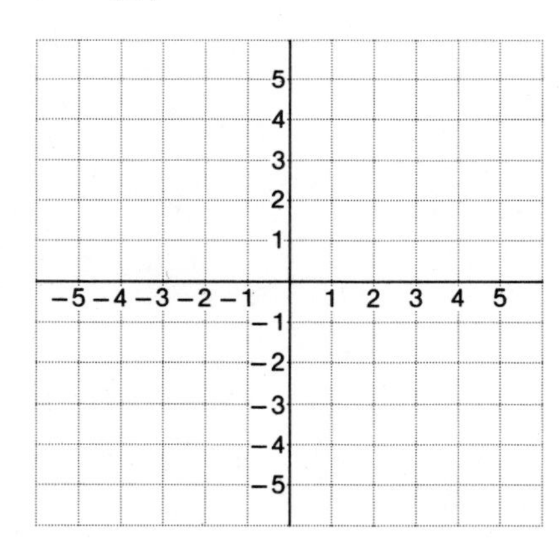

17. $\frac{x}{3} - \frac{y}{2} = 1$

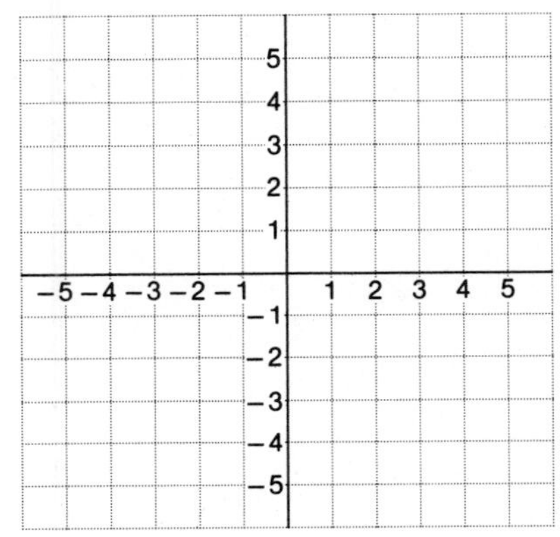

18. $Q - 3P = -2$

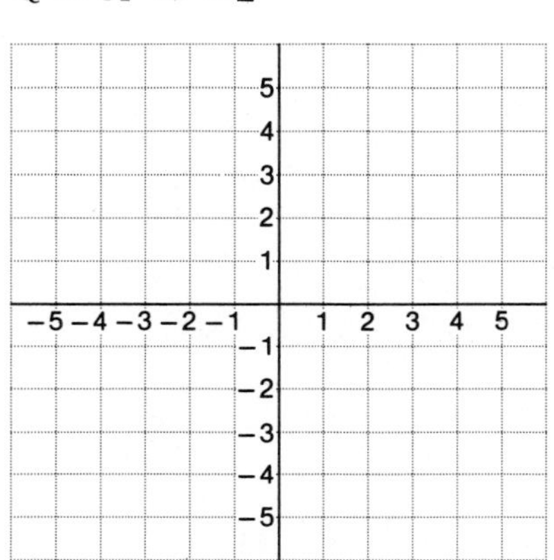

19. See graph.

20. See graph.

19. $y = \left|\frac{x}{2}\right| - 1$

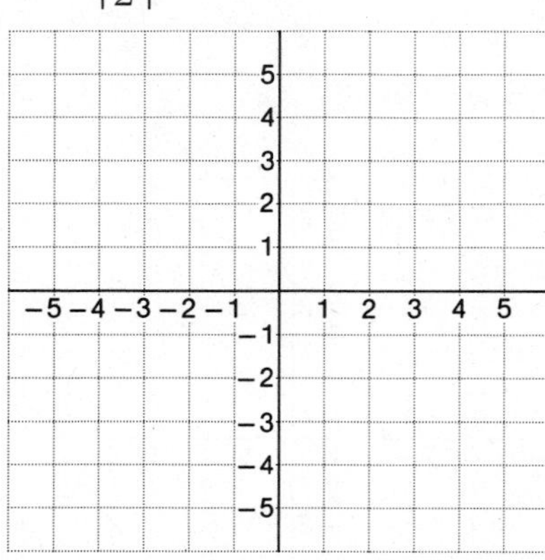

20. $t = (s - 1)^2 + 1$

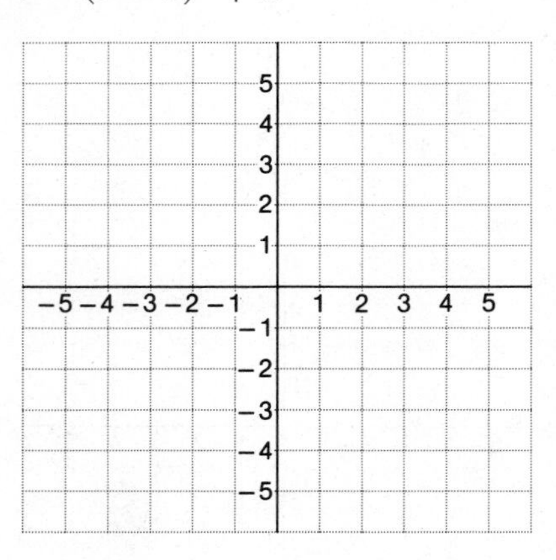

6.4

Graphing Straight Lines

In the last section we noticed that the graph of the solutions to a linear equation in two variables was always a straight line. This is no coincidence. In fact, every time we graph $ax + by = c$ we obtain a straight line. Conversely, if we have a graph that is a straight line, then there are constants a, b, and c such that the equation of the straight line is $ax + by = c$.

DEFINITION 6.3 The *standard form* for the equation of a straight line is $ax + by = c$ (where a, b, and c are arbitrary constants, except that a and b are not both zero). This definition is exactly the same as the one we used for the standard form of a linear equation in two variables.

Example 1 Which of the following are equations of straight lines?

(a) $2x - y = 4$	Yes; it also is in standard form.
(b) $\lvert s\rvert + t = 0$	No; it contains an absolute value term.
(c) $xy = 1$	No; there is an x times y term.
(d) $s = 3t$	Yes. This equation is not in standard form, however.
(e) $2a - 3b + 4 = 0$	Yes. This equation is not in standard form, however.
(f) $-x + y^2 - 1 = 0$	No; it contains the y^2 term.

DO EXERCISE 1.

In the last section, we learned that to graph an equation we must find and plot ordered pairs (points) that satisfy the equation until we have enough points plotted to tell what the entire graph looks like. If we know beforehand that the graph of our equation is going to be a straight line, then we can find and plot just two ordered pairs (if you are cautious you may want to find three ordered pairs) since a straight line is completely determined by any two points on it.

When we graph a straight line, the two (or three) points we choose to plot are completely arbitrary. However, there are two points that quite often are very easy to find, especially when the equation for the straight line is in standard form.

LEARNING OBJECTIVES

After finishing Section 6.4, you should be able to:

- Find the x-intercept and y-intercept of a straight line.
- Graph straight lines.

1. Which of the following are equations of straight lines? If not, why not?
(a) $y = x^2 - 1$
(b) $t = 3s + 4$
(c) $b = \lvert a\rvert - 3$
(d) $-2x + 3y - 1 = 0$
(e) $x + y - z = 4$

2. Find the x-intercept and y-intercept of $-2x + 3y = 12$ and graph the straight line.

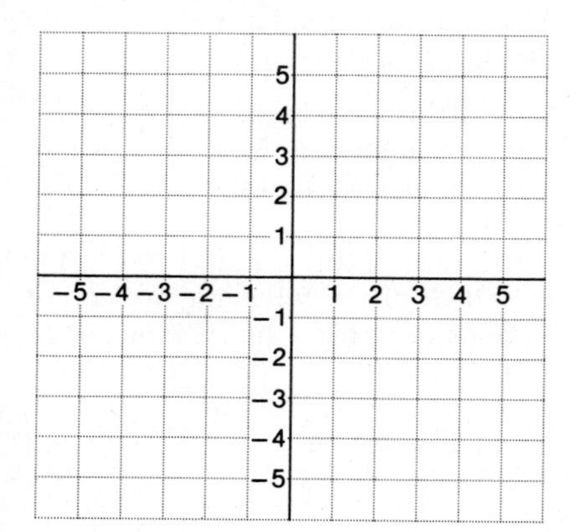

DEFINITION 6.4 The *y-intercept* (or vertical-axis intercept) is the point where the graph of the line crosses the y-axis. The x-coordinate of the y-intercept must be 0.

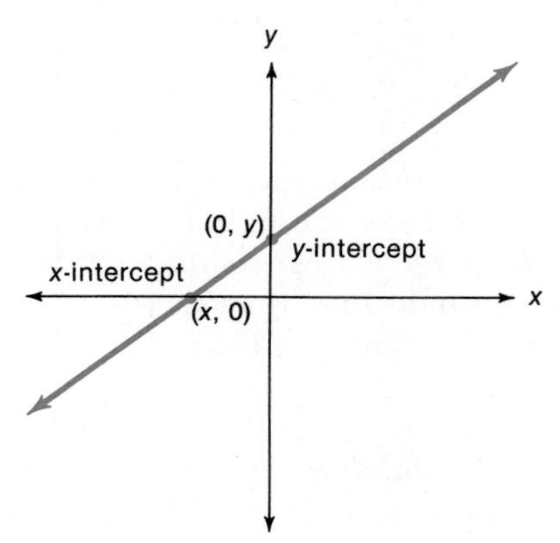

DEFINITION 6.5 The *x-intercept* (or horizontal-axis intercept) is the point where the graph of the line crosses the x-axis. The y-coordinate of the x-intercept must be 0.

Example 2 Graph the straight line $2x + 3y = 6$.

This form of the equation of a straight line lends itself nicely to finding the y-intercept and the x-intercept.

(a) To find the y-intercept, set $x = 0$ and solve for y.

$$\begin{aligned}(2)(0) + 3y &= 6\\ 3y &= 6\\ y &= 2\end{aligned}$$

Thus, (0, 2) is the y-intercept.

(b) To find the x-intercept, set $y = 0$ and solve for x.

$$\begin{aligned}2x + 3(0) &= 6\\ 2x &= 6\\ x &= 3\end{aligned}$$

Thus, (3, 0) is the x-intercept.

(c) Plot the points and draw the graph.

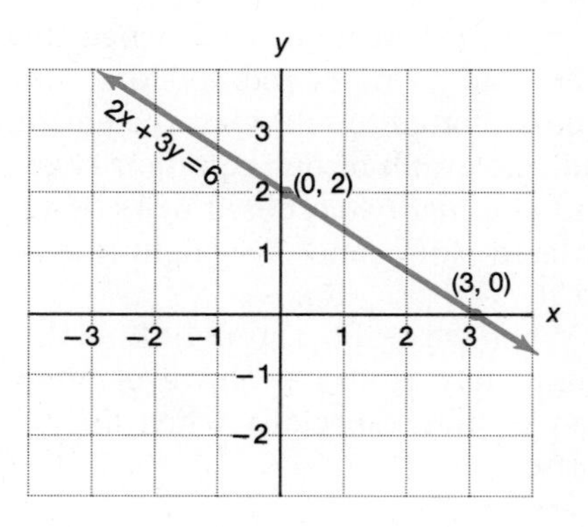

DO EXERCISE 2.

Example 3 Graph the straight line $y = 3x - 2$.

When the equation is in this form, instead of finding the x- and y-intercepts it is easier to pick two values for x and find the corresponding values for y.

(a) If $x = 0$, then $y = -2$. Thus, $(0, -2)$ is a point (actually the y-intercept) on the graph.

(b) If $x = 2$, then $y = 4$. Thus, $(2, 4)$ is on the graph.

(c) Plot the points and draw the graph.

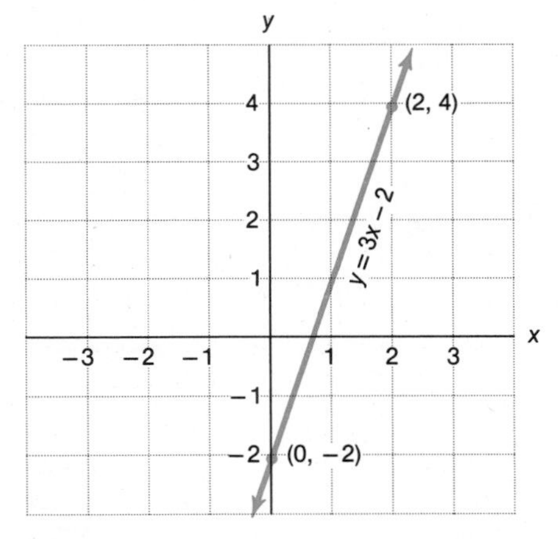

Note that to find the x-intercept of this straight line we set $y = 0$ and solve for x:

$$0 = 3x - 2$$
$$2 = 3x$$
$$\frac{2}{3} = x$$

DO EXERCISE 3.

Example 4 Graph the straight line $y = 4$.

(a) Note that the x variable appears to be missing in the equation of this straight line. Actually the x variable is there, but its coefficient is 0. That is, $y = 4$ is the same equation as $0 \cdot x + y = 4$. If the coefficient of x is 0, we normally do not bother to put the x variable in the equation. What this means for ordered pairs satisfying the equation is that the x-coordinate can be anything, but the y-coordinate must always be 4.

(b) The graph of $y = 4$ is, thus, the horizontal line that is 4 units above the x-axis.

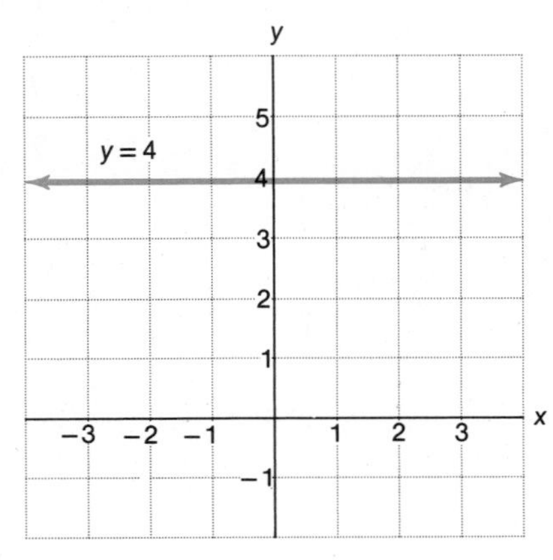

3. Graph $t = \frac{s}{3} + 2$. What is the s-intercept and the t-intercept?

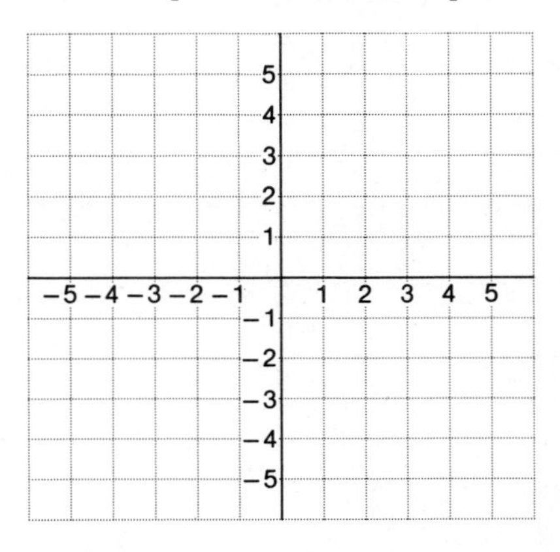

4. Graph $y = -3$.

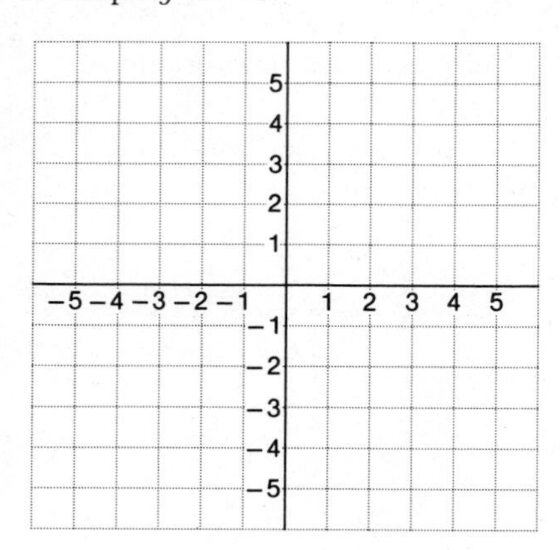

RULE 6.1 The graph of the straight line $y = c$, where c is a constant, is always a horizontal line. Conversely, the equation of any horizontal line has the form $y = c$.

EXAMPLES:

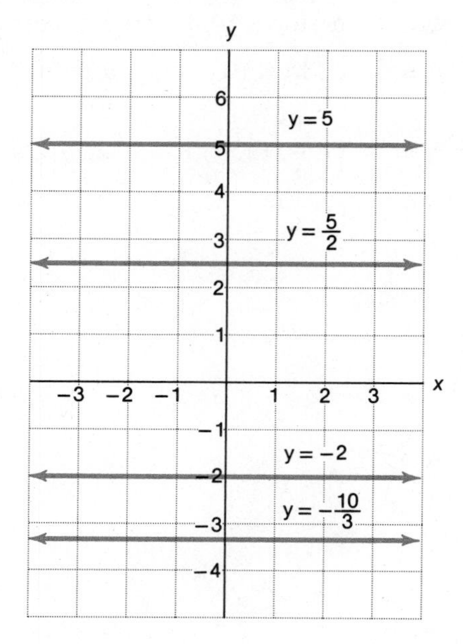

DO EXERCISE 4.

Example 5 Graph the straight line $3x + 4 = -2$.

(a) If we solve $3x + 4 = -2$ for x, we obtain the equivalent equation $3x = -6$ or $x = -2$.

(b) Note that the equation $x = -2$ is the same as the equation $x + 0 \cdot y = -2$. Thus, the y-coordinate can be anything, but the x-coordinate must always be -2.

(c) The graph of $x = -2$ is a vertical line two units to the left of the y-axis.

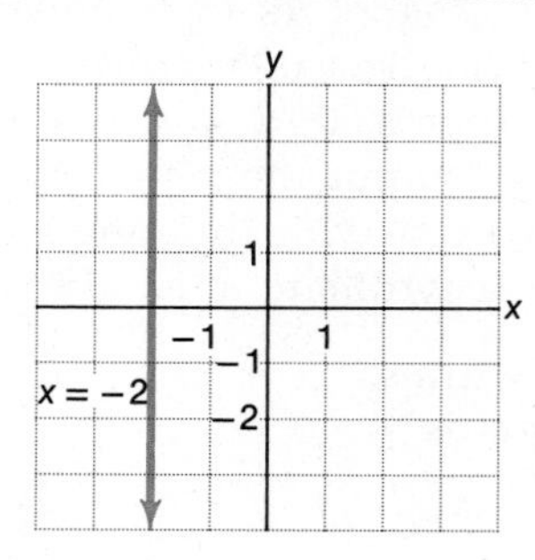

RULE 6.2 The graph of the straight line $x = c$, where c is a constant, is always a vertical line. Conversely, any vertical line has an equation of the form $x = c$.

EXAMPLES:

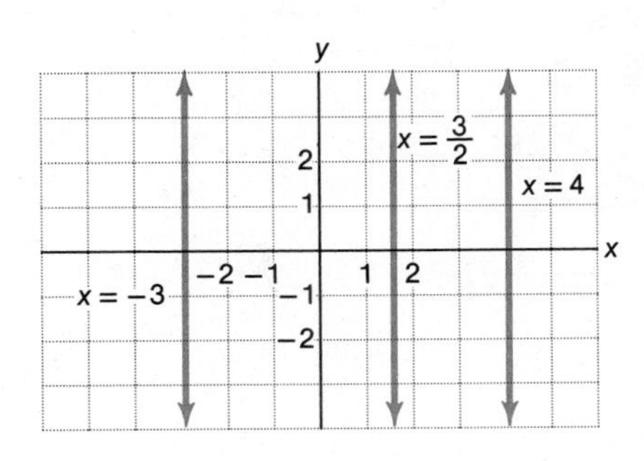

DO EXERCISES 5 AND 6.

DO SECTION PROBLEMS 6.4.

5. Graph $2x - 5 = 0$.

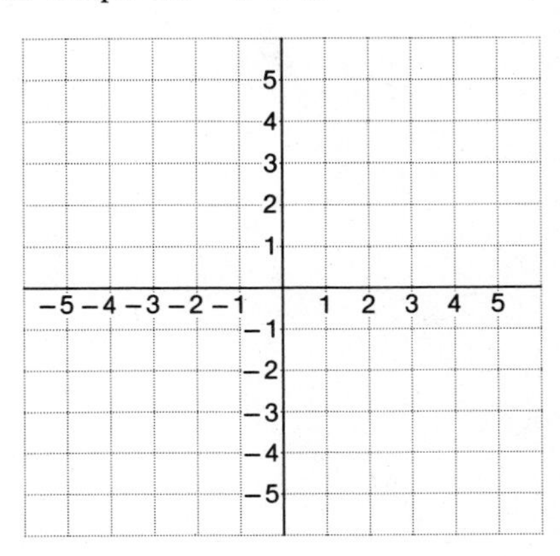

6. Graph $s - 3t = 6$.

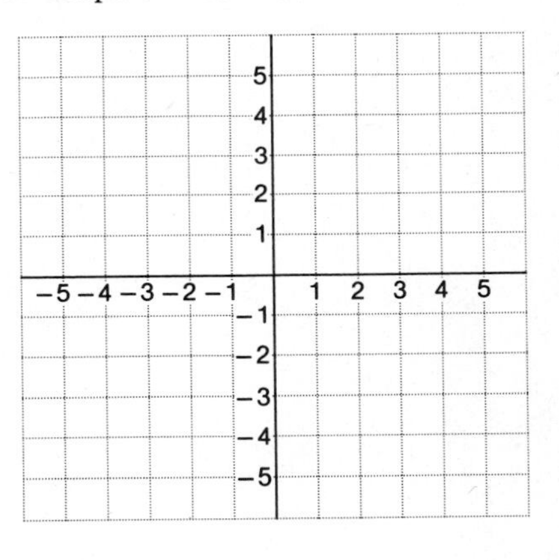

ANSWERS TO MARGINAL EXERCISES

1. (a) No; there is an exponent other than 1 (b) Yes (c) No; there is an absolute value term (d) Yes (e) No; there are three variables

2.

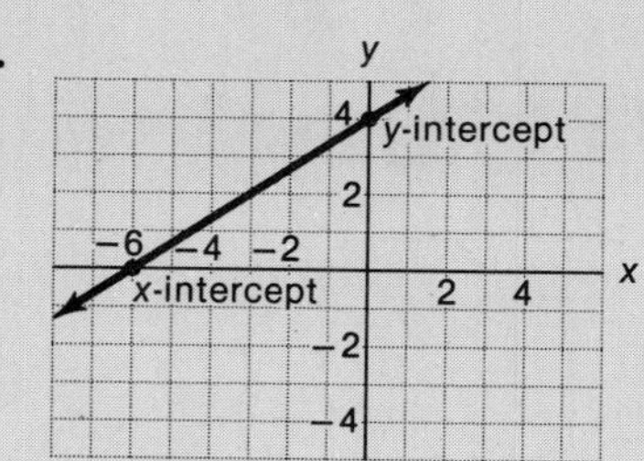

3.

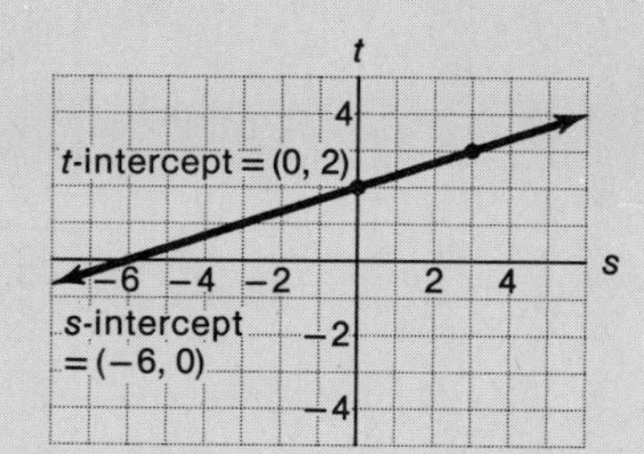

4.

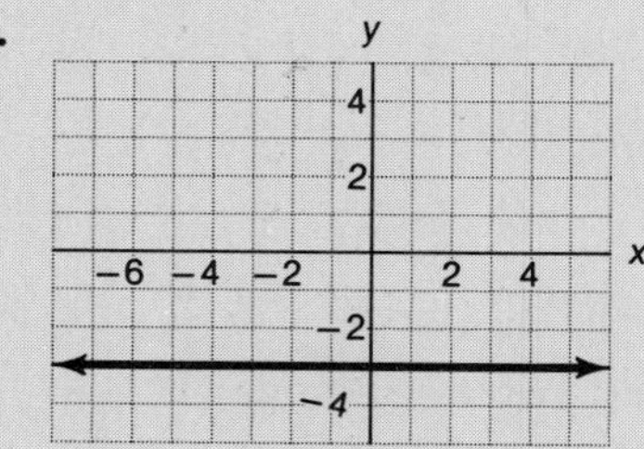

5.

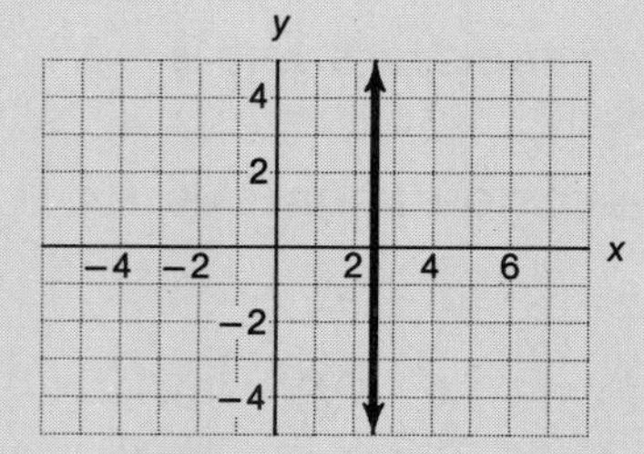

6.

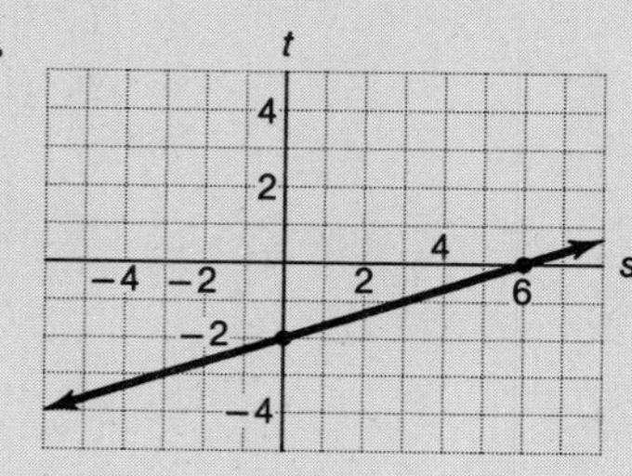

NAME ______ COURSE/SECTION ______ DATE ______

SECTION PROBLEMS 6.4

Are the following equations of straight lines? If not, why not?

1. $2b = -3a + 1$

2. $y = |x - 2|$

3. $x + y = t$

4. $t = s^2 - 1$

5. $\frac{x}{2} - \frac{y}{3} = 4$

6. $3(s - t) = 2st$

Put the equation of the straight line in standard form.

7. $3x - 4y + 7 = 0$

8. $t = -\frac{s}{2} + 5$

9. $x - 1.7 = 0$

10. $a - 6 = 2b$

11. $3m - n = 4$

12. $2(s - 3t) - s = t + 1$

13. Find the x- and y-intercepts for $3x - 4y = 8$.

14. Find the x- and y-intercepts for $\frac{x}{2} - \frac{y}{3} = 4$.

15. Find the s- and t-intercepts for $s - 3t = 6$.

16. Find the s- and t-intercepts for $-4s + t = 6$.

ANSWERS

1. ______
2. ______
3. ______
4. ______
5. ______
6. ______
7. ______
8. ______
9. ______
10. ______
11. ______
12. ______
13. ______
14. ______
15. ______
16. ______

17. ____________

18. ____________

19. ____________

20. ____________

21. See graph.

22. See graph.

23. See graph.

24. See graph.

17. Find the a- and b-intercepts for $-2a + 7b = 10$.

18. Find the x- and y-intercepts for $x = 3$.

19. Find the x- and y-intercepts for $y - 4 = 0$.

20. Find the s- and t-intercepts for $t = 3s - 4$.

Graph the following.

21. $2x - 3y = 6$

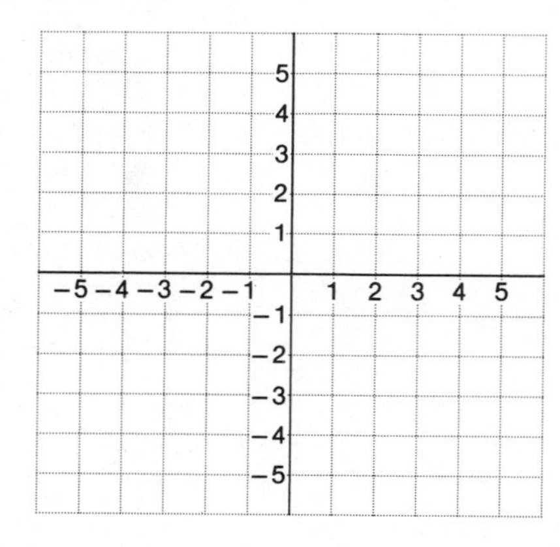

22. $-3a - 4b = 12$

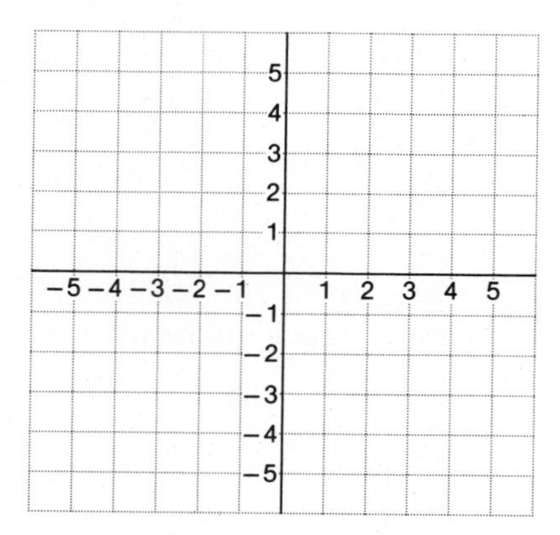

23. $x - 1.5 = 0$

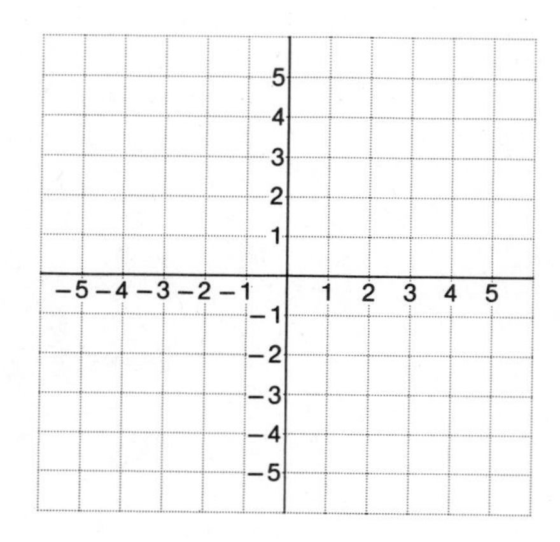

24. $y = -4$

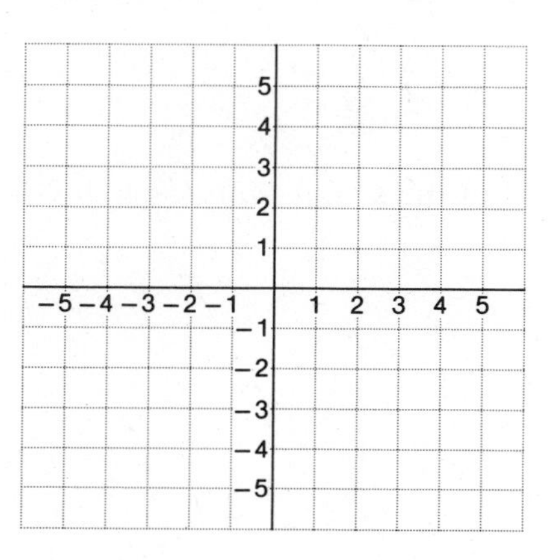

NAME

COURSE/SECTION

DATE

ANSWERS

25. See graph.

26. See graph.

27. See graph.

28. See graph.

29. See graph.

30. See graph.

25. $y = -4x + 2$

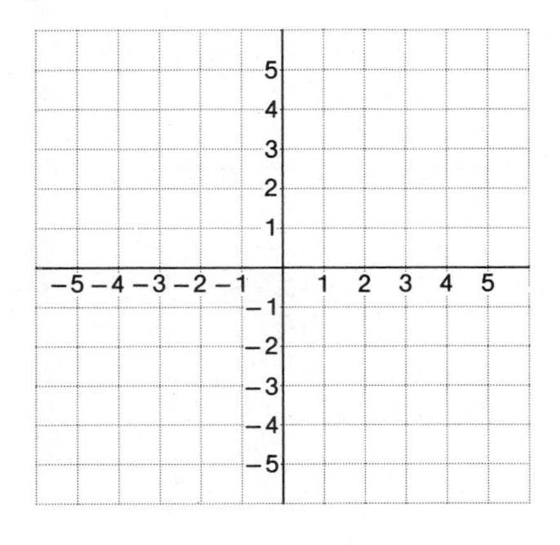

26. $2s - t = 5$

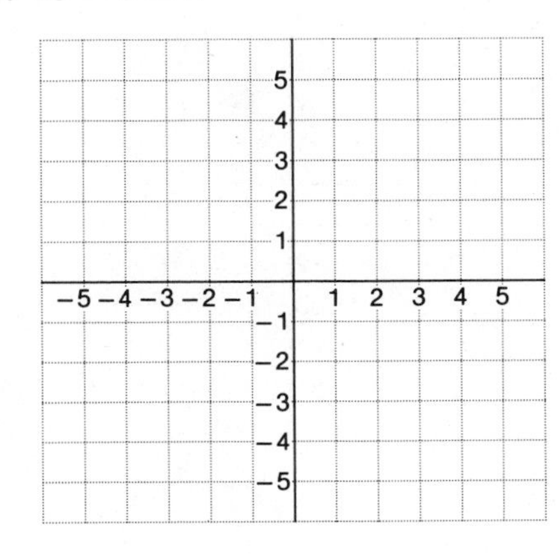

27. $t = \frac{-3s}{2} + 1$

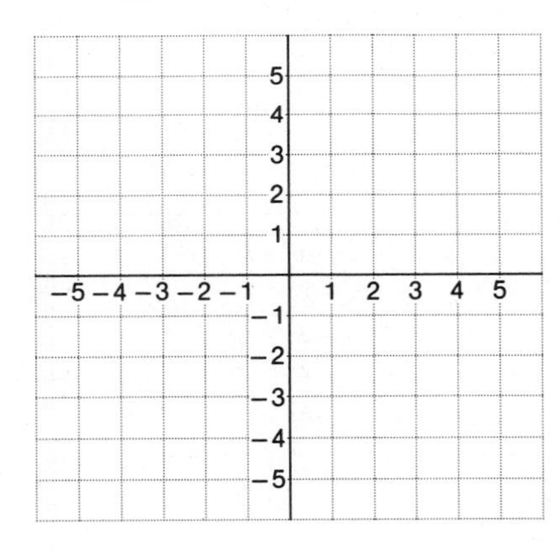

28. $y = \frac{3}{2}x - 1$

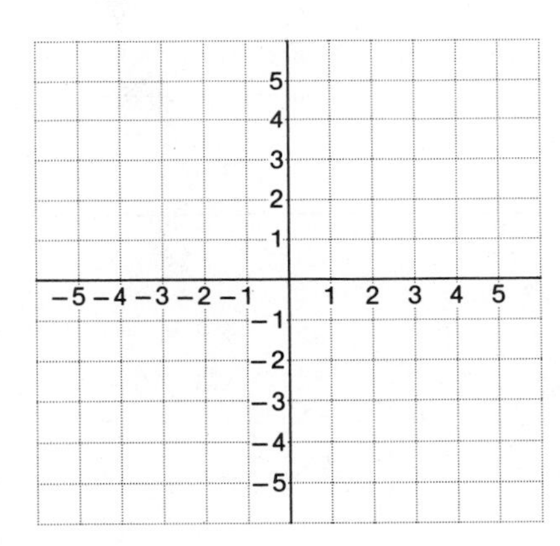

29. $t = -\frac{2}{3}$

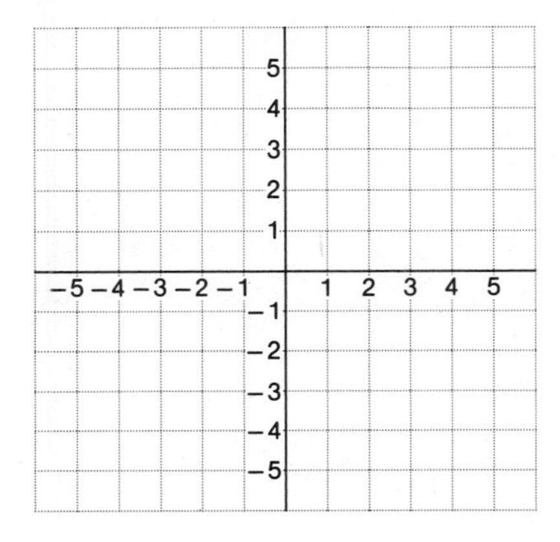

30. $s - 3 = 0$

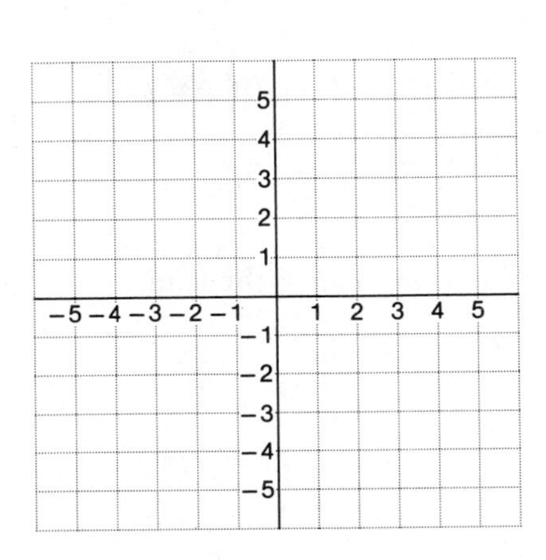

31. See graph.

32. See graph.

33. See graph.

34. See graph.

35. ______

36. ______

37. ______

31. $-3a + 4b = 12$

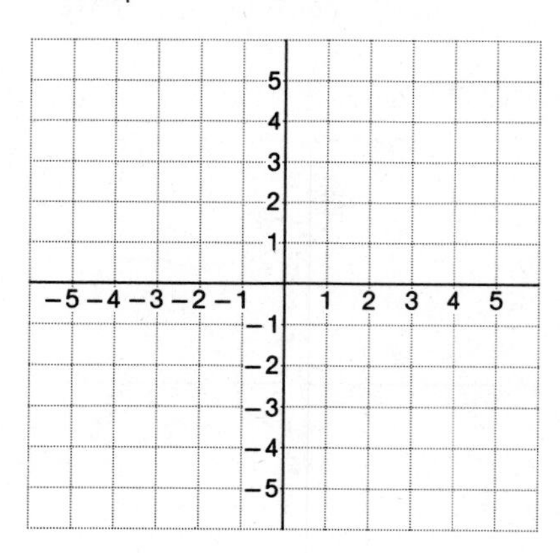

32. $2s - 3t + 6 = 0$

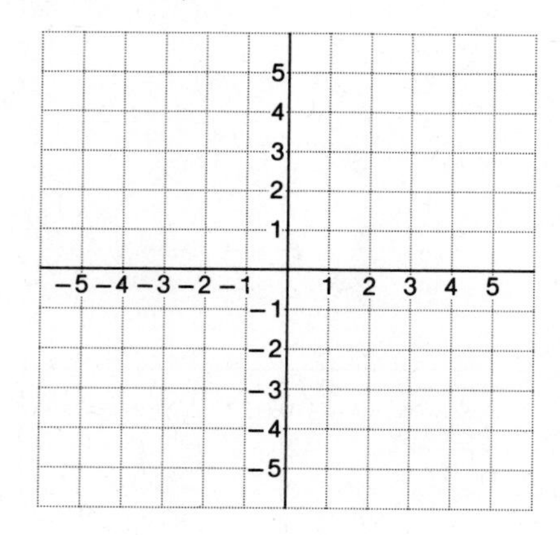

33. $n = -m + 4$

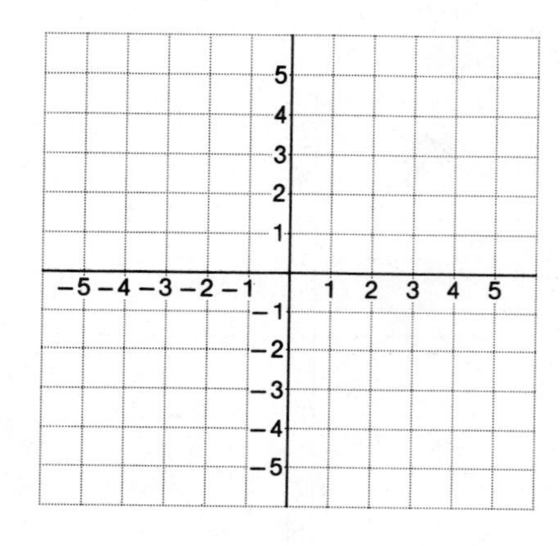

34. $-\frac{x}{2} + \frac{y}{3} = 1$

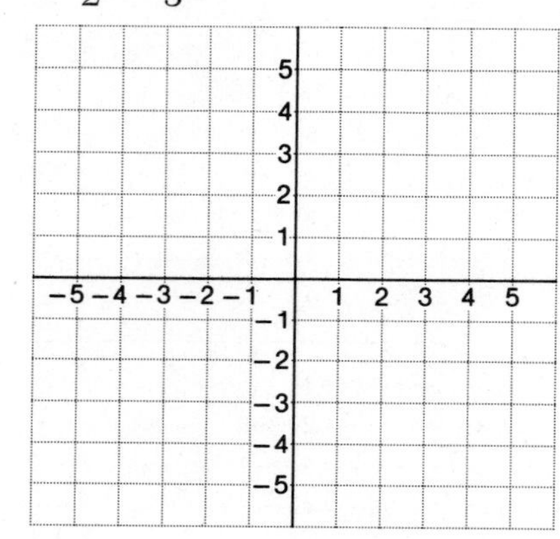

35. What lines do not have x-intercepts?

36. What lines do not have y-intercepts?

37. What lines have the same x- and y-intercepts?

6.5

Slopes of Straight Lines

Next we look at the important concept of the slope of a straight line. First we will give an algebraic definition of slope and then we will look at a geometric interpretation of the slope.

DEFINITION 6.6 If we are given a straight line $ax + by = c$ and two arbitrary points on this line, $P_1 = (x_1, y_1)$ and $P_2 = (x_2, y_2)$, then the *slope* of the line is

$$\frac{\text{difference in } y\text{-coordinates}}{\text{difference in } x\text{-coordinates}} = \frac{y_2 - y_1}{x_2 - x_1}$$

We normally use m to denote the slope of a straight line. Thus,

$$m = \frac{y_2 - y_1}{x_2 - x_1}$$

As can be shown, no matter which two points we choose on the straight line, the ratio of difference in y-coordinates to difference in x-coordinates will always be the same—namely, the slope.

Example 1 Find the slope of the straight line $-2x + y = -1$.

(a) Find two arbitrary points on the line.

$$P_1 = (-2, -5), \qquad P_2 = (1, 1)$$

(b) Find the slope m.

$$m = \frac{y_2 - y_1}{x_2 - x_1} = \frac{1 - (-5)}{1 - (-2)} = \frac{6}{3} = 2$$

(c) Note that the points $P_3 = (0, -1)$ and $P_4 = (3, 5)$ are also on this line. If we find the slope using P_3 and P_4, we have

$$m = \frac{y_2 - y_1}{x_2 - x_1} = \frac{5 - (-1)}{3 - 0} = \frac{6}{3} = 2$$

or using $P_1 = (-2, -5)$ and $P_4 = (3, 5)$, we have

$$m = \frac{y_2 - y_1}{x_2 - x_1} = \frac{5 - (-5)}{3 - (-2)} = \frac{10}{5} = 2$$

In fact, no matter what two points we use, the slope will always be 2.

DO EXERCISE 1.

Example 2 Find the slope of the straight line $y = -3x + 1$.

(a) Find two arbitrary points on the line.

$$P_1 = (1, -2) \quad \text{and} \quad P_2 = (-2, 7)$$

LEARNING OBJECTIVES

After finishing Section 6.5, you should be able to:

- Find the slope of straight lines.

1. Find the slope of the straight line $3x - y = 4$.

2. Find the slope of the straight line $y = \frac{x}{2} + 1$.

3. Find the slope of the straight line $y + 7 = 0$ and graph.

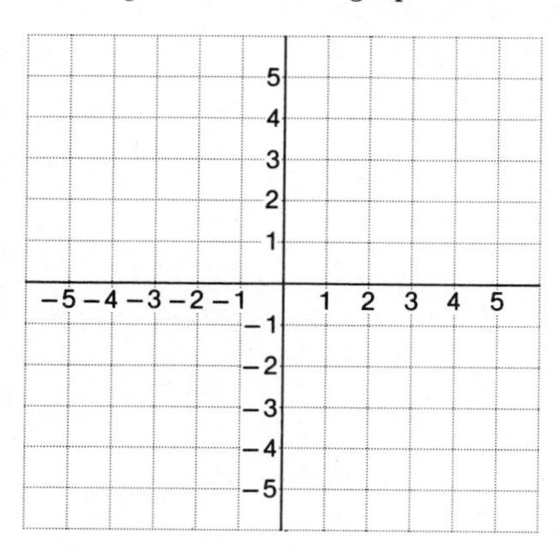

4. Find the slope of the straight line $x = -\frac{3}{4}$ and graph.

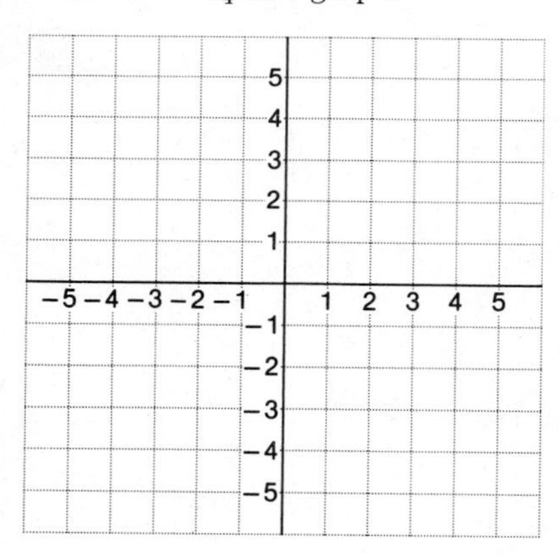

(b) Find m.

$$m = \frac{y_2 - y_1}{x_2 - x_1} = \frac{7 - (-2)}{-2 - 1} = \frac{9}{-3} = -3$$

If we interchange the roles of y_1 and y_2 and of x_1 and x_2, we still obtain the same slope.

$$m = \frac{-2 - 7}{1 - (-2)} = \frac{-9}{3} = -3$$

DO EXERCISE 2.

Example 3 Find the slope of the horizontal line $y = 5$.

(a) Find two arbitrary points on the line.

$$P_1 = (-1, 5) \quad \text{and} \quad P_2 = (2, 5)$$

(b) Find m.

$$m = \frac{y_2 - y_1}{x_2 - x_1} = \frac{5 - 5}{2 - (-1)} = \frac{0}{3} = 0$$

RULE 6.3 All horizontal lines have slope 0. Conversely, any line with slope 0 is a horizontal line.

DO EXERCISE 3.

Example 4 Find the slope of the vertical line $x = -2$.

(a) Find two arbitrary points on the line.

$$P_1 = (-2, 3) \quad \text{and} \quad P_2 = (-2, 7)$$

(b) Find m.

$$m = \frac{y_2 - y_1}{x_2 - x_1} = \frac{7 - 3}{-2 - (-2)} = \frac{4}{0} \quad \textit{Trouble}$$

This is impossible since we can never divide by zero. Thus, the slope of the line is not defined, and we say that the line has no slope.

RULE 6.4 A vertical line has no slope. Conversely, any line that does not have a slope is a vertical line.

DO EXERCISE 4.

Example 5 Find the slope of the straight line $2s - 3t = 4$.

(a) Find two arbitrary points on the line. The s-coordinate will be the first-coordinate.

$$P_1 = (2, 0) \quad \text{and} \quad P_2 = (-1, -2)$$

(b) Find m. Note that a more general definition of slope is

$$m = \frac{\text{difference in vertical axis coordinates}}{\text{difference in horizontal axis coordinates}}$$

Thus, we have

$$m = \frac{t_2 - t_1}{s_2 - s_1} = \frac{-2 - 0}{-1 - 2} = \frac{-2}{-3} = \frac{2}{3}$$

DO EXERCISE 5.

Now we look at a geometric interpretation of slope.

1. Recall that the x-coordinate of a point tells us how far from the y-axis the point is, and the y-coordinate tells us how far from the x-axis the point is. See the examples in the graph at the right.

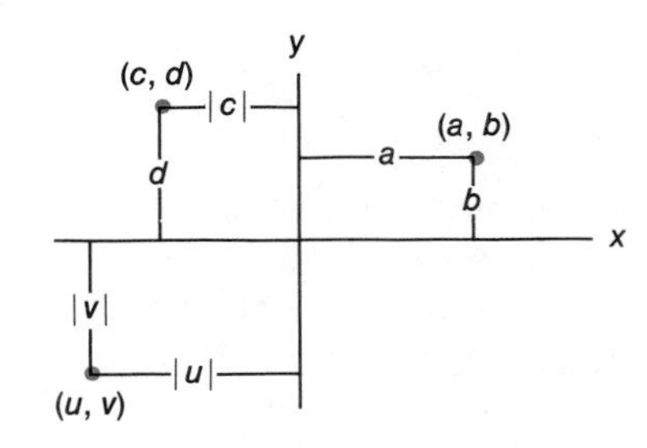

2. Consider the line L at the right and two arbitrary points on L, $P_1 = (x_1, y_1)$ and $P_2 = (x_2, y_2)$. Thus, the slope of L is

$$m = \frac{y_2 - y_1}{x_2 - x_1}$$

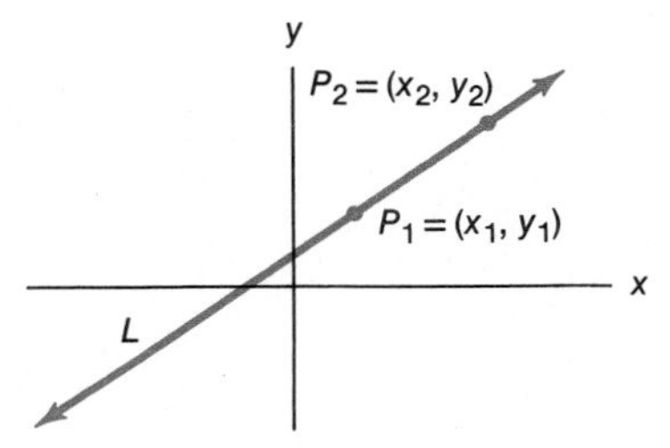

3. Draw the triangle as shown at the right, keeping the sides parallel to the x- and y-axes, respectively. Label the sides a and b. Label the third vertex of the triangle P_3. Then, with the help of item 1, we see that P_3 has coordinates (x_2, y_1).

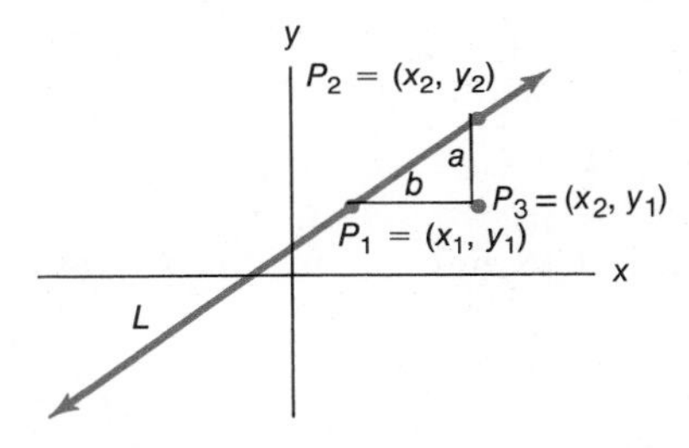

4. The distance from P_2 to P_3 is $|y_2 - y_1| = y_2 - y_1$ (since in our drawing $y_2 - y_1$ is positive), and the distance from P_1 to P_3 is $|x_2 - x_1| = x_2 - x_1$ (once again, $x_2 - x_1$ is positive).

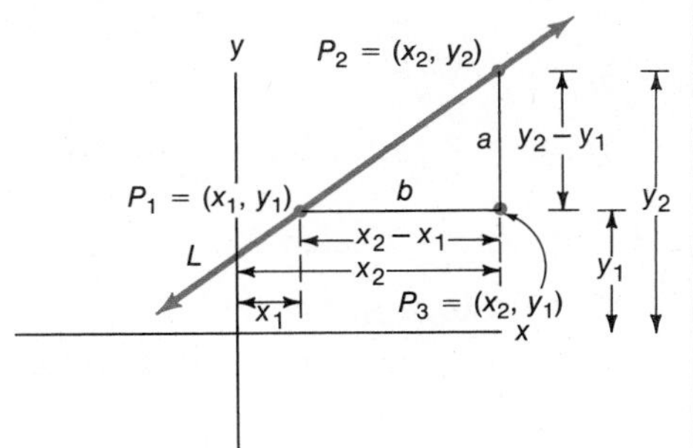

5. Thus, the slope of L,

$$m = \frac{y_2 - y_1}{x_2 - x_1}$$

is the ratio of the length of side a to the length of side b. Also, the length of side a represents a change in the vertical direction in going from P_1 to P_2, and the length of side b represents a change in the horizontal direction in going from P_1 to P_2. Hence, we may also regard the slope as

$$m = \frac{\text{change in vertical direction}}{\text{change in horizontal direction}}$$

5. Find the slope of $2a - b = 3$.

Find the slopes of the lines in the graphs.

6.

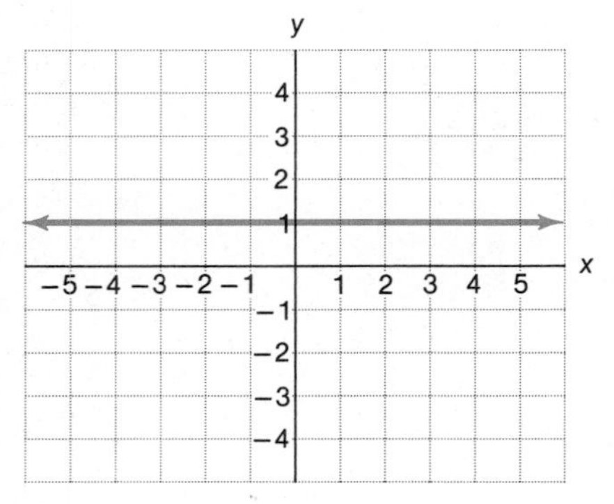

7.

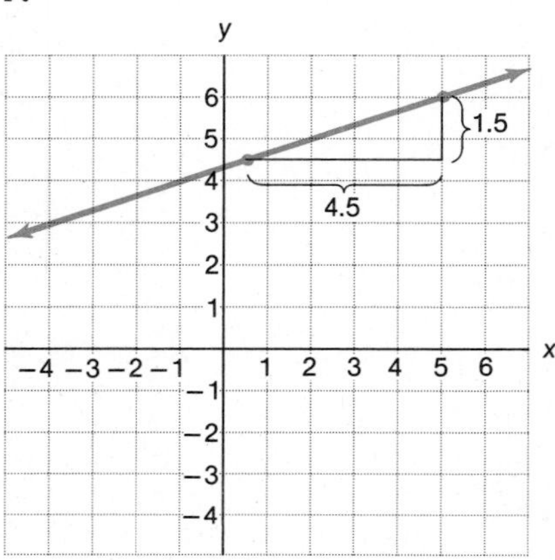

8.

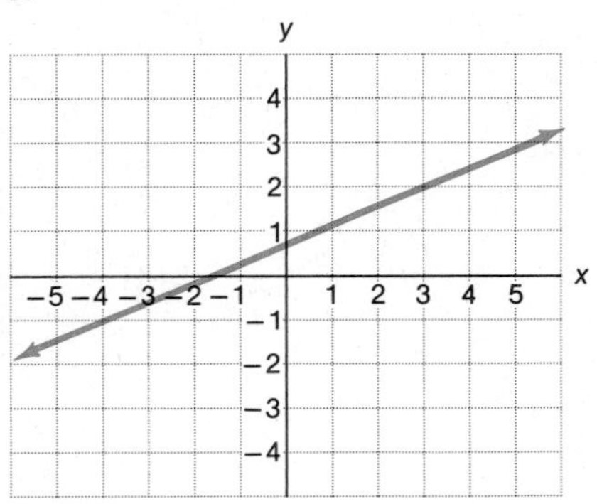

9.

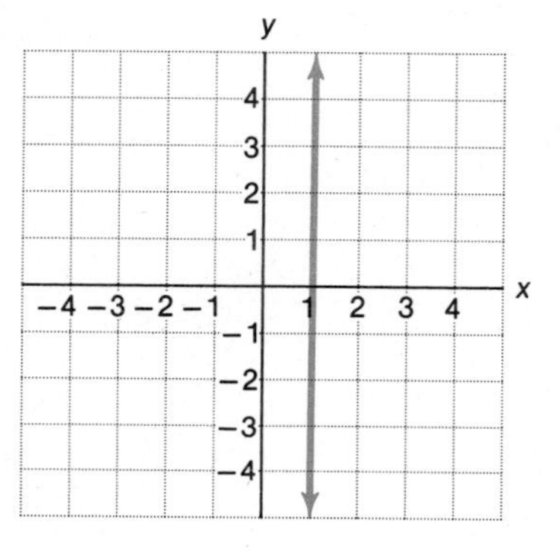

Note

1. All lines that are rising (slope upward) as we go from left to right along the x-axis have *positive* slope. Examples are shown in the graph at the right, where L_1, L_2, and L_3 all have positive slope.

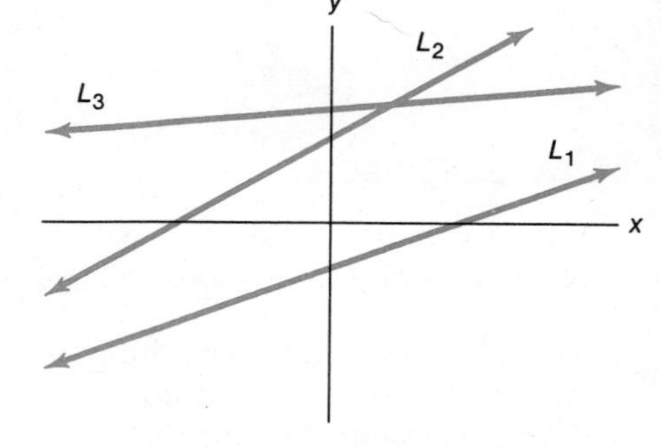

2. The size of a positive slope can tell us how steeply the line is rising. Examples are shown in the graph at the right. We see that the greater the slope, the steeper (faster) the line is rising.

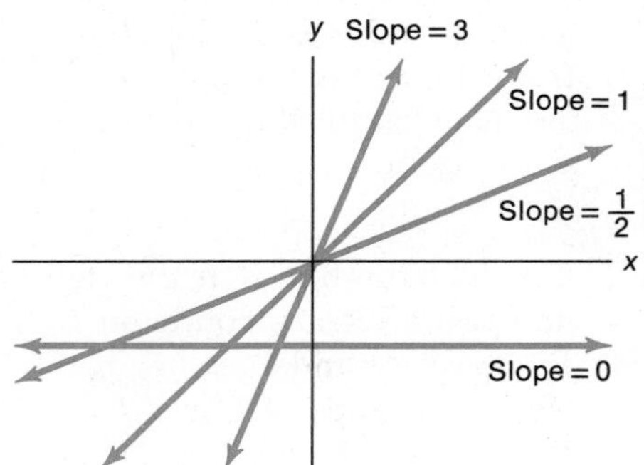

Example 6 Find the slope of the line in the graph.

$$m = \frac{\text{change in vertical direction}}{\text{change in horizontal direction}}$$

Thus, $m = \frac{2}{5}$. (The slope is positive since the line is rising as we go from left to right along the x-axis.)

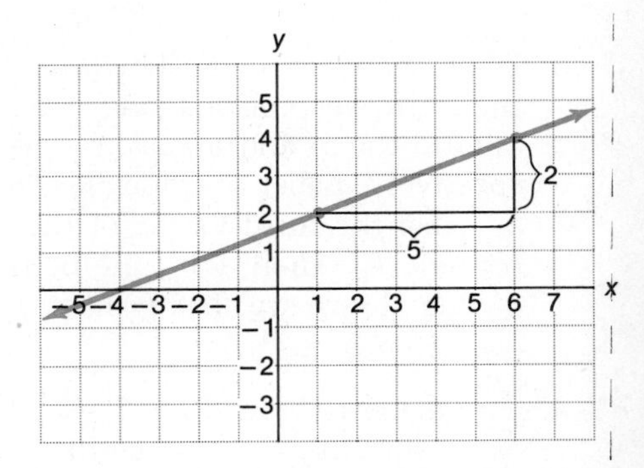

Example 7 Find the slope of the line in the graph.

Using points labeled P_1 and P_2, and

$$m = \frac{\text{change in vertical direction}}{\text{change in horizontal direction}}$$

we have $m = \frac{4}{2} = 2$.

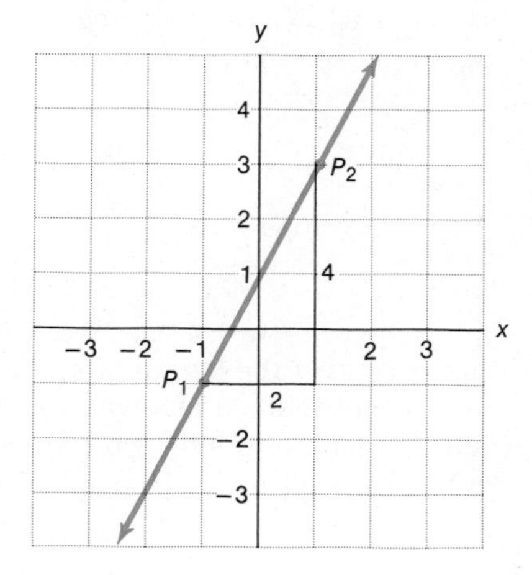

DO EXERCISES 6 THROUGH 9.

Note

1. Consider the line at the right. The slope is

$$m = \frac{y_2 - y_1}{x_2 - x_1}$$

From the figure we see that $y_2 - y_1$ is positive, but $x_2 - x_1$ is negative. Thus, the slope is negative.

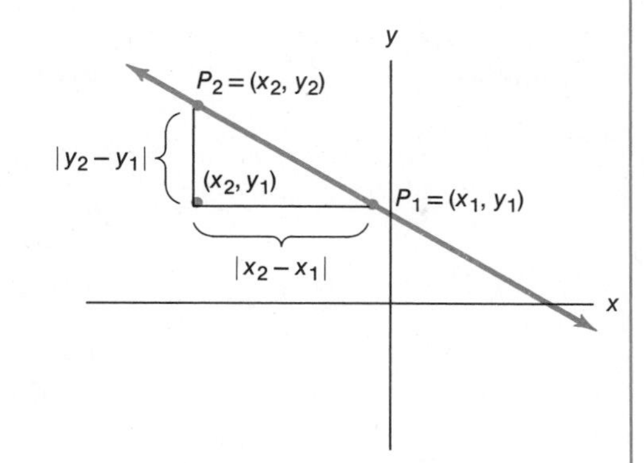

2. All lines that are falling (slope downward) as we go from left to right along the x-axis have *negative* slope. Examples are in the graph at the right where L_1, L_2, and L_3 all have negative slope.

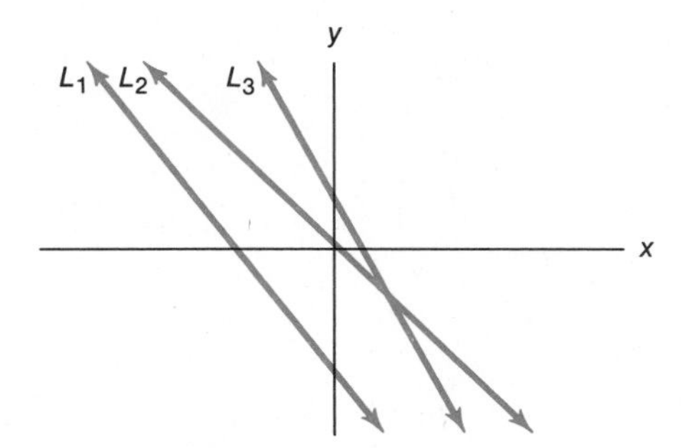

3. The size (in absolute value) of a negative slope can tell us how steeply the line is falling. Examples are in the graph at the right. We see that the greater the slope (in absolute value), the steeper (faster) the line is falling.

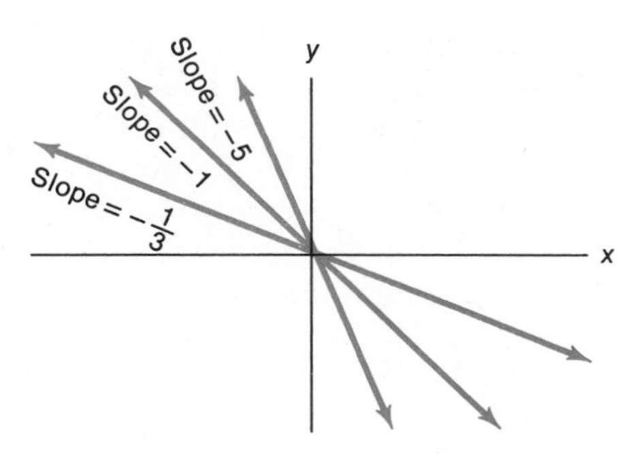

Example 8 Find the slope of the line in the graph.

Using points P_1 and P_2, and

$$m = \frac{\text{change in vertical direction}}{\text{change in horizontal direction}}$$

we have $m = -\frac{4}{8} = -\frac{1}{2}$. The slope is negative since the line is falling.

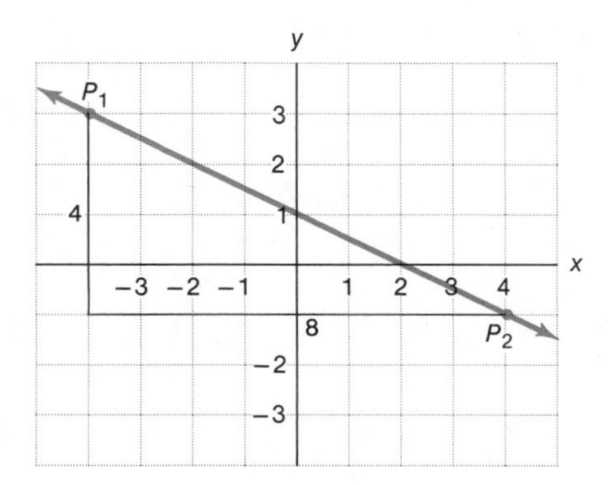

DO EXERCISE 10.

10. Find the approximate slopes of lines L_1, L_2, and L_3.

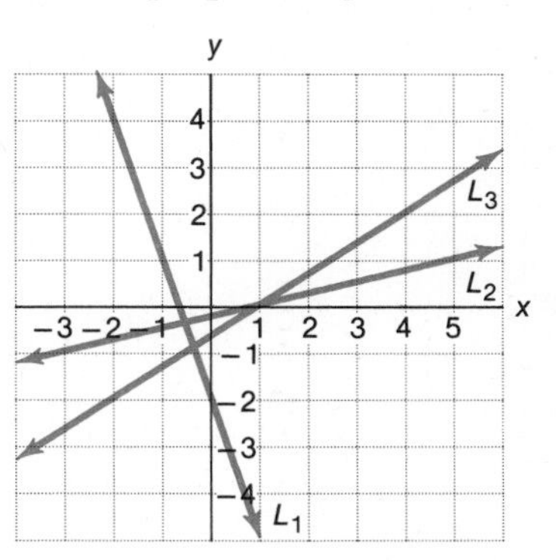

Example 9 Draw the line with slope $\frac{3}{2}$ that goes through the point (1, 1).

(a) Locate the point (1, 1).

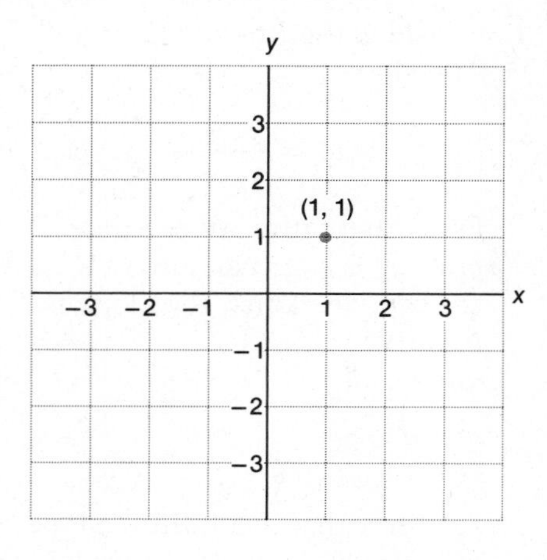

(b) Since

$$m = \frac{3}{2} = \frac{\text{change in vertical direction}}{\text{change in horizontal direction}}$$

and is positive, start at (1, 1) and mark off 2 units to the right in the horizontal direction. Then mark off 3 units upward in the vertical direction. This determines another point on the line.

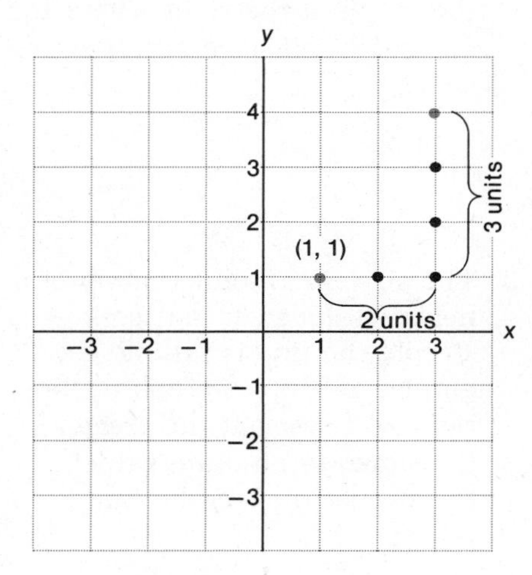

(c) Draw the line.

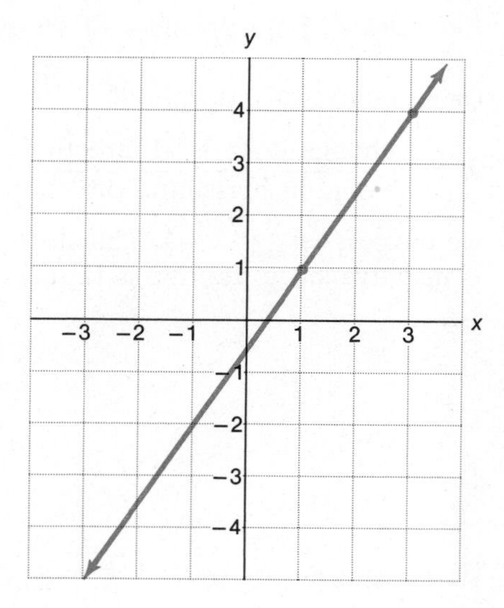

Note

In Example 9, if the slope were $-\frac{3}{2}$, then we could start at (1, 1) and mark off 2 units to the *right* in the horizontal direction and 3 units *downward* in the vertical direction. Look at the graph at the right.

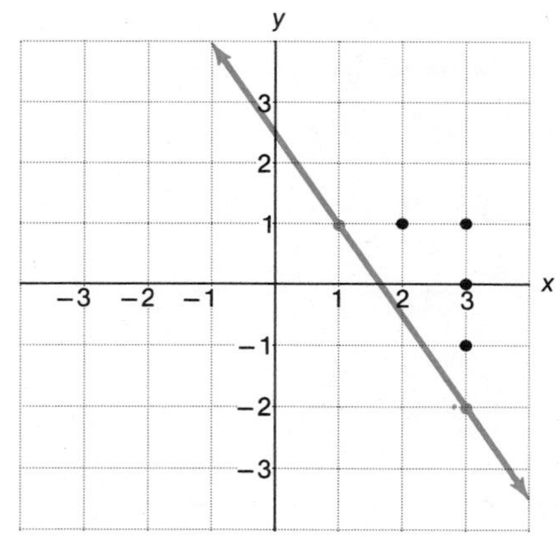
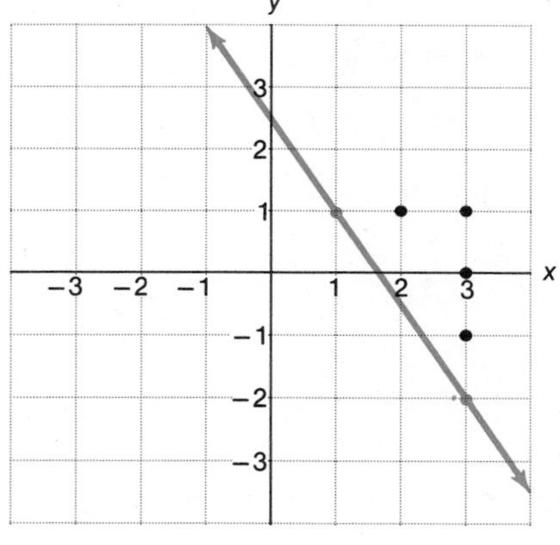

11. Draw lines through (1, 1) that have the following slopes:
(a) $m = \frac{3}{2}$ (b) $m = -2$
(c) $m = 0$

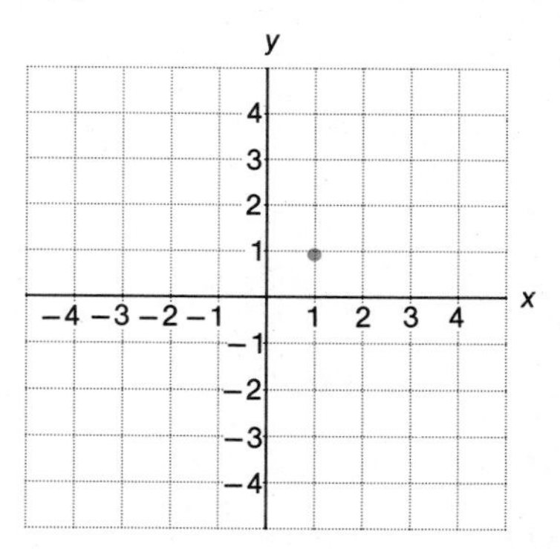

Example 10 Draw the line with slope -3 that goes through the point $(2, -1)$.

(a) Locate the point $(2, -1)$.

(b) Since $m = -3 = -\frac{3}{1}$

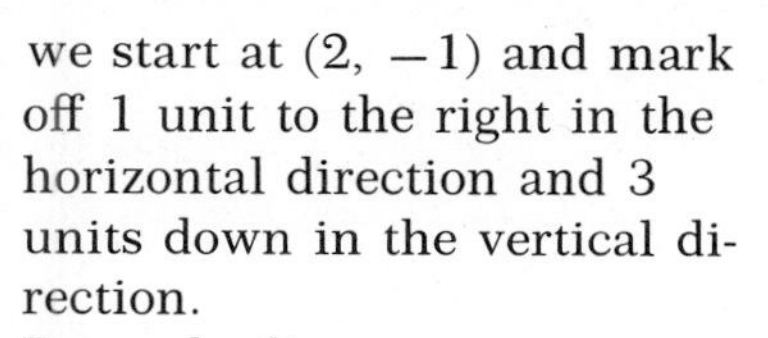

$$= \frac{\text{change in vertical direction}}{\text{change in horizontal direction}}$$

we start at $(2, -1)$ and mark off 1 unit to the right in the horizontal direction and 3 units down in the vertical direction.

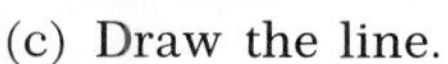

(c) Draw the line.

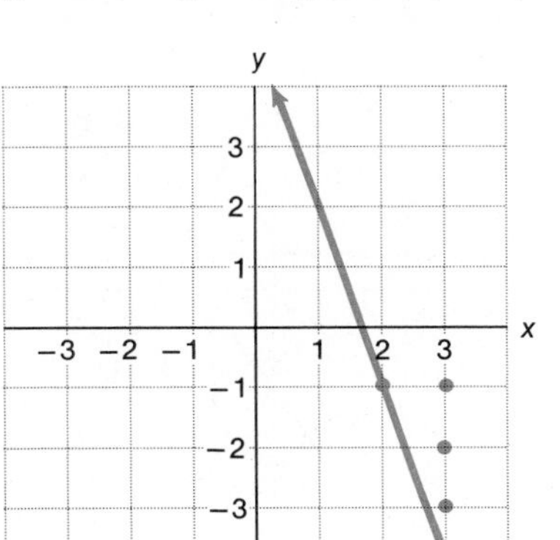

DO EXERCISE 11.

DO SECTION PROBLEMS 6.5.

ANSWERS TO MARGINAL EXERCISES

1. $m = 3$ **2.** $m = \frac{1}{2}$

3. $m = 0$

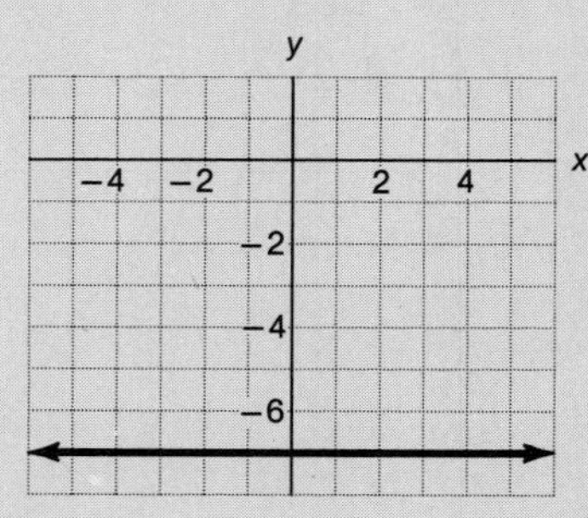

4. No slope

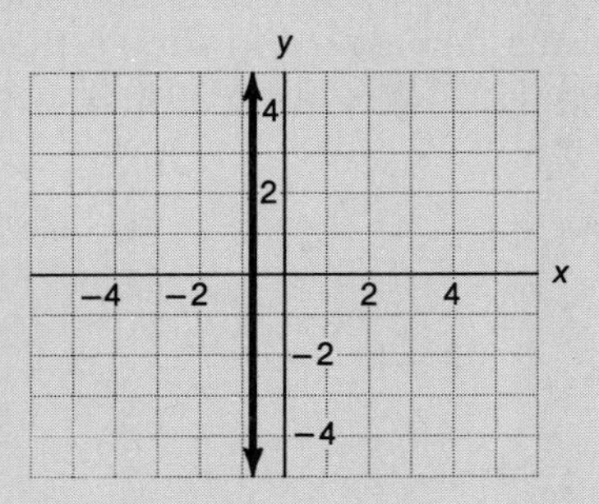

5. $m = 2$ **6.** $m = 0$ **7.** $m = \frac{1}{3}$ **8.** $m = \frac{3}{7}$ **9.** No slope

10. For L_1, $m = -3$; for L_2, $m = \frac{1}{4}$; for L_3, $m = \frac{2}{3}$

11.

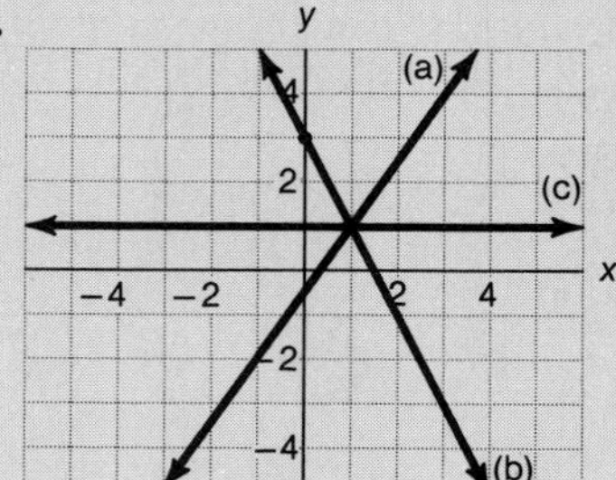

NAME COURSE/SECTION DATE

SECTION PROBLEMS 6.5

Find the slope of the line that goes through the given two points.

1. $P_1 = (2, 6)$, $P_2 = (-3, 4)$

2. $P_1 = (4, 3)$, $P_2 = (-5, 3)$

3. $P_1 = (-1, -6)$, $P_2 = (1, 2)$

4. $P_1 = (0, 4)$, $P_2 = (-2, 0)$

5. $P_1 = (-2, 5)$, $P_2 = (-2, 7)$

6. $P_1 = (\frac{3}{2}, -2)$, $P_2 = (2, 4)$

7. $P_1 = (\frac{1}{2}, 1)$, $P_2 = (\frac{3}{4}, -2)$

8. $P_1 = (0.8, 1.2)$, $P_2 = (0.3, 2.6)$

9. $P_1 = (4, 3)$, $P_2 = (-10, 3)$

10. $P_1 = (-3, 4)$, $P_2 = (-\frac{1}{2}, \frac{1}{2})$

11. Find the slope of the line $2x - 2y = 6$.

12. Find the slope of the line $y = \frac{x}{3} - 1$.

13. Find the slope of the line $y - 3.2 = 0$.

14. Find the slope of the line $4s - t = 1$.

15. Find the slope of the line $a + b = 4$.

16. Find the slope of the line $-3s + 2t = 4$.

17. Find the slope of the line $x - 5 = 0$.

18. Find the slope of the line $-2s - 3t = 6$.

19. Find the slope of the line $7m - 2n = -14$.

20. Find the slope of the line $\frac{x}{2} + \frac{y}{3} = -1$.

ANSWERS

1. ______
2. ______
3. ______
4. ______
5. ______
6. ______
7. ______
8. ______
9. ______
10. ______
11. ______
12. ______
13. ______
14. ______
15. ______
16. ______
17. ______
18. ______
19. ______
20. ______

Find the approximate slope of the following lines.

21. L_1 **22.** L_2

23. L_3 **24.** L_4

25. L_5 **26.** L_6

27. L_7 **28.** The x-axis

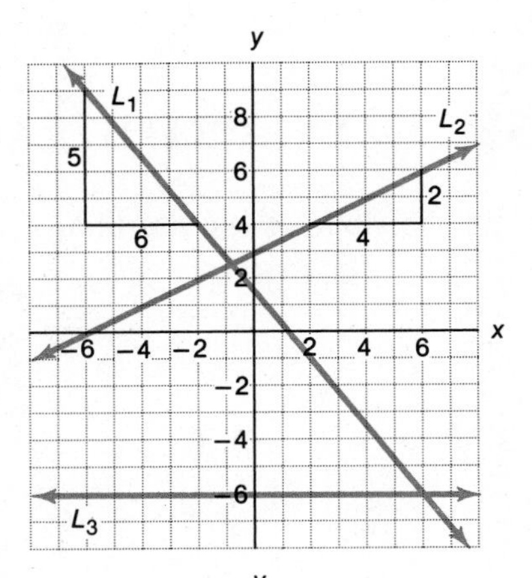

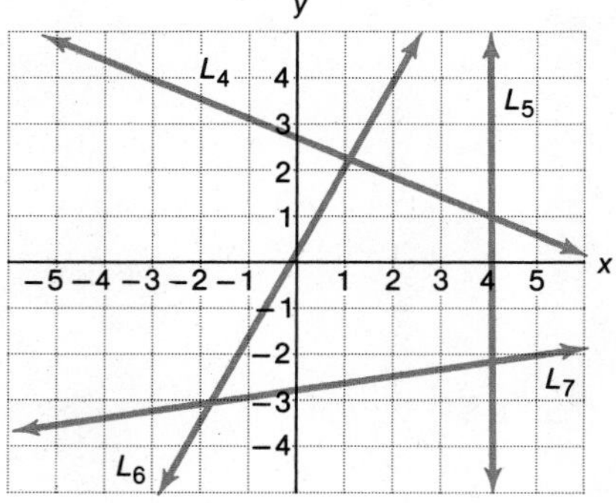

Draw lines through (1, 1) that have the following slopes.

29. $m = 2$ **30.** $m = -3$

31. $m = \frac{3}{4}$ **32.** $m = 0$

33. Slope undefined **34.** $m = \frac{7}{2}$

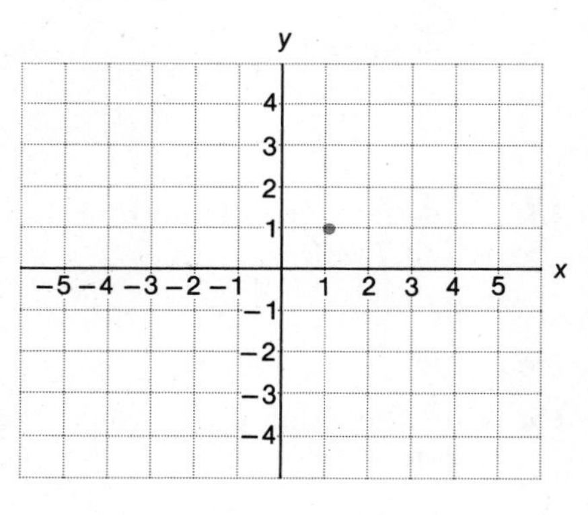

35. What does a positive slope tell us about a line?

36. Which lines do not have a slope?

37. What do the x- and y-coordinates of a point tell us about that point?

38. What does a negative slope tell us about a line?

21. ______________

22. ______________

23. ______________

24. ______________

25. ______________

26. ______________

27. ______________

28. ______________

29. See graph.

30. See graph.

31. See graph.

32. See graph.

33. See graph.

34. See graph.

35. ______________

36. ______________

37. ______________

38. ______________

6.6

Equations of Straight Lines

LEARNING OBJECTIVES

After finishing Section 6.6, you should be able to:

- Find the equation of a line with a given slope that passes through a given point.
- Find the equation of a line through two points.

There is a connection between the slope of certain lines and the equations of those lines. Note the following examples.

1. Consider the straight line $y = 3x - 1$. $P_1 = (0, -1)$ and $P_2 = (1, 2)$ are points on this line. Thus, its slope is

$$m = \frac{y_2 - y_1}{x_2 - x_1} = \frac{2 - (-1)}{1 - 0} = \frac{3}{1} = 3$$

Note that the coefficient of x in the equation is also 3.

2. Consider the straight line $y = -\frac{x}{4} + 2$. $P_1 = (0, 2)$ and $P_2 = (-4, 3)$ are points on this line. Thus, its slope is

$$m = \frac{3 - 2}{-4 - 0} = \frac{1}{-4} = -\frac{1}{4}$$

And, once again, the coefficient of x in the equation is $-\frac{1}{4}$.

3. Consider the straight line $2x + 3y = 4$. $P_1 = (0, \frac{4}{3})$ and $P_2 = (2, 0)$ are points on this line. Thus, its slope is

$$m = \frac{0 - \frac{4}{3}}{2 - 0} = \frac{-\frac{4}{3}}{\frac{2}{1}} = \left(-\frac{4}{3}\right)\left(\frac{1}{2}\right) = -\frac{2}{3}$$

But the coefficient of x in this equation is not $-\frac{2}{3}$. However, let us solve the equation $2x + 3y = 4$ for y in order to put the equation in the same form as the equations in (1) and (2).

$$3y = -2x + 4$$

$$y = -\frac{2}{3}x + \frac{4}{3}$$

Now, once the equation is in the same form as (1) and (2) above, we have the same pattern holding—namely, the coefficient of x is the same as the slope.

DEFINITION 6.7

$$y = mx + b$$

is called the *slope-intercept form* of the equation of a straight line, where m is the slope of the line and $(0, b)$ is the y-intercept.

Note

1. If the equation of a straight line is in the form $y = mx + b$ (solved for a positive one y), then the slope of the line is m and is the coefficient of x, and the y-intercept is the point $(0, b)$.
2. The equation for any straight line that is not vertical can be put in the form $y = mx + b$. Recall that vertical lines do not have a slope. However, we already know that the equation of a vertical line is $x =$ constant.

1. Find the slope and y-intercept of $y = \frac{3x}{4} - 1$.

2. Find the slope and y-intercept of $3x - y = 2$. Graph.

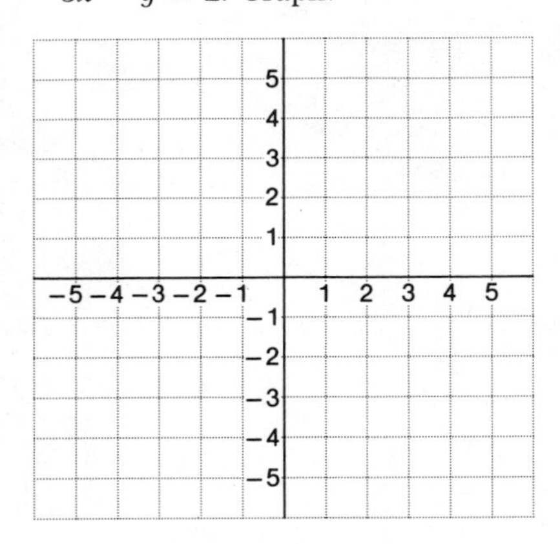

3. Find the slope and t-intercept of $-2s + 4t = 5$. Graph.

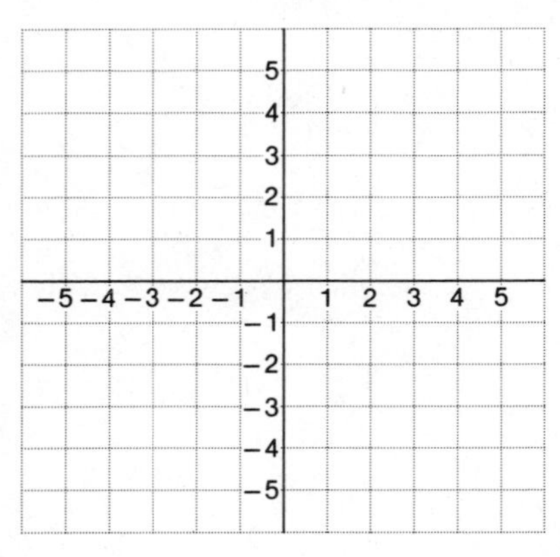

Example 1 Find the slope and y-intercept of the straight line $y = -2x + 5$.

Since the equation is already in the form $y = mx + b$, the slope is -2 (the coefficient of x) and the y-intercept is (0, 5).

DO EXERCISE 1.

Example 2 Find the slope and y-intercept of the straight line $4x - 2y = 6$. There are two ways to find the slope.

1. Find two arbitrary points on the line and use

$$m = \frac{y_2 - y_1}{x_2 - x_1}$$

2. Change the equation to the form $y = mx + b$ by solving for y.

We will use method 2:

$$4x - 2y = 6$$

$$-2y = -4x + 6 \qquad \text{We added } -4x \text{ to each side.}$$

$$y = \frac{-4}{-2}x + \frac{6}{-2} \qquad \text{We multiplied both sides by } -\frac{1}{2}.$$

$$y = 2x - 3$$

Thus, the slope is 2 and the y-intercept is (0, -3).

DO EXERCISE 2.

Example 3 Put the equation $2T + 5R = 1$ in slope-intercept form.

In slope-intercept form, the variable that we solve for is the variable representing the vertical axis. In our problem, T (since it comes after R in the alphabet) will represent the vertical axis, and we will solve the equation for T.

$$2T + 5R = 1$$

$$2T = -5R + 1$$

$$T = -\frac{5}{2}R + \frac{1}{2}$$

Thus, the slope is $-\frac{5}{2}$ and the T-intercept is $(0, \frac{1}{2})$.

DO EXERCISE 3.

Thus far we have been given the equation of a straight line and asked certain information about it. Now we want to start with information concerning a straight line and then find the equation of it.

A common type of problem is one in which the slope of a line and a point that the line goes through are given, and we are asked to find the equation of the line.

Example 4 Find the equation of a straight line with slope 4 that has y-intercept (0, -2).

(a) Since this line obviously has a slope, we know we can write its equation in the form $y = mx + b$. All we have to do is determine the value for m and the value for b.

(b) m is the slope, and hence $m = 4$. b is the y-coordinate of the y-intercept, and hence $b = -2$.

(c) Thus, the equation of the line is $y = 4x - 2$.

DO EXERCISE 4.

Example 5 Find the equation of a straight line with slope 0 that goes through the point $(1, -2)$.

(a) We can put this line in the form $y = mx + b$ since it does have a slope.
(b) The slope is $m = 0$, and thus we have a horizontal line (the equation is $y =$ constant).
(c) Since the line goes through the point $(1, -2)$ and the y-coordinate must be constant, we have the equation $y = -2$.

DO EXERCISES 5 AND 6.

Example 6 Find the equation of a straight line with slope $\frac{2}{3}$ that goes through the point $(-6, 2)$.

(a) The equation can be put in the form $y = mx + b$ where $m = \frac{2}{3}$. Thus, so far, we have $y = \frac{2}{3}x + b$.

(b) To find b we use the fact that if $(-6, 2)$ is on the line $y = \frac{2}{3}x + b$, then the ordered pair $(-6, 2)$ must satisfy the equation $y = \frac{2}{3}x + b$. Thus, substituting $x = -6$ and $y = 2$ in $y = \frac{2}{3}x + b$, we obtain

$$2 = \left(\frac{2}{3}\right)(-6) + b$$
$$2 = -4 + b$$
$$6 = b$$

(c) The equation of the line is $y = \frac{2}{3}x + 6$.

Note

The equation of the line with slope m which goes through the point (x_1, y_1) may also be found by substituting m, x_1, and y_1 in the formula

$$y - y_1 = m(x - x_1)$$

Of course, we may also use the method used in Examples 4–6.

DO EXERCISE 7.

Another common type of problem is to find the equation of a line if we know two points that the line goes through.

Example 7 Find the equation of the straight line that goes through the points $(-1, 3)$ and $(2, -2)$.

(a) We can put this equation in the form $y = mx + b$; that is, it is not a vertical line since the x-coordinates of the two points are different.

4. Find the equation of the straight line with slope -3 and y-intercept $(0, -4)$.

5. Find the equation of a straight line with slope 0 that goes through the point $(3, -4)$.

6. Find the equation of the vertical line that goes through the point $(1, 2)$.

7. Find the equation of the straight line with slope $-\frac{3}{4}$ that goes through the point $(-4, 2)$.

Find the equation of the straight line that goes through the following pairs of points.

8. $(-3, 4)$ and $(5, 4)$

9. $(1, 5)$ and $(-2, 3)$

Find the equation of the straight line that goes through the following pair of points.

10. $(3, -4)$ and $(3, -\frac{1}{2})$

(b) Find m.

$$m = \frac{y_2 - y_1}{x_2 - x_1} = \frac{-2 - 3}{2 - (-1)} = \frac{-5}{3} = -\frac{5}{3}$$

Thus, so far, we have $y = -\frac{5}{3}x + b$.

(c) Find b. Substitute the coordinates for either point that the line goes through in the equation $y = -\frac{5}{3}x + b$, and then solve for b. Substitute $x = 2$ and $y = -2$.

$$-2 = \left(-\frac{5}{3}\right)(2) + b$$

$$-2 = -\frac{10}{3} + b$$

$$-\frac{6}{3} + \frac{10}{3} = b$$

$$\frac{4}{3} = b$$

(d) Thus, the equation is $y = -\frac{5}{3}x + \frac{4}{3}$.

DO EXERCISES 8 AND 9.

Example 8 Find the equation of the straight line through $(2, -3)$ and $(2, 7)$.

(a) This equation cannot be put in the form $y = mx + b$ since if we attempt to find m we have

$$m = \frac{y_2 - y_1}{x_2 - x_1} = \frac{7 - (-3)}{2 - 2} = \frac{10}{0} \quad \textit{Trouble}$$

(b) Thus, the line does not have a slope and must be a vertical line with equation $x =$ constant. Also, note that the x-coordinates of the two points are identical.

(c) The equation of the line is $x = 2$. See the graph at the right.

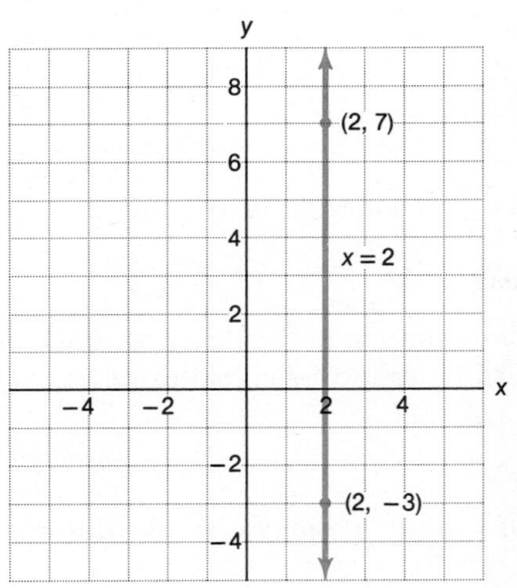

Note

The equation of a nonvertical line that goes through two points (x_1, y_1) and (x_2, y_2) may also be found by substituting x_1, y_1, x_2, and y_2 in the formula

$$y - y_1 = \frac{y_2 - y_1}{x_2 - x_1}(x - x_1)$$

DO EXERCISE 10.

There is a relationship between the slopes of two lines that are parallel.

DEFINITION 6.8 Two straight lines are parallel if they have the same slope. Conversely, parallel lines have identical slopes.

Example 9 What slopes do the following lines have?

(a) Any line parallel to the line $y = 5x - 3$

Since the line $y = 5x - 3$ has slope 5, any line parallel to it must have slope 5.

(b) Any line parallel to the line $2x - 7y = 14$

Change $2x - 7y = 14$ to the form $y = mx + b$.

$$2x - 7y = 14$$
$$-7y = -2x + 14$$
$$y = \frac{2}{7}x - 2$$

Thus, its slope is $\frac{2}{7}$ and any line parallel to it would have slope $\frac{2}{7}$.

DO EXERCISE 11.

Example 10 Find the equation of the straight line L that is parallel to the line $-2x + 3y = 1$ and goes through the point $(3, 4)$.

We will put the equation of L in the form $y = mx + b$.

(a) Find m. Since L is parallel to $-2x + 3y = 1$, it has the same slope. Find the slope of $-2x + 3y = 1$ by solving for y.

$$3y = 2x + 1$$
$$y = \frac{2}{3}x + \frac{1}{3}$$

Thus, L has slope $\frac{2}{3}$, and so far our equation is $y = \frac{2}{3}x + b$.

(b) Find b. Since $(3, 4)$ is on L, substitute $x = 3$ and $y = 4$ in the equation $y = \frac{2}{3}x + b$.

$$4 = \left(\frac{2}{3}\right)(3) + b$$
$$4 = 2 + b$$
$$2 = b$$

(c) Thus, L has the equation $y = \frac{2}{3}x + 2$.

DO EXERCISE 12.

11. Any line parallel to the line $5x - 2y = 1$ would have what slope?

12. Find the equation of the straight line parallel to the line $y = -2x + 3$ that goes through the point $(2, 2)$.

13. Classify the following pairs of lines as parallel, perpendicular, or neither.
(a) $y = 2x - 1$ and $4x - 2y = 3$
(b) $-3a + 2b = 1$ and $6a - 9b = -1$
(c) $y - 2 = 3x$ and $6x + 2y = 0$
(d) $x - 4 = 0$ and $y = -3$

14. Find the equation of the straight line L that is perpendicular to the line that goes through the points $(2, 3)$ and $(-1, 4)$ and itself goes through the point $(-2, -1)$.

There is also a relationship between the slopes of perpendicular lines.

DEFINITION 6.9 If two straight lines are perpendicular and have slopes $m_1 \neq 0$ and $m_2 \neq 0$, respectively, then m_1 and m_2 are negative reciprocals of each other ($m_1 = -1/m_2$ or $m_2 = -1/m_1$). Conversely, if two lines have slopes that are negative reciprocals, then they are perpendicular. A special case is a horizontal line (slope 0) being perpendicular to a vertical line (no slope).

Example 11 Find the slope of any line perpendicular to the line $y = \frac{4}{5}x + 3$.

The line $y = \frac{4}{5}x + 3$ has slope $\frac{4}{5}$ and, thus, any line perpendicular to it must have slope $-\frac{5}{4}$. Note that $\frac{4}{5}$ and $-\frac{5}{4}$ are negative reciprocals.

Example 12 Find the equation of the straight line L that is perpendicular to the line $y = -\frac{3}{2}x - 3$ and goes through the point $(6, -4)$.

We will put the equation of L in the form $y = mx + b$.

(a) Find m. The line $y = -\frac{3}{2}x - 3$ has slope $-\frac{3}{2}$, and since L is perpendicular to this line it must have slope $\frac{2}{3}$ (the negative reciprocal of $-\frac{3}{2}$). Thus, we have $y = \frac{2}{3}x + b$ so far.

(b) Find b. Substitute $x = 6$ and $y = -4$ in $y = \frac{2}{3}x + b$.

$$-4 = \frac{2}{3}(6) + b$$

$$-4 = 4 + b$$

$$-8 = b$$

(c) Thus, L has the equation $y = \dfrac{2}{3}x - 8$.

DO EXERCISES 13 AND 14.

DO SECTION PROBLEMS 6.6.

ANSWERS TO MARGINAL EXERCISES

1. $m = \frac{3}{4}$, y-intercept $= (0, -1)$

2. $m = 3$, y-intercept $= (0, -2)$

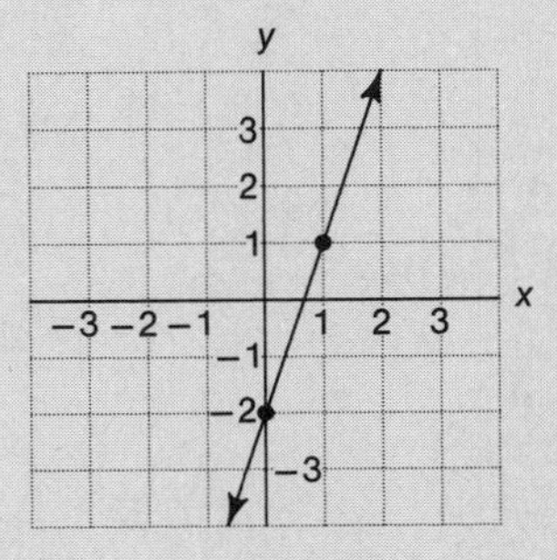

3. $m = \frac{1}{2}$, t-intercept $= (0, \frac{5}{4})$

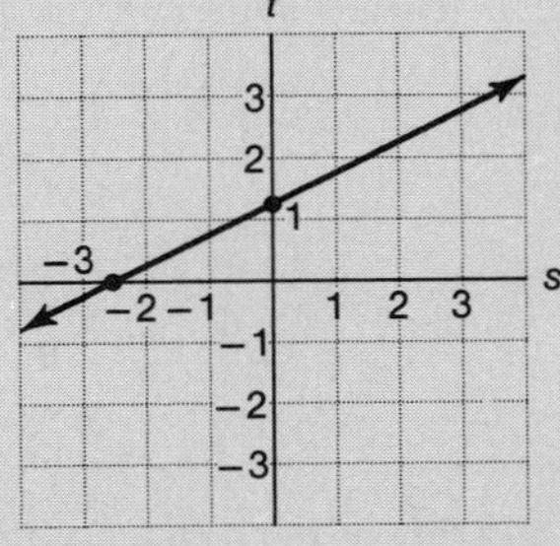

4. $y = -3x - 4$ **5.** $y = -4$ **6.** $x = 1$ **7.** $y = -\frac{3}{4}x - 1$
8. $y = 4$ **9.** $y = \frac{2}{3}x + \frac{13}{3}$ **10.** $x = 3$ **11.** $\frac{5}{2}$ **12.** $y = -2x + 6$
13. (a) Parallel (b) Neither (c) Neither (d) Perpendicular
14. $y = 3x + 5$

NAME ______ COURSE/SECTION ______ DATE ______

SECTION PROBLEMS 6.6

Put in slope-intercept form.

1. $2x - y = 6$
2. $3y + 4 = -x$
3. $2y - 5 = 0$
4. $2s - 3t = -6$
5. $3(a - b) = 2a + 7$
6. $2P + 3Q = -10$
7. $-3x + 2y = 6$
8. $2u + 5v = 4$

Find the slope and vertical axis intercept.

9. $y = -3x + 1$
10. $-2x + 3y = 6$
11. $x - 1 = 0$
12. $t = \frac{3}{2}s + 4$
13. $y + 2 = 0$
14. $-3b + 2a = 1$
15. $5s - 7t = 14$
16. $5y - 3 = 0$
17. $t = \frac{s}{3} + 2$
18. $-x - 2y + 6 = 0$
19. $2(P - 1) - 3(Q + 1) = 0$
20. $\frac{x}{3} - \frac{y}{4} = -1$

ANSWERS

1. ______
2. ______
3. ______
4. ______
5. ______
6. ______
7. ______
8. ______
9. ______
10. ______
11. ______
12. ______
13. ______
14. ______
15. ______
16. ______
17. ______
18. ______
19. ______
20. ______

21. ______________

22. ______________

23. ______________

24. ______________

25. ______________

26. ______________

27. ______________

28. ______________

29. ______________

30. ______________

31. ______________

32. ______________

33. ______________

34. ______________

35. ______________

36. ______________

Find the equation of the straight line with slope m that goes through the given point.

21. $m = \frac{3}{2}$, through (0, 2)

22. $m = -4$, through (1, 3)

23. $m = 0$, through (2, −1)

24. $m = -\frac{3}{4}$, through (2, 4)

25. slope undefined, through (9, −3)

26. $m = 7$, through (2, 14)

27. $m = 4$, through (1, −2)

28. $m = -3$, through (−2, −1)

29. $m = -\frac{3}{2}$, through (−4, −6)

30. $m = 0$, through (4, 5)

Find the equation of the straight line through the given two points.

31. (−2, −5) and (4, −1)

32. ($\frac{1}{2}$, 3) and (3, −1)

33. (−1, 5) and (−1, 2)

34. (4, −3) and (−2, −1)

35. (7.1, 4) and (−1.2, 4)

36. (−4, −4) and (−10, 1)

NAME COURSE/SECTION DATE

ANSWERS

37. ______

38. ______

39. ______

40. ______

41. ______

42. ______

43. ______

44. ______

45. ______

37. $(7, 3)$ and $(-2, 1)$

38. $(1.4, -2.3)$ and $(-0.6, -1.9)$

39. $(-2, -4)$ and $(3, -1)$

40. $(\frac{1}{2}, 4)$ and $(-\frac{3}{2}, 8)$

41. Any line that is parallel to the line $-3x + 5y = 4$ would have what slope?

42. Any line that is perpendicular to the line $-3x + 5y = 4$ would have what slope?

43. Find the equation of the line that is parallel to the line $2x - y = 3$ and goes through the point $(1, 3)$.

44. Find the equation of the line that is parallel to the line $-3a + b = 4$ and goes through the point $(2, 5)$.

45. Find the equation of the line that is perpendicular to the line $y = \frac{2}{5}x + 1$ and goes through the point $(4, -3)$.

46. ______________

47. ______________

48. ______________

49. ______________

50. ______________

51. ______________

52. ______________

46. Find the equation of the line that is perpendicular to the line $y - 3 = 0$ and goes through the point $(-2, 1)$.

47. Find the equation of the line that is parallel to the line $y - 3 = 0$ and goes through the point $(-2, 5)$.

48. Find the equation of the line that is perpendicular to the line $s - 3t = -6$ and goes through the point $(-2, 3)$.

49. Find the equation of the line perpendicular to the line $y = \frac{3x}{5} - 2$ that goes through the point $(12, 4)$.

50. Find the equation of the line perpendicular to the line $-2x - y = 4$ that goes through the point $(2, -5)$.

51. Find the equation of the line that is parallel to the line through the points $(-2, 3)$ and $(1, 7)$ and has y-intercept $(0, -1)$.

52. Find the value of n where the point $(-3, n)$ is on the line L and L is perpendicular to the line $2x - 3y = 1$ and contains the point $(4, 1)$.

6.7

Applications of Linear Equations in Two Variables

LEARNING OBJECTIVES

After finishing Section 6.7, you should be able to:

- Apply linear equations in two variables to some practical problems.

Now we want to look at practical applications of linear equations in two variables.

DEFINITION 6.10 If the relationship between two variables is described by the equation of a straight line, we call this a *linear relationship.*

In a linear relationship, the slope of the straight line tells us something about the relationship between the two variables in the equation of this straight line. For example, consider the equation

$$y = 2x + 1$$

The slope is positive. This means as x increases y also increases (for example, if x increases from 1 to 5, then y increases from 3 to 11). The size of the slope, 2, tells us how fast this increase is. That is, if x increases 1 unit, then y will increase 2 units.

As another example, consider the equation

$$y = -\tfrac{3}{2}x + 1$$

The slope is negative. This means as x increases y must decrease (for example, if x increases from 2 to 6, then y decreases from -2 to -8). The size of the slope in absolute value, $|-\frac{3}{2}| = \frac{3}{2}$, tells us how fast this decrease is. That is, if x increases 1 unit, then y will decrease $\frac{3}{2}$ units.

DO EXERCISE 1.

1. What does the slope tell us about the relationship between x and y in the following equations?
(a) $y = \frac{7}{2}x + 1$
(b) $y = -4x + 1$

As we now look at some practical situations of linear relationships we will see that the slope of a straight line (and sometimes the y-intercept) gives us important information about the interaction of the two variables.

Example 1 For the X and J Company, the relationship between the number of units produced and the profit is a linear relationship. If the profit is \$10,000 when 5000 units are produced and \$20,000 when 9000 units are produced, answer these questions.
(a) What is the equation relating profit and number of units produced?
(b) What is the profit when 10,000 units are produced?

We use the following steps.

1. Let x = number of units produced and
 y = profit when x units are produced.
2. We have two data points that are on our straight line, namely, (5000, 10,000) [that is, the profit y is \$10,000 when 5000 units (x) are

2. At a price of \$30 a shoe store has a demand for 25 oxford striders in a week. At a price of \$40 the demand drops to 17 oxford striders in a week. Assuming a linear relationship between price p and demand q, if the store wants the weekly demand to be 20 oxford striders, what price should it charge?

produced] and (9000, 20,000). We want the equation of the line through these two points.

3. Find m.

$$m = \frac{y_2 - y_1}{x_2 - x_1} = \frac{20{,}000 - 10{,}000}{9000 - 5000} = \frac{10{,}000}{4000} = \frac{5}{2}$$

Thus, $y = \frac{5}{2}x + b$ so far.

4. Find b. (Substitute $x = 5000$ and $y = 10{,}000$ in $y = \frac{5}{2}x + b$.)

$$10{,}000 = \tfrac{5}{2}(5000) + b$$
$$10{,}000 = 12{,}500 + b$$
$$-2500 = b$$

(a) Using Steps 1–4, we see that the equation relating the number of units x and profit y is

$$y = \tfrac{5}{2}x - 2500$$

(b) If 10,000 units are produced, that is, when $x = 10{,}000$, then we have the profit $y = \frac{5}{2}(10{,}000) - 2500 = 25{,}000 - 2500 = \$22{,}500$.

The slope and y-intercept in the equation in Example 1 have interesting physical meanings. The positive slope means that as the number of units x increases, the profit y will also increase; that is, the line $y = \frac{5}{2}x - 2500$ slopes upward. The size of the slope, $\frac{5}{2}$, tells us how fast the increase is. That is, an increase in production of one unit (x) will result in an increase in profit y of $\frac{5}{2}$ units or \$2.50. The y-intercept of $-\$2500$ tells us that if we produce zero units (that is, $x = 0$), then our profit will actually be a loss of \$2500.

DO EXERCISE 2.

Example 2 In straight line depreciation, the relationship between present book value V and the number of years t an item has depreciated is linear. A truck originally cost \$10,000 and is being depreciated at a constant rate of \$2000 a year. Find the equation involving V and t, and find the book value after $2\frac{1}{2}$ years.

(a) We can start with the following word equations.

$$\begin{pmatrix}\text{Amount}\\\text{of}\\\text{depreciation}\end{pmatrix} = \begin{pmatrix}\text{Yearly}\\\text{depreciation}\\\text{rate}\end{pmatrix} \cdot \begin{pmatrix}\text{Number}\\\text{of}\\\text{years}\end{pmatrix} = 2000t$$

$$\begin{pmatrix}\text{Present}\\\text{book value}\end{pmatrix} = \begin{pmatrix}\text{Original}\\\text{cost}\end{pmatrix} - \begin{pmatrix}\text{Amount of}\\\text{depreciation}\end{pmatrix}$$

or
$$V = 10{,}000 - 2000t$$

(b) Thus, when $t = 2.5$ years, we have

$$V = 10{,}000 - 2000(2.5) = 10{,}000 - 5000 = \$5000$$

Note that in the linear equation $V = 10{,}000 - 2000t$, the slope, -2000, is negative, which indicates that as the number of years increases, the book value decreases, that is, the line $V = 10{,}000 - 2000t$ slopes downward. Also, the absolute value of the slope gives the annual rate of depreciation. The V-intercept of \$10,000 gives us the original cost of the truck.

DO EXERCISE 3.

Example 3 Hazeltine Manufacturing can sell all of the lamps it produces for \$60 a lamp. The materials for a lamp cost \$12 and labor per lamp is \$20. The fixed cost for producing the lamps is \$21,000.

(a) Find the revenue, cost, and profit equations.

(b) How many lamps must Hazeltine sell so that their profit is 0 (this is called the *break-even point*)?

(a) We choose the following variables.
x = number of lamps produced/sold
R = revenue from selling x lamps
C = cost of producing x lamps
P = profit from producing and selling x lamps

We have the following linear equations.

(1) Revenue = (price)(demand)
$R = 60x$ (Note the slope is 60.)

(2) Cost = Fixed cost + Variable cost
$C = 21{,}000 +$ Material cost + Labor cost
$C = 21{,}000 + 12x + 20x$
$C = 21{,}000 + 32x$ (Note the slope is 32.)

(3) Profit = Revenue − Cost
$P = 60x - (21{,}000 + 32x)$
$P = 60x - 21{,}000 - 32x$
$P = 28x - 21{,}000$ (Note the slope is 28.)

(b) The break-even point is when $P = 0$ or, in other words, when we have sold enough lamps to recover our fixed cost. Thus let $P = 0$ in $P = 28x - 21{,}000$ and solve for x.

$$0 = 28x - 21{,}000$$
$$21{,}000 = 28x$$
$$\frac{21{,}000}{28} = x$$
$$750 = x$$

Thus we must sell 750 lamps to break even. If we sell less than 750 lamps, we have a loss. If we sell more than 750 lamps, we have a positive profit.

DO EXERCISES 4 AND 5.

DO SECTION PROBLEMS 6.7.

3. (a) If a typewriter originally cost \$500 and we are depreciating it at a constant rate of \$80 a year, find the linear equation involving the book value V after a number of years t of depreciation.
(b) How long will it take the book value to depreciate to \$320?

4. For Example 3, give a physical interpretation of the slope and intercept for the revenue, cost, and profit equation.

5. Jim Mayashi works on a set weekly salary of \$400 plus a commission of 3% of his weekly sales.
(a) Find the linear equation that involves Jim's total weekly pay P and the amount of his weekly sales S.
(b) What physical meaning can you attach to the slope of this line and the P-intercept?
(c) If Jim's sales in a week are \$4,000, what is his total pay for the week?

ANSWERS TO MARGINAL EXERCISES

1. (a) If x increases 1 unit, then y increases $\frac{7}{2}$ units. (b) If x increases 1 unit, then y decreases 4 units. **2.** $36.25 **3.** (a) $V = 500 - 80t$ (b) $2\frac{1}{4}$ years **4.** *Revenue:* With each additional lamp we sell, the revenue increases $60; the intercept means if we sell zero lamps, our revenue is zero. *Cost:* With each additional lamp we produce, the cost increases $32; the intercept means if we produce zero lamps, our cost is still $21,000. *Profit:* Each lamp we sell generates a profit of $28; the intercept means if we sell zero lamps, our profit is a loss of $21,000. **5.** (a) $P = 400 + .03S$ (b) For each dollar of sales, Jim's salary increases 3¢. If Jim's sales are zero, his salary is $400. (c) $520

NAME COURSE/SECTION DATE

ANSWERS

1. ______

2. (a) ______

(b) ______

(c) ______

3. (a) ______

(b) ______

4. (a) ______

(b) ______

5. ______

6. ______

SECTION PROBLEMS 6.7

1. A piece of machinery originally cost \$14,000 and depreciates at a constant rate of \$2500 a year. How long does it take to depreciate to one-half its original value?

2. A desk originally costs \$600 and depreciates at a constant rate of \$80 a year.

(a) Find the linear equation involving the book value V after a number of years t of depreciation.
(b) How long before the desk is worth \$300?
(c) What is the desk worth after $5\frac{1}{2}$ years?

3. Mrs. Larsen receives a salary of \$300 a week plus a commission of 6% of her weekly sales.

(a) Find the equation $y = mx + b$ relating weekly salary y and weekly sales x.
(b) How much should Mrs. Larsen sell for her weekly salary to be \$450?

4. Daniel Howard works on a straight commission of 8% of sales.

(a) Find the equation $y = mx + b$ relating weekly salary y and weekly sales x.
(b) What is Daniel's weekly salary in a week that his sales were \$5,600?

5. At a price of \$35, Mr. Zweig can sell 20 desk lamps a week, but at a price of \$45, he can only sell 15 desk lamps a week. Assume a linear relationship between price and demand. If Mr. Zweig wants to sell 18 desk lamps a week, how much should he charge for each lamp?

6. Tammy's Bar and Grill sells 40 hamburgers a day at a price of \$.50 each. If the price goes up to \$.65, Tammy's only sells 30 hamburgers a day. Assuming a linear relationship between price and number of hamburgers sold, how many hamburgers a day can Tammy's sell at \$.60 each?

7. (a) ________

(b) ________

8. (a) ________

(b) ________

(c) ________

9. (a) ________

(b) ________

10. (a) ________

(b) ________

11. (a) ________

(b) ________

12. (a) ________

(b) ________

7. The cost c to produce 20 units q of a product is \$500 and the cost to produce 50 units is \$1100.

(a) If the cost is linearly related to the number of units produced, find the linear equation $c = mq + b$.
(b) If the cost is \$800, how many units are being produced?

8. (a) If the relationship between number of units produced x and profit P is a linear relationship, find the linear equation $P = mx + b$ involving these two variables if producing 500 units results in a profit of \$2000 and producing 700 units results in a profit of \$3000.
(b) What is the profit if 800 units are produced?
(c) How many units would you have to produce to make a profit of \$2500?

9. The Adam Company sells all the tables it can produce at a price of \$80 a table. Adam's has a fixed cost of \$14,000 and a variable cost of \$25 for labor and \$20 for materials per table.

(a) Find the revenue, cost, and profit equations.
(b) Find the break-even point.

10. The Georgetown Historical Society is staging a concert to raise money. The band charges \$600. Tickets sell for \$4, but \$1 for every ticket sold must go toward renting the hall.

(a) What is the break-even point?
(b) How many tickets must the Society sell to realize a profit of \$450?

11. The WT Ad Agency claims that for every dollar you spend on television advertising, your weekly profit will increase \$12.

(a) If you presently spend \$200 on television advertising and your weekly profit is \$5000, find an equation expressing the linear relationship between profit P and amount x spent on television advertising. Hint: The slope is 12.
(b) What can you expect your profit to be if you spend \$300 on television advertising?

12. Sam currently has \$5,000 invested. The return on his investment is \$800. Sam is told that for every additional dollar he invests he will receive a return of 15¢. Assume a linear relationship between amount invested x and return y on investment.

(a) Find the equation $y = mx + b$.
(b) How much should Sam invest to have a return of \$1000?

NAME ________ COURSE/SECTION ________ DATE ________

ANSWERS

1. ________
2. ________
3. ________
4. ________
5. ________
6. ________
7. ________
8. ________
9. ________
10. ________
11. ________
12. ________
13. See graph.
14. See graph.
15. See graph.
16. See graph.
17. See graph.
18. See graph.

CHAPTER 6 REVIEW PROBLEMS

SECTION 6.1

Are the following linear equations in two variables? If not, why not?

1. $s = 3t^2 + 2t + 1$

2. $y = 2x - 3w$

3. $x - 3y = 4$

4. Which of the following are solutions to $x - 3y = -2$?

(a) $(1, 1)$ (b) $(0, -3)$ (c) $(-1, 2)$ (d) $(4, 2)$

Put in standard form.

5. $\frac{y}{4} = 3x - 1$

6. $2(3x - 4y) = 2x + 5$

7. Find two ordered pairs that are solutions to $2x - y = 5$ and one ordered pair that is not.

8. If $(m, 3)$ is a solution to the equation $x - 4y = 7$, what is the value of m?

SECTION 6.2

Tell which quadrant the following points would be in.

9. x-coordinate positive, y-coordinate negative

10. x-coordinate negative, y-coordinate negative

11. both coordinates positive

12. first-coordinate negative, second-coordinate positive

Plot the following points.

13. $R = (2, -5)$

14. $S = (4, 0)$

15. $T = (-\frac{1}{2}, -3)$

16. $U = (1\frac{1}{4}, 2\frac{1}{2})$

17. $V = (-3, -4)$

18. $W = (0, -\frac{10}{3})$

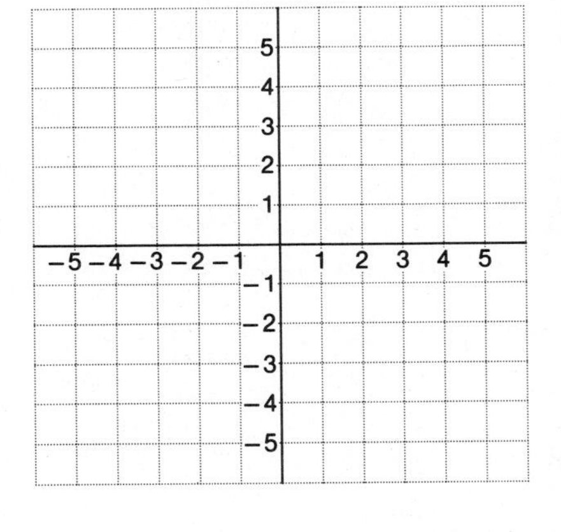

19. ______________

20. ______________

21. ______________

22. ______________

23. ______________

24. See graph.

25. See graph.

26. See graph.

27. ______________

28. ______________

Find the approximate coordinates of the following points.

19. P

20. Q

21. R

22. S

23. T

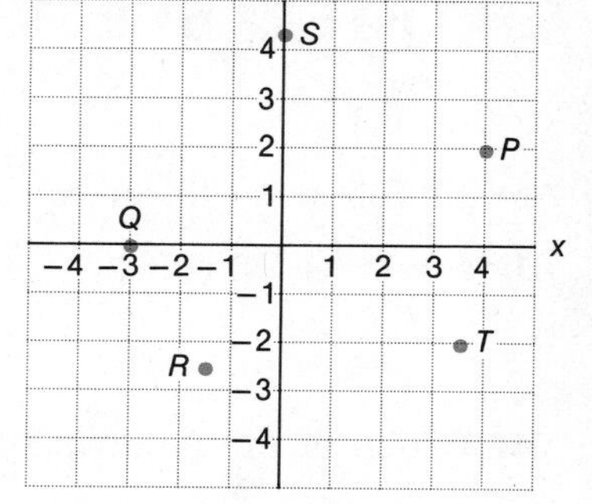

SECTION 6.3

Graph.

24. $y = 2x + 1$

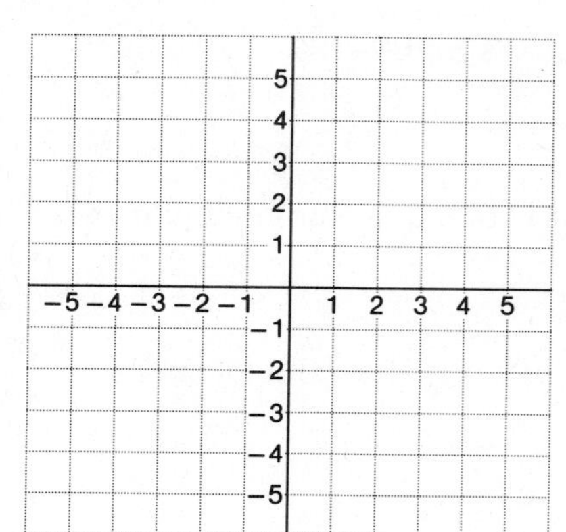

25. $t = |s - 2|$

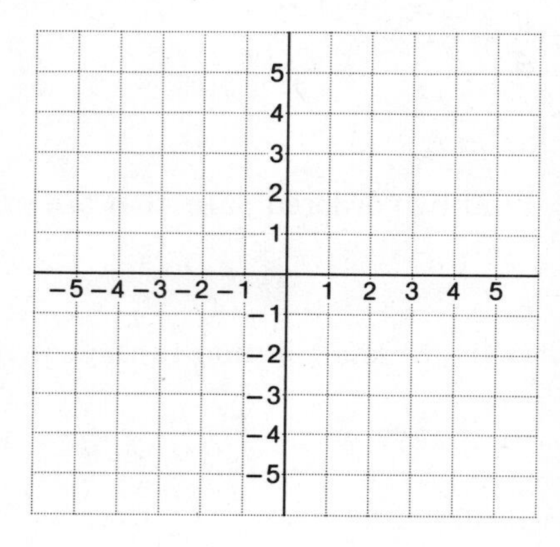

26. $b = (a + 1)^2$

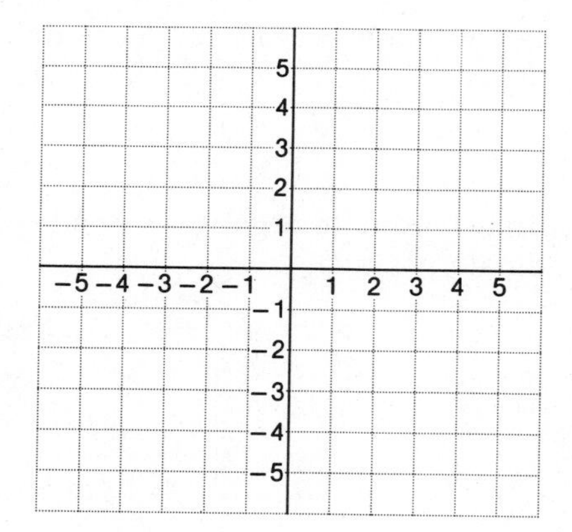

SECTION 6.4

Put in standard form.

27. $y = 3x - 4$

28. $-2a + b - 14 = 0$

NAME COURSE/SECTION DATE ANSWERS

29. Find the x- and y-intercepts for $3x - 4y = -12$.

30. Find the a- and b-intercepts for $b = \dfrac{3a}{4} - 2$.

Graph.

31. $y = \frac{7}{2}x - 2$

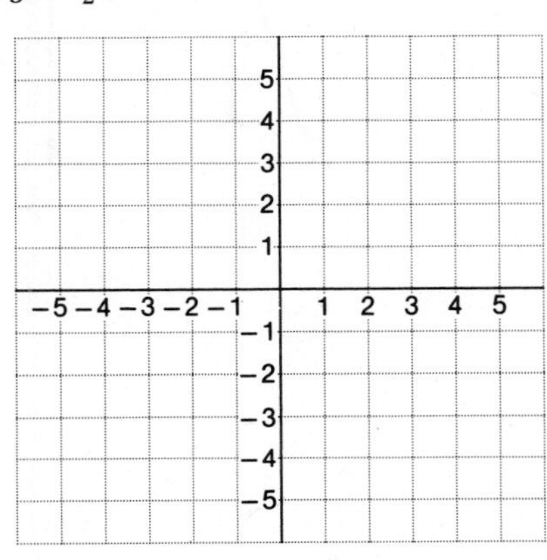

32. $y = -3\frac{1}{3}$

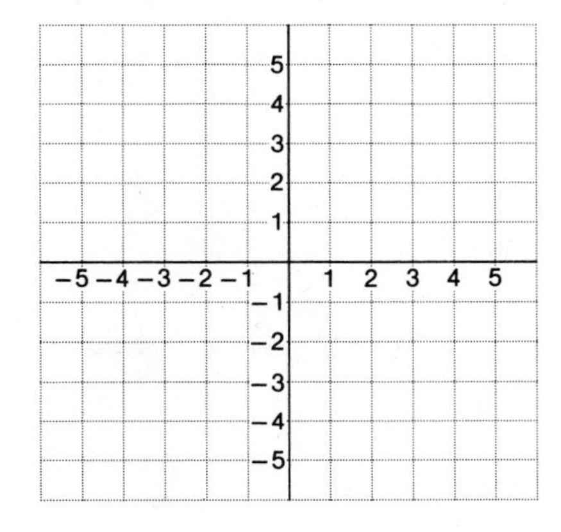

33. $x - 2.5 = 0$

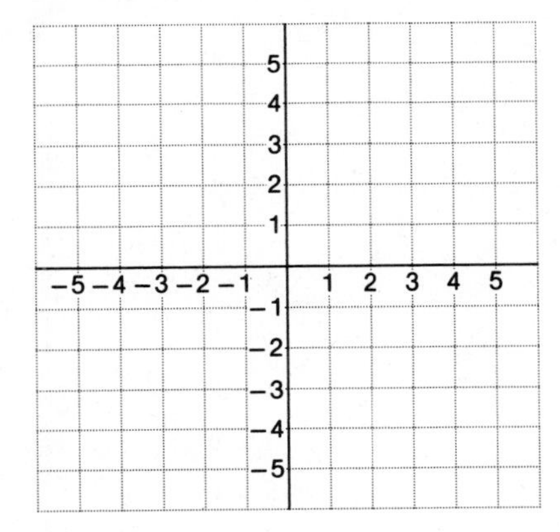

SECTION 6.5

Find the slope of the line that goes through the two given points.

34. $(-2, 4)$ and $(3, 1)$

35. $(2, -5)$ and $(-3, -10)$

36. Find the slope of the line $3x - y = 9$.

29. ______

30. ______

31. See graph.

32. See graph.

33. See graph.

34. ______

35. ______

36. ______

37. ______

38. ______

39. ______

40. ______

41. ______

42. See graph.

43. See graph.

44. See graph.

45. ______

46. ______

47. ______

48. ______

49. ______

50. ______

37. Find the slope of the line $2x - 5 = 7$.

Find the slopes of the following lines.

38. L_1 **39.** L_2

40. L_3 **41.** L_4

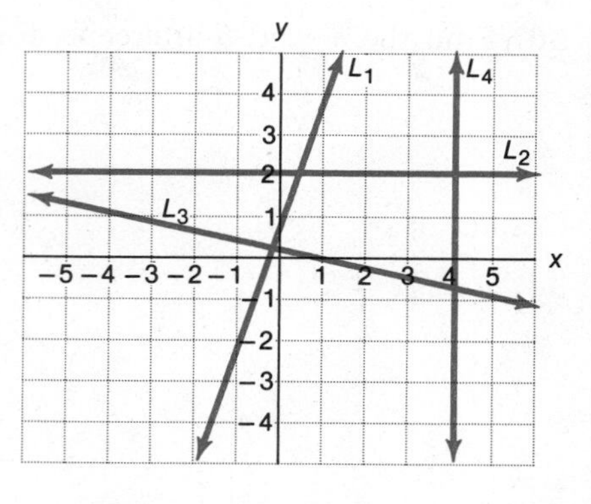

Draw lines through (1, 1) that have the following slopes.

42. $m = 4$ **43.** $m = -\frac{2}{3}$

44. m undefined

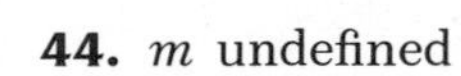

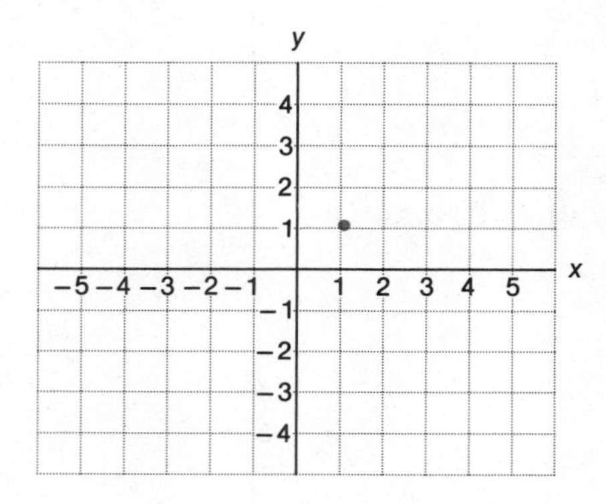

SECTION 6.6

Put in slope-intercept form.

45. $-x + y = 5$ **46.** $s - 5 = t$ **47.** $y - 7 = 0$

Find the slope and vertical axis intercept.

48. $y = \frac{3}{2}x - 4$ **49.** $2x - 5y = 10$ **50.** $s = 2$

NAME COURSE/SECTION DATE

ANSWERS

51. ______
52. ______
53. ______
54. ______
55. ______
56. ______
57. ______
58. ______
59. ______

Find the equation of the line with the given slope (m) that goes through the given point (P).

51. $m = -\frac{1}{3}$, $P = (-6, 1)$

52. $m = 0$, $P = (2, 5)$

53. m is undefined, $P = (3, \frac{2}{3})$

54. $m = 1.2$, $P = (3, 2)$

Find the equation of the line through the given points (P_1 and P_2).

55. $P_1 = (2, 3)$, $P_2 = (4, 7)$

56. $P_1 = (-2, 5)$, $P_2 = (1, -2)$

57. $P_1 = (1, 7)$, $P_2 = (1, -3)$

58. $P_1 = (0, 5)$, $P_2 = (2, 5)$

59. Find the equation of the line parallel to the line $y = -3x + 4$ that goes through the point $(2, 5)$.

60. ____________

61. (a) ____________

(b) ____________

(c) ____________

62. (a) ____________

(b) ____________

60. Find the equation of the line parallel to the line $3x - 2y = 5$ that goes through the point $(-4, 1)$.

SECTION 6.7

61. A truck costs $12,000 and depreciates at a constant rate of $2400 a year.

(a) Find the linear equation $V = mt + b$ involving book value V and length of depreciation t.
(b) What is the truck worth after $3\frac{1}{3}$ years?
(c) When is the truck worth $2,000?

62. At a price of $65, Samuel's sells 20 desk lamps a week, but at a price of $80 they only sell 15 desk lamps a week. Assume a linear relationship between price and number of lamps sold.

(a) Find how many Samuel's will sell if the price is set at $50.
(b) If the price is $75, how many will they sell?

NAME COURSE/SECTION DATE

ANSWERS

1. (a) ______
(b) ______
(c) ______
(d) ______
(e) ______

2. ______

3. See test sheet.

4. See graph.

5. See graph.

CHAPTER 6 TEST

1. Which of the following are linear equations in two variables? If not, why not?

(a) $s - t = 5 + v$ (b) $2y = 3x - 4$ (c) $a - 4b + 5 = 0$
(d) $y = |x| - 3$ (e) $x - xy = 4$

2. If $(7, n)$ is a solution to the equation $x - 3y = 4$, what is the value of n?

3. Match the points on the graph with their coordinates.

P	$(0, 1)$
Q	$(3, 2)$
R	$(1, 0)$
S	$(-\frac{5}{2}, -\frac{3}{2})$
T	$(2, 3)$
	$(-3, 1)$
	$(0, -2)$

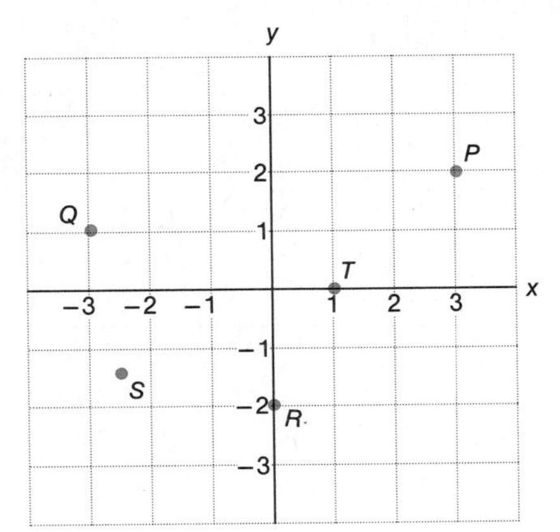

4. Graph $-2x + y = 4$.

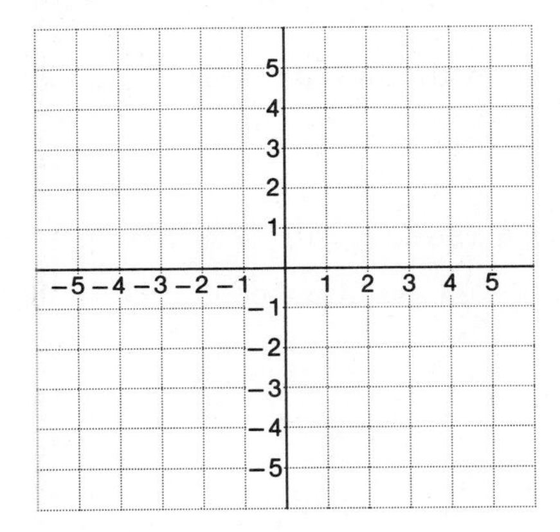

5. Graph $t = -3s - 1$.

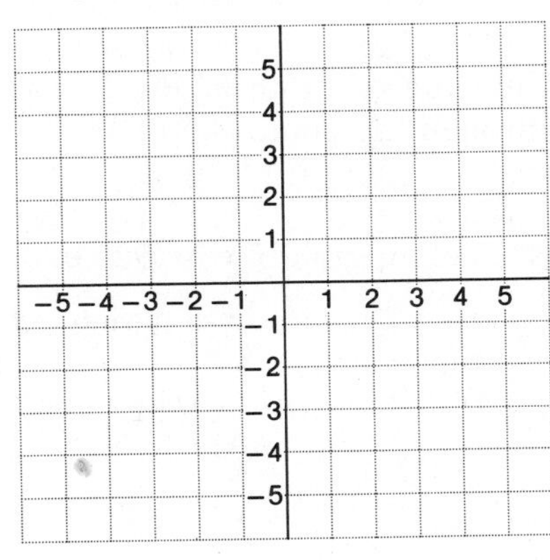

6. See graph.

7. See graph.

8. ______________

9. ______________

10. See graph.

11. ______________

12. ______________

13. ______________

14. ______________

15. (a) ______________

(b) ______________

6. Graph $x + 1 = 4$.

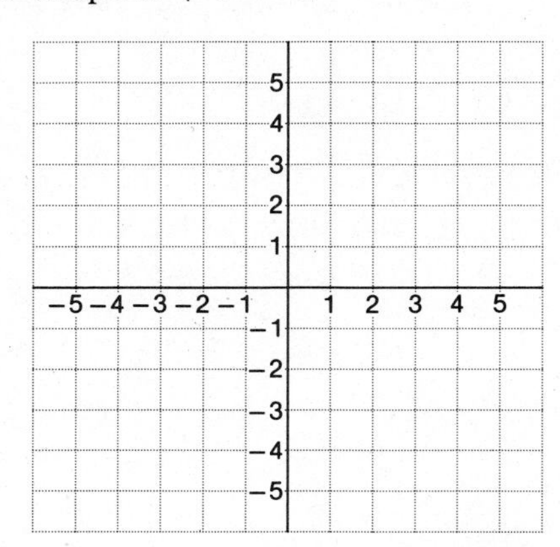

7. Graph $b = |a + 2|$

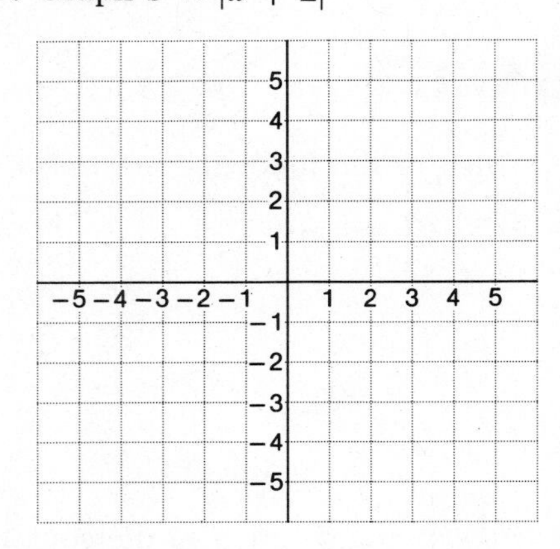

8. Find the slope and y-intercept of the line $y = \frac{3}{2}x - 4$.

9. Find the slope, x-intercept, and y-intercept of the line $2x - 3y = -12$.

10. Draw a line through (1,1) with slope $\frac{4}{3}$.

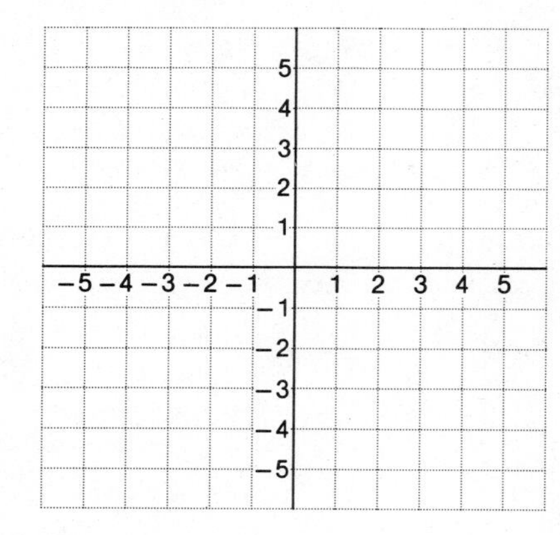

11. Find the slope of lines L_1 and L_2.

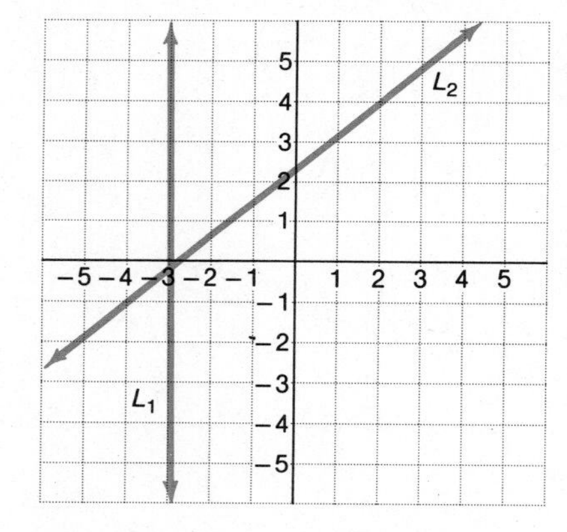

12. Find the equation of the line with slope −3 that goes through the point (1, 4).

13. Find the equation of the line that goes through the points (2, −5) and (4, 3).

14. Find the equation of the line that is perpendicular to the line $2x - 3y = 6$ and goes through the point (4, −1).

15. A truck originally cost $25,000 and depreciates at a constant rate of $6,000 per year.

(a) Find the linear equation relating book value V and the number of years t it has depreciated.

(b) How long will it take to depreciate to $10,000?

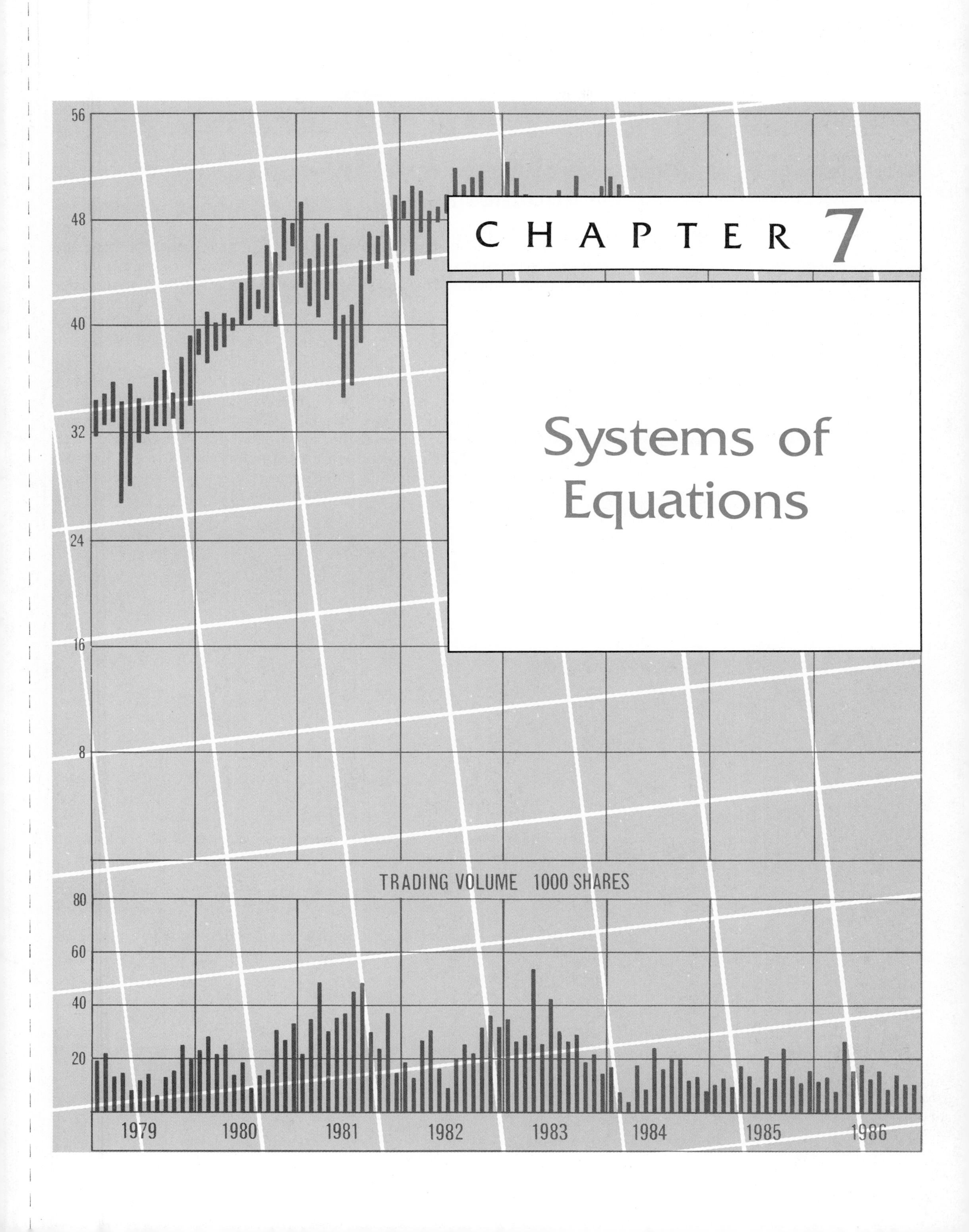

CHAPTER 7

Systems of Equations

LEARNING OBJECTIVES

After finishing Section 7.1, you should be able to:

- Determine if an ordered pair (a, b) is a solution to a given system of two linear equations in two unknowns.
- Identify the different categories of solutions to a system of two linear equations in two unknowns.

7.1

Introduction to Systems of Equations

In this chapter we look at solutions to and practical applications of systems of equations. Systems of equations are useful in many situations, particularly in solving word problems when it is convenient to use more than one variable (unknown).

In Chapter 3 we learned to solve first-degree equations in one variable (such as the equation $2x + 1 = 7$). We also found, with the exception of cases in which there is no solution or infinitely many solutions, that such equations have a unique solution. For example, the solution to $2x + 1 = 7$ is the single number 3. However, in Chapter 6 we learned that there are infinitely many solutions to linear equations in two variables (such as $2x + y = 4$). The solutions are not single numbers, but rather are ordered pairs (points). Geometrically the solutions to such an equation are all points on a straight line. For example, if we plot all the solutions to the equation $2x + y = 4$, we obtain the following graph.

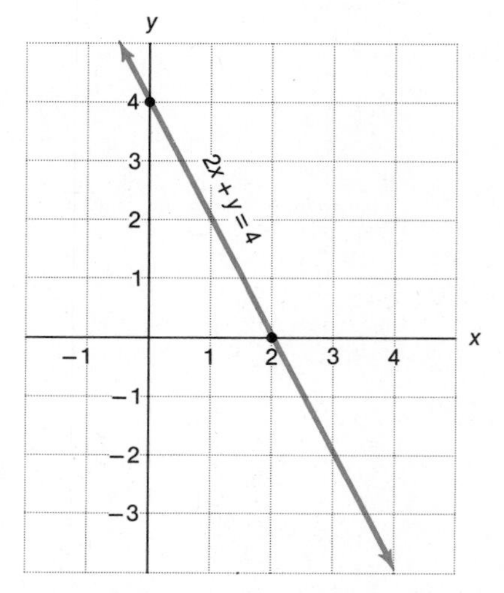

In this chapter we look at pairs of linear equations in two variables (unknowns), such as

$$\begin{Bmatrix} x + y = 3 \\ 2x - y = 0 \end{Bmatrix}$$

which we call a *system of equations*. We are particularly interested in the solutions to such a system.

Let us consider the system

$$\begin{Bmatrix} x + y = 3 \\ 2x - y = 0 \end{Bmatrix}$$

and look separately at the solutions to each of the two equations in the system.

The equation

$$x + y = 3$$

has for solutions all the points on the line at the right.

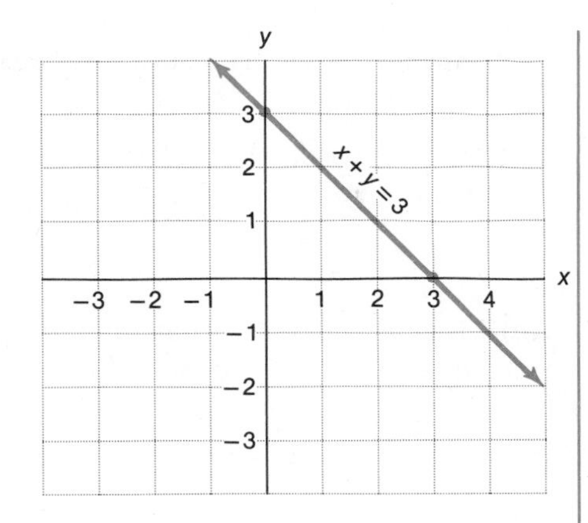

The equation

$$2x - y = 0$$

has for solutions all the points on the line at the right.

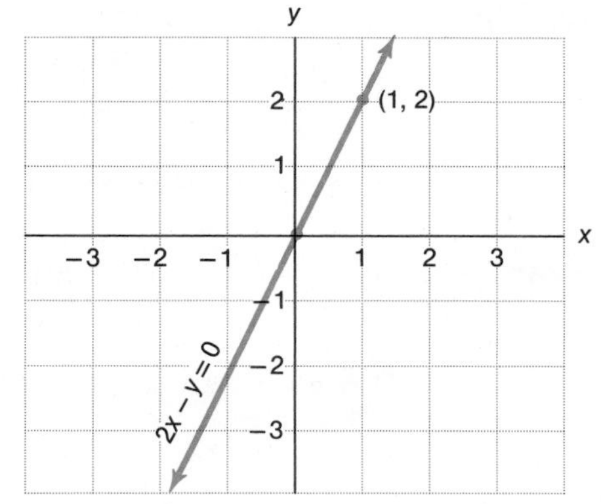

Now, if we draw the graphs of the solutions to both equations on the same coordinate axis, we obtain the following graph.

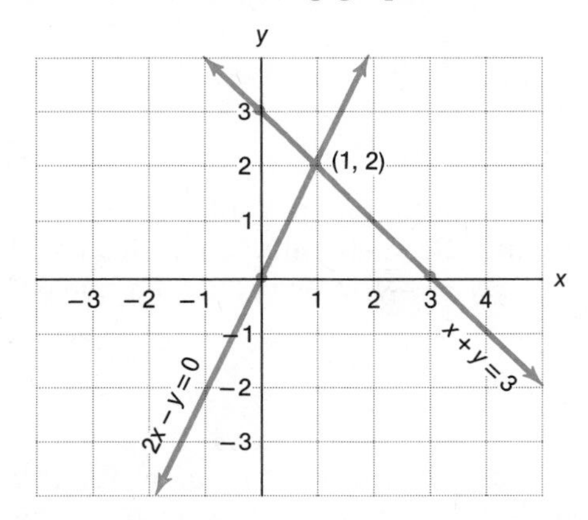

Notice that there is one point, (1, 2), that is common to both lines and hence is a *common solution to both equations.* This leads to the following definition.

DEFINITION 7.1 An ordered pair (a, b) is a *solution* to a system of two equations in two unknowns, such as

$$\begin{Bmatrix} x + y = 3 \\ 2x - y = 0 \end{Bmatrix}$$

if the ordered pair satisfies both equations at the same time. Geometrically this means (a, b) must be a common point on the graph of both equations.

1. Is $(1, -2)$ a solution to the system $\begin{Bmatrix} 2x - y = 4 \\ x + y = 1 \end{Bmatrix}$?

2. (a) Is $(2, 3)$ a solution to the system $\begin{Bmatrix} -x + y = 1 \\ 2x - y = 1 \end{Bmatrix}$?

(b) Geometrically, what does this tell you about the graphs of the lines $-x + y = 1$ and $2x - y = 1$?

Example 1 Is $(4, 2)$ a solution to the system $\begin{Bmatrix} x - y = 2 \\ -x + 3y = -1 \end{Bmatrix}$?

(a) $(4, 2)$ is a solution for the equation $x - y = 2$.

$x - y$	2
$4 - 2$	2
2	2

(b) $(4, 2)$ is not a solution for the equation $-x + 3y = -1$.

$-x + 3y$	-1
$-4 + 3(2)$	-1
$-4 + 6$	-1
2	-1

(c) Thus $(4, 2)$ is not a solution to the system

$$\begin{Bmatrix} x - y = 2 \\ -x + 3y = -1 \end{Bmatrix}$$

since it does not satisfy both equations at the same time.

Example 2 Is $(2, -1)$ a solution to the system $\begin{Bmatrix} 2x - y = 5 \\ x + 3y = -1 \end{Bmatrix}$?

(a) $(2, -1)$ is a solution for the equation $2x - y = 5$.

$2x - y$	5
$2(2) - (-1)$	5
$4 + 1$	5
5	5

(b) $(2, -1)$ is a solution for the equation $x + 3y = -1$.

$x + 3y$	-1
$2 + 3(-1)$	-1
$2 - 3$	-1
-1	-1

(c) Thus $(2, -1)$ is a solution (in fact the only one, as we shall see later) to the system of equations. Geometrically this means that the two lines $2x - y = 5$ and $x + 3y = -1$ intersect at a unique point, namely $(2, -1)$.

DO EXERCISES 1 AND 2.

For convenience, throughout the rest of this chapter when we use the description "a 2-by-2 system," we will always mean a system of two equations where each equation is a linear equation in two variables.

Before we learn how to solve systems of equations, we should look at a very convenient and informative relationship between the various types of solutions to a 2-by-2 system and the graphs of pairs of straight lines. If we graph the solutions to each equation in a 2-by-2 system, we will have two straight lines. If we graph two arbitrary lines, there are three possibilities for the interaction of the lines, and hence there are three possibilities for the solution to a 2-by-2 system. The three possibilities are as follows.

1. The two lines may *intersect* at a point.

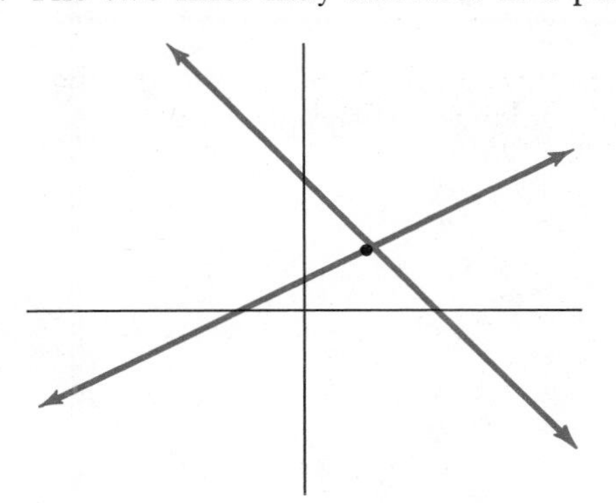

In this case the system has a *unique* (only one) solution (which is the point where they intersect).

2. The two lines may be *parallel*.

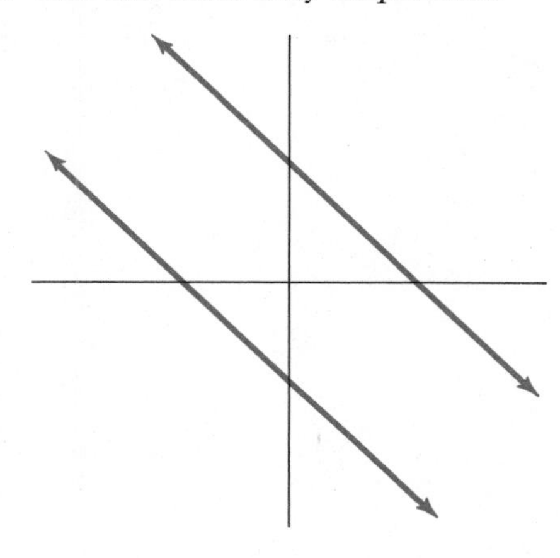

In this case the system has *no solution,* that is, the lines never intersect and hence never have a point in common.

3. The two lines may actually graph as the *same line*.

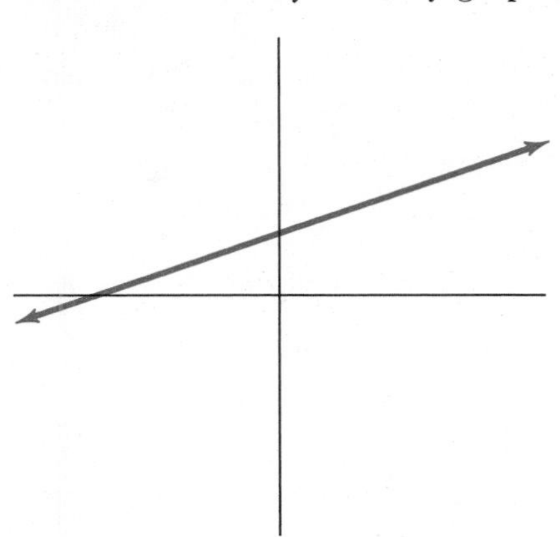

In this case the system has *infinitely many solutions;* that is, the two lines intersect infinitely often and all the points on the line satisfy both equations.

Thus for a 2-by-2 system, we will have a *unique solution, no solution,* or *infinitely many solutions.* Just as in the case of linear equations in one variable, the "no solution" and "infinitely many solutions" cases are not as common as the "unique solution" case.

DO EXERCISE 3.

DO SECTION PROBLEMS 7.1.

3. If the two equations in a 2-by-2 system were to have the following graphs, what could we say about the type of solution to the system?

(a)

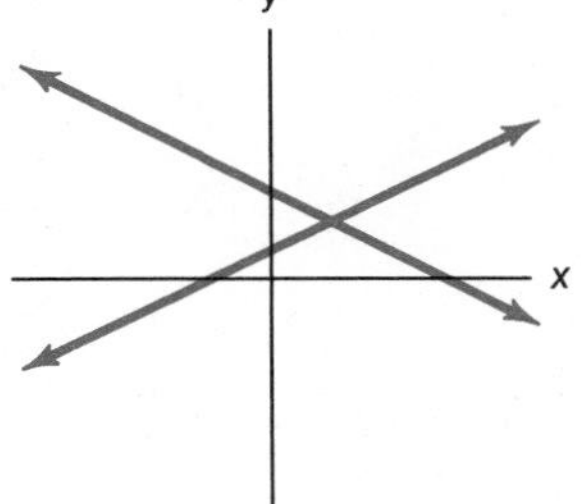

(b)

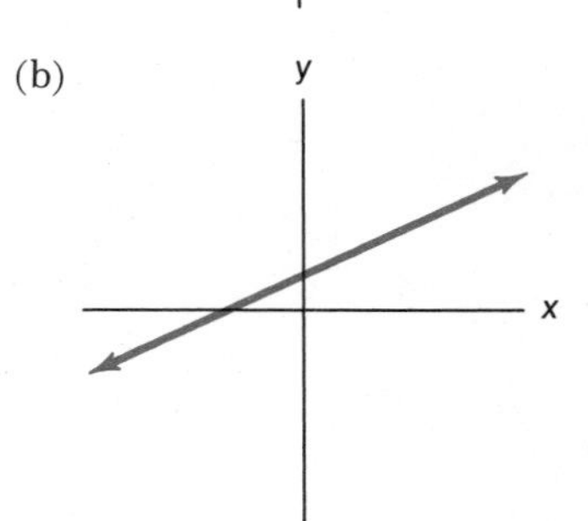

(c)

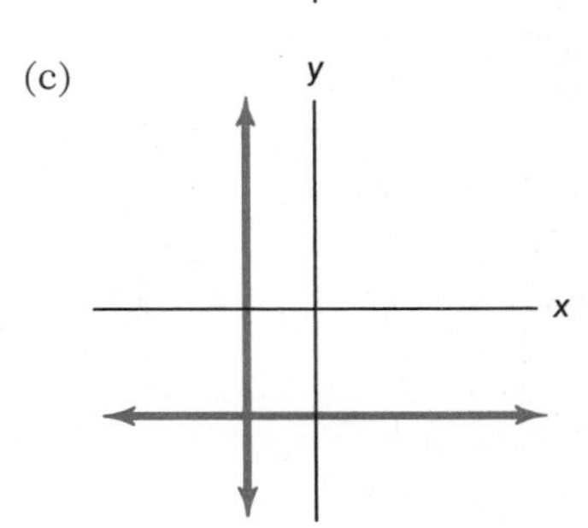

(d)

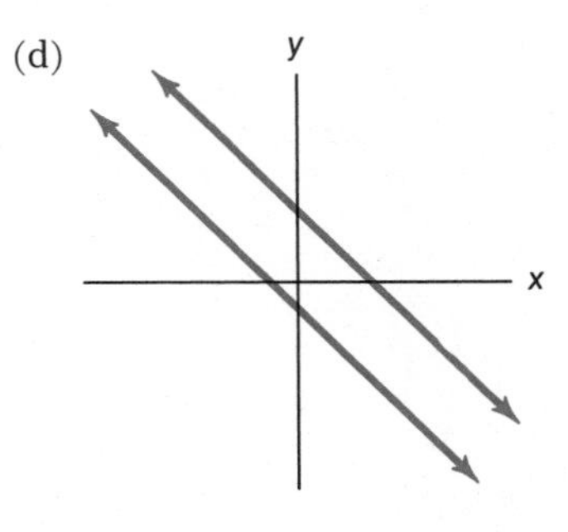

ANSWERS TO MARGINAL EXERCISES

1. No **2.** (a) Yes (b) They intersect at the point (2, 3). **3.** (a) Unique solution (b) Infinitely many solutions (c) Unique solution (d) No solution

NAME ______ COURSE/SECTION ______ DATE ______

SECTION PROBLEMS 7.1

1. Given the systems of equations $\begin{Bmatrix} x + 2y = 1 \\ 2x - y = 2 \end{Bmatrix}$.

(a) Is (3, 1) a solution?
(b) Is (1, 0) a solution?
(c) Is (0, −2) a solution?

2. Given the system of equations $\begin{Bmatrix} 2s - 3t = -8 \\ -s + t = 3 \end{Bmatrix}$.

(a) Is (−4, 0) a solution?
(b) Is (−1, 2) a solution?

3. Given the system of equations $\begin{Bmatrix} -a + 3b = -2 \\ 2a - 6b = 4 \end{Bmatrix}$.

(a) Is (2, 0) a solution?
(b) Is (−1, −1) a solution?
(c) Is $(3, \frac{1}{3})$ a solution?

4. Given the system of equations $\begin{Bmatrix} 3x - y = 2 \\ -2x + y = 1 \end{Bmatrix}$.

(a) Is (1, 1) a solution?
(b) Is (2, 4) a solution?
(c) Is (3, 7) a solution?

ANSWERS

1. (a) ______
(b) ______
(c) ______

2. (a) ______
(b) ______

3. (a) ______
(b) ______
(c) ______

4. (a) ______
(b) ______
(c) ______

5. ________

6. ________

7. ________

8. ________

9. (a) ________

(b) ________

(c) ________

10. ________

Which type of solution (unique, no solution, or infinitely many solutions) would a 2-by-2 system have if the graphs of the two equations in the system are as follows.

5.

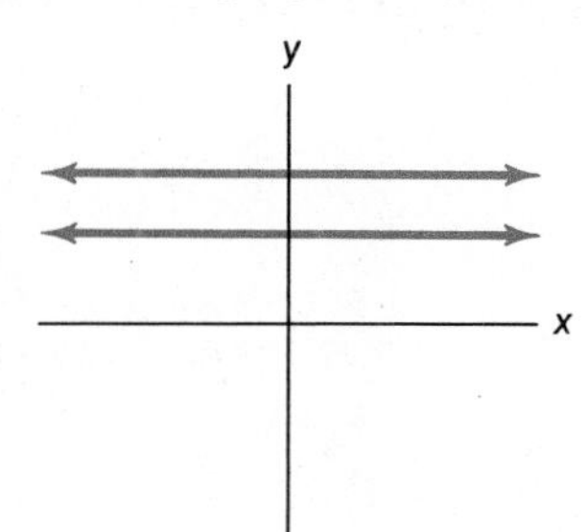

6.

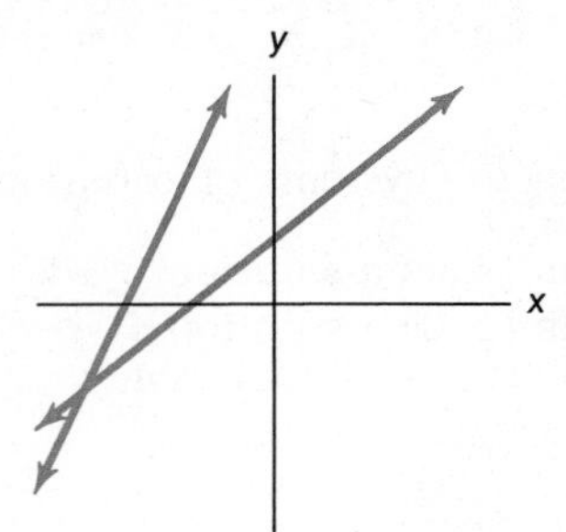

7.

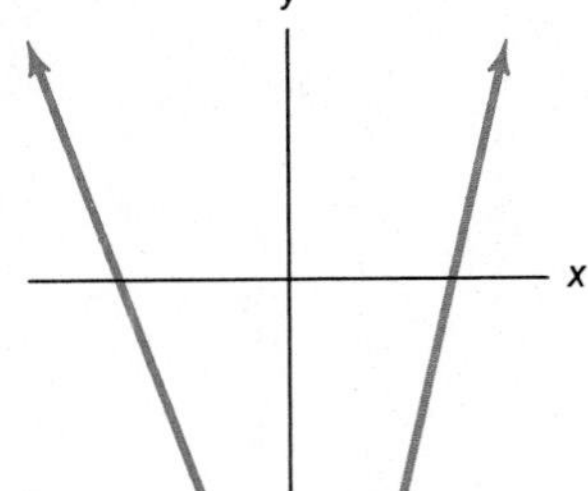

8.

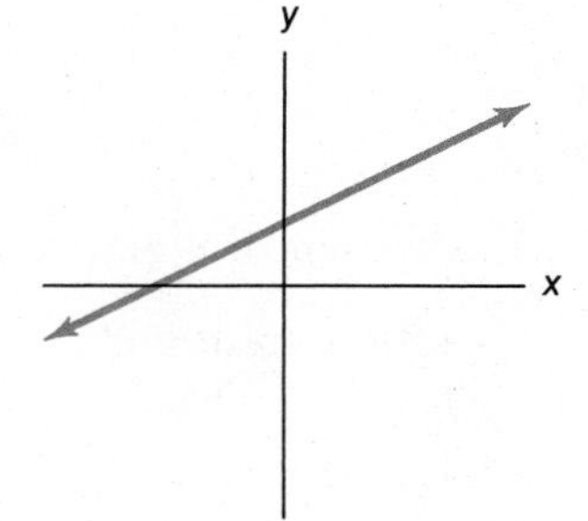

9. Given a 2-by-2 system where the graph of each equation is a line, what is the relationship between the slopes of the two lines and the y-intercepts of the two lines if the system has the following type of solutions?

(a) Unique solution
(b) No solution
(c) Infinitely many solutions

Hint: Recall that parallel lines have the same slope.

10. Is $(\frac{1}{2}, -2)$ a solution to the system $\left\{\begin{aligned} 4x - y &= 4 \\ -2x + 3y &= -7 \end{aligned}\right\}$?

7.2

The Graphing Method and Word Problems

The first method for solving 2-by-2 systems is the graphing method.

STEP 1: GRAPH

Graph both equations in the system on the same coordinate axes.

STEP 2: SOLUTION

Determine the solution to the system from the graph.

1. *Unique solution* The lines *intersect.*
   Read the coordinates at the point where the lines intersect.
2. *No solution* The lines are *parallel.*
3. *Infinitely many solutions* The lines are actually the same line.

STEP 3: CHECK

If there is a unique solution, check the coordinates of the point in both equations, since it is often difficult to read accurately the coordinates of a point where two lines intersect.

Example 1 Use the graphing method to solve $\begin{cases} x - y = 1 \\ 3x - y = -1 \end{cases}$

STEP 1: GRAPH

Graph $x - y = 1$ and $3x - y = -1$ on the same coordinate axis.

STEP 2: SOLUTION

There is a unique solution where the lines intersect. It appears the lines intersect at $(-1, -2)$.

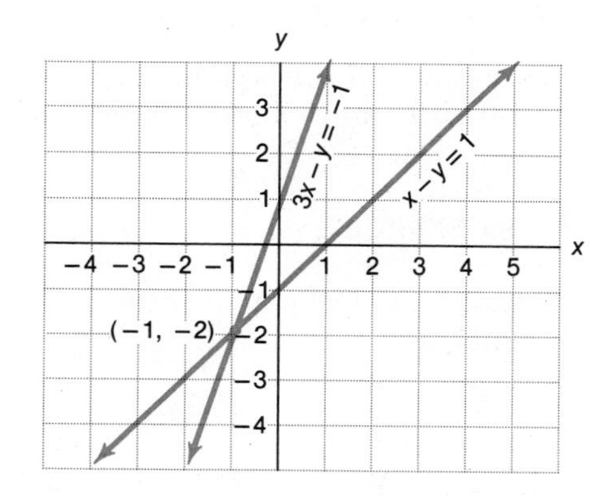

STEP 3: CHECK

Substitute $(-1, -2)$ in both equations.

(1)

$x - y$	1
$-1 - (-2)$	1
$-1 + 2$	1
1	1

(2)

$3x - y$	-1
$3(-1) - (-2)$	-1
$-3 + 2$	-1
-1	-1

Thus the solution to the system is $(-1, -2)$.

DO EXERCISE 1.

LEARNING OBJECTIVES

After finishing Section 7.2, you should be able to:

- Solve a 2-by-2 system by the graphing method.
- Solve word problems requiring the use of a 2-by-2 system.

Solve by the graphing method.

1. $\begin{cases} x + y = 3 \\ -2x + y = 6 \end{cases}$

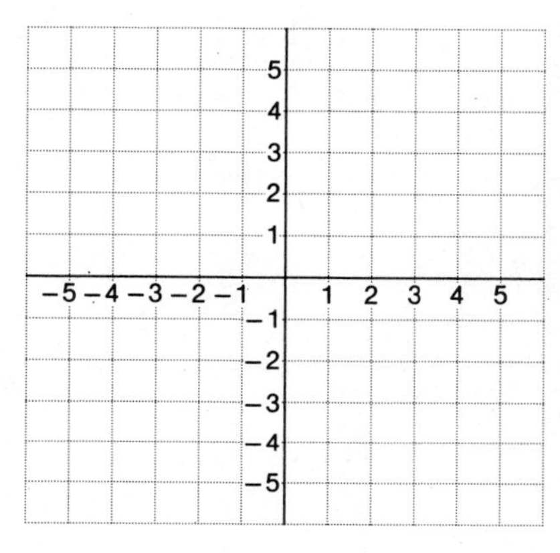

Solve by the graphing method.

2. $\begin{Bmatrix} x - 2y = 1 \\ -2x + 4y = -2 \end{Bmatrix}$

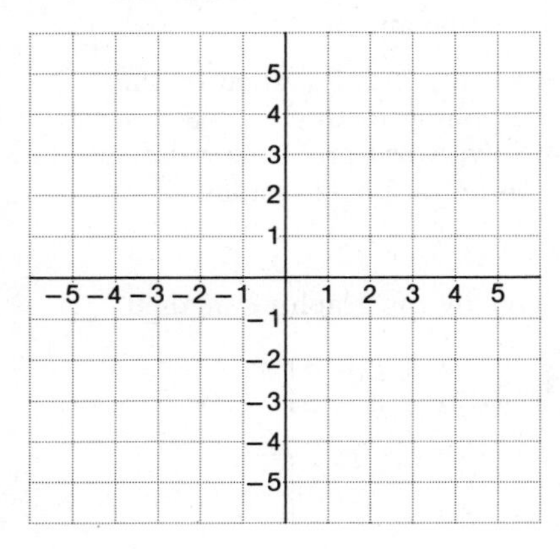

Example 2 Use the graphing method to solve $\begin{Bmatrix} 2s - t = 2 \\ -4s + 2t = 4 \end{Bmatrix}$.

STEP 1: GRAPH

Graph $2s - t = 2$ and $-4s + 2t = 4$ on the same coordinate axis.

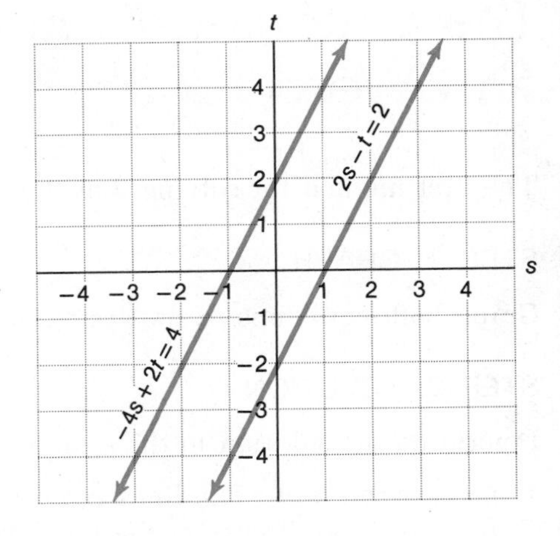

STEP 2: SOLUTION

The lines appear to be parallel, and thus there is no solution to the system.

STEP 3: CHECK

To see if the two lines are actually parallel we find the slope of each line.

(a) $2s - t = 2$

$-t = -2s + 2$

$t = 2s - 2$

Thus, the slope is 2.

(b) $-4s + 2t = 4$

$2t = 4s + 4$

$t = 2s + 2$

Thus, the slope is 2.

We see that both lines have the same slope and hence they must be parallel. (Note that the lines have different t-intercepts, which means that they are not the same line.)

DO EXERCISE 2.

Recall from Chapter 3 the five steps in solving a word problem: Picture (optional, as a picture will not help in all word problems), Variable, Equation, Solve, and Check.

Example 3 Wilfred Brown drops off his five children at the movies. Tickets cost \$2 for anyone under 12 and \$4 for anyone 12 or older. If the total cost of the tickets is \$16, how many children does Wilfred have who are under 12 years of age?

STEP 1: PICTURE

A picture will not help in this problem.

STEP 2: VARIABLE

This problem could be solved with one or two variables. We will use two variables.

Let x = number of children under 12, and

y = number of children 12 or older.

STEP 3: EQUATION

If we solve a problem using one variable, we need only one equation. However, if we solve a problem using two variables, then we need a system of two equations.

(1) The total number of children is 5.

$$x + y = 5$$

(2) $\begin{pmatrix}\text{Total cost}\\ \text{for children}\\ \text{under 12}\end{pmatrix} + \begin{pmatrix}\text{Total cost}\\ \text{for children}\\ \text{12 or older}\end{pmatrix} = \begin{pmatrix}\text{Total}\\ \text{cost}\end{pmatrix}$

$$2x + 4y = 16$$

Thus our system is

$$\begin{Bmatrix} x + y = 5 \\ 2x + 4y = 16 \end{Bmatrix}$$

STEP 4: SOLVE

We will solve the system by using the graphing method.

Graph

Solution

The unique solution is (2, 3).

Check

Substitute (2, 3) in both equations.

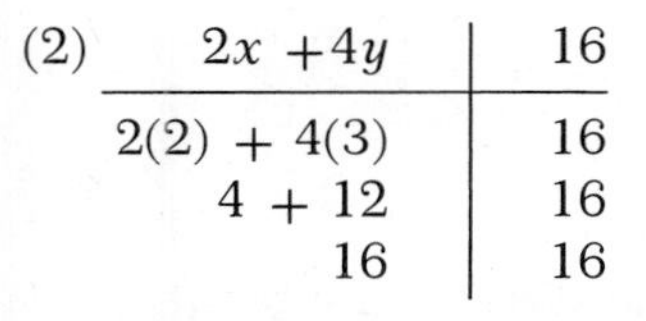

(1) $x + y$	5
$2 + 3$	5
5	5

(2) $2x + 4y$	16
$2(2) + 4(3)$	16
$4 + 12$	16
16	16

Thus Wilfred has 2 children under 12 (and 3 children 12 or older).

DO EXERCISE 3.

Example 4 Kathy and Kim Barnes pick 10 boxes of strawberries. If Kim picks two more than Kathy, how many boxes does each woman pick?

STEP 2: VARIABLE

Let x = number of boxes Kathy picks,
and y = number of boxes Kim picks.

3. Jennifer Ludlow has 7 dogs. Each female dog eats 4 pounds of dog food a week, and each male dog eats 5 pounds a week. If the dogs consume 30 pounds of dog food a week, how many female dogs does Jennifer have?

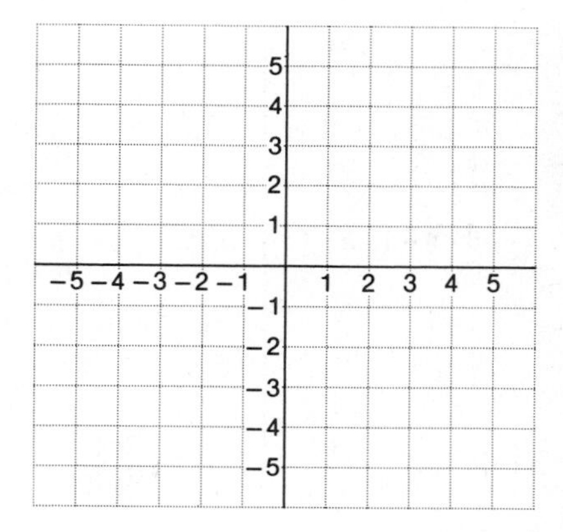

4. Together Paul and Abraham have ten brothers and sisters. If Paul has one more than twice as many as Abraham, how many brothers and sisters do Paul and Abraham each have?

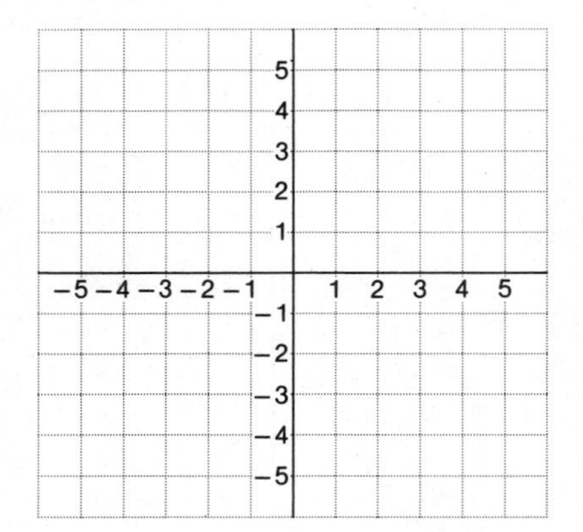

STEP 3: EQUATION

We will need a 2-by-2 system.

$$\begin{pmatrix}\text{Number of boxes}\\ \text{Kathy picks}\end{pmatrix} + \begin{pmatrix}\text{Number of boxes}\\ \text{Kim picks}\end{pmatrix} = \begin{pmatrix}\text{Total number}\\ \text{of boxes picked}\end{pmatrix}$$

$$(1) \qquad x + y = 10$$

Kim picks two more boxes than Kathy.

$$(2) \qquad y = x + 2$$

Thus the system is $\begin{Bmatrix} x + y = 10 \\ y = x + 2 \end{Bmatrix}$.

STEP 4: SOLVE

We will use the graphing method.

Graph

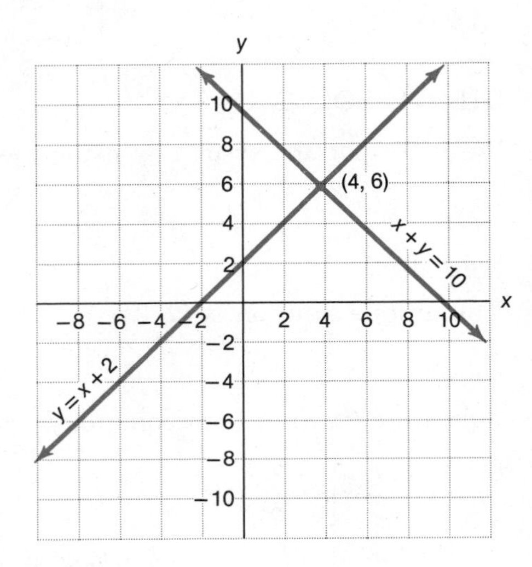

Solution

The solution is (4, 6).

Check

(1)

$x + y$	10
$4 + 6$	10
10	10

(2)

y	$x + 2$
6	$4 + 2$
6	6

Thus Kathy picks 4 boxes, and Kim picks 6 boxes.

DO EXERCISE 4.

Since it is often difficult to read accurately the coordinates of a point where two lines intersect, the graphing method for solving a 2-by-2 system is somewhat limited in its use. The graphing method is all right when a system has no solution (parallel lines), infinitely many solutions (same line), or a unique solution (intersecting lines) for which the coordinates of the solution are integers. In other cases the graphing method is not reliable, and we will develop other methods of solving 2-by-2 systems that will always be reliable.

DO SECTION PROBLEMS 7.2.

ANSWERS TO MARGINAL EXERCISES

1. $(-1, 4)$ **2.** Infinitely many solutions **3.** 5 **4.** Paul has 7 and Abraham has 3.

NAME

COURSE/SECTION

DATE

SECTION PROBLEMS 7.2

Solve the following systems by the graphing method.

1. $\begin{Bmatrix} 2x + y = 5 \\ x - y = 1 \end{Bmatrix}$

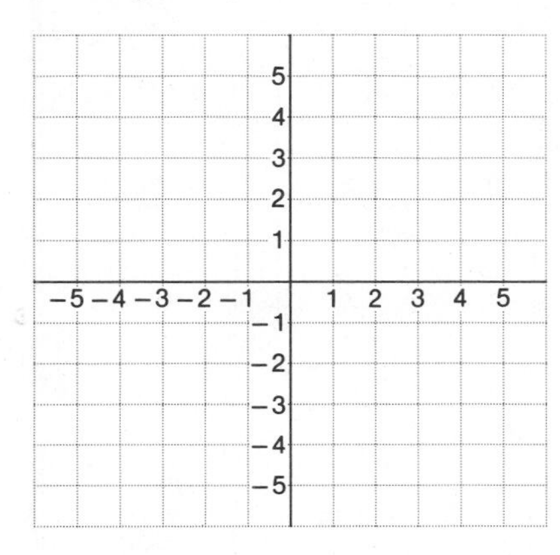

2. $\begin{Bmatrix} s + t = 3 \\ -2s + 2t = 4 \end{Bmatrix}$

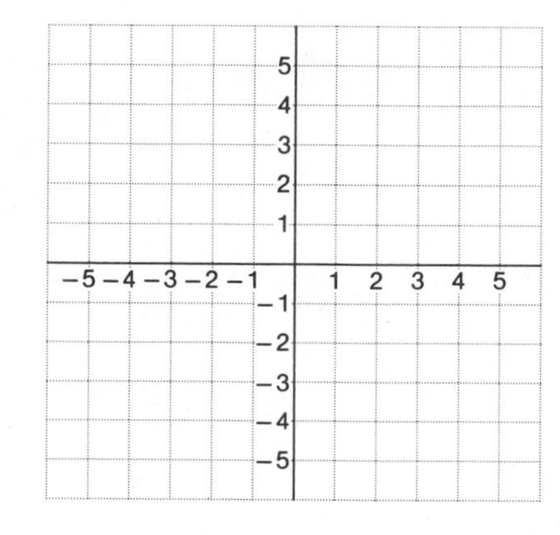

3. $\begin{Bmatrix} 2a + 3 = 0 \\ b - 3 = 0 \end{Bmatrix}$

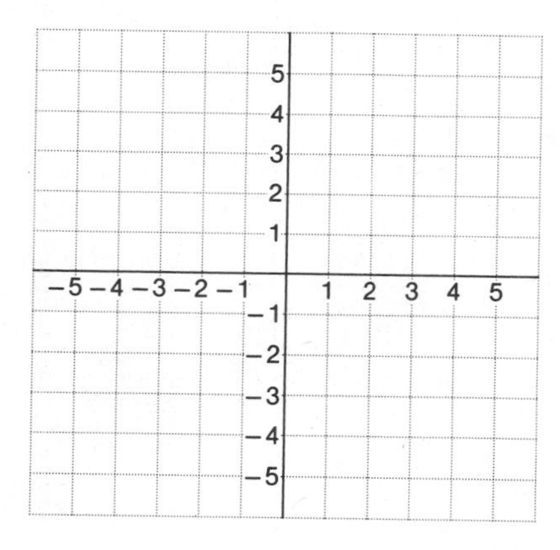

4. $\begin{Bmatrix} x + 2y = -3 \\ -\dfrac{x}{2} - y = \dfrac{7}{2} \end{Bmatrix}$

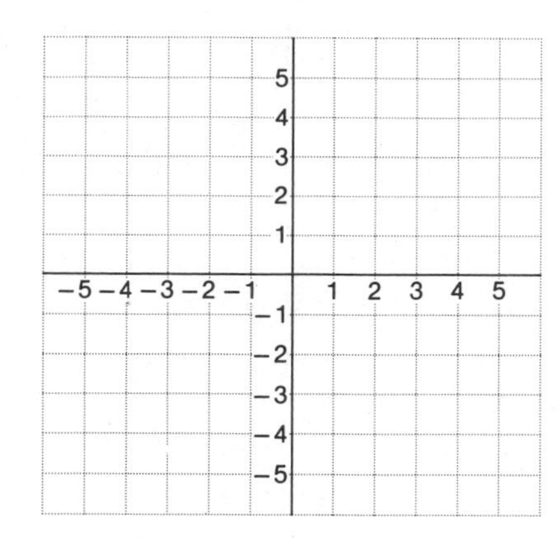

5. $\begin{Bmatrix} y = -x + 2 \\ -x + y = 4 \end{Bmatrix}$

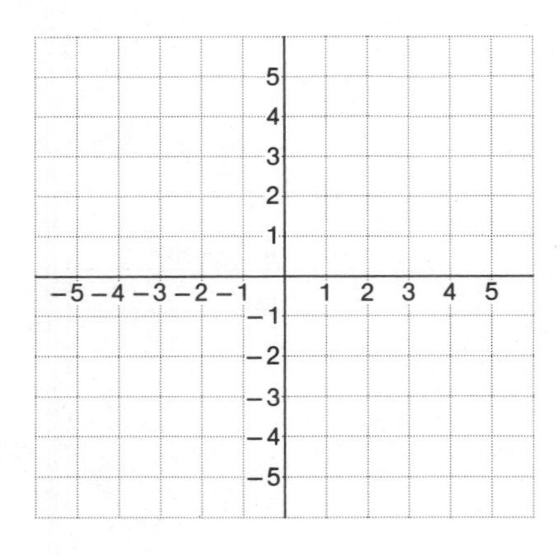

6. $\begin{Bmatrix} -2m + n = 3 \\ 4m - 2n = -6 \end{Bmatrix}$

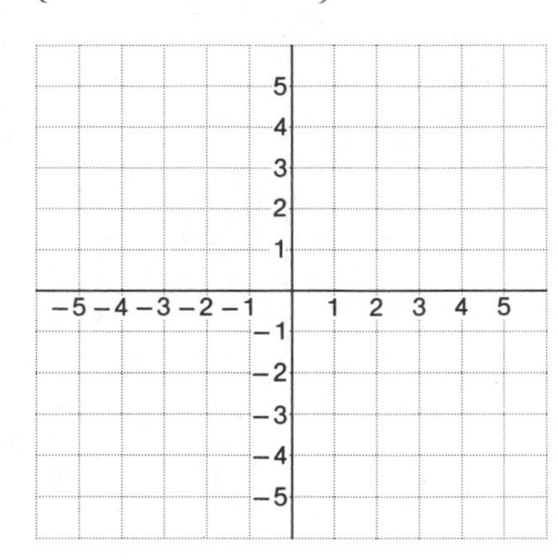

ANSWERS

1. See graph.
2. See graph.
3. See graph.
4. See graph.
5. See graph.
6. See graph.

7. See graph.

8. See graph.

9. See graph.

10. See graph.

11. See graph.

12. See graph.

7. 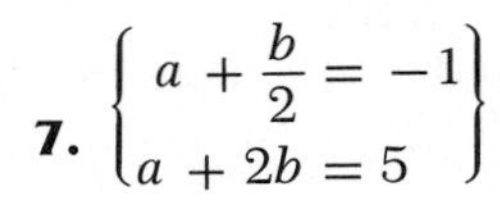
$$\begin{Bmatrix} a + \dfrac{b}{2} = -1 \\ a + 2b = 5 \end{Bmatrix}$$

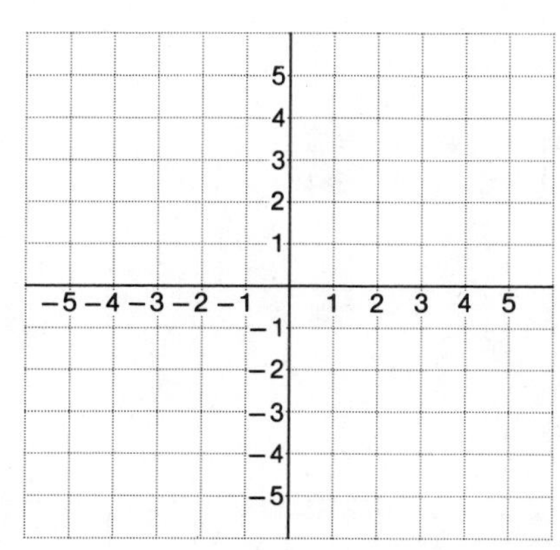

8. $\begin{Bmatrix} 3x - 4y = 12 \\ x - 2 = 0 \end{Bmatrix}$

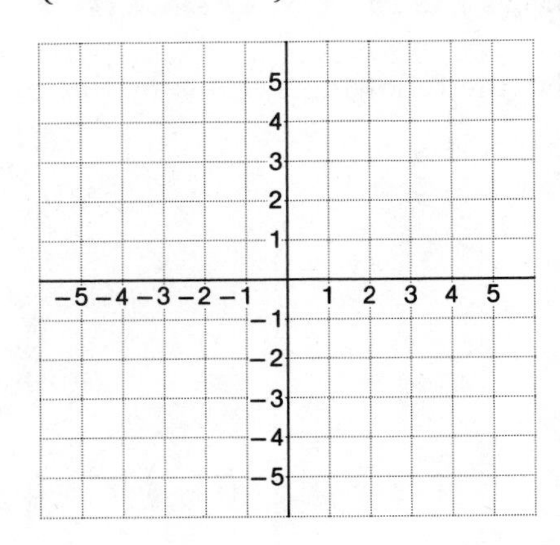

9. $\begin{Bmatrix} a + b = 3 \\ b = -2a + 1 \end{Bmatrix}$

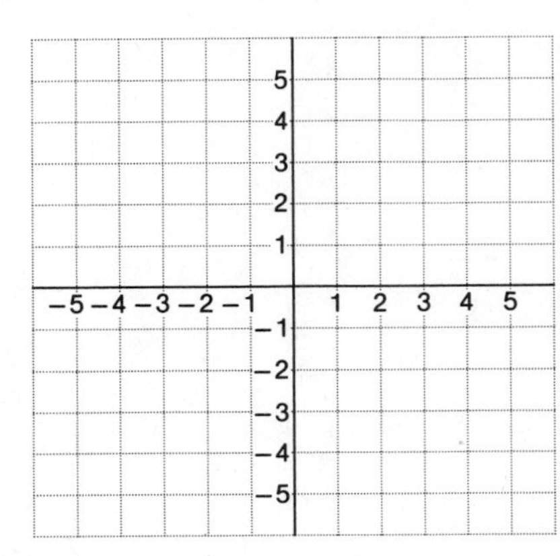

10. $\begin{Bmatrix} 2x - y = -1 \\ -4x + 2y = 0 \end{Bmatrix}$

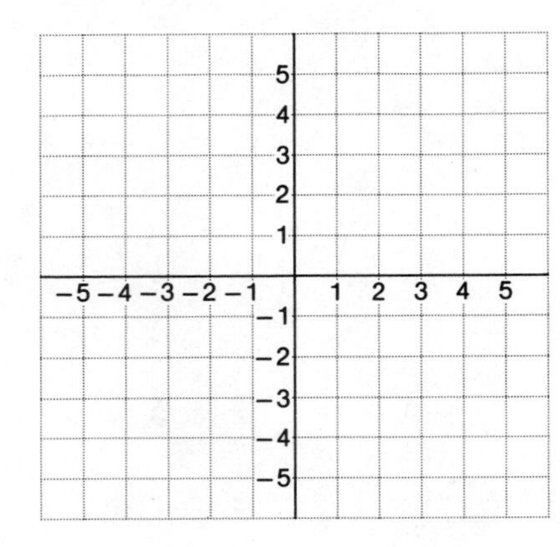

11. $\begin{Bmatrix} x + y = 4 \\ -2x + 4y = -2 \end{Bmatrix}$

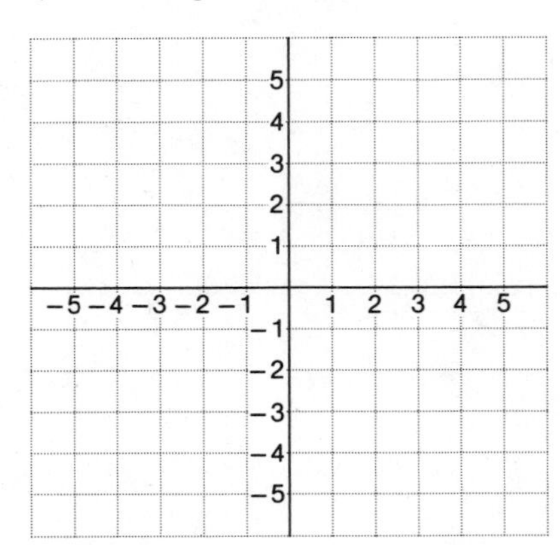

12. $\begin{Bmatrix} x - 3 = 0 \\ x + y = -1 \end{Bmatrix}$

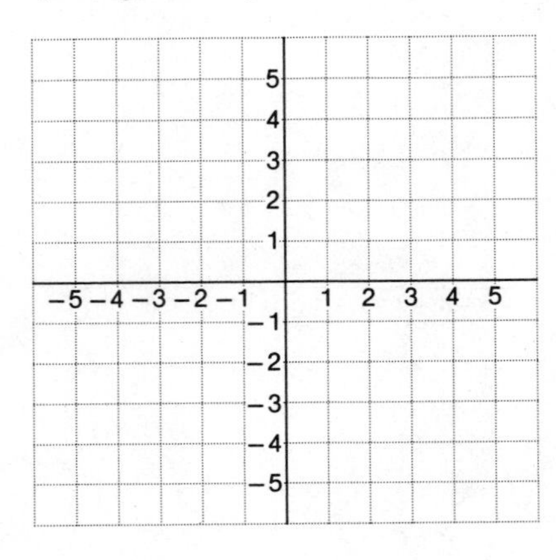

NAME | COURSE/SECTION | DATE | ANSWERS

13. $\begin{Bmatrix} m + n = 1 \\ -2m - n = 0 \end{Bmatrix}$

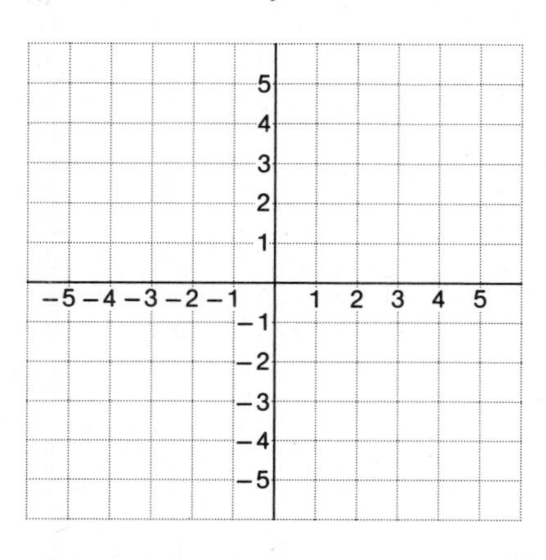

14. $\begin{Bmatrix} -2x + 2y = 8 \\ 3x + y = 0 \end{Bmatrix}$

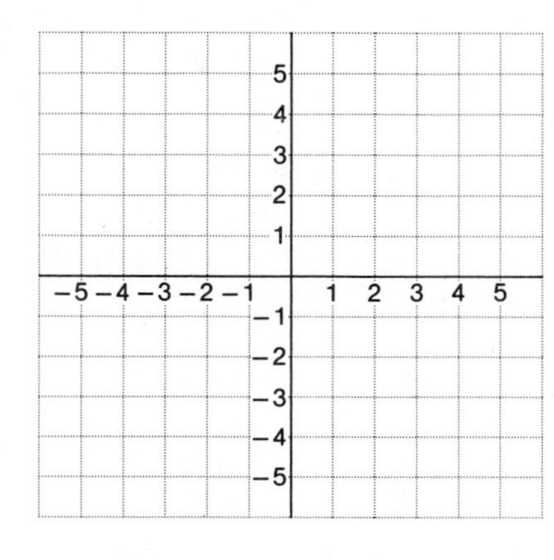

15. If twice a number minus another number is 2, and if twice the sum of the two numbers is 8, find the two numbers.

16. One number is the same as another number minus 5, and twice the first number plus the second number is -1. Find the two numbers.

17. Jeannie and Dena are going to divide 10 pieces of candy between them. If twice what Jeannie receives minus three times what Dena receives is zero, how many more pieces does Jeannie receive than Dena?

ANSWERS

13. See graph.

14. See graph.

15. ______

16. ______

17. ______

18. ______________

19. ______________

20. ______________

18. Tim and Larry together eat 8 chocolate doughnuts. If twice the number Larry eats is the same as the number Tim eats plus one, how many does each person eat?

19. Lori Barber has 8 coins consisting of dimes and nickels. If their total face value is 50¢, how many of each type of coin does Lori have?

20. Laurent Cardeau purchases 6 pieces of fruit, apples and oranges, at the neighborhood fruit stand. Each orange costs 50¢ and each apple costs 40¢. If the total bill was $2.80, how many apples and how many oranges did Laurent buy?

7.3

The Substitution Method and Word Problems

LEARNING OBJECTIVES

After finishing Section 7.3, you should be able to:

- Solve a 2-by-2 system by the substitution method.
- Solve additional word problems.

In this section we look at the substitution method for solving 2-by-2 systems, and in the next section we will look at the addition method. As we mentioned in the last section, the graphing method is sufficient only for certain 2-by-2 systems. However, the substitution and addition methods will always work for any 2-by-2 system. The steps in the substitution method are given as follows.

STEP 1: SELECT

Select either equation in the system and solve for one of the variables in terms of the other variable.

STEP 2: SUBSTITUTE

Substitute the variable you have solved for in the other equation in the system. This gives one equation in only one variable (which we can then solve).

STEP 3: EQUIVALENT SYSTEM

Write an equivalent system. Recall from Chapter 3 that *equivalent equations* were equations that had exactly the same solution. Similarly, *equivalent systems* are systems that have exactly the same solution.

STEP 4: SOLVE

Solve the equivalent system, and hence find the solution for the original system.

STEP 5: CHECK

We can check our solution by substitution in the two original equations.

Example 1 Use the substitution method to solve $\begin{Bmatrix} 4x - y = 2 \\ x - 3y = -5 \end{Bmatrix}$.

STEP 1: SELECT

We arbitrarily pick the second equation to solve for x.

$$x - 3y = -5$$
$$(1) \quad x = 3y - 5$$

STEP 2: SUBSTITUTE

Substitute $3y - 5$ for the x in the first equation, $4x - y = 2$. After this substitution, we obtain an equation in just one variable, which we can then solve.

$$4x - y = 2$$
$$4(3y - 5) - y = 2$$
$$12y - 20 - y = 2$$
$$(2) \quad 11y - 20 = 2$$

STEP 3: EQUIVALENT SYSTEM

To obtain our equivalent system, we use equation (1) from Step 1 and equation (2) from Step 2.

$$\begin{Bmatrix} x = 3y - 5 \\ 11y - 20 = 2 \end{Bmatrix}$$

Solve by the substitution method. (Check your solutions.)

1. $\begin{cases} x - 2y = 3 \\ 2x + y = -4 \end{cases}$

STEP 4: SOLVE

Solve $\begin{cases} x = 3y - 5 \\ 11y - 20 = 2 \end{cases}$ by first solving for y and then for x.

$$11y - 20 = 2$$
$$11y = 22$$
$$y = 2$$
$$x = 3y - 5$$
$$x = 3(2) - 5$$
$$x = 1$$

Thus the solution to the original system is (1, 2).

STEP 5: CHECK

Substitute (1, 2) in the original system.

$4x - y$	2
$4(1) - 2$	2
$4 - 2$	2
2	2

$x - 3y$	-5
$1 - 3(2)$	-5
$1 - 6$	-5
-5	-5

DO EXERCISE 1.

Example 2 Solve $\begin{cases} 2x - y = 4 \\ 3x + 2y = 6 \end{cases}$ by the substitution method.

STEP 1: SELECT

Solve the first equation for y.

$$2x - y = 4$$
$$-y = -2x + 4$$
$$(1) \quad y = 2x - 4$$

STEP 2: SUBSTITUTE

Substitute $2x - 4$ for the y in the second equation.

$$3x + 2y = 6$$
$$3x + 2(2x - 4) = 6$$
$$3x + 4x - 8 = 6$$
$$(2) \quad 7x - 8 = 6$$

STEP 3: EQUIVALENT SYSTEM

Use equation (1) from Step 1 and equation (2) from Step 2.

$$\begin{cases} y = 2x - 4 \\ 7x - 8 = 6 \end{cases}$$

STEP 4: SOLVE

Solve first for x and then for y.

$$7x - 8 = 6$$
$$7x = 14$$
$$x = 2$$
$$y = 2x - 4$$
$$y = 2(2) - 4$$
$$y = 0$$

Thus the solution to the original system is (2, 0).

STEP 5: CHECK

Substitute (2, 0) in the original system.

$2x - y$	4
$2(2) - 0$	4
$4 - 0$	4
4	4

$3x + 2y$	6
$3(2) + 2(0)$	6
$6 + 0$	6
6	6

DO EXERCISES 2 AND 3.

Now we want to see what happens if we use the substitution method to solve a system that has no solution or infinitely many solutions.

Example 3 Use the substitution method to solve $\begin{Bmatrix} 2x - 4y = 1 \\ -x + 2y = -2 \end{Bmatrix}$.

STEP 1: SELECT

Solve the second equation for x.

$$-x + 2y = -2$$
$$-x = -2y - 2$$
$$(1) \quad x = 2y + 2$$

STEP 2: SUBSTITUTE

Substitute $2y + 2$ for the x in the first equation.

$$2x - 4y = 1$$
$$2(2y + 2) - 4y = 1$$
$$4y + 4 - 4y = 1$$
$$(2) \quad 4 = 1$$

Notice that after we made the substitution and simplified the equation, all the variables disappeared and we were left with the false statement $4 = 1$. This means the original system has no solution. The general rule follows.

RULE 7.1 When we solve a system of equations, if all the variables disappear and we are left with

A. a false statement (such as $2 = 1$, $-3 = 2$, $0 = 4$, and so on), then the system has *no solution*.

B. a true statement (such as $2 = 2$, $0 = 0$, $-5 = -5$, and so on), then the system has *infinitely many solutions*.

DO EXERCISE 4.

Example 4 Use the substitution method to solve $\begin{Bmatrix} x - 4y = -2 \\ -3x + 12y = 6 \end{Bmatrix}$

STEP 1: SELECT

Solve the first equation for x.

$$x - 4y = -2$$
$$(1) \quad x = 4y - 2$$

Solve by the substitution method. (Check your solutions.)

2. $\begin{Bmatrix} -2s - t = 3 \\ 3s + 2t = -4 \end{Bmatrix}$

3. $\begin{Bmatrix} x + 3y = 4 \\ y = 2x - 1 \end{Bmatrix}$

Solve the following system by the substitution method.

4. $\begin{Bmatrix} 2x - 4y = -3 \\ -x + 2y = 0 \end{Bmatrix}$

Solve the following systems by the substitution method.

5. $\begin{Bmatrix} -m - 2n = 1 \\ 2m + 4n = -2 \end{Bmatrix}$

6. $\begin{Bmatrix} 4s - t = 2 \\ s = 3t - 5 \end{Bmatrix}$

STEP 2: SUBSTITUTE

Substitute $4y - 2$ for the x in the second equation.

$$-3x + 12y = 6$$
$$-3(4y - 2) + 12y = 6$$
$$-12y + 6 + 12y = 6$$
$$(2) \qquad 6 = 6$$

Since all the variables have disappeared in equation (2) and we are left with a true statement, the original system has infinitely many solutions.

DO EXERCISES 5 AND 6.

Before we look at some word problems, we want to do two more examples.

Example 5 Use the substitution method to solve $\begin{Bmatrix} -3s + 2t = 4 \\ 5s - 3t = 1 \end{Bmatrix}$.

STEP 1: SELECT

Solve the first equation for t.

$$-3s + 2t = 4$$
$$2t = 3s + 4$$
$$(1) \qquad t = \frac{3}{2}s + 2$$

STEP 2: SUBSTITUTE

Substitute $\frac{3}{2}s + 2$ for the t in the second equation.

$$5s - 3t = 1$$
$$5s - 3\left(\frac{3}{2}s + 2\right) = 1$$
$$5s - \frac{9}{2}s - 6 = 1$$
$$10s - 9s - 12 = 2$$
$$(2) \qquad s - 12 = 2$$

STEP 3: EQUIVALENT SYSTEM

Use equation (1) from Step 1 and equation (2) from Step 2.

$$\begin{Bmatrix} t = \frac{3}{2}s + 2 \\ s - 12 = 2 \end{Bmatrix}$$

STEP 4: SOLVE

Solve first for s and then for t.

$$s - 12 = 2$$
$$s = 14$$
$$t = \frac{3}{2}s + 2$$
$$t = \left(\frac{3}{2}\right)(14) + 2$$
$$t = 21 + 2$$
$$t = 23$$

Thus the solution is (14, 23).

STEP 5: CHECK

Substitute (14, 23) in the original system.

$-3s + 2t$	4
$-3(14) + 2(23)$	4
$-42 + 46$	4
4	4

$5s - 3t$	1
$5(14) - 3(23)$	1
$70 - 69$	1
1	1

DO EXERCISE 7.

Solve the following system by the substitution method.

7. $\begin{cases} 2x - 3y = -27 \\ 3x + 4y = 2 \end{cases}$

Example 6 Use the substitution method to solve $\begin{Bmatrix} \dfrac{x}{3} - \dfrac{y}{4} = 1 \\ -\dfrac{x}{6} + \dfrac{3y}{2} = \dfrac{1}{3} \end{Bmatrix}$

OPTIONAL STEP

We may change the fractions in both equations to integers by multiplying both sides of the first equation by 12 and both sides of the second equation by 6.

$$12\left(\frac{x}{3} - \frac{y}{4}\right) = 12(1) \qquad 6\left(-\frac{x}{6} + \frac{3y}{2}\right) = 6\left(\frac{1}{3}\right)$$

$$(1) \quad 4x - 3y = 12 \qquad (2) \quad -x + 9y = 2$$

This gives us the equivalent system

$$\begin{Bmatrix} 4x - 3y = 12 \\ -x + 9y = 2 \end{Bmatrix}$$

which we can solve.

STEP 1: SELECT

Solve the second equation for x.

$$-x + 9y = 2$$
$$-x = -9y + 2$$
$$(1) \quad x = 9y - 2$$

STEP 2: SUBSTITUTE

Substitute $9y - 2$ for the x in the first equation.

$$4x - 3y = 12$$
$$4(9y - 2) - 3y = 12$$
$$36y - 8 - 3y = 12$$
$$(2) \quad 33y - 8 = 12$$

STEP 3: EQUIVALENT SYSTEM

Use equations (1) and (2).

$$\begin{Bmatrix} x = 9y - 2 \\ 33y - 8 = 12 \end{Bmatrix}$$

Solve the following system by the substitution method.

8. $\begin{Bmatrix} -\frac{a}{3} + \frac{2b}{5} = 2 \\ a - 3b = 1 \end{Bmatrix}$

STEP 4: SOLVE

Solve first for y and then for x.

$$33y - 8 = 12$$
$$33y = 20$$
$$y = \frac{20}{33}$$

$$x = 9y - 2$$
$$x = 9\left(\frac{20}{33}\right) - 2$$
$$x = \frac{180}{33} - \frac{66}{33}$$
$$x = \frac{114}{33} = \frac{38}{11}$$

Thus the solution to the original system is $(\frac{38}{11}, \frac{20}{33})$.

STEP 5: CHECK

Substitute $(\frac{38}{11}, \frac{20}{33})$ in the original system.

$\frac{x}{3} - \frac{y}{4}$	1
$\frac{\frac{38}{11}}{3} - \frac{\frac{20}{33}}{4}$	1
$\frac{38}{11}\cdot\frac{1}{3} - \frac{20}{33}\cdot\frac{1}{4}$	1
$\frac{38}{33} - \frac{5}{33}$	1
$\frac{33}{33}$	1
1	1

$-\frac{x}{6} + \frac{3y}{2}$	$\frac{1}{3}$
$\frac{-\frac{38}{11}}{6} + \frac{3\left(\frac{20}{33}\right)}{2}$	$\frac{1}{3}$
$-\frac{38}{11}\cdot\frac{1}{6} + \frac{20}{11}\cdot\frac{1}{2}$	$\frac{1}{3}$
$-\frac{19}{33} + \frac{10}{11}$	$\frac{1}{3}$
$-\frac{19}{33} + \frac{30}{33}$	$\frac{1}{3}$
$\frac{11}{33}$	$\frac{1}{3}$
$\frac{1}{3}$	$\frac{1}{3}$

DO EXERCISE 8.

Example 7 Mr. Trikas is at a dog show. He would like to know how many dogs and how many people are present. He counts 130 legs altogether (people legs and dog legs). He also knows that there are five people there without dogs and everyone else has exactly one dog. How many dogs and how many people are at the dog show?

STEP 2: VARIABLE

Let x = the number of people present, and
y = the number of dogs present.

STEP 3: EQUATION

We will need a 2-by-2 system. In arriving at the first equation, we will assume that each person has 2 legs and each dog has 4 legs.

$$\begin{pmatrix}\text{Total number}\\ \text{of people legs}\end{pmatrix} + \begin{pmatrix}\text{Total number}\\ \text{of dog legs}\end{pmatrix} = \begin{pmatrix}\text{Total number}\\ \text{of legs}\end{pmatrix}$$

$$(1) \qquad 2x + 4y = 130$$

The statement, "There are five people there without dogs and everyone else has exactly one dog," means that the number of people present is 5 more than the number of dogs present. Translating this into a mathematical equation gives us

$$(2) \qquad x = y + 5$$

Thus the system is $\left\{\begin{aligned} 2x + 4y &= 130 \\ x &= y + 5 \end{aligned}\right\}$.

STEP 4: SOLVE

We will use the substitution method.

Select

The second equation is already solved for x.

$$(1)\ x = y + 5$$

Substitute

Substitute $y + 5$ for x in the first equation.

$$\begin{aligned} 2x + 4y &= 130 \\ 2(y + 5) + 4y &= 130 \\ 2y + 10 + 4y &= 130 \\ (2) \qquad 6y + 10 &= 130 \end{aligned}$$

Equivalent system

$$\left\{\begin{aligned} x &= y + 5 \\ 6y + 10 &= 130 \end{aligned}\right\}$$

Solve

$$\begin{aligned} 6y + 10 &= 130 \\ 6y &= 120 \\ y &= 20 \end{aligned}$$

$$\begin{aligned} x &= y + 5 \\ x &= 20 + 5 = 25 \end{aligned}$$

Thus there are 20 dogs and 25 persons at the show.

STEP 5: CHECK

The answer is reasonable, and note that 4(20) plus 2(25) is 130.

DO EXERCISE 9.

Example 8 A nursery owner has clover seed worth 30¢ a pound and alfalfa seed worth 15¢ a pound. How many pounds of each should the nursery owner use to make a 300-pound mixture worth 20¢ a pound?

9. Phyllis and George Cavanaugh both worked overtime this week. Phyllis is paid \$20 an hour for overtime and George is paid \$18 an hour for overtime. Their total overtime pay for the week was \$232, and Phyllis worked twice as many overtime hours as George. How many overtime hours did each person work?

10. Dena has a collection of nickels and dimes worth \$1.90. If the number of nickels she has equals twice the number of dimes minus two, how many of each coin does she have?

STEP 1: PICTURE

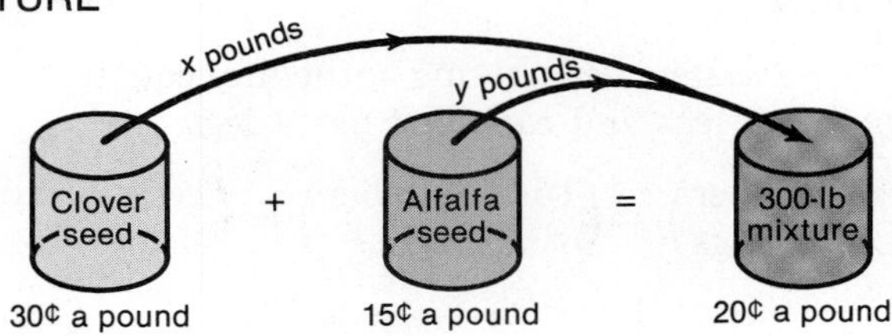

STEP 2: VARIABLE

Let x = number of pounds of clover seed in the mixture, and
y = number of pounds of alfalfa seed in the mixture.

STEP 3: EQUATION

We will need a 2-by-2 system.

$$\begin{pmatrix}\text{Number of}\\ \text{pounds of}\\ \text{clover}\end{pmatrix} + \begin{pmatrix}\text{Number of}\\ \text{pounds of}\\ \text{alfalfa}\end{pmatrix} = \begin{pmatrix}\text{Number of}\\ \text{pounds in}\\ \text{final mixture}\end{pmatrix}$$

$$(1) \qquad x + y = 300$$

$$\begin{pmatrix}\text{Total value}\\ \text{of } x \text{ pounds}\\ \text{of clover}\end{pmatrix} + \begin{pmatrix}\text{Total value}\\ \text{of } y \text{ pounds}\\ \text{of alfalfa}\end{pmatrix} = \begin{pmatrix}\text{Total value}\\ \text{of mixture}\end{pmatrix}$$

$$(2) \qquad 30x + 15y = 20(300)$$

Thus the system is $\left\{\begin{matrix} x + y = 300 \\ 30x + 15y = 6000 \end{matrix}\right\}$.

STEP 4: SOLVE

We will use the substitution method.

Select

Solve the first equation for x.

$$x + y = 300$$
$$(1) \qquad x = 300 - y$$

Substitute

Substitute $300 - y$ for x in the second equation.

$$\begin{aligned} 30x + 15y &= 6000 \\ 30(300 - y) + 15y &= 6000 \\ 9000 - 30y + 15y &= 6000 \\ (2) \qquad 9000 - 15y &= 6000 \end{aligned}$$

Equivalent system

$$\left\{\begin{matrix} x = 300 - y \\ 9000 - 15y = 6000 \end{matrix}\right\}$$

Solve

$$\begin{aligned} 9000 - 15y &= 6000 \\ -15y &= -3000 \\ y &= 200 \\ x &= 300 - y \\ x &= 300 - 200 \\ x &= 100 \end{aligned}$$

Thus the nursery owner wants 100 pounds of clover seed and 200 pounds of alfalfa seed in his mixture.

DO EXERCISE 10.

Example 9 We are going to make two types of cake, cake A and cake B. The recipe for cake A calls for 2 eggs and 1 cup of sugar, and the recipe for cake B calls for 3 eggs and 2 cups of sugar. If we have 76 eggs and 44 cups of sugar available to us, how many of each type of cake should we make in order to use all the eggs and sugar?

STEP 1: PICTURE

A picture will not help, but a table may be useful to record some of the information.

	Eggs	*Cups of sugar*
Cake A	2	1
Cake B	3	2
Total available	76	44

STEP 2: VARIABLE

Let $x =$ the number of cake A to be made,
and $y =$ the number of cake B to be made.

STEP 3: EQUATION

We will need a 2-by-2 system. (Note that to make one cake A we need 2 eggs. Thus to make x cakes of type A, we need 2 times x eggs.)

$$\begin{pmatrix}\text{Number of eggs}\\ \text{needed to make}\\ x\text{ cakes of}\\ \text{type A}\end{pmatrix} + \begin{pmatrix}\text{Number of}\\ \text{eggs needed}\\ \text{to make }y\\ \text{cakes of}\\ \text{type B}\end{pmatrix} = \begin{pmatrix}\text{Number}\\ \text{of eggs}\\ \text{available}\end{pmatrix}$$

$$(1) \qquad 2x + 3y = 76$$

$$\begin{pmatrix}\text{Number of}\\ \text{cups of sugar}\\ \text{to make }x\\ \text{cakes of}\\ \text{type A}\end{pmatrix} + \begin{pmatrix}\text{Number of}\\ \text{cups of sugar}\\ \text{to make }y\\ \text{cakes of}\\ \text{type B}\end{pmatrix} = \begin{pmatrix}\text{Number of}\\ \text{cups of}\\ \text{sugar}\\ \text{available}\end{pmatrix}$$

$$(2) \qquad 1x + 2y = 44$$

$$\begin{Bmatrix} 2x + 3y = 76 \\ x + 2y = 44 \end{Bmatrix}$$

STEP 4: SOLVE

We will use the substitution method.

Select

Solve the second equation for x.

$$x + 2y = 44$$
$$(1) \qquad x = 44 - 2y$$

Substitute

Substitute $44 - 2y$ for the x in the first equation.

$$\begin{aligned} 2x + 3y &= 76 \\ 2(44 - 2y) + 3y &= 76 \\ 88 - 4y + 3y &= 76 \\ (2) \qquad 88 - y &= 76 \end{aligned}$$

11. Before they are considered as completed, tables must go through a preparation time and a finishing time. Table A requires 2 hours for preparation and $1\frac{1}{2}$ hours for finishing. Table B requires 3 hours for preparation and 2 hours for finishing. If we have 65 hours of preparation time available and 45 hours of finishing time, how many of each table should we make in order to use all the hours of preparation and finishing time available to us?

Equivalent system

$$\begin{Bmatrix} x = 44 - 2y \\ 88 - y = 76 \end{Bmatrix}$$

Solve

$$88 - y = 76$$
$$-y = -12$$
$$y = 12$$
$$x = 44 - 2y$$
$$x = 44 - 2(12)$$
$$x = 20$$

Thus the eggs and sugar will be used up if we make 20 of cake A and 12 of cake B.

DO EXERCISE 11.

We note that Example 7 and Example 8 in this section could have been worked using one variable instead of two. However, Example 9 would be very difficult to work using only one variable.

DO SECTION PROBLEMS 7.3.

ANSWERS TO MARGINAL EXERCISES

1. $(-1, -2)$ **2.** $(-2, 1)$ **3.** $(1, 1)$ **4.** No solution **5.** Infinitely many solutions **6.** $(1, 2)$ **7.** $(-6, 5)$ **8.** $\left(-\frac{32}{3}, -\frac{35}{9}\right)$ **9.** 8 for Phyllis and 4 for George **10.** 18 nickels and 10 dimes **11.** 10 of table A and 15 of table B

NAME COURSE/SECTION DATE

SECTION PROBLEMS 7.3

Solve by the substitution method.

1. $\begin{Bmatrix} x = 3y - 1 \\ 2x - y = 4 \end{Bmatrix}$

2. $\begin{Bmatrix} 2x - y = 1 \\ y = 3x + 4 \end{Bmatrix}$

3. $\begin{Bmatrix} s - 5 = 0 \\ 2s - t = 1 \end{Bmatrix}$

4. $\begin{Bmatrix} 3x - y = 4 \\ -2x + y = -2 \end{Bmatrix}$

5. $\begin{Bmatrix} a - 2b = 9 \\ 2a + b = 8 \end{Bmatrix}$

6. $\begin{Bmatrix} s - 2t = 3 \\ -s + 2t = -4 \end{Bmatrix}$

7. $\begin{Bmatrix} 2x - y = 4 \\ y - 3 = 3 \end{Bmatrix}$

8. $\begin{Bmatrix} 2s + 3t = 5 \\ -s + 2t = 1 \end{Bmatrix}$

9. $\begin{Bmatrix} x - 4 = 0 \\ y + 1 = 0 \end{Bmatrix}$

10. $\begin{Bmatrix} 3x + 2y = -2 \\ 4x - 5y = 5 \end{Bmatrix}$

11. $\begin{Bmatrix} -6x + 4y = 2 \\ 3x - 2y = -1 \end{Bmatrix}$

12. $\begin{Bmatrix} \frac{x}{2} = \frac{y}{3} + 1 \\ -3x + 2y = -6 \end{Bmatrix}$

13. $\begin{Bmatrix} -3m + 2n = 4 \\ 5m - 3n = 1 \end{Bmatrix}$

14. $\begin{Bmatrix} \frac{a}{2} - b = 1 \\ -a + 2b = 3 \end{Bmatrix}$

15. $\begin{Bmatrix} x + y = 2 \\ 2x - 5y = 11 \end{Bmatrix}$

16. $\begin{Bmatrix} 0.5s - 3t = 4 \\ s = 2t + 4 \end{Bmatrix}$

17. $\begin{Bmatrix} 2x - y = 4 \\ -3x + y = -4 \end{Bmatrix}$

18. $\begin{Bmatrix} x - 1.5y = 4 \\ x + 2 = 0 \end{Bmatrix}$

19. $\begin{Bmatrix} 0.5x - 2y = 1 \\ x = 6y - 4 \end{Bmatrix}$

20. $\begin{Bmatrix} 7s - 3t = 1 \\ 2s + 5t = 12 \end{Bmatrix}$

21. $\begin{Bmatrix} x - y = 2 \\ 2x + 3y = -26 \end{Bmatrix}$

22. $\begin{Bmatrix} a = b - 3 \\ 2a - b = 4 \end{Bmatrix}$

23. $\begin{Bmatrix} x - 2 = 3 \\ 2y + 1 = -3 \end{Bmatrix}$

24. $\begin{Bmatrix} \frac{s}{2} = 3t + 1 \\ s - 2t = 3 \end{Bmatrix}$

25. $\begin{Bmatrix} -3x + 2y = 4 \\ 4x - 5y = 9 \end{Bmatrix}$

26. $\begin{Bmatrix} \frac{m}{2} - \frac{n}{3} = 1 \\ 3m + n = 4 \end{Bmatrix}$

27. $\begin{Bmatrix} 0.5x - 3y = 4 \\ x + 4y = 6 \end{Bmatrix}$

28. $\begin{Bmatrix} 2x - 4y = 1 \\ -x + 2y = 0 \end{Bmatrix}$

29. $\begin{Bmatrix} -3P - Q = 4 \\ 2P - 2Q = -3 \end{Bmatrix}$

30. $\begin{Bmatrix} \frac{s}{3} - \frac{t}{2} = 1 \\ \frac{2s}{5} + t = 2 \end{Bmatrix}$

ANSWERS

1. ______
2. ______
3. ______
4. ______
5. ______
6. ______
7. ______
8. ______
9. ______
10. ______
11. ______
12. ______
13. ______
14. ______
15. ______
16. ______
17. ______
18. ______
19. ______
20. ______
21. ______
22. ______
23. ______
24. ______
25. ______
26. ______
27. ______
28. ______
29. ______
30. ______

31. ______

32. ______

33. ______

34. ______

35. ______

36. ______

37. ______

38. ______

39. ______

40. ______

31. One number is 8 more than another number. If their sum is 15, what are the two numbers?

32. One number is twice as large as another number. If 3 times the smaller number minus 4 times the larger number is 10, what are the two numbers?

33. A tea dealer has one brand worth 50¢ an ounce and another worth 80¢ an ounce. How many ounces of each kind must he use to make 120 ounces of a mixture worth 72¢ an ounce?

34. Lynda has 22 coins consisting of dimes and quarters. If the total value of her coins is $3.40, how many quarters does she have?

35. A car's radiator holds 12 liters. At present the liquid in the radiator is 40% antifreeze. How much pure (100%) antifreeze must you add to bring the level to 60% antifreeze?

36. A tea grower has a mixture of tea made from tea worth $1.40 a pound and tea worth $1.70 a pound. The mixture is valued at $20.60. If the amount of $1.40-a-pound tea is doubled and the amount of $1.70-a-pound tea is tripled, the mixture is valued at $54.80. How many pounds of each tea are in the final mixture?

37. The perimeter of a rectangle is 40 inches. If the width is $\frac{2}{3}$ of the length, find the dimensions of the rectangle.

38. Mrs. O'Brien wants to break up her 28-day vacation into two parts—one part consisting of a trip to Las Vegas, and the other part of a trip to her home town of Tacoma, Washington. She wants the time spent on the trip to Las Vegas to be $\frac{2}{5}$ of the amount of time she spends in Tacoma. How many days will she spend in Las Vegas?

39. Huckster McPete is brewing some of his infamous white lightning and blue lightning. It takes 5 pounds of mash and 1 pound of sugar to make one gallon of white lightning. It takes 4 pounds of mash and $\frac{3}{4}$ pound of sugar to make one gallon of blue lightning. Huckster has 83 pounds of mash and 16 pounds of sugar available. How many gallons of white lightning and how many gallons of blue lightning should Huckster brew to use up all the mash and all the sugar?

40. BuDord Realty sells land (for which their commission is 10%) and homes (for which their commission is 8%). In May they received a total of $17,800 in commissions on total sales of $210,000 (land and homes). What was the total sales for land in May?

7.4

The Addition Method

The last method we will look at for solving a 2-by-2 system is the addition method. The basic idea in the addition method is that we can add (possibly after multiplying one or both equations by a nonzero number) our two equations together so that one of the variables will sum to zero. This leaves us with one equation in one unknown, which we can then solve.

The steps in the addition method are as follows.

STEP 1: PROPER FORM

Put both equations in standard form, $ax + by = c$.

STEP 2: CHOOSE

Decide which variable you wish to eliminate.

STEP 3: EQUIVALENT SYSTEM

If necessary, multiply one or both equations in the system by a nonzero number to obtain an equivalent system where the coefficients of the variable you want to eliminate are additive inverses.

STEP 4: SOLVE

Solve this equivalent system by adding the two equations, solving the resulting equation for the remaining variable, and then substituting the value for this variable into either of the two equations in the equivalent or original system to find the value of the other variable.

STEP 5: CHECK

LEARNING OBJECTIVES

After finishing Section 7.4, you should be able to:

- Solve a 2-by-2 system by the addition method.
- Solve additional word problems.

Example 1 Use the addition method to solve $\begin{Bmatrix} 2x - 2y = 3 \\ x + 2y = -6 \end{Bmatrix}$.

STEP 1: PROPER FORM

The equations are already in proper form.

STEP 2: CHOOSE

We will eliminate the y variable.

STEP 3: EQUIVALENT SYSTEM

Since the coefficients of y are already additive inverses (2 and -2), we do not need to do anything to obtain an equivalent system.

$$\begin{Bmatrix} 2x - 2y = 3 \\ x + 2y = -6 \end{Bmatrix}$$

STEP 4: SOLVE

Add the two equations.

$$\begin{aligned} 2x - 2y &= 3 \\ x + 2y &= -6 \\ \hline 3x \qquad &= -3 \end{aligned}$$

Solve by the addition method.

1. $\begin{Bmatrix} 3x + y = 4 \\ -3x + 2y = 2 \end{Bmatrix}$

Solve resulting equation for x.

$$3x = -3$$
$$x = -1$$

Substitute -1 for x in $x + 2y = -6$ and solve for y.

$$x + 2y = -6$$
$$-1 + 2y = -6$$
$$2y = -5$$
$$y = -\frac{5}{2}$$

Thus the solution to the system is $(-1, -\frac{5}{2})$.

STEP 5: CHECK

Substitute $(-1, \frac{5}{2})$ in the original system.

$2x - 2y$	3
$2(-1) - 2\left(-\frac{5}{2}\right)$	3
$-2 + 5$	3
3	3

$x + 2y$	-6
$-1 + 2\left(-\frac{5}{2}\right)$	-6
$-1 - 5$	-6
-6	-6

DO EXERCISE 1.

Example 2 Use the addition method to solve $\begin{Bmatrix} a = 3b + 4 \\ 2a + 2b = 1 \end{Bmatrix}$.

STEP 1: PROPER FORM

Add $-3b$ to each side of the first equation.

$$a = 3b + 4$$
$$a + (-3b) = 4$$

Thus our system becomes $\begin{Bmatrix} a - 3b = 4 \\ 2a + 2b = 1 \end{Bmatrix}$.

STEP 2: CHOOSE

We will eliminate the a variable.

STEP 3: EQUIVALENT SYSTEM

To obtain additive inverses for the coefficients of a, we multiply the first equation by -2 and keep the second equation as it is.

$$a - 3b = 4$$
$$-2(a - 3b) = -2(4)$$
$$(1) \quad -2a + 6b = -8$$
$$(2) \quad 2a + 2b = 1$$

$$\begin{Bmatrix} -2a + 6b = -8 \\ 2a + 2b = 1 \end{Bmatrix}$$

STEP 4: SOLVE

Add.

$$\begin{array}{r} -2a + 6b = -8 \\ 2a + 2b = 1 \\ \hline 8b = -7 \end{array}$$

Solve for b.

$$b = -\frac{7}{8}$$

Substitute $-\frac{7}{8}$ for b in $2a + 2b = 1$ and solve for a.

$$2a + 2b = 1$$
$$2a + 2\left(-\frac{7}{8}\right) = 1$$
$$2a - \frac{14}{8} = \frac{8}{8}$$
$$2a = \frac{22}{8}$$
$$a = \frac{11}{8}$$

Thus the solution is $(\frac{11}{8}, -\frac{7}{8})$.

STEP 5: CHECK

Substitute $(\frac{11}{8}, -\frac{7}{8})$ in the original system.

$a - 3b$	4
$\frac{11}{8} - 3\left(-\frac{7}{8}\right)$	4
$\frac{11}{8} + \frac{21}{8}$	4
$\frac{32}{8}$	4
4	4

$2a + 2b$	1
$2\left(\frac{11}{8}\right) + 2\left(-\frac{7}{8}\right)$	1
$\frac{22}{8} - \frac{14}{8}$	1
$\frac{8}{8}$	1
1	1

DO EXERCISE 2.

Example 3 Use the addition method to solve $\begin{Bmatrix} 2x - 5y = 4 \\ -3x + 7y = -2 \end{Bmatrix}$.

STEP 2: CHOOSE

We will eliminate the x variable.

STEP 3: EQUIVALENT SYSTEM

To obtain a coefficient of 6 for the x in the first equation, multiply by 3.

$$2x - 5y = 4$$
$$3(2x - 5y) = 3(4)$$
$$(1) \quad 6x - 15y = 12$$

To obtain a coefficient of -6 for the x in the second equation, multiply by 2.

$$-3x + 7y = -2$$
$$2(-3x + 7y) = 2(-2)$$
$$(2) \quad -6x + 14y = -4$$

$$\begin{Bmatrix} 6x - 15y = 12 \\ -6x + 14y = -4 \end{Bmatrix}$$

Solve by the addition method.

2. $\begin{Bmatrix} -2x - 2y = 4 \\ 3x + y = -6 \end{Bmatrix}$

Solve by the addition method.

3. $\begin{Bmatrix} \frac{x}{2} - y = 3 \\ 3x + 2y = 2 \end{Bmatrix}$

STEP 4: SOLVE

Add.

$$\begin{array}{r} 6x - 15y = 12 \\ -6x + 14y = -4 \\ \hline -y = 8 \end{array}$$

Solve for y. $$y = -8$$

Substitute and solve for x.

$$\begin{aligned} 6x - 15y &= 12 \\ 6x - 15(-8) &= 12 \\ 6x + 120 &= 12 \\ 6x &= -108 \\ x &= -18 \end{aligned}$$

Thus the solution is $(-18, -8)$.

DO EXERCISE 3.

Example 4 Use the addition method to solve $\begin{Bmatrix} 4s - 10t = 1 \\ -2s + 5t = 0 \end{Bmatrix}$

STEP 2: CHOOSE

We will eliminate the t variable.

STEP 3: EQUIVALENT SYSTEM

Keep the first equation as is; multiply the second equation by 2.

$$\begin{aligned} (1) \qquad 4s - 10t &= 1 \\ 2(-2s + 5t) &= 2(0) \\ (2) \qquad -4s + 10t &= 0 \end{aligned}$$

Thus the equivalent system is $\begin{Bmatrix} 4s - 10t = 1 \\ -4s + 10t = 0 \end{Bmatrix}$.

STEP 4: SOLVE

Add.

$$\begin{array}{r} 4s - 10t = 1 \\ -4s + 10t = 0 \\ \hline 0 = 1 \end{array} \quad \textit{False statement}$$

Since all the variables have disappeared and we are left with a false statement, we have *no solution* for our system.

Example 5 Use the addition method to solve $\begin{Bmatrix} -x + 3y = 4 \\ y = -2x - 1 \end{Bmatrix}$

Note

This system would probably be more easily solved by the substitution method. However, we can do it by the addition method.

STEP 1: PROPER FORM

The second equation has to be changed.

(1) $-x + 3y = 4$

$y = -2x - 1$

(2) $2x + y = -1$

$$\begin{cases} -x + 3y = 4 \\ 2x + y = -1 \end{cases}$$

STEP 2: CHOOSE

We will eliminate the x variable.

STEP 3: EQUIVALENT SYSTEM

Multiply the first equation by 2 and keep the second as is.

$-x + 3y = 4$

$2(-x + 3y) = 2(4)$

(1) $-2x + 6y = 8$

(2) $2x + y = -1$

$$\begin{cases} -2x + 6y = 8 \\ 2x + y = -1 \end{cases}$$

STEP 4: SOLVE

Add.

$$\begin{array}{r} -2x + 6y = 8 \\ 2x + y = -1 \\ \hline 7y = 7 \end{array}$$

Solve for y.

$y = 1$

Solve for x.

$2x + y = -1$

$2x + 1 = -1$

$2x = -2$

$x = -1$

Thus the solution is $(-1, 1)$.

STEP 5: CHECK

Substitute $(-1, 1)$ in the original system.

$-x + 3y$	4
$-(-1) + 3(1)$	4
$1 + 3$	4
4	4

$2x + y$	-1
$2(-1) + 1$	-1
$-2 + 1$	-1
-1	-1

DO EXERCISES 4 THROUGH 6.

Now we will do some word problems. Again, recall the five steps we use to solve word problems: Picture, Variable, Equation, Solve, and Check.

Example 6 Harold McDiver has just bought two parcels of land for a total of \$30,000. Parcel A has 10 acres in it and Parcel B has 15 acres. The average cost of an acre from Parcel A and an acre from Parcel B is \$1250. Find the cost per acre of the land in Parcel A and the land in Parcel B.

Solve by the addition method.

4. $\begin{cases} -2a + 3b = 3 \\ -4a - 6b = -6 \end{cases}$

5. $\begin{cases} x = 2y - 1 \\ -3x + y = 4 \end{cases}$

6. In Chapter 3, we learned that there are two rules for changing one equation to an equivalent equation (the Addition Principle and the Multiplication Principle). Which (if any) of these two rules do we use in solving a 2-by-2 system by the addition method?

7. The sum of the ages of Shane and his mother is 38. Twice Shane's age in 10 years plus his mother's age in 10 years will be 76. How old is each now?

STEP 2: VARIABLE

Let a = cost per acre in Parcel A, and
b = cost per acre in Parcel B.

STEP 3: EQUATION

We need a 2-by-2 system.

$$\begin{pmatrix}\text{Total cost}\\ \text{of Parcel A}\end{pmatrix} + \begin{pmatrix}\text{Total cost}\\ \text{of Parcel B}\end{pmatrix} = \begin{pmatrix}\text{Total}\\ \text{cost}\end{pmatrix}$$

$$(1) \qquad 10a + 15b = 30{,}000$$

Note: $\begin{pmatrix}\text{Total cost}\\ \text{of Parcel A}\end{pmatrix} = \begin{pmatrix}\text{Number of acres}\\ \text{in Parcel A}\end{pmatrix}\begin{pmatrix}\text{Cost per}\\ \text{acre}\end{pmatrix} = 10a$

Average cost of an acre from Parcel A and an acre from Parcel B is \$1250.

$$(2) \qquad \frac{a+b}{2} = 1250$$

Our system is $\begin{Bmatrix} 10a + 15b = 30{,}000 \\ \dfrac{a+b}{2} = 1250 \end{Bmatrix}$ or equivalently $\begin{Bmatrix} 10a + 15b = 30{,}000 \\ a + b = 2500 \end{Bmatrix}$.

STEP 4: SOLVE

We will use the addition method.

Choose

We will eliminate the a variable.

Equivalent system

Keep the first equation as is and multiply the second equation by -10.

$$\begin{aligned} (1) \qquad 10a + 15b &= 30{,}000 \\ a + b &= 2500 \\ -10(a+b) &= -10(2500) \\ (2) \qquad -10a - 10b &= -25{,}000 \end{aligned}$$

$$\begin{Bmatrix} 10a + 15b = 30{,}000 \\ -10a - 10b = -25{,}000 \end{Bmatrix}$$

Solve

Add and solve for b.

$$\begin{aligned} 10a + 15b &= 30{,}000 \\ -10a - 10b &= -25{,}000 \\ \hline 5b &= 5000 \\ b &= 1000 \end{aligned}$$

Solve for a.

$$\begin{aligned} a + b &= 2500 \\ a + 1000 &= 2500 \\ a &= 1500 \end{aligned}$$

Thus each acre in Parcel A cost \$1500 and each acre in Parcel B cost \$1000.

■ **DO EXERCISE 7.**

Example 7 John Michos has money invested in stocks and in real estate. During the last year, the return on the money he had invested in stocks was 8%, and the return on his real estate investments was 15%. He had a total of \$32,000 invested in stocks and real estate, and the total return on his money was \$4,100. How much does he have invested in stocks and how much in real estate?

We omit Step 1 since a picture would not help us solve this problem.

STEP 2: VARIABLE

Let x = amount invested in stocks,
and y = amount invested in real estate.

STEP 3: EQUATION

We need a 2-by-2 system.

$$\begin{pmatrix}\text{Amount invested}\\ \text{in stocks}\end{pmatrix} + \begin{pmatrix}\text{Amount invested}\\ \text{in real estate}\end{pmatrix} = \begin{pmatrix}\text{Total}\\ \text{amount}\end{pmatrix}$$

$$(1) \qquad x + y = 32{,}000$$

$$\begin{pmatrix}\text{Return on}\\ \text{stocks}\end{pmatrix} + \begin{pmatrix}\text{Return on}\\ \text{real estate}\end{pmatrix} = \begin{pmatrix}\text{Total}\\ \text{return}\end{pmatrix}$$

$$8\% \text{ of } x + 15\% \text{ of } y = 4100$$

$$(2) \qquad 0.08x + 0.15y = 4100$$

$$\begin{Bmatrix} x + y = 32{,}000 \\ 0.08x + 0.15y = 4100 \end{Bmatrix}$$

STEP 4: SOLVE

We will use the addition method.

Choose

We will eliminate the x variable.

Equivalent system

Multiply the first equation by -0.08 and keep the second equation as is.

$$x + y = 32{,}000$$

$$-0.08(x + y) = -0.08(32{,}000)$$

$$(1) \qquad -0.08x - 0.08y = -2560$$

$$(2) \qquad 0.08x + 0.15y = 4100$$

$$\begin{Bmatrix} -0.08x - 0.08y = 2560 \\ 0.08x + 0.15y = 4100 \end{Bmatrix}$$

Solve

Add.

$$\begin{aligned} -0.08x - 0.08y &= -2560 \\ 0.08x + 0.15y &= 4100 \\ \hline 0.07y &= 1540 \end{aligned}$$

$$y = 22{,}000$$

$$x + y = 32{,}000$$

$$x = 10{,}000$$

Thus John has \$10,000 invested in stocks and \$22,000 invested in real estate.

DO EXERCISE 8.

8. The sum of one number and twice another number is 220. If 20% of the first number plus 25% of the second number is 32, what are the two numbers?

Example 8 Phil Gilbert drove from Tacoma to Omaha, a distance of 1860 miles, on 90 gallons of gasoline. Part of the time he used regular gasoline, for which he got 20 miles per gallon (mpg), and the rest of the time he used gasohol, with which he got 22 mpg. How many gallons of regular and how many gallons of gasohol did he use on the trip?

We omit Step 1 since a picture would not help us solve this problem.

STEP 2: VARIABLE

Let x = number of gallons of regular gas
and y = number of gallons of gasohol.

STEP 3: EQUATION

We need a 2-by-2 system.

$$\begin{pmatrix}\text{Number of}\\ \text{gallons of}\\ \text{regular}\end{pmatrix} + \begin{pmatrix}\text{Number of}\\ \text{gallons of}\\ \text{gasohol}\end{pmatrix} = \begin{pmatrix}\text{Total}\\ \text{number}\\ \text{of gallons}\\ \text{used}\end{pmatrix}$$

$$(1) \qquad x + y = 90$$

(*Hint:* If you can travel 20 miles on one gallon of regular, then with x number of gallons you can travel 20 times x miles.)

$$\begin{pmatrix}\text{Number of miles}\\ \text{traveled on } x \text{ gallons}\\ \text{of regular gas}\end{pmatrix} + \begin{pmatrix}\text{Number of}\\ \text{miles traveled}\\ \text{on } y \text{ gallons}\\ \text{of gasohol}\end{pmatrix} = \begin{pmatrix}\text{Total number}\\ \text{of miles}\\ \text{traveled}\end{pmatrix}$$

$$(2) \qquad 20x + 22y = 1860$$

$$\begin{Bmatrix} x + y = 90 \\ 20x + 22y = 1860 \end{Bmatrix}$$

STEP 4: SOLVE

We will use the substitution method.

Select

Solve the first equation for x.

$$x + y = 90$$
$$(1) \qquad x = 90 - y$$

Substitution

Substitute $90 - y$ for x in the second equation.

$$20x + 22y = 1860$$
$$20(90 - y) + 22y = 1860$$
$$1800 - 20y + 22y = 1860$$
$$(2) \qquad 1800 + 2y = 1860$$

Equivalent system

$$\begin{Bmatrix} x = 90 - y \\ 1800 + 2y = 1860 \end{Bmatrix}$$

Solve

$$1800 + 2y = 1860$$
$$2y = 60$$
$$y = 30$$
$$x = 90 - y$$
$$x = 60$$

Thus Phil used 60 gallons of regular gas and 30 gallons of gasohol.

DO EXERCISE 9.

Example 9 In 1982 and 1983, the Fletchers heated their home by using wood and oil. The total amount of money spent for fuel in 1982 was \$650.00. In 1983, they added an additional wood stove and spent twice as much on wood as in 1982 and one-half the amount on oil as in 1982. They spent a total of \$625.00 for fuel in 1983. How much did they spend for wood and oil each in 1982 and 1983?

We omit Step 1 since a picture would not help us solve this problem.

STEP 2: VARIABLE

Let x = amount spent on wood in 1982
and y = amount spent on oil in 1982.

STEP 3: EQUATION

We need a 2-by-2 system.

$$\begin{pmatrix}\text{Amount spent}\\ \text{on wood in}\\ 1982\end{pmatrix} + \begin{pmatrix}\text{Amount spent}\\ \text{on oil in}\\ 1982\end{pmatrix} = \begin{pmatrix}\text{Total amount}\\ \text{spent in 1982}\end{pmatrix}$$

$$(1) \qquad x + y = 650$$

$$\begin{pmatrix}\text{Amount spent}\\ \text{on wood in}\\ 1983\end{pmatrix} + \begin{pmatrix}\text{Amount spent}\\ \text{on oil in}\\ 1983\end{pmatrix} = \begin{pmatrix}\text{Total amount}\\ \text{spent in 1983}\end{pmatrix}$$

$$(2) \qquad 2x + \frac{y}{2} = 625$$

$$\begin{Bmatrix} x + y = 650 \\ 2x + \dfrac{y}{2} = 625 \end{Bmatrix}$$

STEP 4: SOLVE

We will use the addition method.

Choose

We will eliminate the y variable.

9. Trudy Carter buys two building lots at a total cost of \$29,500. Building site #1 cost \$3500 per acre and site #2 cost \$4000 per acre. If there are a total of 8 acres in the 2 building sites, how many acres are in each site?

10. Jeannie Eldest takes the tens and twenties from her shop at the end of the day to the bank. There were a total of 33 bills with a total value of $510. How many tens and how many twenties did Jeannie deposit?

Equivalent system

Keep the first equation as is and multiply the second equation by -2.

$$(1) \qquad x + y = 650$$

$$-2\left(2x + \frac{y}{2}\right) = -2(625)$$

$$(2) \qquad -4x - y = -1250$$

$$\begin{Bmatrix} x + y = 650 \\ -4x - y = -1250 \end{Bmatrix}$$

Solve

Add.

$$\begin{array}{rl} x + y = & 650 \\ -4x - y = & -1250 \\ \hline -3x \quad = & -\ 600 \end{array}$$

$$x = 200$$

$$x + y = 650$$

$$y = 450$$

Thus in 1982 the Fletchers spent $200 for wood and $450 for oil. In 1983 they spent $400 for wood and $225 for oil.

DO EXERCISE 10.

DO SECTION PROBLEMS 7.4.

ANSWERS TO MARGINAL EXERCISES

1. $\left(\frac{2}{3}, 2\right)$ **2.** $(-2, 0)$ **3.** $(2, -2)$ **4.** $(0, 1)$ **5.** $\left(-\frac{7}{5}, -\frac{1}{5}\right)$ **6.** Addition Principle is always used and sometimes the Multiplication Principle. **7.** 8 and 30 **8.** 60 and 80 **9.** 5 acres in site #1 and 3 in site #2 **10.** 15 tens and 18 twenties

NAME COURSE/SECTION DATE

SECTION PROBLEMS 7.4

Solve by the addition method.

1. $\begin{Bmatrix} x - y = 4 \\ 3x + y = 0 \end{Bmatrix}$

2. $\begin{Bmatrix} -m + 3n = -2 \\ m + 2n = 7 \end{Bmatrix}$

3. $\begin{Bmatrix} 2a = b + 5 \\ -3a + b = -4 \end{Bmatrix}$

4. $\begin{Bmatrix} x + 6 = 0 \\ 2x - y = 1 \end{Bmatrix}$

5. $\begin{Bmatrix} x - 2y = 11 \\ 2x + y = 7 \end{Bmatrix}$

6. $\begin{Bmatrix} 3s - t = 4 \\ -2s + t = 1 \end{Bmatrix}$

7. $\begin{Bmatrix} 3s + 2t = 10 \\ -2s + 4t = 4 \end{Bmatrix}$

8. $\begin{Bmatrix} P - 2Q = 2 \\ 2P + Q = 4 \end{Bmatrix}$

9. $\begin{Bmatrix} 3x - 4y = 1 \\ -6x + 8y = -1 \end{Bmatrix}$

10. $\begin{Bmatrix} 3x - 5y = 4 \\ -2x + 2y = 4 \end{Bmatrix}$

11. $\begin{Bmatrix} -7a + 2b = 1 \\ 3a - 4b = 9 \end{Bmatrix}$

12. $\begin{Bmatrix} 2x - 5 = y \\ 3y - 1 = 2x \end{Bmatrix}$

13. $\begin{Bmatrix} \frac{P}{2} - Q = -8 \\ 2P + \frac{Q}{3} = -6 \end{Bmatrix}$

14. $\begin{Bmatrix} 2(x - 3y) = 1 \\ -3x + 6y = -4 \end{Bmatrix}$

15. $\begin{Bmatrix} 2x - (x - y) = 4 \\ -3(x + y) = -2 \end{Bmatrix}$

16. $\begin{Bmatrix} 7a - 5b = -2 \\ -3a - 2b = 5 \end{Bmatrix}$

17. $\begin{Bmatrix} s - 2t = 3 \\ -s + 3t = -1 \end{Bmatrix}$

18. $\begin{Bmatrix} a - 3b = 4 \\ 2a + b = 7 \end{Bmatrix}$

ANSWERS

1. ____
2. ____
3. ____
4. ____
5. ____
6. ____
7. ____
8. ____
9. ____
10. ____
11. ____
12. ____
13. ____
14. ____
15. ____
16. ____
17. ____
18. ____

19. ____________

20. ____________

21. ____________

22. ____________

23. ____________

24. ____________

25. ____________

26. ____________

19. $\begin{Bmatrix} s = 2t - 1 \\ 2s - t = 3 \end{Bmatrix}$

20. $\begin{Bmatrix} -\dfrac{x}{3} + y = 7 \\ x - \dfrac{y}{2} = 4 \end{Bmatrix}$

21. $\begin{Bmatrix} 7a + 5b = 4 \\ -3a + 4b = -2 \end{Bmatrix}$

22. $\begin{Bmatrix} x = -3y + 10 \\ y = -2x - 4 \end{Bmatrix}$

23. John is twice as old as Mike. Two-thirds of John's age in five years will equal Mike's age in five years. How old will John be in fifteen years?

24. The difference between Beth's age and that of her father is 30. Five times Beth's age in 5 years plus 2 will equal her father's age in 5 years. How old is each now?

25. We have a line that goes through the points (1, 2) and (−2, −4) and another line that has slope −3 and goes through the point (1, 6). Where do the lines intersect?

26. A line with slope $\frac{5}{2}$ goes through the point (2, −1). Another line with slope −2 goes through the point (−1, 4). Where do the lines intersect?

NAME COURSE/SECTION DATE

ANSWERS

27. ______

28. ______

29. ______

30. ______

27. Adult tickets for the pancake supper cost \$1.50 each and children's tickets cost \$.50 each. If the total receipts for the pancake supper were \$132.50, and if three times the number of adults is the same as twice the number of children buying tickets plus 10, how many people were at the pancake supper?

28. There were 13 B's given in a physics course. 10% of the men in the class received B's, and 15% of the women received B's. If the number of men in the class doubled is the same as the number of women in the class plus 20, how many men are in the class?

29. The Jones family eats out at the local hamburger stand. They order hamburgers at 75¢ each and french fries at 50¢ an order. If they order two more hamburgers than they do orders of french fries and the total bill is \$9.00, how many hamburgers and how many orders of french fries do they order?

30. At the beginning of 1986 the Novaks' oil burning furnace collapsed, and they spent \$3600 installing a new furnace. Their combined oil bill for 1985 and 1986 was \$3800. The new furnace was more efficient. In fact, if their 1986 oil bill was \$60 less, then it would be exactly 70% of their 1985 bill. How much did the Novaks spend on oil in 1985 and in 1986?

31. ____________

32. ____________

33. ____________

34. ____________

35. ____________

36. ____________

37. ____________

38. ____________

39. ____________

40. ____________

41. ____________

42. ____________

43. ____________

44. ____________

45. ____________

Solve the following using the addition or substitution method.

31. $\begin{Bmatrix} x + 3y = 2 \\ 2x - 4y = -16 \end{Bmatrix}$

32. $\begin{Bmatrix} 3a - b = 6 \\ 2a + b = 4 \end{Bmatrix}$

33. $\begin{Bmatrix} s = 2t - 1 \\ 2s - 3t = 4 \end{Bmatrix}$

34. $\begin{Bmatrix} a - 3 = 0 \\ b + 2a = 7 \end{Bmatrix}$

35. $\begin{Bmatrix} 5x - 7y = -14 \\ -3x - 2y = -4 \end{Bmatrix}$

36. $\begin{Bmatrix} \dfrac{x}{3} - \dfrac{y}{4} = 2 \\ -4x + 3y = 0 \end{Bmatrix}$

37. $\begin{Bmatrix} -s + 4t = 1 \\ 2s = 3t \end{Bmatrix}$

38. $\begin{Bmatrix} 9x - 4y = -2 \\ -6x + 15y = 63 \end{Bmatrix}$

39. $\begin{Bmatrix} 2s = t + 5 \\ -3s + t = -2 \end{Bmatrix}$

40. $\begin{Bmatrix} \dfrac{x}{2} - \dfrac{y}{3} = 1 \\ -x + \dfrac{y}{2} = -4 \end{Bmatrix}$

41. $\begin{Bmatrix} -7a + 6b = 15 \\ 4a - 3b = -12 \end{Bmatrix}$

42. $\begin{Bmatrix} s - 2t = 3 \\ 3s + 2t = 5 \end{Bmatrix}$

43. $\begin{Bmatrix} 2a + b = 4 \\ -a + 2b = 3 \end{Bmatrix}$

44. $\begin{Bmatrix} 14x - y = 3 \\ -10x - 5y = 16 \end{Bmatrix}$

45. $\begin{Bmatrix} 7P - 3 = 5Q \\ -2Q + 4 = 6P \end{Bmatrix}$

7.5

Motion Problems

LEARNING OBJECTIVES

After finishing Section 7.5, you should be able to:

- Apply linear equations in two variables to practical problems.

Now we will look at some problems that come under the general description of motion problems. We first need the motion formula, which describes the relationship among distance d, rate or average speed r, and the time t:

$$d = r \cdot t$$

Note

1. This formula says that distance is equal to the rate times the time. For example, if we travel 50 miles per hour for 3 hours we will go a distance of $d = rt = (50 \text{ mph})(3 \text{ hr}) = 150$ miles.
2. In using this formula be careful that the units match. For example, if r is in miles per hour, then d must be in miles and t must be in hours, but if r is in feet per second, then d must be in feet, and t must be in seconds.

Example 1 At 12:00 noon a train traveling at 80 mph leaves Omaha going west. An hour later another train leaves Omaha traveling east at 70 mph. At what time will they be 380 miles apart?

STEP 1: PICTURE

A diagram showing each trip is essential.

West ← Omaha → East

STEP 2: VARIABLE

We need a variable for distance, rate, and time for each trip.

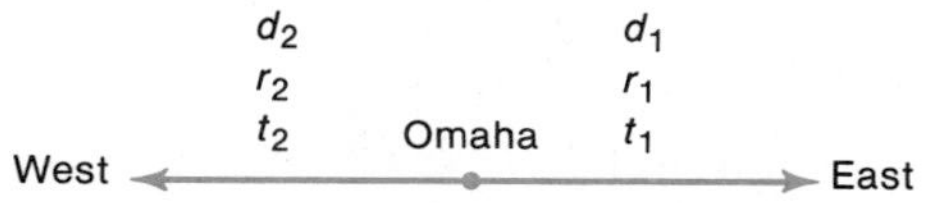

	d_2		d_1
	r_2		r_1
	t_2	Omaha	t_1
West ←		•	→ East

Fill in what is known about the variable, including the relationship between t_1 and t_2. Note that the train traveling west travels for one hour longer than the train traveling east.

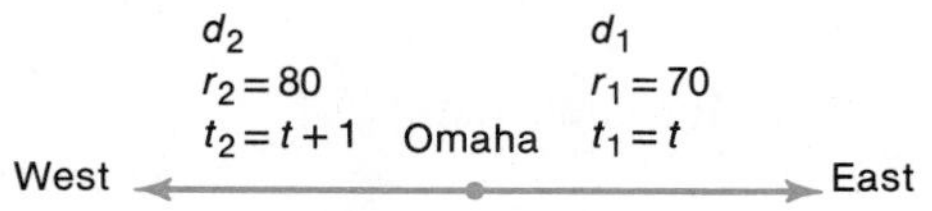

	d_2		d_1
	$r_2 = 80$		$r_1 = 70$
	$t_2 = t + 1$	Omaha	$t_1 = t$
West ←		•	→ East

STEP 3: EQUATION

Write the motion formula $d = r \cdot t$ for each trip.

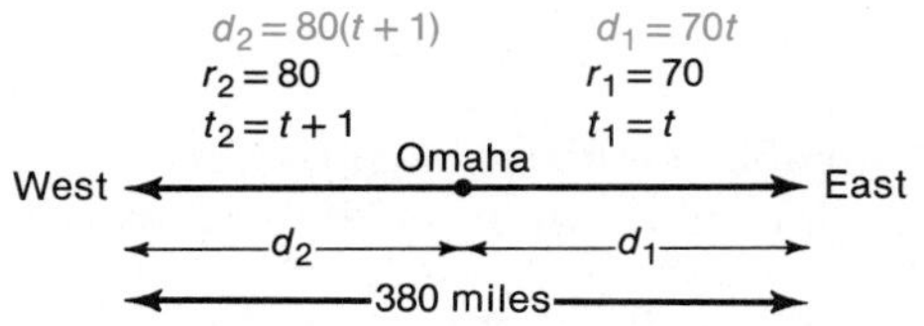

	$d_2 = 80(t + 1)$		$d_1 = 70t$
	$r_2 = 80$		$r_1 = 70$
	$t_2 = t + 1$	Omaha	$t_1 = t$
West ←		•	→ East
	← d_2 →		← d_1 →
	← 380 miles →		

Note that the relationship between the distances is $d_2 + d_1 = 380$.

1. Cheryl leaves Salt Lake City traveling south at 50 mph. An hour and a half later Marilyn leaves Salt Lake City traveling north at 60 mph. When the distance between them is 295 miles, how far has Cheryl traveled?

Instead of constructing a 2-by-2 system, such as

$$\begin{cases} d_2 = 80(t + 1) \\ 380 - d_2 = 70t \end{cases}$$

if we make a "double substitution" (that is, substitute for d_2 and d_1 what they are equal to in terms of t), we can obtain one equation in the variable t immediately.

$$d_2 + d_1 = 380$$
$$80(t + 1) + 70t = 380$$

STEP 4: SOLVE

$$80t + 80 + 70t = 380$$
$$150t = 300$$
$$t = 2 \text{ hours}$$

Thus the train traveling east travels for two hours, and since it left at 1:00, it will be 3:00 P.M. when the two trains are 380 miles apart.

DO EXERCISE 1.

Example 2 A plane flew from Brownsport to Titleville, a distance of 450 miles, in $2\frac{1}{2}$ hours with a tailwind. The return trip, against the same wind, took $3\frac{3}{4}$ hours. Find the speed of the wind and the speed of the plane in still air.

STEP 1: PICTURE

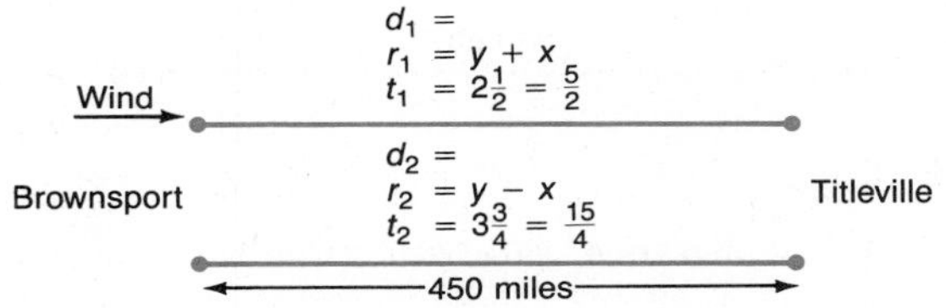

Wind
$d_1 =$
$r_1 = y + x$
$t_1 = 2\frac{1}{2} = \frac{5}{2}$
Brownsport
$d_2 =$
$r_2 = y - x$
$t_2 = 3\frac{3}{4} = \frac{15}{4}$
Titleville
450 miles

STEP 2: VARIABLE

Let x = speed of the wind and
y = speed of the plane in still air. Then
$y + x$ = resultant speed of the plane with tailwind, and
$y - x$ = resultant speed of the plane against headwind.

STEP 3: EQUATION

Write the motion formula for each trip.

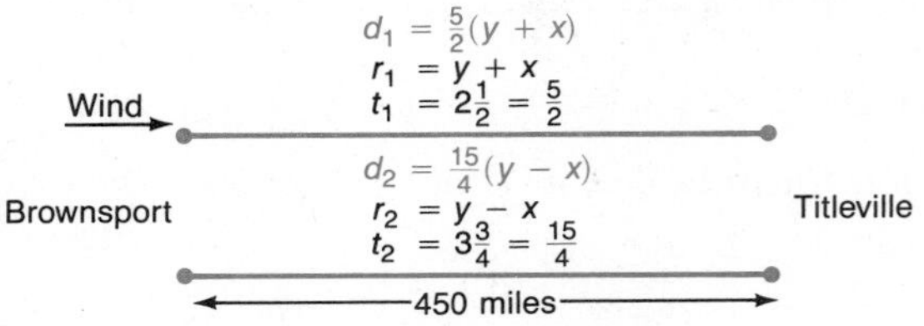

Wind
$d_1 = \frac{5}{2}(y + x)$
$r_1 = y + x$
$t_1 = 2\frac{1}{2} = \frac{5}{2}$
Brownsport
$d_2 = \frac{15}{4}(y - x)$
$r_2 = y - x$
$t_2 = 3\frac{3}{4} = \frac{15}{4}$
Titleville
450 miles

Now d_1 and d_2 are both 450 (see above diagram). If we substitute 450 for d_1 and d_2 in the two motion formulas, we have the following system to solve.

$$\begin{cases} \frac{5}{2}(y + x) = 450 \\ \frac{15}{4}(y - x) = 450 \end{cases}$$

STEP 4: SOLVE

We first clear fractions in our two equations by multiplying equation (1) by 2 and equation (2) by 4. This gives us the following equivalent system.

$$\begin{Bmatrix} 5(y + x) = 900 \\ 15(y - x) = 1800 \end{Bmatrix} \quad \text{or} \quad \begin{Bmatrix} 5y + 5x = 900 \\ 15y - 15x = 1800 \end{Bmatrix}$$

We will use the addition method to solve.

Choose

We will eliminate the x variable.

Equivalent system

Multiply the first equation by 3 and keep the second equation as is.

$$3(5y + 5x) = 3(900)$$

$$(1) \quad 15y + 15x = 2700$$

$$(2) \quad 15y - 15x = 1800$$

$$\begin{Bmatrix} 15y + 15x = 2700 \\ 15y - 15x = 1800 \end{Bmatrix}$$

Solve

Add and solve for y and then x.

$$\begin{aligned} 15y + 15x &= 2700 \\ 15y - 15x &= 1800 \\ \hline 30y &= 4500 \\ y &= \frac{4500}{30} \\ y &= 150 \\ 5y + 5x &= 900 \\ 5(150) + 5x &= 900 \\ 750 + 5x &= 900 \\ 5x &= 150 \\ x &= 30 \end{aligned}$$

Thus the speed of the wind is 30 mph, and the speed of the plane in still air is 150 mph.

DO EXERCISE 2.

Example 3 Mary Golden leaves the house and travels west on Interstate 9 at 50 mph. Twenty minutes later her husband Saul discovers she left her wallet at home and rushes after Mary to catch up with her. If Saul drives at 70 mph, how long will it take him to catch Mary?

STEP 1: PICTURE

Mary: d_1, $r_1 = 50$, $t_1 = t + \frac{1}{3}$

Saul: d_2, $r_2 = 70$, $t_2 = t$

2. It takes Ed Rosenberg 60 minutes to run his boat upriver from his house to Twome's Landing. To come back takes only 45 minutes. If the distance from his house to Twome's Landing is 18 miles, what is the speed of the current in the river?

3. Two cars leave South Bend at the same time and are traveling north on Highway 102. If one car travels at 60 mph and the other at 45 mph, how long before they are 20 miles apart?

STEP 2: VARIABLE

Note that 20 min. $= \frac{1}{3}$ hour.

STEP 3: EQUATION

Write the motion formula for each trip.

$d_1 = 50\left(t + \frac{1}{3}\right)$
$r_1 = 50$
$t_1 = t + \frac{1}{3}$
Mary

$d_2 = 70t$
$r_2 = 70$
$t_2 = t$
Saul

We have $d_1 = d_2$ or, making appropriate substitutions, we have

$$50\left(t + \frac{1}{3}\right) = 70t$$

STEP 4: SOLVE

$$50t + \frac{50}{3} = 70t$$

$$\frac{50}{3} = 20t$$

$$\left(\frac{50}{3}\right)\left(\frac{1}{20}\right) = t$$

$$\frac{50}{60} = t$$

Thus $t = \frac{5}{6}$ of an hour, or it takes Saul 50 minutes to catch Mary.

DO EXERCISE 3.

Note that in motion problems we sometimes obtain a 2-by-2 system to solve (as in Example 2), but other times (see Examples 1 and 3) it is easier to make appropriate substitutions and obtain our one equation in one unknown immediately.

DO SECTION PROBLEMS 7.5.

ANSWERS TO MARGINAL EXERCISES

1. 175 miles **2.** 3 mph **3.** $1\frac{1}{3}$ hours

NAME COURSE/SECTION DATE

ANSWERS

1. ____________

2. ____________

3. ____________

4. ____________

5. ____________

6. ____________

SECTION PROBLEMS 7.5

1. Larry Parker flies his plane from Potluck to Sapper in 3 hours with the help of a 30-mph tail wind. The return flight against a 20-mph head wind takes 4 hours. What is the speed of Larry's plane in still air?

2. Traveling at 600 mph, a plane leaves Tacoma at 1:00 to fly to Omaha, a distance of 1600 miles. At 2:00 a plane leaves Omaha traveling at 500 mph to fly to Tacoma. At what time will they cross paths?

3. A plane leaves Kansas City at 2:00 traveling east at 250 mph. An hour later a plane leaves Kansas City traveling west at 300 mph. At what time will they be 1075 miles apart?

4. Two trains leave Denver on parallel tracks traveling east. One train travels at 80 mph and the other at 50 mph. How long before they are 75 miles apart?

5. John can run at a steady pace of 12 mph, and his roommate Mike can run at a steady pace of 10 mph. If John gives Mike $\frac{3}{4}$ of a mile head start and they both start running at the same time, how many minutes will it take John to catch up with Mike?

6. Barbara leaves Boston traveling south at 50 mph. Thirty minutes later, Helen leaves Boston traveling at 70 mph in order to catch Barbara. How far from Boston is Barbara when Helen catches her?

7. ____________

8. ____________

9. ____________

10. ____________

11. ____________

12. ____________

7. A train leaves Des Moines traveling east at 100 mph. An hour later, another train leaves Des Moines traveling west at 120 mph. When they are 870 miles apart, how far from Des Moines is the train that is going west?

8. Mrs. Crowley leaves Boston traveling south in her car at 40 mph. Forty-five minutes later, her husband leaves Boston traveling south in his car at 60 mph. How long before Mrs. Crowley's husband passes her?

9. A plane flying with a tail wind of 20 mph makes a trip in 7 hours. The return trip against the 20-mph wind takes an hour longer. What is the speed of the plane in still air?

10. Ruth leaves North Platt traveling north at 40 mph. Thirty minutes later, Betty leaves North Platt, also traveling north, at 60 mph. How long will Ruth have been traveling when Betty is 10 miles in front of her?

11. John Cooper wants to mix $1.00-per-ounce tea with $.70-per-ounce tea to make a mixture of 30 ounces worth $.90 an ounce. How much of each kind should be in the mixture?

12. Mrs. Perrotta has $50,000 to invest. She invests some in the stock market, for which she earns an 8% return on her money, and the rest in real estate, which returns her 10% on her investment. If the total amount of money she earned was $4,560, how much did Mrs. Perrotta have invested in stocks and how much in real estate?

NAME COURSE/SECTION DATE

CHAPTER 7 REVIEW PROBLEMS

SECTION 7.1

1. Is (2, −3) a solution to the system $\begin{Bmatrix} 2x + y = 1 \\ -x - y = -5 \end{Bmatrix}$?

2. Is (−1, 2) a solution to the system $\begin{Bmatrix} x + y = 1 \\ -2x - y = 0 \end{Bmatrix}$?

3. What are the three possibilities for solutions to a 2-by-2 system?

SECTION 7.2

Solve by the graphing method.

4. $\begin{Bmatrix} 4s - t = 4 \\ -2s + 3t = -2 \end{Bmatrix}$

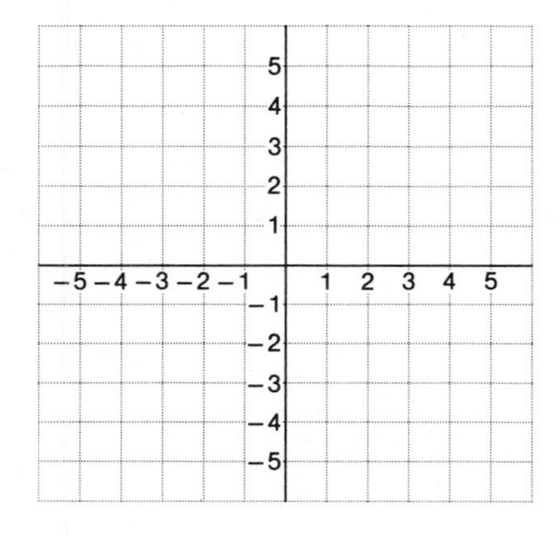

5. $\begin{Bmatrix} -2a + b = 3 \\ 4a - 2b = 1 \end{Bmatrix}$

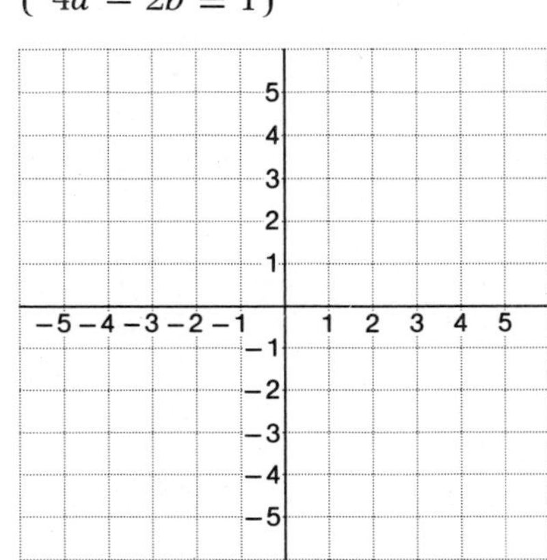

ANSWERS

1. ____________

2. ____________

3. ____________

4. See graph.

5. See graph.

6. See graph.

7. See graph.

8. ______

9. ______

10. ______

11. ______

12. ______

13. ______

6. $\begin{cases} \dfrac{x}{2} + y = -4 \\ \dfrac{-3x}{2} - 3y = 12 \end{cases}$

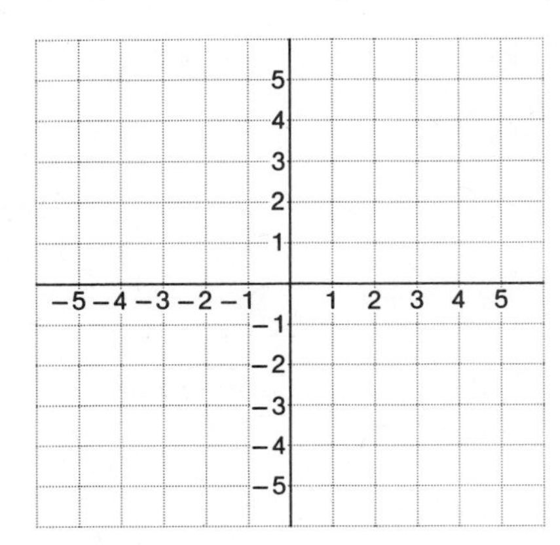

7. $\begin{cases} 3s + t = 4 \\ 2t + 4 = 0 \end{cases}$

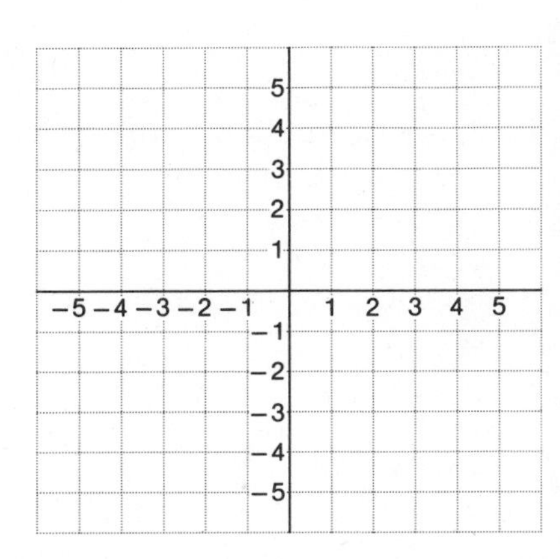

8. The sum of two numbers is -1 and their difference is 5. Find the two numbers.

SECTION 7.3

Solve by the substitution method.

9. $\begin{cases} m - 2n = -2 \\ -2m - n = -11 \end{cases}$

10. $\begin{cases} 2s - t = 3 \\ -s + 2t = 4 \end{cases}$

11. $\begin{cases} x - 3y = -1 \\ 2x - 6y = -2 \end{cases}$

12. $\begin{cases} -3a + b = 4 \\ 2a - 5 = 0 \end{cases}$

13. A line goes through the points (4, 3) and (2, -1), and another line that has slope -3 goes through the point (-1, 2). Where do the lines intersect?

NAME | COURSE/SECTION | DATE

SECTION 7.4

Solve by the addition method.

14. $\begin{Bmatrix} 3x - y = 4 \\ 2x + y = 1 \end{Bmatrix}$

15. $\begin{Bmatrix} 2x - y = 3 \\ -4x + 2y = 0 \end{Bmatrix}$

16. $\begin{Bmatrix} 2m - 3n = 2 \\ -3m + 5n = 4 \end{Bmatrix}$

17. $\begin{Bmatrix} 5x - 7y = 4 \\ -3x + 4y = -6 \end{Bmatrix}$

Solve by any method.

18. $\begin{Bmatrix} x = 2y - 5 \\ 3x - y = 4 \end{Bmatrix}$

19. $\begin{Bmatrix} \dfrac{x}{2} = y - 5 \\ -3x + 2y = 4 \end{Bmatrix}$

20. $\begin{Bmatrix} 10x - 5y = 4 \\ -3x + 4y = 2 \end{Bmatrix}$

21. The sum of two numbers is 2 and their difference is 18. Find the two numbers.

22. Helen McGuinnes has a solution that is 40% alcohol and another solution that is 60% alcohol. If Helen wants 12 pints of a solution that is 45% alcohol, how many pints of the 40% solution and of the 60% solution should she mix together to obtain her 45% solution?

ANSWERS

14. ____________

15. ____________

16. ____________

17. ____________

18. ____________

19. ____________

20. ____________

21. ____________

22. ____________

23. ____________

24. ____________

25. ____________

26. ____________

27. ____________

SECTION 7.5

23. A train leaves Fargo traveling east at 100 mph. An hour later another train leaves Fargo traveling west at 120 mph. When they are 980 miles apart, how far from Fargo will each train be?

24. Henry Farlow runs at a speed of 10 mph and his friend Stacy at a speed of 8 mph. If Henry gives Stacy a one-half-mile head start, how long will it take Henry to catch her?

25. The sum of two numbers is 19 and their difference is 5. Find the two numbers.

26. The average of two numbers is 34. If one number is one less than twice the other, find the numbers.

27. Mr. Hauser calculates he needs 30 gallons of primer and overcoat to paint his house. If he plans to use four more gallons of overcoat than of primer, how many gallons of each does he need?

NAME ______ COURSE/SECTION ______ DATE ______

ANSWERS

1. ______

2. See graph.

3. ______

4. ______

5. ______

6. ______

7. ______

8. ______

CHAPTER 7 TEST

1. Is $(-3, 2)$ a solution to the system $\begin{Bmatrix} 2x - 5y = -16 \\ -x + 3y = 9 \end{Bmatrix}$?

2. Solve by the graphing method.
$\begin{Bmatrix} s - t = -5 \\ 2s + 2t = 2 \end{Bmatrix}$

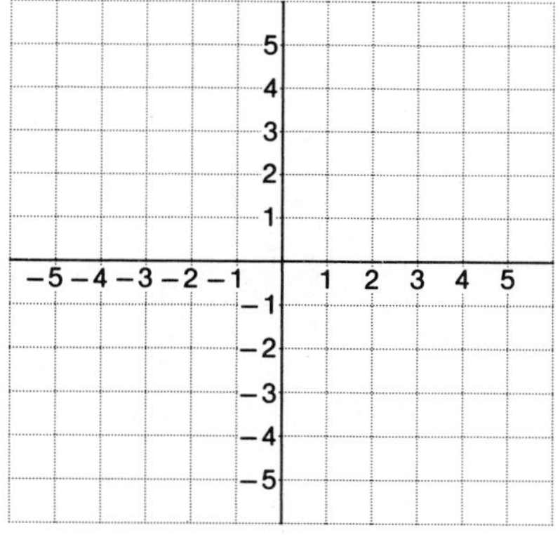

3. Solve by the addition method.
$\begin{Bmatrix} x - 3y = 6 \\ -2x + y = -7 \end{Bmatrix}$

4. Solve by the substitution method.
$\begin{Bmatrix} 2s - t = 2 \\ -s + 3t = -11 \end{Bmatrix}$

Solve by any method.

5. $\begin{Bmatrix} -3a - b = 4 \\ 2a + b = -1 \end{Bmatrix}$

6. $\begin{Bmatrix} 2x - 3y = -7 \\ -3x + 5y = 13 \end{Bmatrix}$

7. $\begin{Bmatrix} 2m = 3 + n \\ -4m + 2 = -2n \end{Bmatrix}$

8. $\begin{Bmatrix} -P + 2Q = 6 \\ 2P = 3Q \end{Bmatrix}$

9. ______________

10. ______________

11. ______________

12. ______________

13. ______________

14. ______________

15. ______________

9. Given the three general cases for solutions to a 2-by-2 system (unique solution, no solution, and infinitely many solutions), give a geometric interpretation for each case.

10. The sum of two numbers is 1. If 3 times one number plus twice the other is 4, what are the numbers?

11. Kathy Schleyer has a solution that is 40% salt and another solution that is 60% salt. If Kathy wants 12 liters of a 45% salt solution, how many liters of the 40% solution and of the 60% solution should she mix together?

12. Carolyn Jury invested $15,000 in stocks and bonds. Her annual return on the stocks was 15% and 9% on the bonds. The total amount of her annual return was $1890. How much did Carolyn invest in stocks?

13. At 2:00 Pete Hunter starts from Livermore and drives at 60 mph towards Milo. Thirty minutes later, Steve Mooreland starts in Milo and drives at 50 mph towards Livermore. If Livermore and Milo are 195 miles apart, what time is it when Pete and Steve meet?

14. Susan and Larry Ferrara are hunting for agates. If together they find 55 agates, and Susan finds 5 less than Larry, how many agates does each find?

15. Give an example of a 2-by-2 system that has infinitely many solutions.

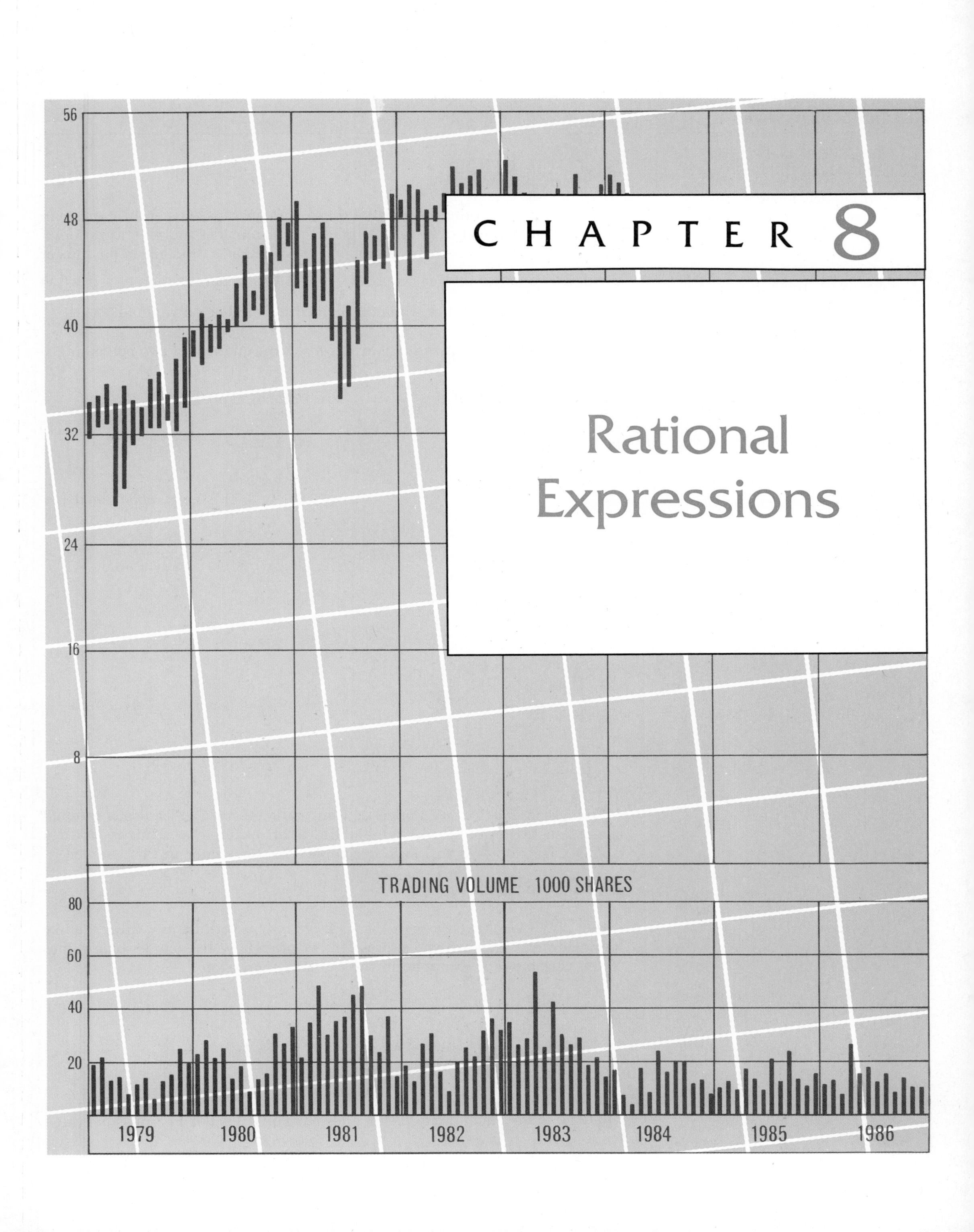
CHAPTER 8
Rational Expressions
56
48
40
32
24
16
8
TRADING VOLUME 1000 SHARES
80
60
40
20
1979
1980
1981
1982
1983
1984
1985
1986

8.1

Introduction

As we mentioned at the very start of this text, algebra is a generalization or an extension of arithmetic. In this chapter we will consider the generalization of the concept of a fraction in arithmetic. Recall that a fraction is an integer divided by another integer $\left(\text{such as } \frac{4}{5}, \frac{10}{21}\right)$. Our generalization of a fraction will take the form of a polynomial divided by another polynomial.

DEFINITION 8.1 A *rational expression* is the quotient of two polynomials.

Example 1 Which of the following are rational expressions?

(a) $\dfrac{x + 2}{x^2 + 3x + 4}$ Yes

(b) $\dfrac{1}{y - 2}$ Yes

(c) $\dfrac{x}{|x - 1|}$ No; $|x - 1|$ is not a polynomial.

(d) $\dfrac{\sqrt{t}}{t^2 + 1}$ No; $\sqrt{t}$ is not a polynomial.

(e) $\dfrac{4}{5}$ Yes

(f) $\dfrac{\frac{1}{a^2} + 3}{a^2 + 1}$ No; $\frac{1}{a^2} + 3$ is not a polynomial.

(g) $\dfrac{x^2 + y^2}{x - y}$ Yes

From Example 1(e) we notice that any rational number is also a rational expression. Also note that Example 1(f) is an example of what is called a complex fraction. We will look at complex fractions later in this chapter.

DO EXERCISE 1.

In a polynomial such as $x^2 - 3x + 4$, we may substitute any value we like for x. However, in a rational expression, we often have to exclude some values of the variable. For example, in the rational expression

$$\frac{x + 2}{x - 3}$$

we cannot let x be 3 since if x is 3, then we have

$$\frac{3 + 2}{3 - 3} = \frac{5}{0}$$

and the 0 in the denominator is not allowed. Thus, 3 is not a permissible value for x.

LEARNING OBJECTIVES

After finishing Section 8.1, you should be able to:

- Identify rational expressions.
- Find poles of rational expressions.

1. Which of the following are rational expressions? If not, why not?

(a) $\dfrac{x^2 - 3}{4x + 1}$ (b) $\dfrac{t - 1}{3}$

(c) $\dfrac{\sqrt{y}}{y^2 + 2y - 3}$ (d) $\dfrac{2}{|a|}$

(e) $\dfrac{3}{7}$ (f) $\dfrac{x(x + 3)}{x^2 + 4}$

In general, any value of the variable that makes the denominator zero in a rational expression will not be a permissible value for the variable. We call a number that makes the denominator zero in a rational expression a *pole* of the rational expression.

Example 2 Find the poles of the following rational expressions.

Rational expression	*Pole*
(a) $\frac{1}{y + 4}$	$y = -4$
(b) $\frac{2x^2}{(x + 2)(x + 1)}$	$x = -2$ and $x = -1$

Note that if $(x + 2)(x + 1) = 0$, then $x + 2 = 0$ (or $x = -2$), or $x + 1 = 0$ (or $x = -1$).

(c) $\frac{t - 4}{t^2 + 1}$	There are no poles.

Note that $t^2 + 1$ is always positive and can never be zero.

(d) $\frac{x - 1}{2x^3 + 5x^2 - 3x}$	$x = 0$, $x = -3$, and $x = \frac{1}{2}$

Note that

$$\begin{aligned} 2x^3 + 5x^2 - 3x &= x(2x^2 + 5x - 3) \\ &= (x)(2x - 1)(x + 3) \end{aligned}$$

Now, if

$$(x)(2x - 1)(x + 3) = 0$$

then

$$(1)\ x = 0 \quad \text{or} \quad (2)\ 2x - 1 = 0 \quad \text{or} \quad (3)\ x + 3 = 0$$

$$2x = 1 \qquad\qquad x = -3$$

$$x = \frac{1}{2}$$

We will assume for the remainder of this chapter that whenever we have a rational expression that has poles, we will exclude the poles as values for the variable in the rational expression.

DO EXERCISE 2.

The operations we perform with rational expressions are the same operations we perform with fractions—namely, addition, subtraction, multiplication, division, and reducing to lowest terms. Moreover, the methods and procedures we use to perform the various operations on rational expressions are identical to the methods and procedures we use when working with fractions. This should not seem very surprising since fractions are also rational expressions. Thus, we will use what we know about adding, subtracting, multiplying, and dividing fractions to help perform the same operations on rational expressions.

DO SECTION PROBLEMS 8.1.

2. Find the poles for the following rational expressions.

(a) $\frac{x}{2x - 1}$ (b) $\frac{t - 4}{t^2 - 2t - 3}$

(c) $\frac{4y^2 - 5}{y^2 + 4}$ (d) $\frac{5}{a^2 - 16}$

(e) $\frac{3x - 1}{2x^3 + 5x^2 - 12x}$

ANSWERS TO MARGINAL EXERCISES

1. (a) Yes (b) Yes (c) No; the numerator is not a polynomial. (d) No; the denominator is not a polynomial. (e) Yes (f) Yes **2.** (a) $\frac{1}{2}$ (b) 3, -1 (c) None (d) -4, 4 (e) 0, $\frac{3}{2}$, -4

NAME ______ COURSE/SECTION ______ DATE ______

SECTION PROBLEMS 8.1

Are the following rational expressions? If not, why not?

1. $\dfrac{x-2}{y}$

2. $\dfrac{\sqrt{x}}{x}$

3. $\dfrac{t^2}{t+\frac{3}{2}}$

4. 5

5. $\dfrac{y^{-1}}{y^{-2}-1}$

6. $\dfrac{1-\frac{x}{y}}{1+\frac{x}{y}}$

7. $\dfrac{t^2-3t}{2t}$

8. $\dfrac{|x-1|}{5y}$

Find the poles.

9. $\dfrac{x-2}{x}$

10. $\dfrac{3y^3}{y^2-1}$

11. $\dfrac{t}{t^2-t-20}$

ANSWERS

1. ______
2. ______
3. ______
4. ______
5. ______
6. ______
7. ______
8. ______
9. ______
10. ______
11. ______

12. ______________

13. ______________

14. ______________

15. ______________

16. ______________

17. ______________

18. ______________

19. ______________

20. ______________

12. $\dfrac{x}{x^2 + 1}$

13. $\dfrac{4}{x(x^2 - 16)}$

14. $\dfrac{4}{5a^3 + 9a^2 - 2a}$

15. $\dfrac{2}{3 - y}$

16. $\dfrac{3}{x(x - 1)(x + 2)(2x - 3)}$

17. $\dfrac{y^2 + 3}{2y^2 + 4}$

18. $\dfrac{1}{x - 2}$

19. $\dfrac{x - 2}{x^3 - 4x^2 + 3x}$

20. $\dfrac{7}{(t - 3)^2}$

8.2

Simplification of Rational Expressions

The first operation we look at is reducing (simplification of) rational expressions. If we multiply the two fractions

$$\frac{2}{3} \cdot \frac{6}{5} = \frac{12}{15}$$

we do not leave our answer as $\frac{12}{15}$. Instead, we reduce $\frac{12}{15}$ to lowest terms and obtain $\frac{12}{15} = \frac{4}{5}$. As we might expect, we also want to be able to reduce rational expressions to lowest terms. We recall how we reduce fractions in the following example.

Example 1 Reduce $\frac{12}{15}$.

(a) Factor the numerator and denominator and look for common factors in the numerator and denominator (a *factor* is part of a product).

$$\frac{12}{15} = \frac{4 \cdot 3}{3 \cdot 5} = \frac{2 \cdot 2 \cdot 3}{3 \cdot 5}$$

We see that 3 is a common factor.

(b) Simplify.

$$\frac{12}{15} = \frac{2 \cdot 2 \cdot 3}{5 \cdot 3} = \frac{3}{3} \cdot \frac{2 \cdot 2}{5}$$

$$= 1 \cdot \frac{4}{5} = \frac{4}{5}$$

We can also use the Fundamental Property of Fractions, namely,

$$\frac{m \cdot c}{n \cdot c} = \frac{m}{n} \qquad \text{where } c \neq 0$$

which we first saw in Section 1.2. This allows us to simplify a fraction any time we have a common factor in the numerator and denominator.

DO EXERCISES 1 THROUGH 4.

Note

1. In doing Example 1, we could use the shortcut of "canceling" the 3 in the numerator and denominator, and we may write

$$\frac{12}{15} = \frac{2 \cdot 2 \cdot \not{3}}{\not{3} \cdot 5} = \frac{2 \cdot 2}{5} = \frac{4}{5}$$

This is certainly acceptable as long as we understand why we can cancel the common factors and when we can cancel in a problem. Also, remember it is never wrong to write out all the steps in working a problem.

LEARNING OBJECTIVES

After finishing Section 8.2, you should be able to:

- Simplify rational expressions.

Reduce the following fractions by the method used in Example 1.

1. $\frac{24}{36}$

2. $\frac{14}{21}$

3. $\frac{15}{28}$

4. $\frac{18}{12}$

What is *wrong* (if anything) with the following problems involving reducing?

5. $\frac{2+10}{2} = \frac{\not{2}+10}{\not{2}} = 10$

6. $\frac{3+7}{3+2} = \frac{\not{3}+7}{\not{3}+2} = \frac{7}{2}$

7. $\frac{4 \cdot 5}{3 \cdot 4} = \frac{\not{4} \cdot 5}{3 \cdot \not{4}} = \frac{5}{3}$

8. $\frac{16}{64} = \frac{1\not{6}}{\not{6}4} = \frac{1}{4}$

Reduce.

9. $\frac{4a-8}{3a-6}$

2. The fact that the numbers we cancel must be common factors is very important. For example,

$$\frac{3+15}{3} \neq \frac{\not{3}+15}{\not{3}} = 15$$

Note that

$$\frac{3+15}{3} = \frac{18}{3} = 6$$

The reason we cannot cancel the 3 in the numerator and denominator of $\frac{3+15}{3}$ is that the 3 in the numerator is not a factor; that is, it is not 3 times something. The canceling in

$$\frac{3 \cdot 15}{3} = \frac{\not{3} \cdot 15}{\not{3}} = 15$$

is all right, however, because here the 3 *is* a common factor.

3. Reducing (or canceling) incorrectly is one of the more common mistakes in algebra, and we must be careful not to make this mistake.

DO EXERCISES 5 THROUGH 8.

Now we want to reduce rational expressions by duplicating the steps involved in reducing a fraction.

Example 2 Reduce $\frac{3a-9}{a-3}$.

(a) Factor the numerator and denominator and note any common factors.

$$\frac{3a-9}{a-3} = \frac{3(a-3)}{a-3}$$

(b) Since we have a common factor, namely $a - 3$, we can simplify.

$$\frac{3a-9}{a-3} = \frac{3(a-3)}{a-3} = \frac{3}{1} \cdot \frac{a-3}{a-3} = \frac{3}{1} \cdot 1 = 3$$

DO EXERCISE 9.

Example 3 Reduce $\frac{x^2+2x+1}{x^2-x-2}$.

(a) Factor the numerator and denominator, and note any common factors.

$$\frac{x^2+2x+1}{x^2-x-2} = \frac{(x+1)(x+1)}{(x-2)(x+1)}$$

(b) Since there is a common factor, $(x + 1)$, we can simplify.

$$\frac{x^2 + 2x + 1}{x^2 - x - 2} = \frac{(x + 1)(x + 1)}{(x - 2)(x + 1)} = \frac{x + 1}{x - 2} \cdot \frac{x + 1}{x + 1} = \frac{x + 1}{x - 2} \cdot 1 = \frac{x + 1}{x - 2}$$

DO EXERCISE 10.

Example 4 Reduce $\dfrac{x^3 + x^2 - 12x}{x^3 + 4x^2}$.

(a) Factor the numerator and denominator and note any common factors.

$$\frac{x^3 + x^2 - 12x}{x^3 + 4x^2} = \frac{x(x^2 + x - 12)}{x^2(x + 4)} = \frac{x(x + 4)(x - 3)}{x \cdot x(x + 4)}$$

(b) Since there are common factors, x and $x + 4$, we can simplify.

$$\frac{x^3 + x^2 - 12x}{x^3 + 4x^2} = \frac{x(x + 4)(x - 3)}{x \cdot x(x + 4)} = \frac{x}{x} \cdot \frac{x + 4}{x + 4} \cdot \frac{x - 3}{x} = 1 \cdot 1 \cdot \frac{x - 3}{x} = \frac{x - 3}{x}$$

DO EXERCISES 11 AND 12.

There are some other rational expressions we want to reduce.

Example 5 Reduce $\dfrac{2x - 2y}{y - x}$.

(a) Factor.

$$\frac{2x - 2y}{y - x} = \frac{2(x - y)}{y - x}$$

Note there are no apparent common factors. However, $x - y$ and $y - x$ differ only by a sign; that is, in $x - y$ the x is positive and the y is negative and in $y - x$ just the opposite is true. We can do something to reduce further in this case.

(b) Factor -1 from the denominator and then note that we will have a common factor.

$$\frac{2x - 2y}{y - x} = \frac{2(x - y)}{y - x} = \frac{2(x - y)}{-1(-y + x)} = \frac{2(x - y)}{-1(x - y)}$$

Reduce.

10. $\dfrac{t - 3}{t^2 - 9}$

Reduce.

11. $\dfrac{x^2 + x - 6}{x^2 - 6x + 8}$

12. $\dfrac{y^4 - 16y^2}{y^3 - 8y^2 + 16y}$

Reduce.

13. $\dfrac{t^2 + 4t + 4}{2 - t}$

Reduce.

14. $\dfrac{4x^3y^2z}{-12x^2y^3z^2}$

15. $\dfrac{(x - 1)^3(x + 3)(x - 2)}{4(x^2 - 3x + 2)}$

(c) Simplify.

$$\frac{2x - 2y}{y - x} = \frac{2(x - y)}{-1(x - y)} = \frac{2}{-1} \cdot \frac{x - y}{x - y} = \frac{2}{-1} = -2$$

DO EXERCISE 13.

Example 6 Reduce $\dfrac{20a^3b}{12a^2b^2}$.

(a) Factor.

$$\frac{20a^3b}{12a^2b^2} = \frac{4 \cdot 5 \cdot a^2 \cdot a \cdot b}{4 \cdot 3 \cdot a^2 \cdot b \cdot b}$$

Note that we factor a^3 in the numerator as $a^2 \cdot a$ (instead of $a \cdot a \cdot a$) to match the a^2 in the denominator.

(b) Simplify.

$$\frac{20a^3b}{12a^2b^2} = \frac{4 \cdot 5 \cdot a^2 \cdot a \cdot b}{4 \cdot 3 \cdot a^2 \cdot b \cdot b} = \frac{4}{4} \cdot \frac{a^2}{a^2} \cdot \frac{b}{b} \cdot \frac{5 \cdot a}{3 \cdot b} = \frac{5a}{3b}$$

Note we could also use the exponential laws of Chapter 2 to reduce $20a^3b/12a^2b^2$; that is,

$$\frac{20a^3b}{12a^2b^2} = \frac{5a^{3-2}b^{1-2}}{3} = \frac{5ab^{-1}}{3} = \frac{5a}{3b}$$

Recall that we do not leave negative exponents in final answers.

DO EXERCISES 14 AND 15.

Finally, we want to look again at the shortcut method of reducing called canceling. However, first we give three notes of caution and advice.

1. It is never wrong to write out all the steps in a reducing problem, or in any type of problem, for that matter.
2. In using any shortcut method, such as canceling, it is vital to know *why* the method works and *when* (on what type of problem) it can be used.
3. We can only cancel when we have a common factor.

Example 7 Reduce $\dfrac{35}{49}$.

$$\frac{35}{49} = \frac{7 \cdot 5}{7 \cdot 7} = \frac{\cancel{7} \cdot 5}{\cancel{7} \cdot 7} = \frac{5}{7}$$

Example 8 Reduce $\dfrac{x^2(x-2)^2}{x^3-4x}$.

$$\frac{x^2(x-2)^2}{x^3-4x} = \frac{x \cdot x(x-2)(x-2)}{x(x^2-4)} = \frac{x \cdot x(x-2)(x-2)}{x(x-2)(x+2)}$$

$$= \frac{\not{x} \cdot x(\cancel{x-2})(x-2)}{\not{x}(\cancel{x-2})(x+2)}$$

$$= \frac{x(x-2)}{x+2}$$

DO EXERCISES 16 THROUGH 22.

DO SECTION PROBLEMS 8.2.

Reduce.

16. $\dfrac{21x^2(y-1)}{7x(y-1)^2}$

17. $\dfrac{2t^2-5t-3}{(2t+1)(t^2-9)}$

18. $\dfrac{(x+1)(x^2-x)}{1-x^2}$

19. $\dfrac{15x^2y^3z^2}{-3xy^4z^2}$

What is *wrong* with the reducing in the following problems?

20. $\dfrac{a+ab}{a} = \dfrac{\not{a}+ab}{\not{a}} = ab$

21. $\dfrac{x^2-3}{x^2+x-4} = \dfrac{\cancel{x^2}-3}{\cancel{x^2}+x-4} = \dfrac{-3}{x-4}$

22. $\dfrac{x-5}{x} = \dfrac{\not{x}-5}{\not{x}} = -5$

ANSWERS TO MARGINAL EXERCISES

1. $\frac{2}{3}$ **2.** $\frac{2}{3}$ **3.** $\frac{15}{28}$ **4.** $\frac{3}{2}$ **5.** The 2 is not a factor. **6.** The 3 is not a factor. **7.** Nothing wrong **8.** The 6 is not a factor. **9.** $\frac{4}{3}$ **10.** $\frac{1}{t + 3}$ **11.** $\frac{x + 3}{x - 4}$ **12.** $\frac{y(y + 4)}{y - 4}$ **13.** This is already reduced. **14.** $-\frac{x}{3yz}$ **15.** $\frac{(x - 1)^2(x + 3)}{4}$ **16.** $\frac{3x}{y - 1}$ **17.** $\frac{1}{t + 3}$ **18.** $-x$ **19.** $-\frac{5x}{y}$ **20.** The a is not a factor. **21.** The x^2 is not a factor. **22.** The x is not a factor.

NAME COURSE/SECTION DATE

SECTION PROBLEMS 8.2

Reduce.

1. $\frac{20}{48}$

2. $\frac{15}{28}$

3. $\frac{30}{140}$

4. $\frac{51}{85}$

5. $\frac{24}{36}$

6. $\frac{125}{200}$

7. $\frac{63}{55}$

8. $\frac{84}{147}$

9. $\frac{a^2b}{ab^2}$

10. $\frac{-18xy^2z}{4xy}$

11. $\frac{24(ab)^2}{20a^4b^2}$

12. $\frac{x^2y}{xy^2z}$

13. $\frac{3x^2y^4}{12xy^2z}$

14. $\frac{a^2bc^3}{ab^2c}$

15. $\frac{4a-4}{3a-3}$

16. $\frac{3t-3}{4t-4}$

17. $\frac{3xy-9y}{3y-12yz}$

18. $\frac{4a^2b-6ab}{6ab-3b^2}$

ANSWERS

1. ________
2. ________
3. ________
4. ________
5. ________
6. ________
7. ________
8. ________
9. ________
10. ________
11. ________
12. ________
13. ________
14. ________
15. ________
16. ________
17. ________
18. ________

19. ______________

20. ______________

21. ______________

22. ______________

23. ______________

24. ______________

25. ______________

26. ______________

27. ______________

28. ______________

29. ______________

30. ______________

31. ______________

32. ______________

33. ______________

34. ______________

35. ______________

36. ______________

19. $\dfrac{x-1}{1-x}$

20. $\dfrac{4x-10}{2-6x}$

21. $\dfrac{3y-3}{5-5y}$

22. $\dfrac{x-y}{y-x}$

23. $\dfrac{x(x-1)^2}{(x+1)(x-1)}$

24. $\dfrac{x^2+3x+2}{x^2+4}$

25. $\dfrac{y^2-4y+3}{y^2-1}$

26. $\dfrac{x^2-9}{x^2-3x}$

27. $\dfrac{t^2-16}{t+4}$

28. $\dfrac{6x^2+7x-3}{3x^2+5x-2}$

29. $\dfrac{9-n^2}{n^2-3n}$

30. $\dfrac{4x^2+4x}{4x^3-2x^2-6x}$

31. $\dfrac{y^2(y-3)^2(y+1)}{y(y+1)^2(y-3)^2}$

32. $\dfrac{2x^2+3x-2}{x^2+x-2}$

33. $\dfrac{y^2-5y}{y^2+9y+20}$

34. $\dfrac{t^3-t^2-2t}{t^2-4}$

35. $\dfrac{t-1}{t^2-6t+5}$

36. $\dfrac{x^2-9}{x^2+6x+9}$

8.3

Multiplication and Division of Rational Expressions

LEARNING OBJECTIVES

After finishing Section 8.3, you should be able to:

- Multiply rational expressions.
- Divide rational expressions.

The next operation we look at is multiplication of rational expressions. Recall that to multiply two fractions, we multiply their numerators and then we multiply their denominators,

$$\frac{a}{b} \cdot \frac{c}{d} = \frac{a \cdot c}{b \cdot d}$$

For example,

$$\frac{2}{3} \cdot \frac{4}{5} = \frac{2 \cdot 4}{3 \cdot 5} = \frac{8}{15}$$

We multiply rational expressions exactly the same way.

Example 1 Multiply the following rational expressions.

(a) $$\frac{x}{x+1} \cdot \frac{3}{x+4} = \frac{(x)(3)}{(x+1)(x+4)} = \frac{3x}{x^2+5x+4}$$

(b) $$\frac{t-1}{t+2} \cdot t = \frac{t-1}{t+2} \cdot \frac{t}{1} = \frac{(t-1)t}{(t+2)1} = \frac{t^2-t}{t+2}$$

(c) $$\frac{1}{y} \cdot \frac{x}{x^2+1} = \frac{1 \cdot x}{y(x^2+1)} = \frac{x}{yx^2+y}$$

(d) $$x^2 \cdot \frac{2}{3} \cdot \frac{x+1}{x^2+2} = \frac{x^2}{1} \cdot \frac{2}{3} \cdot \frac{x+1}{x^2+2} = \frac{(x^2)(2)(x+1)}{(1)(3)(x^2+2)} = \frac{2x^2(x+1)}{3(x^2+2)} = \frac{2x^3+2x^2}{3x^2+6}$$

DO EXERCISES 1 THROUGH 4.

Multiply.

1. $\frac{x}{x+2} \cdot \frac{3x+1}{x-1}$

2. $t \cdot \frac{2}{t^2+1}$

3. $\frac{3}{4} \cdot \frac{2x}{y} \cdot \frac{x+1}{5}$

4. $(-2)\left(\frac{1}{x+5}\right)\left(\frac{-x}{2-x}\right)$

Since we always want our final answers to be in reduced form, we look at the problem of multiplying and reducing rational expressions as a single process. We begin by looking at a problem with fractions:

$$\frac{21}{70} \cdot \frac{25}{12} = \frac{21 \cdot 25}{70 \cdot 12} = \frac{3 \cdot 7 \cdot 5 \cdot 5}{7 \cdot 2 \cdot 5 \cdot 3 \cdot 4}$$

$$= \frac{3}{3} \cdot \frac{7}{7} \cdot \frac{5}{5} \cdot \frac{5}{2 \cdot 4} = \frac{5}{8}$$

Multiply and reduce.

5. $\frac{15}{28} \cdot \frac{14}{9}$

6. $\frac{3x^2y}{2z} \cdot \frac{10z^2x}{9xy^2}$

Multiply and reduce.

7. $\frac{2t-2}{15} \cdot \frac{3}{8t-8}$

8. $\frac{t}{t^2-9} \cdot \frac{t+3}{t-1}$

9. $\frac{x^2-2x+1}{x^2-2x-3} \cdot \frac{x^2-3x}{x^2+x-2}$

In this problem, instead of multiplying $21 \cdot 25$ and $70 \cdot 12$ and then reducing the resulting fraction $\frac{525}{840}$, we started by factoring the 21, 25, 70, and 12, reducing what we could, and then multiplying. This process is quite efficient, and we will use it with rational expressions.

Example 2 Multiply and reduce $\frac{3x^2y}{10z} \cdot \frac{5z^3}{xy^2}$.

$$\frac{3x^2y}{10z} \cdot \frac{5z^3}{xy^2} = \frac{3 \cdot 5 \cdot x^2 \cdot y \cdot z^3}{10 \cdot x \cdot y^2 \cdot z}$$
$$= \frac{3 \cdot 5 \cdot x \cdot x \cdot y \cdot z \cdot z^2}{2 \cdot 5 \cdot x \cdot y \cdot y \cdot z}$$
$$= \frac{5}{5} \cdot \frac{x}{x} \cdot \frac{y}{y} \cdot \frac{z}{z} \cdot \frac{3xz^2}{2y} = \frac{3xz^2}{2y}$$

Note that the exponential laws could also be used to do this problem.

DO EXERCISES 5 AND 6.

Example 3 Multiply and reduce $\frac{2x-4}{x+1} \cdot \frac{x^2+x}{10x-20}$.

$$\frac{2x-4}{x+1} \cdot \frac{x^2+x}{10x-20} = \frac{(2x-4)(x^2+x)}{(x+1)(10x-20)}$$
$$= \frac{2(x-2)(x)(x+1)}{10(x+1)(x-2)}$$
$$= \frac{2}{2} \cdot \frac{x-2}{x-2} \cdot \frac{x+1}{x+1} \cdot \frac{x}{5}$$
$$= 1 \cdot 1 \cdot 1 \cdot \frac{x}{5} = \frac{x}{5}$$

Example 4 Multiply and reduce $\frac{t}{t^2+2t+1} \cdot \frac{t^2-1}{t^2+2t}$.

$$\frac{t}{t^2+2t+1} \cdot \frac{t^2-1}{t^2+2t} = \frac{t(t^2-1)}{(t^2+2t+1)(t^2+2t)}$$
$$= \frac{t(t-1)(t+1)}{(t+1)(t+1)(t)(t+2)}$$
$$= \frac{t}{t} \cdot \frac{t+1}{t+1} \cdot \frac{t-1}{(t+1)(t+2)}$$
$$= \frac{t-1}{(t+1)(t+2)} \quad \left(\text{or } \frac{t-1}{t^2+3t+2}\right)$$

DO EXERCISES 7 THROUGH 9.

We will do two more examples before we look at division of rational expressions.

Example 5 Multiply and reduce.

$$\frac{2x^2+3x-2}{x}\cdot\frac{x+3}{x-3}\cdot\frac{2x-6}{x^2+5x+6}=\frac{(2x^2+3x-2)(x+3)(2x-6)}{x(x-3)(x^2+5x+6)}$$
$$=\frac{2(2x-1)(x+2)(x+3)(x-3)}{x(x-3)(x+2)(x+3)}$$
$$=\frac{x+2}{x+2}\cdot\frac{x+3}{x+3}\cdot\frac{x-3}{x-3}\cdot\frac{2(2x-1)}{x}$$
$$=\frac{2(2x-1)}{x}$$

We can also use the shortcut method of canceling to do Example 5, as we will show in Example 6.

Example 6 Multiply and reduce.

$$\frac{2x^2+3x-2}{x}\cdot\frac{x+3}{x-3}\cdot\frac{2x-6}{x^2+5x+6}=\frac{2(2x-1)(x+2)(x+3)(x-3)}{x(x-3)(x+2)(x+3)}$$
$$=\frac{2(2x-1)\cancel{(x+2)}\cancel{(x+3)}\cancel{(x-3)}}{x\cancel{(x-3)}\cancel{(x+2)}\cancel{(x+3)}}$$
$$=\frac{2(2x-1)}{x}$$

DO EXERCISES 10 THROUGH 13.

The last operation we will look at in this section is division of rational expressions. As before, first we review what happens with fractions.

Recall from Chapter 1 that when we divide fractions, we change the division to multiplication and then multiply by the reciprocal of the divisor; or, in symbols,

$$\frac{a}{b}\div\frac{c}{d}=\frac{a}{b}\cdot\frac{d}{c}=\frac{a\cdot d}{b\cdot c}$$

Thus, the problem $\frac{2}{3}\div\frac{4}{5}$ becomes $\frac{2}{3}\div\frac{4}{5}=\frac{2}{3}\cdot\frac{5}{4}=\frac{10}{12}=\frac{5}{6}$.

DO EXERCISE 14.

The procedure for dividing two rational expressions is exactly the same as that for dividing two fractions.

Example 7 Divide $\frac{x}{x+1}$ by $\frac{2}{x}$.

$$\frac{x}{x+1}\div\frac{2}{x}=\frac{x}{x+1}\cdot\frac{x}{2}=\frac{x\cdot x}{2(x+1)}\quad\left(\text{or }\frac{x^2}{2x+2}\right)$$

DO EXERCISE 15.

Multiply and reduce.

10. $y\cdot\frac{1}{y-1}\cdot\frac{y^2-1}{4y^2}$

11. $\frac{4-x^2}{x}\cdot\frac{3x}{x-2}$

12. $\frac{x}{x+2}\cdot\frac{2x+1}{x^2}\cdot\frac{x^2+x-2}{2x^2-5x-3}$

13. What are the poles for each of the rational expressions in Exercise 12?

Perform the indicated operation.

14. $\frac{3}{5}\div\frac{9}{2}$

Perform the indicated operation.

15. $\frac{6}{x}\div\frac{4}{x+2}$

Perform the indicated operation.

16. $\dfrac{t^2 - 25}{t^2} \div \dfrac{t + 5}{t}$

Example 8 Divide $\dfrac{t + 3}{t^2 - 1}$ by $\dfrac{t^2 - 9}{t + 1}$.

$$\frac{t + 3}{t^2 - 1} \div \frac{t^2 - 9}{t + 1} = \frac{t + 3}{t^2 - 1} \cdot \frac{t + 1}{t^2 - 9} = \frac{(t + 3)(t + 1)}{(t^2 - 1)(t^2 - 9)}$$
$$= \frac{(t + 3)(t + 1)}{(t - 1)(t + 1)(t - 3)(t + 3)}$$
$$= \frac{t + 3}{t + 3} \cdot \frac{t + 1}{t + 1} \cdot \frac{1}{(t - 1)(t - 3)}$$
$$= \frac{1}{(t - 1)(t - 3)}$$

DO EXERCISE 16.

Example 9 Divide $\dfrac{4a^2t}{c}$ by $\dfrac{-8as}{c^2}$.

$$\frac{4a^2t}{c} \div \frac{-8as}{c^2} = \frac{4a^2t}{c} \cdot \frac{c^2}{-8as} = \frac{4a^2tc^2}{-8asc}$$
$$= \frac{4}{4} \cdot \frac{a}{a} \cdot \frac{c}{c} \cdot \frac{atc}{-2s} = -\frac{atc}{2s}$$

Perform the indicated operation.

17. $\dfrac{2y^2 - 5y - 12}{y^2 - 36} \div \dfrac{2y^2 + 5y + 3}{y^2 + 7y + 6}$

Example 10 Divide $\dfrac{x^2 - x - 2}{x - 3}$ by $x^2 - 4x + 4$.

$$\frac{x^2 - x - 2}{x - 3} \div x^2 - 4x + 4 = \frac{x^2 - x - 2}{x - 3} \div \frac{x^2 - 4x + 4}{1}$$
$$= \frac{x^2 - x - 2}{x - 3} \cdot \frac{1}{x^2 - 4x + 4}$$
$$= \frac{x^2 - x - 2}{(x - 3)(x^2 - 4x + 4)}$$
$$= \frac{(x - 2)(x + 1)}{(x - 3)(x - 2)(x - 2)}$$
$$= \frac{x - 2}{x - 2} \cdot \frac{x + 1}{(x - 3)(x - 2)} = \frac{x + 1}{(x - 3)(x - 2)}$$

18. Are $x \div \dfrac{1}{x - 1}$ and $\dfrac{1}{x - 1} \div x$ equal?

DO EXERCISES 17 AND 18.

DO SECTION PROBLEMS 8.3.

ANSWERS TO MARGINAL EXERCISES

1. $\dfrac{3x^2 + x}{x^2 + x - 2}$ **2.** $\dfrac{2t}{t^2 + 1}$ **3.** $\dfrac{3x^2 + 3x}{10y}$ **4.** $\dfrac{2x}{-x^2 - 3x + 10}$
5. $\dfrac{5}{6}$ **6.** $\dfrac{5x^2z}{3y}$ **7.** $\dfrac{1}{20}$ **8.** $\dfrac{t}{(t - 3)(t - 1)}$ **9.** $\dfrac{x(x - 1)}{(x + 1)(x + 2)}$
10. $\dfrac{y + 1}{4y}$ **11.** $-3(x + 2)$ **12.** $\dfrac{x - 1}{x(x - 3)}$ **13.** $-2, 0, -\dfrac{1}{2}, 3$
14. $\dfrac{2}{15}$ **15.** $\dfrac{3(x + 2)}{2x}$ **16.** $\dfrac{t - 5}{t}$ **17.** $\dfrac{y - 4}{y - 6}$ **18.** No

NAME COURSE/SECTION DATE

SECTION PROBLEMS 8.3

Perform the indicated operation and reduce where possible.

1. $\frac{3}{5} \cdot \frac{4}{7}$

2. $\frac{7}{3} \cdot \frac{12}{5}$

3. $\frac{20}{21} \cdot \frac{35}{8}$

4. $\frac{8}{3} \cdot \frac{15}{4} \cdot \frac{5}{6}$

5. $y \cdot \frac{1}{y} \cdot \frac{1}{y^2}$

6. $\frac{2}{3} \cdot \frac{x}{x+1}$

7. $(-3)\left(\frac{x}{x+1}\right)\left(\frac{y-1}{y}\right)$

8. $\frac{3ab}{2c} \cdot \frac{8c^2}{9a}$

9. $x \cdot \frac{1}{x+1} \cdot \frac{x}{x+4}$

10. $4y \cdot \frac{2y-1}{y^2-4}$

ANSWERS

1. __________
2. __________
3. __________
4. __________
5. __________
6. __________
7. __________
8. __________
9. __________
10. __________

11. ______________

12. ______________

13. ______________

14. ______________

15. ______________

16. ______________

17. ______________

18. ______________

19. ______________

20. ______________

11. $\dfrac{t}{t^2 - 4t - 5} \cdot 0$

12. $\dfrac{a^2}{bc} \cdot \dfrac{ab^2c}{a^2c}$

13. $\dfrac{(x - 1)^2}{3x} \cdot \dfrac{-9x}{x^2 + 5x - 6}$

14. $\dfrac{y^2 - 16}{y} \cdot \dfrac{y^2 + 2y}{y^2 - 5y + 4}$

15. $\dfrac{2x^2 + 7x + 3}{x} \cdot \dfrac{x^2 + 9x}{x^2 + 3x}$

16. $1 \cdot \dfrac{x^3}{x^2 - 3x - 4}$

17. $x \cdot \dfrac{3 - x}{x + 2}$

18. $a \cdot \dfrac{a + 1}{a^2 - a} \cdot \dfrac{2a - 2}{a^2 + 2a + 1}$

19. $\dfrac{(y - 1)^2}{4y^2 + 4y - 3} \cdot \dfrac{(2y - 1)^2}{(y - 1)^3}$

20. $\dfrac{t - 1}{t^2 + 4t + 4} \cdot \dfrac{t + 2}{1 - t^2}$

NAME COURSE/SECTION DATE

ANSWERS

21. $\frac{9}{10} \div \frac{3}{5}$

22. $\frac{3}{5} \div 6$

23. $2a \div \frac{a}{b}$

24. $3a \div \frac{2}{a^2}$

25. $\frac{y}{4} \div 3y^2$

26. $\frac{x - 3}{x + 4} \div 1$

27. $0 \div \frac{y^2 - y}{1 - y}$

28. $0 \div \frac{3x}{y}$

29. $\frac{4x^2y}{z} \div 1$

30. $(2x - 1) \div (4x^2 - 4x + 1)$

21. ____________

22. ____________

23. ____________

24. ____________

25. ____________

26. ____________

27. ____________

28. ____________

29. ____________

30. ____________

31. ______________

32. ______________

33. ______________

34. ______________

35. ______________

36. ______________

37. ______________

38. ______________

39. ______________

31. $\dfrac{x^2 - 3x + 2}{x + 4} \div \dfrac{x^2 - 3x + 2}{x^2 + 3x}$

32. $\dfrac{t^2 - 4}{3t + 12} \div \dfrac{(t + 2)^2}{12t^2 - 48t}$

33. $\dfrac{x^2}{x^2 - 1} \div \dfrac{x}{x + 1}$

34. $\dfrac{y + 2}{y - 1} \div \dfrac{y - 2}{y - 1}$

35. $\dfrac{a}{a + 1} \div \dfrac{a}{a - 2}$

36. $\dfrac{y^2 - 3y}{y - 1} \cdot \dfrac{y^2 - 1}{y^2 - 6y + 9}$

37. $\dfrac{a^2 - 3a + 2}{a^2 - 7a + 12} \cdot \dfrac{a^2 - 5a + 4}{a^2 + a - 2}$

38. $\dfrac{y^2 - 7y + 12}{y + 1} \div \dfrac{y - 4}{y + 1}$

39. $\left[\dfrac{x^2 + 2x}{x - 3} \cdot \dfrac{x^2 - 3x}{x - 1}\right] \div \dfrac{x}{x^2 + x - 2}$

8.4

Addition and Subtraction with Common Denominators, LCD, and Building Up Rational Expressions

To add or subtract fractions we must have a common denominator, and when we do have a common denominator we just add or subtract the numerators. For example,

$$\frac{3}{7} + \frac{2}{7} = \frac{3+2}{7} = \frac{5}{7}$$

We add or subtract rational expressions that have a common denominator in exactly the same way.

Example 1 $\frac{2x}{x+1} + \frac{3}{x+1} = \frac{2x+3}{x+1}$

Example 2 $\frac{4t}{t-2} - \frac{8}{t-2} = \frac{4t-8}{t-2} = \frac{4(t-2)}{t-2}$

$$= \frac{4}{1} \cdot \frac{t-2}{t-2} = 4$$

DO EXERCISES 1 AND 2.

Example 3 $\frac{4x}{x^2-9} + \frac{3}{x^2-9} + \frac{7x}{x^2-9} = \frac{4x+3+7x}{x^2-9} = \frac{11x+3}{x^2-9}$

Example 4 $\frac{2y}{x^2+1} - \frac{3x}{x^2+1} + \frac{4}{x^2+1} = \frac{2y-3x+4}{x^2+1}$

DO EXERCISES 3 AND 4.

The real challenge in adding or subtracting rational expressions begins when the rational expressions do not have a common denominator. To understand the steps involved in adding such rational expressions, we will look at a similar problem involving the addition of two fractions.

Example 5 Add $\frac{2}{3} + \frac{3}{4}$.

(a) Find the least common denominator (LCD). Later we will review the method for finding the LCD that we used in Chapter 1.

The LCD of 3 and 4 is 12.

LEARNING OBJECTIVES

After finishing Section 8.4, you should be able to:

- Add rational expressions with common denominators.
- Subtract rational expressions with common denominators.
- Find Least Common Denominators of rational expressions.

Perform the indicated operation.

1. $\frac{3x}{2x-1} + \frac{x}{2x-1}$

2. $\frac{t^2}{t-1} + \frac{t-2}{t-1}$

Perform the indicated operation.

3. $\frac{3}{a-1} - \frac{4a}{a-1}$

4. $\frac{2m}{m^2+4} - \frac{m^2}{m^2+4} + \frac{3m^2}{m^2+4}$

Find the LCD for the following problems.

5. $\frac{1}{15} + \frac{3}{20}$

6. $\frac{1}{12} - \frac{5}{18}$

7. $\frac{1}{24} + \frac{7}{90} - \frac{3}{22}$

8. $\frac{3}{8} + \frac{5}{24} + \frac{1}{6}$

9. $1 + \frac{1}{5}$

(b) Change $\frac{2}{3}$ and $\frac{3}{4}$ to equivalent fractions that have 12 for a denominator. This is called *building up* the fraction.

$$\frac{2}{3} = \frac{8}{12}$$

$$\frac{3}{4} = \frac{9}{12}$$

(c) Perform the addition.

$$\frac{2}{3} + \frac{3}{4} = \frac{8}{12} + \frac{9}{12}$$
$$= \frac{8+9}{12} = \frac{17}{12} \left(\text{or } 1\frac{5}{12}\right)$$

In step (a), we used the least common denominator, but actually any common denominator would work. The reason for using the LCD instead of a larger common denominator is not because it necessarily makes adding fractions easier, but because it will certainly make adding rational expressions easier (as we will see later).

Now, before we can proceed to add or subtract rational expressions, we must look at two procedures.

1. How do we find a least common denominator (LCD) for two or more rational expressions?
2. How do we build up rational expressions?

Before we start to find the LCD for rational expressions, we will review the procedure for fractions. The LCD of two fractions is the smallest whole number that both denominators divide into evenly. Often it is possible to see what the LCD is going to be by inspection. For example, the LCD for $\frac{5}{12} + \frac{1}{4}$ would be 12 and the LCD for $\frac{3}{5} + \frac{1}{3}$ would be 15. However, in those cases where the LCD is not easily recognizable $\left(\text{such as } \frac{5}{18} + \frac{1}{15}\right)$ we need a procedure to follow. The following is a repeat of Rule 1.4 from Chapter 1.

To find the LCD of two or more denominators, we follow these steps:

1. Write the prime factorization for each denominator.
2. Include each prime in the LCD the most number of times it occurs in any one of the individual factorizations.

Example 6 Find the LCD for the problem $\frac{5}{18} + \frac{1}{15}$.

$$18 = 2 \cdot 9 = 2 \cdot 3 \cdot 3$$
$$15 = 3 \cdot 5 = \quad 3 \cdot 5$$
$$\text{LCD} = 2 \cdot 3 \cdot 3 \cdot 5 = 90$$

Example 7 Find the LCD for the problem $\frac{1}{63} + \frac{3}{8} + \frac{1}{42}$.

$$63 = 9 \cdot 7 = 3 \cdot 3 \cdot 7 = 3 \cdot 3 \cdot 7$$
$$8 = 2 \cdot 4 = 2 \cdot 2 \cdot 2 = \quad 2 \cdot 2 \cdot 2$$
$$42 = 6 \cdot 7 = 2 \cdot 3 \cdot 7 = \quad 3 \cdot 7 \cdot 2$$
$$\text{LCD} = 3 \cdot 3 \cdot 7 \cdot 2 \cdot 2 \cdot 2 = 504$$

DO EXERCISES 5 THROUGH 9.

We find the LCD for rational expressions by using exactly the same procedure except that instead of writing the prime factorization of integers we will be factoring polynomials.

Example 8 Find the LCD for the problem

$$\frac{1}{x^2 - 2x + 1} + \frac{x}{x^2 - 3x + 2}$$

(a) Factor the denominators.

$$x^2 - 2x + 1 = (x - 1)(x - 1)$$
$$x^2 - 3x + 2 = (x - 1)(x - 2)$$

The LCD will have only the factors $(x - 1)$ and $(x - 2)$ in it since these are the only different factors occurring in either factorization.

(b) The LCD will have the factor $x - 1$ twice (since twice is the most number of times $x - 1$ occurs in either factorization) and the factor $x - 2$ once. Thus, we have

$$\begin{aligned} x^2 - 2x + 1 &= (x - 1)(x - 1) \\ x^2 - 3x + 2 &= (x - 1)(x - 2) \\ \text{LCD} &= (x - 1)(x - 1)(x - 2) \quad (\text{or} \quad (x - 1)^2(x - 2)) \end{aligned}$$

We could arrive at a common denominator (but not the least common denominator) of

$$x^2 - 2x + 1 \quad \text{and} \quad x^2 - 3x + 2$$

by just multiplying the two polynomials and obtaining

$$(x^2 - 2x + 1)(x^2 - 3x + 2), \quad \text{or} \quad (x - 1)^3(x - 2)$$

Later we will see there are advantages to using the least common denominator instead of just a common denominator.

DO EXERCISE 10.

Example 9 Find the LCD for the problem

$$\frac{1}{x^2 - 2x} + \frac{3}{x^3 - 3x^2} - \frac{2x}{x^2 - 5x + 6}$$

$$\begin{aligned} x^2 - 2x &= x(x - 2) &&= x \cdot (x - 2) \\ x^3 - 3x^2 &= x^2(x - 3) = x \cdot x(x - 3) &&= x \cdot x \cdot (x - 3) \\ x^2 - 5x + 6 &= (x - 2)(x - 3) &&= (x - 2) \cdot (x - 3) \\ & & \text{LCD} &= x \cdot x \cdot (x - 2)(x - 3) \\ & & &(\text{or} \quad x^2(x - 2)(x - 3)) \end{aligned}$$

Example 10 Find the LCD for the problem

$$\frac{1}{9} - \frac{5}{6a} + \frac{3}{2a^2 + 6a}$$

$$\begin{aligned} 9 &= 3 \cdot 3 &&= 3 \cdot 3 \\ 6a &= 2 \cdot 3 \cdot a &&= 2 \cdot 3 \cdot a \\ 2a^2 + 6a &= 2a(a + 3) &&= 2 \cdot a \cdot (a + 3) \\ & & \text{LCD} &= 2 \cdot 3 \cdot 3 \cdot a \cdot (a + 3) \quad (\text{or} \quad 18a(a + 3)) \end{aligned}$$

DO EXERCISES 11 THROUGH 13.

Find the LCD.

10. $\dfrac{1}{x^2 - 16} + \dfrac{1}{x^2 - 5x + 4}$

Find the LCD for the following problems.

11. $\dfrac{1}{y + 3} - \dfrac{2}{y^2 - 9} + \dfrac{3}{y^2 - 3y}$

12. $\dfrac{1}{a^3 - a^2} + \dfrac{4}{a^3 - 2a^2 + a} + \dfrac{3}{a^2 + a}$

13. $\dfrac{x}{1 - x^2} + \dfrac{3}{4x - 4} + \dfrac{3x}{2x^2 - 2}$

Find the missing numerator.

14. $\frac{2}{3} = \frac{?}{27}$

Find the missing numerator.

15. $\frac{3x}{x-2} = \frac{?}{x^2-4}$

The second procedure we must look at is "building up" rational expressions, or, in other words, taking a rational expression and changing it to an equivalent rational expression that has a different denominator. As before, first we will review what happens with fractions.

Example 11 Change $\frac{3}{4}$ to an equivalent fraction having 24 for a denominator.

$$\frac{3}{4} = \frac{?}{24}$$

In changing the denominator 4 to the denominator 24, we multiplied the 4 by 6. Therefore, because we multiplied the denominator by 6, we must also multiply the numerator by 6 to obtain an equivalent fraction.

$$\frac{3}{4} = \frac{3}{4} \cdot 1 = \frac{3}{4} \cdot \frac{6}{6} = \frac{3 \cdot 6}{4 \cdot 6} = \frac{18}{24}$$

Thus,

$$\frac{3}{4} = \frac{18}{24}$$

DO EXERCISE 14.

Now we apply this procedure to "building up" rational expressions.

Example 12 Change $\frac{3}{x-3}$ to an equivalent rational expression that has $x^2 - 2x - 3$ as its denominator. [Note: $x^2 - 2x - 3 = (x-3)(x+1)$]

$$\frac{3}{x-3} = \frac{?}{x^2 - 2x - 3} = \frac{?}{(x-3)(x+1)}$$

In changing the denominator $x - 3$ to the denominator $(x-3)(x+1)$, we multiplied the $x - 3$ by the factor $x + 1$. Therefore, we must multiply the numerator by this same factor $x + 1$ to obtain an equivalent rational expression.

$$\frac{3}{x-3} = \frac{3}{x-3} \cdot \frac{x+1}{x+1} = \frac{3(x+1)}{(x-3)(x+1)}$$

Thus,

$$\frac{3}{x-3} = \frac{3(x+1)}{x^2 - 2x - 3}$$

DO EXERCISE 15.

Example 13 Find the missing numerator in $a = \frac{?}{a+2}$.

$$a = \frac{a}{1} = \frac{?}{a+2}$$

In changing the denominator 1 to the denominator $a + 2$, we multiplied the 1 by the factor $a + 2$. Thus, we must multiply the numerator by the same factor.

$$a = \frac{a}{1} = \frac{a}{1} \cdot \frac{a+2}{a+2} = \frac{a(a+2)}{a+2}$$

Hence,

$$a = \frac{a(a+2)}{a+2}$$

and the missing numerator is $a(a + 2)$.

Example 14 Find the missing numerator in

$$\frac{4x}{x^2 + x - 2} = \frac{?}{(x+2)^2(x-1)}$$

$$\frac{4x}{x^2 + x - 2} = \frac{4x}{(x+2)(x-1)} = \frac{?}{(x+2)^2(x-1)}$$

In changing the denominator $(x + 2)(x - 1)$ to the denominator

$$(x+2)^2(x-1) = (x+2)(x+2)(x-1)$$

we multiplied the denominator $(x + 2)(x - 1)$ by the factor $x + 2$. Thus, we have

$$\begin{aligned} \frac{4x}{x^2 + x - 2} &= \frac{4x}{(x+2)(x-1)} \cdot \frac{x+2}{x+2} \\ &= \frac{4x(x+2)}{(x+2)(x+2)(x-1)} \\ &= \frac{4x(x+2)}{(x+2)^2(x-1)} \end{aligned}$$

Hence,

$$\frac{4x}{x^2 + x - 2} = \frac{4x(x+2)}{(x+2)^2(x-1)}$$

and the missing numerator is $4x(x + 2)$.

DO EXERCISES 16 THROUGH 18.

Now that we know how to find the LCD for rational expressions and how to "build up" rational expressions, we can add or subtract any rational expression. This is what we will do in the next section.

DO SECTION PROBLEMS 8.4.

Find the missing numerator for the following problems.

16. $\frac{a-1}{a+1} = \frac{?}{(a+1)^2(a+3)}$

17. $3x = \frac{?}{ab}$

18. $\frac{-4}{y^2 - y} = \frac{?}{y^4 - 3y^3 + 2y^2}$

ANSWERS TO MARGINAL EXERCISES

1. $\frac{4x}{2x-1}$ **2.** $t+2$ **3.** $\frac{3-4a}{a-1}$ **4.** $\frac{2m^2+2m}{m^2+4}$ **5.** 60
6. 36 **7.** 3960 **8.** 24 **9.** 5 **10.** $(x-4)(x+4)(x-1)$
11. $y(y-3)(y+3)$ **12.** $a^2(a-1)^2(a+1)$ **13.** $4(x-1)(x+1)$
14. 18 **15.** $3x(x+2)$ **16.** $(a-1)(a+1)(a+3)$ **17.** $(3x)(ab)$
18. $-4(y)(y-2)$

NAME ______ COURSE/SECTION ______ DATE ______

SECTION PROBLEMS 8.4

Perform the indicated operation.

1. $\dfrac{2}{x-1} + \dfrac{3x}{x-1}$

2. $\dfrac{t}{t+4} - \dfrac{2t}{t+4}$

3. $\dfrac{2y+5}{2y-1} + \dfrac{-6}{2y-1}$

4. $\dfrac{3x-4}{x} + \dfrac{2x+1}{x}$

5. $\dfrac{3}{t} + \dfrac{4-t}{t}$

6. $\dfrac{x-1}{x^2+4} + \dfrac{x^2-x}{x^2+4}$

7. $\dfrac{x^2}{x^2-9} - \dfrac{x+6}{x^2-9}$

8. $\dfrac{-2y}{y^2-36} + \dfrac{3y-6}{y^2-36}$

9. $\dfrac{2t}{t+3} + \dfrac{t+1}{t+3} - \dfrac{4t}{t+3}$

10. $\dfrac{3x}{x-2} + \dfrac{2x+1}{x-2} - \dfrac{4x+3}{x-2}$

Find the LCD. (Do not perform the operation.)

11. $\dfrac{3}{20} + \dfrac{3}{28}$

12. $\dfrac{1}{4} + \dfrac{3}{10} - \dfrac{1}{15}$

13. $\dfrac{1}{x} + \dfrac{1}{x-3}$

14. $\dfrac{1}{y^2-16} + \dfrac{1}{y^2-5y+4}$

ANSWERS

1. ______
2. ______
3. ______
4. ______
5. ______
6. ______
7. ______
8. ______
9. ______
10. ______
11. ______
12. ______
13. ______
14. ______

15. ______
16. ______
17. ______
18. ______
19. ______
20. ______
21. ______
22. ______
23. ______
24. ______
25. ______

15. $\dfrac{1}{x} + \dfrac{1}{x^2 - x} + \dfrac{2}{x^2 - 2x + 1}$

16. $\dfrac{y}{y^2 - 3y + 2} + \dfrac{1}{(y - 2)^2} + \dfrac{3}{y + 3}$

17. $\dfrac{1}{(x - 1)^2(x + 4)} + \dfrac{1}{x^2 + 3x - 4} + \dfrac{1}{x^2 + 8x + 16}$

18. $\dfrac{5}{18x - 36} + \dfrac{1}{15x^2 - 30x} + \dfrac{3}{2x^2}$

Build up the rational expression by finding the missing numerator.

19. $\dfrac{x}{x - 1} = \dfrac{?}{x^2 - 2x + 1}$

20. $\dfrac{3}{y} = \dfrac{?}{y^2 - 4y}$

21. $\dfrac{-2}{t^2 - 1} = \dfrac{?}{t(t + 1)^2(t - 1)}$

22. $5 = \dfrac{?}{3x - 7}$

23. $x = \dfrac{?}{x^2 - 3x}$

24. $\dfrac{4}{x^2 - 3x + 2} = \dfrac{?}{(x^2 - 6x + 8)(x - 1)}$

25. $7x = \dfrac{?}{3(x - 4)}$

8.5

Addition and Subtraction of Rational Expressions

LEARNING OBJECTIVES

After finishing Section 8.5, you should be able to:

- Add and subtract rational expressions.
- Simplify complex fractions.

Since all the preliminary work was done in the last section, we are now ready to add and subtract rational expressions that do not have common denominators. Once again, the first example will be done with fractions. Notice that the steps used in adding or subtracting rational expressions are exactly the same as those used in adding or subtracting fractions.

Example 1 Add $\frac{4}{15} + \frac{7}{9}$.

(a) Find the LCD.

$$15 = 3 \cdot 5 = \quad 3 \cdot 5$$
$$9 = 3 \cdot 3 = 3 \cdot 3$$
$$\text{LCD} = 3 \cdot 3 \cdot 5 = 45$$

(b) Change $\frac{4}{15}$ and $\frac{7}{9}$ to equivalent fractions that have 45 as a denominator.

$$\frac{4}{15} = \frac{4 \cdot 3}{15 \cdot 3} = \frac{12}{45}$$
$$\frac{7}{9} = \frac{7 \cdot 5}{9 \cdot 5} = \frac{35}{45}$$

(c) Add.

$$\frac{4}{15} + \frac{7}{9} = \frac{12}{45} + \frac{35}{45}$$
$$= \frac{12 + 35}{45}$$
$$= \frac{47}{45} \quad \left(\text{or} \quad 1\frac{2}{45}\right)$$

DO EXERCISE 1.

Perform the indicated operation.

1. $\frac{5}{12} + \frac{7}{15}$

Example 2 Add $\frac{3}{x-1} + \frac{4x}{x^2-1}$.

(a) Find the LCD.

$$x - 1 = (x - 1)$$
$$x^2 - 1 = (x - 1)(x + 1)$$
$$\text{LCD} = (x - 1)(x + 1)$$

(b) Change $\frac{3}{x-1}$ and $\frac{4x}{x^2-1}$ to equivalent fractions that have $(x - 1)(x + 1)$ as a denominator.

$$\frac{3}{x-1} = \frac{3}{x-1} \cdot \frac{x+1}{x+1}$$
$$= \frac{3(x+1)}{(x-1)(x+1)}$$

$\frac{4x}{x^2-1}$ already has the desired denominator.

Perform the indicated operation.

2. $\dfrac{3}{x-2}+\dfrac{4}{x+5}$

(c) Add.

$$\frac{3}{x-1}+\frac{4x}{x^2-1}=\frac{3(x+1)}{(x-1)(x+1)}+\frac{4x}{(x-1)(x+1)}$$

$$=\frac{3(x+1)+4x}{(x-1)(x+1)}=\frac{7x+3}{(x-1)(x+1)}$$

DO EXERCISES 2 AND 3.

Example 3 Find $1+\dfrac{1}{x}+\dfrac{1}{x-4}$. Note that $1=\dfrac{1}{1}$.

(a) Find the LCD.

$$1=1$$
$$x=x$$
$$x-4=(x-4)$$
$$\text{LCD}=(1)(x)(x-4)$$

3. $\dfrac{2}{x-2}+\dfrac{x}{x^2-4}$

(b) Change each rational expression to an equivalent one that has $x(x-4)$ as a denominator.

$$1=\frac{1}{1}\cdot\frac{x(x-4)}{x(x-4)}=\frac{x(x-4)}{x(x-4)}$$

$$\frac{1}{x}=\frac{1}{x}\cdot\frac{x-4}{x-4}=\frac{x-4}{x(x-4)}$$

$$\frac{1}{x-4}=\frac{1}{x-4}\cdot\frac{x}{x}=\frac{x}{x(x-4)}$$

(c) Add.

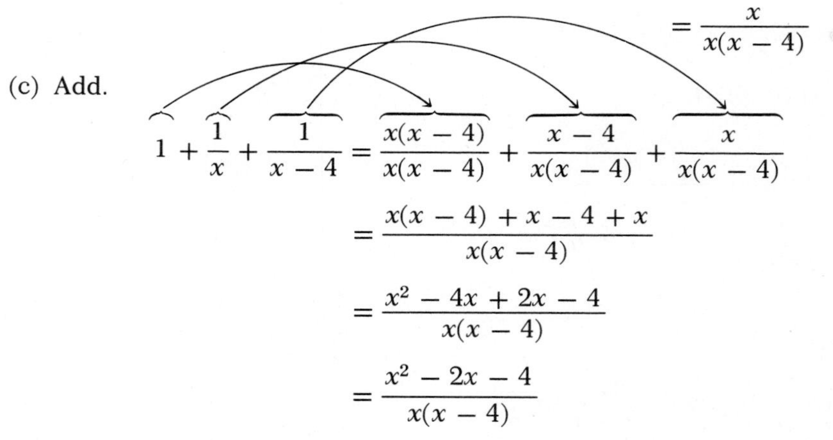

$$1+\frac{1}{x}+\frac{1}{x-4}=\frac{x(x-4)}{x(x-4)}+\frac{x-4}{x(x-4)}+\frac{x}{x(x-4)}$$

$$=\frac{x(x-4)+x-4+x}{x(x-4)}$$

$$=\frac{x^2-4x+2x-4}{x(x-4)}$$

$$=\frac{x^2-2x-4}{x(x-4)}$$

Perform the indicated operation.

4. $\dfrac{x}{x^2-16}\cdot\dfrac{x+4}{x^2-3x}$

Note that $\dfrac{x^2-2x-4}{x(x-4)}$ is reduced since x^2-2x-4 does not have a factor of x or $x-4$.

DO EXERCISE 4.

In the following examples, we have combined steps (b) and (c) of the three previous examples into one step.

Example 4 Find $\dfrac{x-1}{x^2+x-6} - \dfrac{x+1}{2x^2+5x-3}$.

(a) Find the LCD.

$$\begin{aligned} x^2+x-6 &= (x+3)(x-2) = (x+3)(x-2) \\ 2x^2+5x-3 &= (2x-1)(x+3) = (x+3)\qquad\quad (2x-1) \\ \text{LCD} &= (x+3)(x-2)(2x-1) \end{aligned}$$

(b) Subtract.

$$\begin{aligned} &\frac{x-1}{x^2+x-6} - \frac{x+1}{2x^2+5x-3} \\ &= \frac{x-1}{(x+3)(x-2)} \cdot \frac{2x-1}{2x-1} - \frac{x+1}{(2x-1)(x+3)} \cdot \frac{x-2}{x-2} \\ &= \frac{(x-1)(2x-1)}{(x+3)(x-2)(2x-1)} - \frac{(x+1)(x-2)}{(x+3)(x-2)(2x-1)} \\ &= \frac{(x-1)(2x-1)-(x+1)(x-2)}{(x+3)(x-2)(2x-1)} \\ &= \frac{2x^2-3x+1-(x^2-x-2)}{(x+3)(x-2)(2x-1)} \\ &= \frac{2x^2-3x+1-x^2+x+2}{(x+3)(x-2)(2x-1)} \\ &= \frac{x^2-2x+3}{(x+3)(x-2)(2x-1)} \quad \text{which is reduced.} \end{aligned}$$

In Example 4 we could have used

$$(x^2+x-6)(2x^2+5x-3)$$

as a common denominator instead of using the LCD. However, if we had done this, the building up of our rational expressions would have been more difficult and our final answer would not have been reduced. This is why it is easier to use the LCD and not just any common denominator when building up rational expressions.

DO EXERCISES 5 AND 6.

Example 5 Simplify $\dfrac{5}{2} + \dfrac{1}{y} - \dfrac{2y}{y^2-3y}$.

(a) Find the LCD.

$$\begin{aligned} 2 &= 2 \\ y &= y \\ y^2-3y &= y(y-3) \\ \text{LCD} &= 2y(y-3) \end{aligned}$$

Perform the indicated operation.

5. $\dfrac{1}{x-5} - \dfrac{2}{x+4}$

6. $3 - \dfrac{1}{t} + \dfrac{3}{t-2}$

Perform the indicated operation.

7. $\dfrac{-4}{x^2 - 4} - \dfrac{x}{x^2 + 5x + 6}$

8. $\dfrac{y}{y^2 + 2y + 1} + \dfrac{y + 2}{y^2 + y}$

Simplify.

9. $\dfrac{1 + \dfrac{1}{x + 1}}{1 + \dfrac{2}{x}}$

(b) Perform the operations.

$$\frac{5}{2} + \frac{1}{y} - \frac{2y}{y^2 - 3y} = \frac{5}{2} \cdot \frac{y(y-3)}{y(y-3)} + \frac{1}{y} \cdot \frac{2(y-3)}{2(y-3)} - \frac{2y}{y(y-3)} \cdot \frac{2}{2}$$

$$= \frac{5y(y-3)}{2y(y-3)} + \frac{2(y-3)}{2y(y-3)} - \frac{4y}{2y(y-3)}$$

$$= \frac{5y(y-3) + 2(y-3) - 4y}{2y(y-3)}$$

$$= \frac{5y^2 - 15y + 2y - 6 - 4y}{2y(y-3)}$$

$$= \frac{5y^2 - 17y - 6}{2y(y-3)}$$

The final answer is reduced since $5y^2 - 17y - 6$ does not have 2, y, or $y - 3$ as a factor.

DO EXERCISES 7 AND 8.

A *complex fraction* is an algebraic expression for which the numerator or denominator, or both, contains a rational expression. For example,

$$\frac{1 + \dfrac{1}{x}}{1 - \dfrac{1}{x}}$$

is a complex fraction. One way to simplify complex fractions is to combine the numerator and denominator into single rational expressions and then divide the numerator by the denominator.

Example 6 Simplify $\dfrac{1 + \dfrac{1}{x}}{1 - \dfrac{1}{x}}$.

$$\frac{1 + \dfrac{1}{x}}{1 - \dfrac{1}{x}} = \frac{\dfrac{x}{x} + \dfrac{1}{x}}{\dfrac{x}{x} - \dfrac{1}{x}} = \frac{\dfrac{x+1}{x}}{\dfrac{x-1}{x}}$$

$$= \frac{x+1}{x} \div \frac{x-1}{x} = \frac{x+1}{x} \cdot \frac{x}{x-1}$$

$$= \frac{(x+1)(x)}{(x)(x-1)} = \frac{x+1}{x-1} \cdot \frac{x}{x}$$

$$= \frac{x+1}{x-1}$$

DO EXERCISE 9.

Example 7 Simplify $\dfrac{1 + \dfrac{2}{y}}{1 + \dfrac{1}{y} - \dfrac{2}{y^2}}$

$$\frac{1 + \dfrac{2}{y}}{1 + \dfrac{1}{y} - \dfrac{2}{y^2}} = \frac{\dfrac{y}{y} + \dfrac{2}{y}}{\dfrac{y^2}{y^2} + \dfrac{y}{y^2} - \dfrac{2}{y^2}}$$

$$= \frac{\dfrac{y + 2}{y}}{\dfrac{y^2 + y - 2}{y^2}}$$

$$= \frac{y + 2}{y} \div \frac{y^2 + y - 2}{y^2}$$

$$= \frac{y + 2}{y} \cdot \frac{y^2}{y^2 + y - 2}$$

$$= \frac{y^2(y + 2)}{y(y^2 + y - 2)}$$

$$= \frac{y \cdot y(y + 2)}{y(y + 2)(y - 1)}$$

$$= \frac{y}{y} \cdot \frac{y + 2}{y + 2} \cdot \frac{y}{y - 1}$$

$$= \frac{y}{y - 1}$$

DO EXERCISE 10.

DO SECTION PROBLEMS 8.5.

Simplify.

10. $\dfrac{x - \dfrac{y^2}{x}}{1 + \dfrac{y}{x}}$

ANSWERS TO MARGINAL EXERCISES

1. $\frac{53}{60}$ **2.** $\frac{7x + 7}{(x - 2)(x + 5)}$ **3.** $\frac{3x + 4}{(x - 2)(x + 2)}$ **4.** $\frac{1}{(x - 4)(x - 3)}$
5. $\frac{-x + 14}{(x - 5)(x + 4)}$ **6.** $\frac{3t^2 - 4t + 2}{t(t - 2)}$ **7.** $\frac{-x^2 - 2x - 12}{(x - 2)(x + 2)(x + 3)}$
8. $\frac{2y^2 + 3y + 2}{y(y + 1)^2}$ **9.** $\frac{x}{x + 1}$ **10.** $x - y$

NAME COURSE/SECTION DATE

ANSWERS

SECTION PROBLEMS 8.5

Perform the indicated operation.

1. $\frac{3}{20} + \frac{3}{28}$

2. $\frac{1}{4} + \frac{3}{10} - \frac{1}{15}$

3. $\frac{1}{15} - \frac{5}{18} + \frac{7}{30}$

4. $\frac{5}{21} - \frac{1}{9} - \frac{3}{14}$

5. $\frac{3}{x + 4} + \frac{x}{x + 4}$

6. $\frac{2t}{t - 2} - \frac{4}{t - 2}$

7. $\frac{2}{x - 5} + \frac{-3}{5 - x}$

8. $\frac{x}{x - y} + \frac{y}{y - x}$

9. $\frac{1}{a - 1} + a$

10. $x + \frac{2}{x}$

11. $\frac{2}{t - 1} + \frac{5}{t + 1}$

12. $x - \frac{1}{x - 1}$

13. $\frac{3}{t - 5} - \frac{4}{t + 1}$

14. $4 - \frac{1}{3 - x} + \frac{2}{x - 3}$

15. $1 + \frac{1}{a} + \frac{1}{a^2}$

16. $4 - \frac{1}{1 - a} - \frac{a}{1 + a}$

1. ______
2. ______
3. ______
4. ______
5. ______
6. ______
7. ______
8. ______
9. ______
10. ______
11. ______
12. ______
13. ______
14. ______
15. ______
16. ______

17. ____________

18. ____________

19. ____________

20. ____________

21. ____________

22. ____________

23. ____________

24. ____________

25. ____________

26. ____________

27. ____________

28. ____________

17. $\dfrac{2}{y} + \dfrac{y+1}{y^2-y} + \dfrac{5}{y-1}$

18. $\dfrac{x+1}{x-1} - \dfrac{x}{1-x}$

19. $\dfrac{1}{t^2-1} + \dfrac{2}{t+1}$

20. $\dfrac{3}{x-5} - \dfrac{2}{x^2-4x-5}$

21. $\dfrac{2y}{y^2-4} + \dfrac{1}{y^2+6y+8}$

22. $\dfrac{3}{(x-1)^2} + \dfrac{4}{x^2-1}$

23. $\dfrac{-3}{x^2-x} + \dfrac{4}{x-1}$

24. $\dfrac{2x}{x^2-4} + \dfrac{3}{x-2}$

25. $\dfrac{x}{x-4} - \dfrac{4}{x^2-5x+4}$

26. $\dfrac{y}{y^2+3y-4} + \dfrac{3}{y^2+5y+4}$

27. $\dfrac{1}{t-2} + \dfrac{3}{t-3} + \dfrac{-2}{t+1}$

28. $\dfrac{y}{y-5} - \dfrac{2}{y^2-25}$

NAME COURSE/SECTION DATE

ANSWERS

29. ______

30. ______

31. ______

32. ______

33. ______

34. ______

35. ______

29. $\dfrac{y-3}{y+2} \div \dfrac{y^2-9}{y^2+2y}$

30. $\left(x \div \dfrac{x-1}{x}\right) + \dfrac{2}{x-1}$

31. $\left(1+\frac{1}{x}\right)\left(1-\frac{1}{x}\right)$

32. $\dfrac{6}{x^2-3x+2} - \dfrac{1}{(x-2)^2} + \dfrac{2}{x^2+x-6}$

33. $\dfrac{t}{t^2-9} \cdot \dfrac{t+3}{t^2-2t}$

34. $3 - \dfrac{2}{1-x}$

35. $\dfrac{4}{(x-1)^2(x+4)} + \dfrac{3}{x^2+3x-4} - \dfrac{2}{x^2+8x+16}$

36. ______________

37. ______________

38. ______________

39. ______________

40. ______________

41. ______________

42. ______________

43. ______________

44. ______________

36. $\dfrac{t-3}{2t^2+3t-2}-\dfrac{t+2}{2t^2-7t+3}$

37. $\left(\dfrac{x-4}{x^2+3x-4}\cdot\dfrac{x^2+8x+16}{2x^2-5x-12}\right)+\dfrac{x}{x^2-1}$

38. $\dfrac{1}{t}\div\left(1+\dfrac{1}{t}\right)$

Simplify.

39. $\dfrac{1}{1-\dfrac{1}{x}}$

40. $2-\dfrac{1}{1-\dfrac{1}{x+1}}$

41. $\dfrac{1-\dfrac{1}{t-2}}{\dfrac{1}{t-2}-1}$

42. $\dfrac{1-\dfrac{1}{x}-\dfrac{2}{x^2}}{1-\dfrac{2}{x}}$

43. $\dfrac{1-\dfrac{3}{x}}{1+\dfrac{3}{x}}$

44. $\dfrac{1-\dfrac{2}{t}+\dfrac{1}{t^2}}{1-\dfrac{1}{t^2}}$

8.6

Rational Equations

LEARNING OBJECTIVES

After finishing Section 8.6, you should be able to:

- Solve rational equations.
- Solve literal equations for a variable that appears in the denominator.

Now we would like to solve rational equations such as

$$\frac{2}{x-2} = \frac{-4}{x+1}$$

The steps used to solve a rational equation are listed next.

STEP 1: LCD

Find and multiply both sides of the equation by the LCD of all the denominators. This is called "clearing the fractions" and performs the necessary step of getting the variables out of the denominator.

STEP 2: SOLVE

Solve the equation that results from doing Step 1.

STEP 3: CHECK

Checking the solutions in a rational equation consists of two parts:

1. Recall that there are possible values (called poles) for which a rational expression will not be defined. If a solution obtained in Step 2 is a pole, then we have to disregard it as a solution.
2. If a solution is not a pole, then we can substitute it in the original equation to see if we obtain a true statement.

Example 1 Solve $\frac{3}{x} = -4$.

STEP 1: LCD

Multiply both sides by x.

$$x\left(\frac{3}{x}\right) = x(-4)$$

$$\frac{3x}{x} = -4x$$

$$3 = -4x$$

STEP 2: SOLVE

$$-\frac{3}{4} = x$$

STEP 3: CHECK

The number $-\frac{3}{4}$ is not a pole (the only pole is $x = 0$). If we substitute $-\frac{3}{4}$ in the original equation

$$\frac{3}{x} = -4$$

Solve.

1. $\dfrac{3}{x-1} = -2$

we obtain

$\dfrac{3}{x}$	-4
$\dfrac{3}{-\frac{3}{4}}$	-4
$3 \div \left(-\frac{3}{4}\right)$	-4
$\left(\frac{3}{1}\right)\cdot\left(-\frac{4}{3}\right)$	-4
-4	-4

DO EXERCISE 1.

Example 2 Solve $\dfrac{2}{x-2} = \dfrac{-4}{x+1}$.

STEP 1: LCD

LCD = $(x-2)(x+1)$

$$\frac{(x-2)}{1}\frac{(x+1)}{1}\left(\frac{2}{x-2}\right) = \frac{(x-2)}{1}\frac{(x+1)}{1}\left(\frac{-4}{x+1}\right)$$

$$\frac{2(x+1)}{1}\cdot\frac{x-2}{x-2} = \frac{-4(x-2)}{1}\cdot\frac{x+1}{x+1}$$

$$2(x+1) = -4(x-2)$$

STEP 2: SOLVE

$$2x + 2 = -4x + 8$$
$$6x + 2 = 8$$
$$6x = 6$$
$$x = 1$$

STEP 3: CHECK

The number 1 is not a pole. (The only poles are $x = 2$ and $x = -1$.) Substitute $x = 1$ in the original equation.

$\dfrac{2}{x-2}$	$\dfrac{-4}{x+1}$
$\dfrac{2}{1-2}$	$\dfrac{-4}{1+1}$
$\dfrac{2}{-1}$	$\dfrac{-4}{2}$
-2	-2

Solve.

2. $\dfrac{-2}{x-5} = \dfrac{4}{x-2}$

DO EXERCISE 2.

Example 3 Solve $\frac{x}{x-4} + \frac{3}{x-2} = 0$.

STEP 1: LCD

LCD $= (x-4)(x-2)$

$$(x-4)(x-2)\left[\frac{x}{x-4} + \frac{3}{x-2}\right] = (x-4)(x-2)(0)$$

$$\frac{(x-4)(x-2)}{1}\left(\frac{x}{x-4}\right) + \frac{(x-4)(x-2)}{1}\left(\frac{3}{x-2}\right) = 0$$

$$\frac{x(x-2)}{1}\cdot\frac{x-4}{x-4} + \frac{3(x-4)}{1}\cdot\frac{x-2}{x-2} = 0$$

$$x(x-2) + 3(x-4) = 0$$

STEP 2: SOLVE

$$x^2 - 2x + 3x - 12 = 0$$

$$x^2 + x - 12 = 0$$

$$(x+4)(x-3) = 0$$

Thus, (1) $x + 4 = 0$ gives us $x = -4$, or
(2) $x - 3 = 0$ gives us $x = 3$.

STEP 3: CHECK

The numbers -4 and 3 are not poles. (The only poles are 4 and 2.)

$\frac{x}{x-4} + \frac{3}{x-2}$	0
$\frac{-4}{-4-4} + \frac{3}{-4-2}$	0
$\frac{-4}{-8} + \frac{3}{-6}$	0
$\frac{1}{2} + \left(-\frac{1}{2}\right)$	0
0	0

$\frac{x}{x-4} + \frac{3}{x-2}$	0
$\frac{3}{3-4} + \frac{3}{3-2}$	0
$\frac{3}{-1} + \frac{3}{1}$	0
$-3 + 3$	0
0	0

DO EXERCISE 3.

Here are some more examples.

Example 4 Solve $\frac{1}{x+5} - 1 = \frac{-x}{x+2}$.

STEP 1: LCD

The LCD is $(x+5)(x+2)$.

$$\frac{(x+5)(x+2)}{1}\left(\frac{1}{x+5} - 1\right) = \frac{(x+5)(x+2)}{1}\left(\frac{-x}{x+2}\right)$$

$$\frac{x+5}{x+5}\cdot\frac{x+2}{1} - 1(x+5)(x+2) = \frac{x+2}{x+2}\cdot\frac{(-x)(x+5)}{1}$$

$$x + 2 - (x^2 + 7x + 10) = -x(x+5)$$

Solve.

3. $\frac{-x}{x+1} = \frac{-2}{x+3}$

Solve.

4. $\dfrac{1}{x-1} - \dfrac{2}{x+2} = \dfrac{-1}{x}$

Solve.

5. $\dfrac{3}{1 + \dfrac{1}{x}} = 2$

STEP 2: SOLVE

$$\begin{aligned} x + 2 - x^2 - 7x - 10 &= -x^2 - 5x \\ -x^2 - 6x - 8 &= -x^2 - 5x \\ -6x - 8 &= -5x \\ -x - 8 &= 0 \\ -x &= 8 \\ x &= -8 \end{aligned}$$

STEP 3: CHECK

The number -8 is not a pole. (The poles are -5 and -2.)

$\dfrac{1}{x+5} - 1$	$\dfrac{-x}{x+2}$
$\dfrac{1}{-8+5} - 1$	$\dfrac{-(-8)}{-8+2}$
$\dfrac{1}{-3} - 1$	$\dfrac{8}{-6}$
$-\dfrac{4}{3}$	$-\dfrac{4}{3}$

DO EXERCISE 4.

Example 5 Solve $\dfrac{1}{x-3} = \dfrac{2}{x+4} + \dfrac{7}{x^2 + x - 12}$.

STEP 1: LCD

Note that $x^2 + x - 2 = (x + 4)(x - 3)$. Thus, the LCD $= (x - 3)(x + 4)$.

$$\frac{(x-3)(x+4)}{1}\left(\frac{1}{x-3}\right) = \frac{(x-3)(x+4)}{1}\left(\frac{2}{x+4} + \frac{7}{(x-3)(x+4)}\right)$$

$$\frac{x-3}{x-3}\cdot\frac{x+4}{1} = \frac{x+4}{x+4}\cdot\frac{2(x-3)}{1} + \frac{x-3}{x-3}\cdot\frac{x+4}{x+4}\cdot\frac{7}{1}$$

$$x + 4 = 2(x - 3) + 7$$

STEP 2: SOLVE

$$\begin{aligned} x + 4 &= 2x - 6 + 7 \\ x + 4 &= 2x + 1 \\ -x + 4 &= 1 \\ -x &= -3 \\ x &= 3 \end{aligned}$$

STEP 3: CHECK

The poles are $x = 3$ and $x = -4$. Since our solution is a pole, we have to disregard it as a solution. In this case, the original equation has no solution.

DO EXERCISE 5.

Example 6 Solve $\frac{3x}{x^2 - 4} - \frac{2x}{x^2 - 3x + 2} = 0.$

STEP 1: LCD

$$x^2 - 4 = (x - 2)(x + 2)$$
$$x^2 - 3x + 2 = (x - 2)(x - 1)$$
$$\text{LCD} = (x - 2)(x + 2)(x - 1)$$

$$\frac{(x - 2)(x + 2)(x - 1)}{1}\left[\frac{3x}{(x - 2)(x + 2)}\right] - \frac{(x - 2)(x + 2)(x - 1)}{1}\left[\frac{2x}{(x - 2)(x - 1)}\right] = 0$$

$$\frac{x - 2}{x - 2} \cdot \frac{x + 2}{x + 2} \cdot \frac{3x(x - 1)}{1} - \frac{x - 2}{x - 2} \cdot \frac{x - 1}{x - 1} \cdot \frac{2x(x + 2)}{1} = 0$$

$$3x(x - 1) - 2x(x + 2) = 0$$

STEP 2: SOLVE

$$3x^2 - 3x - 2x^2 - 4x = 0$$
$$x^2 - 7x = 0$$
$$x(x - 7) = 0$$

Thus,

$$(1) \quad x = 0 \qquad \text{or} \qquad (2) \quad x = 7$$

STEP 3: CHECK

The numbers 0 and 7 are not poles. (The poles are 2, −2, and 1.)

$\frac{3x}{x^2 - 4} - \frac{2x}{x^2 - 3x + 2}$	0
$\frac{3(0)}{0^2 - 4} - \frac{2(0)}{0 - 3(0) + 2}$	0
$0 + 0$	0
0	0

$\frac{3x}{x^2 - 4} - \frac{2x}{x^2 - 3x + 2}$	0
$\frac{3(7)}{7^2 - 4} - \frac{2(7)}{7^2 - 3(7) + 2}$	0
$\frac{21}{45} - \frac{14}{30}$	0
$\frac{7}{15} - \frac{7}{15}$	0
0	0

DO EXERCISE 6.

Before we leave this section, we would like to look at solving certain formulas (literal equations) for a designated variable, much as we did in Chapter 3.

Example 7 Solve $\frac{I}{PT} = R$ for T.

This is solved the same way we solve a rational equation. Hence, we first clear fractions.

Solve.

6. $\frac{x}{x - 4} = \frac{4}{x - 4} + 2$

7. Solve $t = \frac{d}{r}$ for r.

8. Solve $\frac{1}{P} + \frac{1}{Q} = 1$ for P.

9. Solve $\frac{b+x}{x} = \frac{b+1}{2}$ for x.

(a) Multiply each side by PT.

$$\frac{PT}{1}\left(\frac{I}{PT}\right) = (PT)R$$

$$\frac{PT}{PT} \cdot \frac{I}{1} = PTR$$

$$I = PTR$$

(b) Multiply each side by $\frac{1}{PR}$.

$$\frac{1}{PR} \cdot \frac{I}{1} = \frac{1}{PR} \cdot \frac{PTR}{1}$$

$$\frac{I}{PR} = \frac{PTR}{PR} = \frac{P}{P} \cdot \frac{R}{R} \cdot \frac{T}{1}$$

$$\frac{I}{PR} = T$$

Thus, $T = \frac{I}{PR}$.

DO EXERCISE 7.

Example 8 Solve $A = \frac{B-C}{D+C}$ for C.

(a) Multiply each side by $D + C$ to clear fractions.

$$(D+C)A = \frac{D+C}{1} \cdot \frac{B-C}{D+C}$$

$$AD + AC = \frac{D+C}{D+C} \cdot \frac{B-C}{1}$$

$$AD + AC = B - C$$

(b) Gather all the terms that include C on one side of the equality sign and everything else on the other side. Thus, add C to each side and subtract AD from each side.

$$AD + AC + C = B - C + C$$

$$AD + AC + C - AD = B - AD$$

$$AC + C = B - AD$$

(c) Factor out C.

$$C(A+1) = B - AD$$

(d) Multiply each side by $\frac{1}{A+1}$.

$$\frac{1}{A+1} \cdot \frac{(C)(A+1)}{1} = \frac{1}{A+1} \cdot \frac{B-AD}{1}$$

$$\frac{C}{1} \cdot \frac{A+1}{A+1} = \frac{B-AD}{A+1}$$

$$C = \frac{B-AD}{A+1}$$

DO EXERCISES 8 AND 9.

DO SECTION PROBLEMS 8.6.

ANSWERS TO MARGINAL EXERCISES

1. $-\frac{1}{2}$ **2.** 4 **3.** 1, −2 **4.** $\frac{2}{5}$ **5.** 2 **6.** No solution
7. $r = \frac{d}{t}$ **8.** $P = \frac{Q}{Q-1}$ **9.** $x = \frac{2b}{b-1}$

NAME COURSE/SECTION DATE

ANSWERS

1. ______

2. ______

3. ______

4. ______

5. ______

6. ______

7. ______

8. ______

9. ______

10. ______

SECTION PROBLEMS 8.6

Solve.

1. $\frac{3}{x} = \frac{1}{2}$

2. $\frac{4}{1 - x} = -2$

3. $3 = \frac{12}{y^2}$

4. $4 + \frac{2}{y} = 3$

5. $\frac{2}{x - 1} = x$

6. $\frac{-3}{a + 4} = \frac{2}{a - 1}$

7. $\frac{6}{2x + 1} - \frac{1}{x + 4} = 0$

8. $\frac{x}{x + 1} = \frac{-1}{x - 3}$

9. $\frac{y}{y + 2} - \frac{4}{y - 3} = 0$

10. $\frac{12}{y} - 3 = \frac{5}{8}$

11. ______________

12. ______________

13. ______________

14. ______________

15. ______________

16. ______________

17. ______________

18. ______________

19. ______________

20. ______________

11. $\dfrac{3}{t-2} = \dfrac{-5}{t+1}$

12. $\dfrac{-4}{t-2} = \dfrac{3}{5-t}$

13. $\dfrac{12}{t+4} = t$

14. $\dfrac{x}{x+2} = 4$

15. $\dfrac{-7}{y} = \dfrac{3}{4}$

16. $\dfrac{3x}{4-x} = -1$

17. $\dfrac{x}{x-3} = \dfrac{3}{x-3} + 2$

18. $3 + \dfrac{1}{x} - \dfrac{2}{x^2} = 0$

19. $\dfrac{2}{x-5} + \dfrac{3}{x-4} = \dfrac{1}{x-5}$

20. $\dfrac{x-2}{x+3} - \dfrac{x+1}{x+2} = 0$

NAME

COURSE/SECTION

DATE

ANSWERS

21. $\dfrac{3}{x^2 - x - 2} + \dfrac{2}{x^2 - 1} = 0$

22. $\dfrac{1}{x} + \dfrac{7}{x^2 - 3x} = 2$

23. $\dfrac{y - 1}{y} + \dfrac{y + 1}{y - 1} = 2$

24. $\dfrac{7}{1 - 2y} = -3$

25. $\dfrac{x}{x^2 - x - 6} = \dfrac{x - 3}{x^2 - 4}$

26. $\dfrac{y}{2y^2 - 5y - 3} - \dfrac{y - 3}{2y + 1} = \dfrac{1}{2}$

27. $\dfrac{1 - \dfrac{1}{x}}{1 + \dfrac{1}{x}} = \dfrac{1}{2}$

28. $1 + \dfrac{1}{t - 1} = 2$

29. $\dfrac{3}{x^2 - 1} = \dfrac{2}{x + 1}$

30. $\dfrac{y + 1}{y} + 2 = 1 + \dfrac{1}{y}$

21. ____________

22. ____________

23. ____________

24. ____________

25. ____________

26. ____________

27. ____________

28. ____________

29. ____________

30. ____________

31. ______________

32. ______________

33. ______________

34. ______________

35. ______________

36. ______________

37. ______________

38. ______________

39. ______________

40. ______________

31. $-\frac{2}{3} + \frac{1}{a} = \frac{7}{8}$

32. $\frac{x}{1 + x} = 1 + \frac{1}{x}$

33. Solve $t^2 = \frac{2s}{g}$ for g.

34. Solve $r = \frac{d}{t}$ for t.

35. Solve $P = \frac{I}{RT}$ for T.

36. Solve $\pi = \frac{A}{2rh}$ for h.

37. Solve $\frac{1}{P} + \frac{1}{Q} = 1$ for Q.

38. Solve $\frac{2A}{h} - b = B$ for h.

39. Solve $C = \frac{ax}{A + Bx}$ for x.

40. Solve $x = y - \frac{2}{Ax}$ for A.

8.7

Word Problems

LEARNING OBJECTIVES

After finishing Section 8.7, you should be able to:

- Solve word problems that involve rational equations.

Now that we can solve rational equations, there are additional types of word problems that we can tackle. Recall the five steps we use to solve a word problem: Picture, Variable, Equation, Solve, Check.

Example 1 Bill Fletcher buys a large bag of cashews for \$30 and an even larger bag of pistachios for \$44. There are 3 more pounds of pistachios than cashews and the cashews cost 50¢ a pound more than the pistachios. How many pounds of cashews did Bill buy and how much a pound did he pay? We omit Step 1 since a picture will not help us solve this problem.

STEP 2: VARIABLE

Two choices for our variable might be the number of pounds of cashews or the cost per pound of the cashews. Either choice is fine, but we will choose the former.

Let x = number of pounds of cashews, then
$x + 3$ = number of pounds of pistachios.

STEP 3: EQUATION

Note

The total cost for the cashews is \$30. If there are x number of pounds of cashews then

$\frac{30}{x}$ = cost per pound for the cashews (in dollars). Similarly,

$\frac{44}{x+3}$ = cost per pound for the pistachios (in dollars).

Now, "The cashews cost 50¢ (or \$.50) a pound more than the pistachios." This translates as

$$\begin{pmatrix}\text{Cost per}\\ \text{pound for}\\ \text{the cashews}\end{pmatrix} = \begin{pmatrix}\text{Cost per}\\ \text{pound for}\\ \text{the pistachios}\end{pmatrix} + 0.50$$

$$\frac{30}{x} = \frac{44}{x+3} + \frac{1}{2} \qquad \left(\text{We let } 0.5 = \frac{1}{2}.\right)$$

STEP 4: SOLVE

$$\frac{30}{x} = \frac{44}{x+3} + \frac{1}{2}$$

$$2(x)(x+3)\left[\frac{30}{x}\right] = 2(x)(x+3)\left[\frac{44}{x+3} + \frac{1}{2}\right]$$

$$\frac{60(x)(x+3)}{x} = \frac{88(x)(x+3)}{x+3} + \frac{2(x)(x+3)}{2}$$

$$60(x+3) = 88x + x(x+3)$$

$$60x + 180 = 88x + x^2 + 3x$$

$$0 = x^2 + 31x - 180$$

$$0 = (x-5)(x+36)$$

1. (a) In Example 1, why did we disregard the solution -36?
(b) Rework Example 1 by letting x = cost per pound of cashews. *Hint:* If x = cost per pound of cashews, then $x - 0.50$ = cost per pound of pistachios, and, also, $\frac{30}{x}$ = number of pounds of cashews.

Thus, (1) $x = 5$ or (2) $x = -36$ (which we may disregard). Hence $x = 5$ is the solution, Bill buys 5 pounds of cashews, and the cost per pound $\left(\frac{30}{x}\right)$ is $\frac{30}{5} = \$6$.

STEP 5: CHECK

You may check that 5 satisfies the rational equation in Step 4. Also note that the number of pounds of pistachios, $x + 3$, is 8 and the cost, $\frac{44}{x+3}$, is $\frac{44}{8} = \$5.50$.

DO EXERCISE 1.

The following example is a type of word problem that normally comes under the general description of a "work" problem.

Example 2 Louise Driscol can write the Labresh Company's annual report by herself in 5 days, and Cathy Biglow can do it by herself in 4 days. If they work together, how long will it take to write the annual report?

We omit Step 1 since a picture will not help us solve this problem.

STEP 2: VARIABLE

Let t = time required to write the annual report (in days) if they work together.

STEP 3: EQUATION

Before we can write an equation for this problem, we need to understand the following.

If Louise can write the report in 5 days, then in one day she can complete $\frac{1}{5}$ of the job.

If Cathy can write the report in 4 days, then in one day she can complete $\frac{1}{4}$ of the job.

In general, if t is the time required (in days) to complete a job, then in one day $\frac{1}{t}$ of the job will be completed.

Now we can write the equation.

$$\left(\begin{array}{l}\text{Amount of job}\\ \text{that will be}\\ \text{completed in}\\ \text{one day with}\\ \text{both working}\\ \text{together}\end{array}\right) = \left(\begin{array}{l}\text{Amount of job}\\ \text{that Louise can}\\ \text{do in one day}\end{array}\right) + \left(\begin{array}{l}\text{Amount of job}\\ \text{that Cathy can}\\ \text{do in one day}\end{array}\right)$$

$$\frac{1}{t} = \frac{1}{5} + \frac{1}{4}$$

STEP 4: SOLVE

$$20t\left(\frac{1}{t}\right) = 20t\left(\frac{1}{5}\right) + 20t\left(\frac{1}{4}\right)$$

$$\frac{20t}{t} = \frac{20t}{5} + \frac{20t}{4}$$

$$20 = 4t + 5t$$

$$20 = 9t$$

$$\frac{20}{9} = t$$

Thus, working together it will take $\frac{20}{9}$ or $2\frac{2}{9}$ days to write the report.

STEP 5: CHECK

The answer is certainly reasonable since if they could both do it alone in 4 days, then together it would take them 2 days, or if they could both do it alone in 5 days, then together it would take them $2\frac{1}{2}$ days. Notice that our answer is between the 2 days and $2\frac{1}{2}$ days, which is what we would expect.

DO EXERCISE 2.

Recall that in Chapter 7 we did some word problems called motion problems. In this process we used the motion formula $d = r \cdot t$ [distance equals rate (average speed) times time]. We do another motion problem here.

Example 3 Alice Gibson drives 40 miles to work. On the way home, in similar driving conditions, she drives 10 mph faster and it takes her 8 minutes $\left(\frac{8}{60} = \frac{2}{15}\text{ hour}\right)$ less travel time. What is Alice's average speed going to and coming from work?

STEP 1: PICTURE

$d_1 = r \cdot t$
$r_1 = r$
$t_1 = t$

Home ———————— Work

$d_2 = (r + 10)\left(t - \frac{2}{15}\right)$
$r_2 = r + 10$
$t_2 = t - \frac{2}{15}$

STEP 2: VARIABLE

Let r = rate going to work; then
$r + 10$ = rate going home.
Let t = time (in hours) going to work; then
$t - \frac{8}{60} = t - \frac{2}{15}$ = time going home.

2. Pipe A can fill a pool in 8 hours, pipe B can fill the pool in 7 hours, and pipe C can empty the pool in 12 hours.
(a) If pipes A and B work together, how long will it take to fill the pool if pipe C is closed?
(b) If someone accidentally leaves pipe C open, then how long will it take pipes A and B to fill the pool?

3. A camper rows her boat 2 miles upstream and then returns downstream to her starting point. The trip upstream takes three times as long as the trip downstream. If the camper rows 2 mph in still water, how fast is the stream moving?

STEP 3: EQUATION

Since $d_1 = d_2 = 40$, we have two equations:

$$(1) \qquad 40 = r \cdot t$$

$$(2) \qquad 40 = (r + 10)\left(t - \frac{2}{15}\right)$$

in the unknowns r and t.

STEP 4: SOLVE

We will solve equation (1) for t and substitute this value in equation (2).

$$(1) \qquad \frac{40}{r} = t$$

$$(2) \qquad 40 = (r + 10)\left(\frac{40}{r} - \frac{2}{15}\right)$$

$$40 = (r + 10)\left(\frac{600}{15r} - \frac{2r}{15r}\right)$$

$$40 = (r + 10)\left(\frac{600 - 2r}{15r}\right)$$

$$(15r)(40) = \left(\frac{15r}{1}\right)\left(\frac{r + 10}{1}\right)\frac{(600 - 2r)}{15r}$$

$$600r = (r + 10)(600 - 2r)$$

$$600r = 600r - 2r^2 + 6000 - 20r$$

$$2r^2 + 20r - 6000 = 0$$

$$2(r^2 + 10r - 3000) = 0$$

$$2(r + 60)(r - 50) = 0$$

We have the solution:

(1) $r = -60$ mph (which we may disregard since it is negative), or
(2) $r = 50$ mph.

Thus, Alice drove 50 mph going to work and $r + 10 = 60$ mph going home.

DO EXERCISE 3.

Example 4 A group of editors decides to buy a woodlot for \$4800. After they have started the purchase, two more editors decide to join the group. If the additional two editors decrease the cost per individual of each of the original editors by \$200, how many persons were in the original group?

Since a picture won't help us solve this problem, we omit Step 1.

STEP 2: VARIABLE

Let x = number in original group;
then $x + 2$ = number in new group.

STEP 3: EQUATION

Note the following:

1. $\dfrac{\text{Total cost}}{\text{Total number of members}} = \text{Cost per individual}$
2. $\dfrac{4800}{x} = \text{Cost per individual in original group}$
3. $\dfrac{4800}{x+2} = \text{Cost per individual in new group}$

Our equation will be as follows.

$$\begin{pmatrix}\text{Cost per individual}\\ \text{in original group}\end{pmatrix} - \$200 = \begin{pmatrix}\text{Cost per individual}\\ \text{in new group}\end{pmatrix}$$

$$\frac{4800}{x} - \$200 = \frac{4800}{x+2}$$

STEP 4: SOLVE

$$\frac{x(x+2)}{1}\left(\frac{4800}{x}\right) - (x)(x+2)(200) = \frac{x(x+2)}{1}\left(\frac{4800}{x+2}\right)$$

$$4800(x+2) - (200x^2 + 400x) = 4800(x)$$

$$4800x + 9600 - 200x^2 - 400x = 4800x$$

$$-200x^2 - 400x + 9600 = 0$$

$$-200(x^2 + 2x - 48) = 0$$

$$-200(x+8)(x-6) = 0$$

Thus, $x = -8$ (which we may disregard) or $x = 6$. Hence, there were 6 original members.

DO EXERCISE 4.

DO SECTION PROBLEMS 8.7.

4. Tom has four times as much money as Steve. If Tom gives Steve \$1, Steve will then have one-third as much money as Tom will have. How much does each man have?

ANSWERS TO MARGINAL EXERCISES

1. (a) The variable x cannot be negative since it represents number of pounds. (b) 5 pounds of cashews that cost $6 per pound. **2.** (a) $3\frac{11}{15}$ hr (b) $5\frac{13}{31}$ hr **3.** 1 mph **4.** $4, $16

NAME COURSE/SECTION DATE

ANSWERS

1. ____
2. ____
3. ____
4. ____
5. ____

SECTION PROBLEMS 8.7

1. Three times the reciprocal of a number is the same as the number plus two. Find the number. *Hint:* If $x =$ the number, then the reciprocal $= \frac{1}{x}$.

2. Twice the square of the reciprocal of a number is 18. What is the number?

3. Three times a number divided by the sum of the number and two is the same as the number minus two. Find the number.

4. The sum of a number and 4 divided by the number equals 6 divided by the result of subtracting 2 from the number plus $\frac{1}{2}$. Find the number.

5. A plane travels from Seattle to Omaha, a distance of 1800 miles, with a tailwind in three hours. The return flight against the same wind takes three hours and twenty minutes. Find the speed of the wind and speed of the plane in still air.

6. ______

7. ______

8. ______

9. ______

10. ______

6. A passenger train travels 10 mph faster than a freight train. If the passenger train travels 280 miles during the same time the freight train travels 240 miles, find the speed of each train.

7. Dawn Lacey can do a piece of work in 3 hours and Sue Abrams can do the same piece of work in $2\frac{1}{2}$ hours. If they work together, how long will it take to do this piece of work?

8. Doc can type five pages in 30 minutes and Slim can type five pages in 40 minutes. If they work together on a 25-page report, how long will it take them to complete the job?

9. Mr. Kiner, a purchasing agent, buys a particular lot of tables for a total of $2000. If Mr. Kiner had bought 30 more tables, the individual price per table would have gone down $5 and the new total would have been $2800. How many tables did Mr. Kiner buy?

10. A group of college students decides to buy a television set for $600. They realize the cost for each person is going to be too much; however, they calculate that if they can find two more students to go in with them, they can decrease the cost per individual by $50 each. What will the individual cost be for this enlarged group?

NAME COURSE/SECTION DATE ANSWERS

11. A six-inch pipe can fill a pool by itself in 5 hours and a five-inch pipe can fill the same pool by itself in $5\frac{1}{2}$ hours. If they are turned on together, how long will it take to fill the pool?

11. ____________

12. ____________

13. ____________

14. ____________

15. ____________

12. If in Problem 11 the six-inch pipe is on for an hour by itself and then the five-inch pipe is also turned on, how long will it take to fill the pool?

13. The McCorkle Production Company buys some company cars for a total of $37,500 and also buys some company pickups for a total of $24,000. The number of pickups the company buys is one less than the number of cars, and it pays $1500 more for a car than a pickup. How much did the company pay for each pickup and how much for each car?

14. Doug Nassar buys a tin of Peppermint tea for $9.60 and a tin of Comfrey tea for $10.00. There are 2 more ounces of the Comfrey tea than of the Peppermint tea, and the Peppermint tea costs 20¢ an ounce more than the Comfrey. How many ounces of each kind of tea did Doug buy?

15. Saul drives 10 mph faster than William. If Saul covers 330 miles in the same time that William covers 270 miles, find the speed of each.

16. ______________

17. ______________

18. ______________

19. ______________

20. ______________

16. One plane travels 50 mph faster than another plane. If the first plane travels 2000 miles in the same time that the second plane travels 1800 miles, find the speed of each.

17. Gingerbread Nursery has a rectangular herb bed that contains 120 square feet. They calculate that if they increase the width of the bed by 2 feet and decrease the length by 1 foot, they can obtain an area of 140 square feet. Find the dimensions of the original bed. (*Hint:* Use length $= \dfrac{\text{area}}{\text{width}}$, x = width of original bed, and $x + 2$ = width of new bed.)

18. Joan Cassidy wants to make \$100 in interest by putting money in the bank and leaving it for one year. Her bank has a fixed interest rate, but she calculates that if she can find somewhere to invest her money at an interest rate that is 2% higher than the bank's rate, the amount of money she would have to invest to make the \$100 in interest would decrease by \$250. What is her bank's present interest rate? Note that $I = PRT$ where I = interest, P = principal, R = interest rate, and T = time in years ($T = 1$ for this problem).

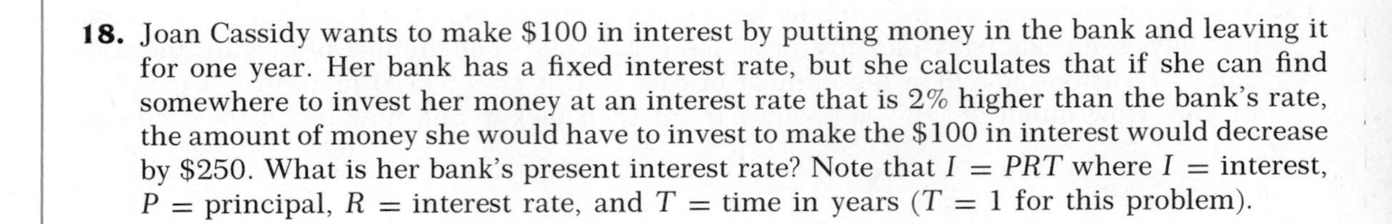

19. A car is traveling 20 mph slower than a train. The car travels 225 miles while the train travels 315 miles. How fast is the train going?

20. It takes Dan 16 hours to wax all the floors in the Sunshine Mission. Larry can do all the floors in 14 hours. If they work together, how long will it take to wax all the floors?

8.8

Proportion

Before we look at proportion problems, we need to discuss ratio briefly. A *ratio* expresses a numerical relation between two quantities.

Example 1 Consider the following rectangle.

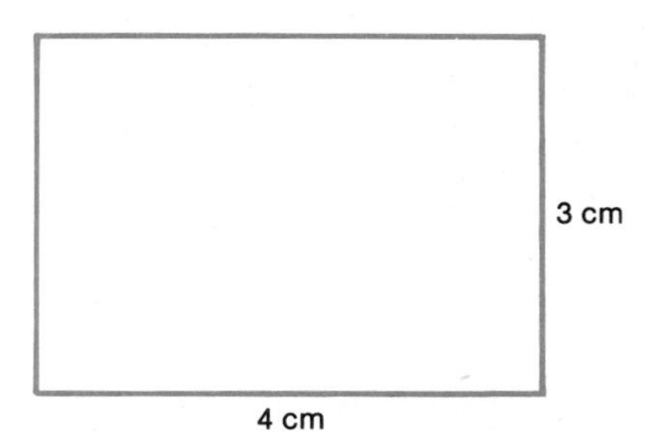

1. The ratio of the length of the rectangle to the width is 4 to 3. Often we write a ratio of 4 to 3 as 4:3 or as a fraction, $\frac{4}{3}$.
2. The ratio of the width to the length is 3:4 $\left(\text{or } \frac{3}{4}\right)$.
3. The ratio of the length to the perimeter is 4:14 $\left(\text{or } \frac{4}{14}\right)$.

Example 2 If we consider the sets of numbers {5, 10}, {15, 30}, and {$2\frac{1}{2}$, 5}, then the ratio of the first number in each set to the second number is 1:2.

DO EXERCISE 1.

We form a *proportion* by taking two ratios that are expressed as fractions and setting them equal to each other. There are two types of proportion problems—direct and indirect—that we are going to consider. To understand the difference between a direct proportion and an indirect proportion, we must first realize that we are always comparing just two quantities when we work a proportion problem. These two quantities may act the same way in the problem: if one increases, the other increases; or if one decreases, the other decreases. Or they may act in opposite ways: as one increases, the other decreases. How they act distinguishes a direct proportion from an indirect proportion.

DEFINITION 8.2 A. In a *direct proportion,* the two quantities we are comparing will act the same way, that is, both increase or both decrease together.

B. In an *indirect proportion,* the two quantities we are comparing will act in opposite ways, that is, one will increase and the other will decrease.

Example 3 If it takes 300 bricks to build 15 feet of wall, how many bricks does it take to build 20 feet of wall?

This is an example of a direct proportion problem. We notice that the two quantities (number of bricks and feet of wall) act the same way. In other words, if the number of feet of wall increases, then the number of bricks must increase; or if the number of bricks decreases, then the amount of wall we can build will also decrease.

LEARNING OBJECTIVES

After finishing Section 8.8, you should be able to:

- Solve proportion problems.

1. In the following rectangle, what is the ratio of (a) length to width? (b) width to length?

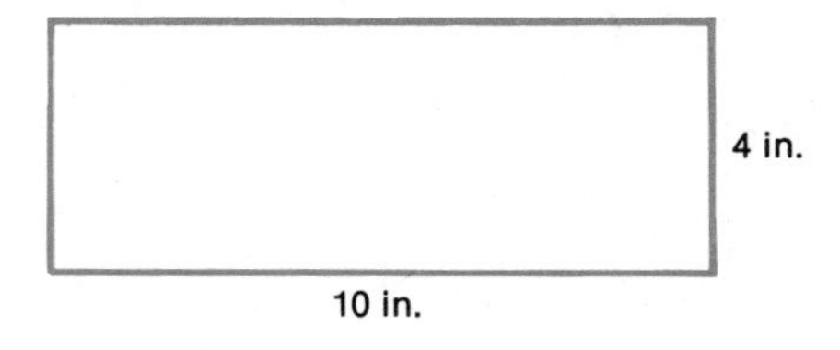

Example 4 If 3 men can rake a yard in 7 hours, how long will it take 5 men to rake the same yard?

This is an example of an indirect proportion problem. Notice that the two quantities (men and hours) act in opposite ways. In other words, as we increase the number of men raking the yard, the number of hours required will decrease.

The only difference in solving a direct proportion and an indirect proportion is in the actual setting up of the proportion. As we work the problems in Examples 3 and 4 we will note this difference.

Example 5 If it takes 300 bricks to build 15 feet of wall, how many bricks does it take to build 20 feet of wall?

STEP 2: VARIABLE

Let x = number of bricks needed to build 20 feet of wall.

STEP 3: EQUATION

Form the two ratios. When forming ratios, we always compare same units to same units, such as bricks to bricks and feet to feet—not bricks to feet. This is very critical for indirect proportions and, while it is not necessary for direct proportions, we may as well get in the habit of doing it the same way for both types of proportion problems.

$$\frac{300 \text{ bricks}}{x \text{ bricks}} = \frac{300}{x} \quad \text{and} \quad \frac{15 \text{ feet}}{20 \text{ feet}} = \frac{15}{20}$$

Note

1. We express the ratios as fractions.
2. The 300 bricks and 15 feet do go together in the problem. That is, it takes 300 bricks to build 15 feet. Thus, since we have the 300 bricks in the numerator of the first ratio, we have to be consistent and put the 15 feet in the numerator of the second ratio.

Form the direct proportion by setting the ratios equal.

$$\frac{300}{x} = \frac{15}{20}$$

STEP 4: SOLVE

$$\frac{300}{x} = \frac{15}{20}$$

$$\frac{20x}{1}\left(\frac{300}{x}\right) = \frac{20x}{1}\left(\frac{15}{20}\right)$$

$$\frac{x}{x} \cdot \frac{(20)(300)}{1} = \frac{20}{20} \cdot \frac{15x}{1}$$

$$6000 = 15x$$

$$400 = x$$

Thus, we need 400 bricks.

STEP 5: CHECK

The answer is certainly reasonable since we know we need more than 300 bricks to build 20 feet of wall.

DO EXERCISE 2.

Example 6 If 3 men can rake a yard in 7 hours, how long will it take 5 men to rake the same yard?

STEP 2: VARIABLE

Let t = number of hours needed for the 5 men.

STEP 3: EQUATION

Form the two ratios (note that we set up the ratios same units to same units).

$$\frac{3 \text{ men}}{5 \text{ men}} = \frac{3}{5} \qquad \frac{7 \text{ hours}}{x \text{ hours}} = \frac{7}{x}$$

To form the indirect proportion, we first invert $\frac{3}{5}$ to $\frac{5}{3}$ (that is, we take the reciprocal of $\frac{3}{5}$) and then set the ratios equal. This is the only place solving an indirect proportion differs from solving a direct proportion.

$$\frac{5}{3} = \frac{7}{x}$$

$\left(\text{We could also invert } \frac{7}{x} \text{ to } \frac{x}{7}, \text{ set the ratios equal, and obtain } \frac{x}{7} = \frac{3}{5}.\right)$

STEP 4: SOLVE

$$\frac{3x}{1} \cdot \frac{5}{3} = \frac{3x}{1} \cdot \frac{7}{x}$$

$$5x = 21$$

$$x = \frac{21}{5} \quad \text{or} \quad 4\frac{1}{5}$$

Thus, it takes the five men $4\frac{1}{5}$ hours to rake the yard.

STEP 5: CHECK

Once again, the answer is reasonable since we realize it should take less time for the 5 men than for the 3 men.

DO EXERCISES 3 AND 4.

In order to work proportion problems correctly, we need to be able to distinguish between direct and indirect proportions. There are basically three ways of doing this.

2. A pump can empty a 3500-gallon swimming pool in 4 hours. How long will it take the same pump to empty a 4900-gallon pool?

3. A gear with 12 teeth is in mesh with a gear with 8 teeth. If the speed of the 12-tooth gear is 450 rpm, what is the speed of the 8-tooth gear? (*Hint:* This is an indirect proportion.)

4. If it takes four 6-inch pipes 2 hours 15 minutes to empty a tank, how long would it take three 6-inch pipes to empty the same tank?

5. If the distance a spring is stretched by a 20-pound weight is 30 cm, how much weight is on a spring that is stretched 40 cm?

1. By reading the problem carefully we can determine whether it is a direct or an indirect proportion.
2. The problem will tell us that one quantity varies directly (direct proportion) or indirectly (indirect proportion) with another quantity.
3. From the experience of working a similar problem in the past we can recognize this problem as a direct or indirect proportion.

In the last two examples we could tell what type of proportion problem each was by reading the problem carefully. In the next two examples, we will be told what type they are.

Example 7 Hooke's law states that the distance d a spring is stretched when a weight is placed on it varies directly as the weight w of the object. (In other words, d and w are directly proportional to each other.) If the distance a spring is stretched is 40 inches when the weight on the spring is 30 pounds, how far is the spring stretched when a 20-pound weight is attached?

STEP 2: VARIABLE

Let x = distance spring is stretched when a 20-pound weight is attached.

STEP 3: EQUATION

Form the two ratios.

$$\frac{40 \text{ inches}}{x \text{ inches}} = \frac{40}{x}$$

$$\frac{30 \text{ pounds}}{20 \text{ pounds}} = \frac{30}{20}$$

Form the direct proportion.

$$\frac{40}{x} = \frac{30}{20}$$

STEP 4: SOLVE

$$\frac{20x}{1} \cdot \frac{40}{x} = \frac{20x}{1} \cdot \frac{30}{20}$$

$$(20)(40) = 30x$$

$$800 = 30x$$

$$\frac{800}{30} = x$$

$$x = \frac{80}{3} = 26\frac{2}{3} \text{ inches}$$

Thus, the spring is stretched $26\frac{2}{3}$ inches when the weight is 20 pounds.

DO EXERCISE 5.

Example 8 The fundamental frequency of vibration of a string and its length are indirectly proportional to each other. If the fundamental frequency of a 30-cm string is 200 cycles per second, what is the length of the string when it has a fundamental frequency of 250 cycles per second?

STEP 2: VARIABLE

Let x = length of a string with fundamental frequency of 250 cycles per second.

STEP 3: EQUATION

Form the ratios.

$$\frac{30 \text{ cm}}{x \text{ cm}} = \frac{30}{x}$$

$$\frac{200 \text{ cycles/second}}{250 \text{ cycles/second}} = \frac{200}{250}$$

Invert $\frac{200}{250}$.

$$\frac{250}{200}$$

Form the indirect proportion.

$$\frac{30}{x} = \frac{250}{200}$$

STEP 4: SOLVE

$$\frac{200x}{1} \cdot \frac{30}{x} = \frac{200x}{1} \cdot \frac{250}{200}$$

$$(200)(30) = 250x$$

$$6000 = 250x$$

$$\frac{6000}{250} = x$$

$$x = 24$$

Thus, the length of the string is 24 cm.

DO EXERCISE 6.

DO SECTION PROBLEMS 8.8.

6. If the fundamental frequency of a 25-cm string is 150 cycles per second, what is the fundamental frequency of a 40-cm string?

ANSWERS TO MARGINAL EXERCISES

1. (a) 10:4 (b) 4:10 **2.** $5\frac{3}{5}$ hr **3.** 675 rpm **4.** 3 hr **5.** $26\frac{2}{3}$ pounds **6.** $93\frac{3}{4}$ cycles per second

NAME COURSE/SECTION DATE

ANSWERS

1. ____________

2. ____________

3. ____________

4. ____________

5. ____________

6. ____________

7. ____________

SECTION PROBLEMS 8.8

1. If 0.75 inch on a map represents 10 miles on the ground, how many miles on the ground will be represented by 6 inches on the map?

2. A pump can empty a 3500-gallon swimming pool in $4\frac{1}{2}$ hours. How long will it take the same pump to empty an 8800-gallon swimming pool?

3. If a family of four uses 120 gallons of water a day, how many gallons a day would Georgetown use if its population were 5000?

4. A 12-inch pulley and an 8-inch pulley are connected by a belt. If the 12-inch pulley has an rpm of 320, what is the rpm of the 8-inch pulley?

5. If eight apples cost $1.00, how much would fifteen such apples cost?

6. It takes 2 hours to cover a certain distance at a speed of 50 mph. How long would it take to cover the same distance at a speed of 70 mph?

7. Henry can type five pages in 20 minutes and Joe can type five pages in 25 minutes. If they work together on a twenty-page report, how long will it take them to complete the job?

8. ______________

9. ______________

10. ______________

11. ______________

12. ______________

13. ______________

8. The sum of the reciprocal of a number and the reciprocal squared of a number is six. Find the number.

9. If three students working together take 8 hours to do a math assignment, how long would it take five students working together to complete the same assignment?

10. If the distance a spring is stretched by a 10-pound weight is 8 inches, how far would a 14-pound weight stretch the same spring?

11. If the fundamental frequency of a 50-cm string is 200 cycles/sec, then what is the length of a string with a fundamental frequency of 250 cycles/sec?

12. A gear with 20 teeth is in mesh with a gear with 30 teeth. If the speed of the 20-tooth gear is 250 rpm, find the speed of the 30-tooth gear.

13. Chris knows that it takes him $1\frac{1}{2}$ hours to type a five-page report. If he has an 18-page report to type, how much time should he allow himself?

NAME ______ COURSE/SECTION ______ DATE ______

ANSWERS

14. ______

15. ______

16. ______

17. ______

18. ______

19. ______

14. For $1.39, Mrs. Holland can buy 5 oranges. How many oranges can she buy with a five-dollar bill?

15. If Mr. Parker employs 5 persons, he can set up all the chairs for the Pilot's concert in 6 hours. He is under pressure to complete the job in 4 hours or less. What is the minimum number of persons Mr. Parker should employ to set up the chairs?

16. If 8 students can consume 18 cans of soda in a night, how many cans of soda can 14 students consume?

17. A driver has a given distance to travel by car. If he drives 50 mph, it will take him $2\frac{1}{2}$ hours to cover the distance. If he wants to cover the distance in 2 hours, how fast will he have to drive?

18. Joan can walk 3 miles in 50 minutes. How far can she walk in $2\frac{1}{2}$ hours?

19. The fundamental frequency of a 30-cm string is 350 cycles/sec. What is the fundamental frequency of a 35-cm string?

20. ____________

21. ____________

22. ____________

23. ____________

24. ____________

25. ____________

20. It takes $1\frac{1}{4}$ hours for 3 pipes to empty a pool. How long will it take 5 pipes to empty the same pool?

21. A 30-tooth gear with a speed of 200 rpm is in mesh with a 50-tooth gear. Find the speed of the 50-tooth gear.

22. A 12-gram weight stretches a spring 10 cm. If the spring is stretched 16 cm, how much weight is on the spring?

23. If 15 persons eat 27 hot dogs, how many hot dogs will 25 persons eat?

24. If 3 inches on a map represents 30 miles on the ground, then 55 miles on the ground is represented by how many inches on the map?

25. If it takes 16 persons 5 hours to pick 250 bushels of apples, how long will it take 20 persons to pick 250 bushels of apples?

NAME _______________ COURSE/SECTION _______________ DATE _______________

CHAPTER 8 REVIEW PROBLEMS

SECTION 8.1

Are the following rational expressions? If not, why not?

1. $\dfrac{x^2 - x}{y}$

2. $\dfrac{x}{\sqrt{x + 1}}$

Find the poles.

3. $\dfrac{3y}{y^2 - 16}$

4. $\dfrac{4}{t^2 + 3}$

SECTION 8.2

Reduce.

5. $\dfrac{2x^2 + 6x}{6x + 18}$

6. $\dfrac{x^2 + x - 6}{x^2 - 3x + 2}$

7. $\dfrac{t(t - 1)^2(t + 2)^2}{t^3(t - 1)(t + 1)(t + 2)^3}$

SECTION 8.3

Perform the indicated operation.

8. $\dfrac{x - 1}{x} \cdot \dfrac{3y}{y + 2}$

9. $\dfrac{(x - 1)^2}{x} \cdot \dfrac{3}{(x - 1)(x + 2)}$

10. $\dfrac{x}{3} \cdot \dfrac{6x - 24}{x^2} \cdot \dfrac{x^2 + x}{x - 2}$

11. $\dfrac{12a^3bc^2}{3b^2c} \cdot \dfrac{6ab^3c^4}{8a^2c}$

ANSWERS

1. _______________
2. _______________
3. _______________
4. _______________
5. _______________
6. _______________
7. _______________
8. _______________
9. _______________
10. _______________
11. _______________

12. ______________

13. ______________

14. ______________

15. ______________

16. ______________

17. ______________

18. ______________

19. ______________

20. ______________

21. ______________

12. $\dfrac{x^2 - 3x + 2}{x^2 - 12x + 36} \cdot 0$

13. $\dfrac{2x^2 - 8}{x^2 - 5x - 6} \div \dfrac{x^3 - 3x^2 + 2x}{x^2 - 1}$

SECTION 8.4

Perform the indicated operation.

14. $\dfrac{2}{x - 4} - \dfrac{3}{x - 4}$

15. $\dfrac{3a}{a - 5} + \dfrac{4}{a - 5} - \dfrac{2a}{a - 5}$

Find the LCD.

16. $\dfrac{1}{4} + \dfrac{13}{15} + \dfrac{1}{6}$

17. $\dfrac{1}{x^2 - x - 6} + \dfrac{1}{x^2 + 4x + 4} + \dfrac{1}{x^2 - 2x - 3}$

SECTION 8.5

Perform the indicated operation.

18. $\dfrac{2}{x + 3} + \dfrac{4}{x - 2}$

19. $y + \dfrac{3}{y} - \dfrac{1}{y - 1}$

20. $\dfrac{1}{x^2 - x} + \dfrac{3}{x^2 + x - 2} - \dfrac{4}{x^2 + 2x}$

21. $\left[\dfrac{x}{x^2 - 3x + 2} \cdot \dfrac{x - 1}{x^2 + x}\right] + \dfrac{3}{x^2 + 2x + 1}$

NAME _______ COURSE/SECTION _______ DATE _______

ANSWERS

22. ________

23. ________

24. ________

25. ________

26. ________

27. ________

28. ________

29. ________

30. ________

22. Simplify $\dfrac{x - \dfrac{1}{x}}{x + \dfrac{1}{x}}$.

SECTION 8.6

Solve.

23. $\dfrac{7}{x} = \dfrac{3}{5}$

24. $\dfrac{6}{y} - \dfrac{1}{y-1} = 0$

25. $\dfrac{x-1}{x} = \dfrac{x-3}{x-1} - \dfrac{5}{x^2 - x}$

26. Solve $\dfrac{S}{P} = (1 + i)^2$ for P.

27. Solve $A = \dfrac{C}{B - C}$ for C.

SECTION 8.7

28. The reciprocal of a number plus six times the number is five. Find the number.

29. A builder buys a parcel of land for \$7000. If he had bought 10 additional acres, the price per acre would have gone down \$30 and the total price would have been \$9600. How many acres did the builder buy for the \$7000?

30. Sharon can do a piece of work in 4 hours. The same piece of work takes Doris 3 hours and takes Rick $2\frac{1}{2}$ hours. How long will it take Doris, Rick, and Sharon to complete the job if they work together?

31. ______________

32. ______________

33. ______________

34. ______________

31. David and John both teach a section of Freshman Calculus. If two of John's students changed to David's class, then the classes would be the same size. If two of David's students changed to John's class, then the ratio of students in John's class to students in David's class would be $\frac{7}{5}$. How many students are in John's class?

SECTION 8.8

32. Pump A can empty a 2000-gallon tank in 2 hours and pump B can empty a 3000-gallon tank in $3\frac{1}{2}$ hours. How long would it take pump A and pump B working together to empty a 6000-gallon tank?

33. If Kate Stern walks 4 miles in one hour, how far can she walk in 40 minutes?

34. A driver drives a given distance at 40 mph in $3\frac{1}{2}$ hours. If he drives 50 mph, how long will it take the driver to drive the same distance?

NAME COURSE/SECTION DATE

ANSWERS

1. ______
2. ______
3. ______
4. ______
5. ______
6. ______
7. ______
8. ______
9. ______
10. ______
11. ______
12. ______

CHAPTER 8 TEST

1. Simplify $\dfrac{6a - 4a^2}{4ab + 4ab^2}$.

2. Simplify $\dfrac{x^2 - 25}{x^2 - 6x + 5}$.

Perform the indicated operation and simplify.

3. $\dfrac{-3}{t - 1} \cdot \dfrac{4s}{t}$

4. $\dfrac{(x - 3)^2}{x + 4} \cdot \dfrac{x^2 - 16}{2x - 6}$

5. $\dfrac{21ab^2c}{4a^2b} \cdot \dfrac{2abc}{7bc^2}$

6. $2b \div \dfrac{3b}{b - 1}$

7. $\dfrac{x^2 - x}{x^2 - 4} \div \dfrac{x^2 + 3x - 4}{x^2 - 4x + 4}$

8. $\dfrac{t}{t - 1} + \dfrac{3}{t + 4}$

9. $2 - \dfrac{1}{y + 1} + \dfrac{2y}{y - 3}$

10. $\dfrac{2a}{a^2 + 2a - 3} + \dfrac{3}{a^2 + a - 6}$

11. $2 - \dfrac{1}{3 - \dfrac{1}{t}}$

12. Solve $\dfrac{3 - x}{x} = 2$.

13. ______________

14. ______________

15. ______________

16. ______________

17. ______________

18. ______________

19. ______________

20. ______________

13. Solve $\frac{2t + 4}{t^2 - 1} = \frac{3}{t - 1}$.

14. Solve $\frac{3}{2a - 1} - \frac{4}{a + 2} = 0$.

15. Solve $\frac{3}{y} + 2 = \frac{-2}{y - 2}$.

16. Simplify $\frac{2 - \frac{1}{x + 1}}{4x - \frac{1}{x}}$.

17. Solve $\frac{1}{R} = \frac{1}{R_1} + \frac{1}{R_2}$ for R.

18. One plus the quotient of three divided by a number is the same as two less than the quotient of four divided by the number. Find the number.

19. Harry Drwyer can split a cord of wood in 4 hours. Peg Crafts can split a cord of wood in 6 hours. If they work together, how long will it take to split a cord of wood?

20. In 9 holes of golf, Kim Little has a total of 21 putts. If Kim plays 30 holes of golf, what is the total number of putts she should have?

CHAPTER 9

Radicals

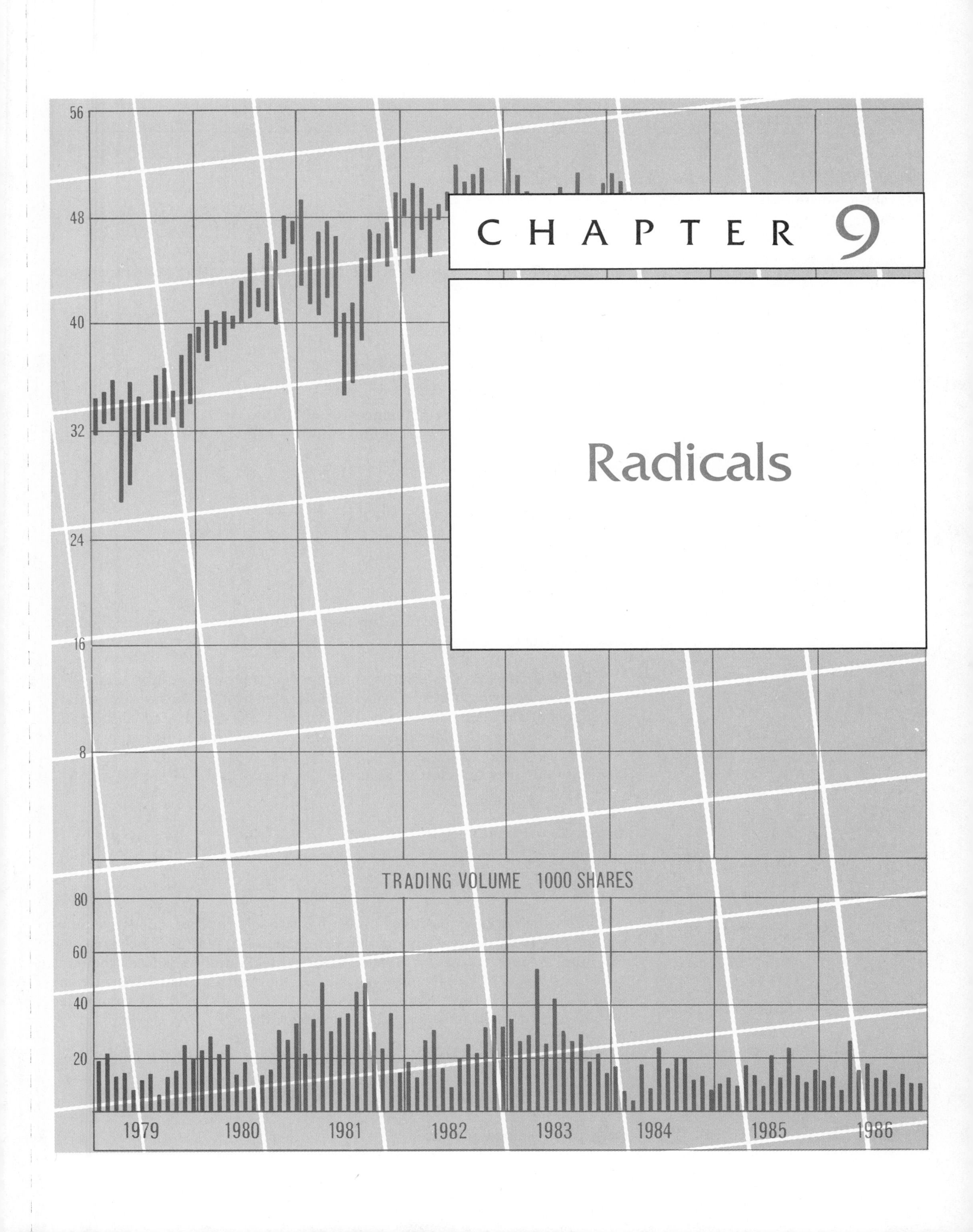

LEARNING OBJECTIVES

After finishing Section 9.1, you should be able to:

- Evaluate square roots.
- Identify irrational numbers.

1. Find both square roots of the following numbers.

(a) 64 (b) 1

(c) $\frac{1}{16}$ (d) 0.04

9.1

Square Roots and Irrational Numbers

The first thing we are going to look at in this section is the square root of a number.

DEFINITION 9.1 The number b is a square root of the number a if and only if $b^2 = a$.

Example 1

(a) 4 is a square root of 16 since $4^2 = 16$.
-4 is also a square root of 16 since $(-4)^2 = 16$.

(b) 10 is a square root of 100 since $10^2 = 100$.
-10 is also a square root of 100 since $(-10)^2 = 100$.

(c) $\frac{1}{2}$ is a square root of $\frac{1}{4}$ since $\left(\frac{1}{2}\right)^2 = \frac{1}{4}$.
$-\frac{1}{2}$ is a square root of $\frac{1}{4}$ since $\left(-\frac{1}{2}\right)^2 = \frac{1}{4}$.

DO EXERCISE 1.

In general, every positive number has two square roots—a positive root and a negative root.

DEFINITION 9.2 The positive square root of a positive number a is called the *principal square root of a* and is denoted by $\sqrt{a}$. The negative square root of a is denoted by $-\sqrt{a}$. The symbol $\sqrt{\ }$ is called a *radical sign* and the expression under the radical sign is called the *radicand*.

If we use our new notation in Example 1, we have:

(a) $\sqrt{16} = 4$ and $-\sqrt{16} = -4$

(b) $\sqrt{100} = 10$ and $-\sqrt{100} = -10$

(c) $\sqrt{\frac{1}{4}} = \frac{1}{2}$ and $-\sqrt{\frac{1}{4}} = -\frac{1}{2}$

Note that while every positive number has two square roots, a negative number does not have any real square root. For example, $\sqrt{-49}$ does not exist as a real number since we *cannot* find two identical real numbers whose product is -49 [note that $(7)(7) = 49$ and $(-7)(-7) = 49$]. Also note that 0 has only one square root, namely itself, since $\sqrt{0} = 0$ and $-\sqrt{0} = -0 = 0$. We summarize the above discussion by stating the following rule.

RULE 9.1 $\sqrt{a}$ exists as a real number if and only if a is nonnegative.

Example 2 Find the following square roots.

(a) $\sqrt{25} = 5$ since $5^2 = 25$.

(b) $-\sqrt{49} = -7$ since $(-7)^2 = 49$.

(c) $\sqrt{-4}$ does not exist as a real number since -4 is negative.

(d) $\sqrt{\frac{1}{9}} = \frac{1}{3}$ since $\left(\frac{1}{3}\right)^2 = \frac{1}{9}$.

DO EXERCISE 2.

2. Evaluate the following, if possible.

(a) $\sqrt{36}$ (b) $-\sqrt{81}$

(c) $\sqrt{-25}$ (d) $\sqrt{\frac{4}{9}}$

(e) $-\sqrt{\frac{25}{36}}$

Some whole numbers have principal square roots that are also whole numbers, such as $\sqrt{0} = 0$, $\sqrt{1} = 1$, $\sqrt{4} = 2$, and $\sqrt{9} = 3$.

DEFINITION 9.3 A whole number whose principal square root is also a whole number is called a *perfect square.*

The first ten perfect squares are 0, 1, 4, 9, 16, 25, 36, 49, 64, and 81.

This brings us to whole numbers that are not perfect squares and the square roots of such numbers. For example, $\sqrt{2}$ does exist but it is not a whole number. In fact, it can be shown that $\sqrt{2}$ is not even a rational number (that is, it does not have the form m/n where m and n are integers). The number $\sqrt{2}$ is an example of what we call an irrational number. We interrupt our study of square roots to take a brief look at this new set of numbers called the *irrational numbers.*

Consider the rational number m/n where m and n are integers and $n \neq 0$. If we change any rational number into its decimal representation, one of two things will happen.

1. We obtain a *terminating* (or finite) *decimal.* This means that when we divide the denominator into the numerator of our fraction, we eventually have a remainder of zero.

EXAMPLES: $\frac{1}{2} = 0.5$ $\frac{1}{4} = 0.25$ $\frac{3}{8} = 0.375$

$$\begin{array}{r} .5 \\ 2\overline{)1.0} \\ \underline{1\,0} \\ 0 \end{array} \qquad \begin{array}{r} .25 \\ 4\overline{)1.00} \\ \underline{8} \\ 20 \\ \underline{20} \\ 0 \end{array} \qquad \begin{array}{r} .375 \\ 8\overline{)3.000} \\ \underline{2\,4} \\ 60 \\ \underline{56} \\ 40 \\ \underline{40} \\ 0 \end{array}$$

2. We obtain a nonterminating decimal, which will have a series of numbers (called the *repetend*) that will repeat over and over again. This is often called a *repeating decimal.*

EXAMPLES: $\frac{1}{3} = 0.\overline{3}$ The bar over the 3 means that the 3 keeps repeating. The number 3 is the repetend.

$$\begin{array}{r} .33\ldots \\ 3\overline{)1.00} \\ \underline{9} \\ 10 \\ \underline{9} \\ 1 \end{array}$$

Express the following fractions as terminating decimals or, if not a terminating decimal, find the repetend.

3. $\frac{3}{4}$

4. $\frac{4}{9}$

5. $\frac{1}{8}$

6. $\frac{5}{6}$

$\frac{3}{11} = 0.\overline{27}$ The series of numbers 27 is the repetend.

```
     .2727 . . .
11)3.00
    2 2
     80
     77
      30
      22
       80
       77
        3
```

$\frac{1}{6} = 0.1\overline{6}$ The number 6 is the repetend.

$4\frac{1}{7} = 4.\overline{142857}$ The series of numbers 142857 is the repetend.

DO EXERCISES 3 THROUGH 6.

We can have a third kind of decimal that does not terminate and never sets up a repeating series of numbers.

EXAMPLES: (a) 0.1011011101110 . . .
(b) 5.12112111211112 . . .
(c) 3.102003000400005 . . .

DEFINITION 9.4 An *irrational number* is a number that, when it is expressed as a decimal, will neither terminate nor set up a repeating series of numbers.

The square root of any whole number that is not a perfect square is an irrational number. Thus, $\sqrt{2}$, $\sqrt{3}$, $\sqrt{10}$, and so on, are examples of irrational numbers. An example of a familiar irrational number is π.

If we plot all the rational numbers on a number line,

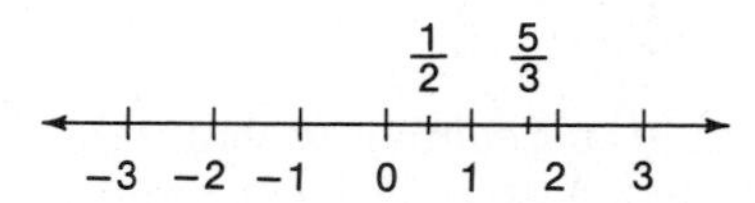

we do not fill up the number line. In other words, there are points on the number line that are not represented by rational numbers. However, if we also plot all the irrational numbers on the number line, then every point on the number line is represented by a rational or an irrational number.

DEFINITION 9.5 The set of *real numbers* (denoted by R) is the set we obtain by combining together all rational numbers and all irrational numbers. We could also characterize R as the set of all decimal numbers. There is a one-to-one correspondence between the set R and all the points on a number line.

We summarize the various sets of numbers we have studied thus far.

$\{1, 2, 3, \ldots\}$	Natural numbers
$\{0, 1, 2, \ldots\}$	Whole numbers
$\{\ldots, -2, -1, 0, 1, 2, \ldots\}$	Integers
$\left\{\frac{m}{n} \mid m \text{ and } n \text{ are integers, } n \neq 0\right\}$	Rational numbers
{Decimals that do not terminate or have a repetend}	Irrational numbers
{All rational and irrational numbers}	Real numbers

DO EXERCISES 7 THROUGH 14.

We now return to square roots. We know that $\sqrt{2}$ is an irrational number. However, we may approximate $\sqrt{2}$ by a rational number to any degree of accuracy we desire. For example,

$\sqrt{2} \cong 1.4$ (to one decimal place)
$\sqrt{2} \cong 1.41$ (to two decimal places)
$\sqrt{2} \cong 1.414$ (to three decimal places)
$\sqrt{2} \cong 1.414213$ (to six decimal places).

There are a number of methods for finding rational approximations to square roots.

1. Use a table (see Appendix D). This is what we will do in this book.
2. Use a slide rule.
3. Work it out in long hand. This involves only pencil and paper and can be done, but it is a bit tedious.
4. Use a hand calculator, many of which have a $\sqrt{\ }$ key.

The Table of Square Roots in Appendix D gives rational approximations of square roots for the numbers 1 through 100 to three-decimal-place accuracy.

Example 3 Using Appendix D, find rational approximations for the following.

(a) $\sqrt{15} \cong 3.873$ (b) $-\sqrt{68} \cong -8.246$

(c) $\sqrt{81} = 9$ (81 is a perfect square.)

(d) $\sqrt{-13}$ does not exist as a real number since -13 is negative.

DO EXERCISES 15 THROUGH 20.

DO SECTION PROBLEMS 9.1.

Identify the following as a rational number or an irrational number.

7. 1.43

8. $\sqrt{5}$

9. $\frac{7}{12}$

10. $\sqrt{49}$

11. 0.21211211121111 . . .

12. -5

13. $-\sqrt{17}$

14. $0.37\overline{07707}$

Use Appendix D to find the following.

15. $\sqrt{38}$

16. $-\sqrt{52}$

17. $\sqrt{64}$

18. $\sqrt{3}$

19. $\sqrt{-16}$

20. $\sqrt{98}$

ANSWERS TO MARGINAL EXERCISES

1. (a) 8, −8 (b) 1, −1 (c) $\frac{1}{4}$, $-\frac{1}{4}$ (d) 0.2, −0.2 **2.** (a) 6 (b) −9 (c) Does not exist. (d) $\frac{2}{3}$ (e) $-\frac{5}{6}$ **3.** 0.75 **4.** $0.\overline{4}$ **5.** 0.125 **6.** $0.8\overline{3}$ **7.** Rational **8.** Irrational **9.** Rational **10.** Rational **11.** Irrational **12.** Rational **13.** Irrational **14.** Rational **15.** 6.164 **16.** −7.211 **17.** 8 **18.** 1.732 **19.** Does not exist. **20.** 9.899

NAME ______ COURSE/SECTION ______ DATE ______

SECTION PROBLEMS 9.1

Find both square roots of the following numbers.

1. 25
2. 144
3. $\frac{1}{36}$
4. 0.09

Evaluate (if possible).

5. $-\sqrt{100}$
6. $\sqrt{-3}$
7. $\sqrt{\frac{9}{16}}$
8. $-\sqrt{64}$
9. $\sqrt{-1}$
10. $\sqrt{1.21}$
11. $-\sqrt{-25}$
12. $\sqrt{0.0004}$
13. $\sqrt{\frac{81}{64}}$
14. $\sqrt{0.16}$

Express as a terminating decimal or find the repetend.

15. $\frac{7}{8}$
16. $\frac{1}{12}$
17. $\frac{5}{11}$
18. $\frac{1}{4}$
19. $\frac{2}{3}$
20. $\frac{7}{5}$
21. $\frac{1}{9}$
22. $\frac{17}{20}$
23. $\frac{1}{16}$
24. $\frac{5}{4}$

Identify as a rational number or an irrational number.

25. $\sqrt{7}$
26. 3.42
27. $-\sqrt{12}$

ANSWERS

1. ______
2. ______
3. ______
4. ______
5. ______
6. ______
7. ______
8. ______
9. ______
10. ______
11. ______
12. ______
13. ______
14. ______
15. ______
16. ______
17. ______
18. ______
19. ______
20. ______
21. ______
22. ______
23. ______
24. ______
25. ______
26. ______
27. ______

28. ____________

29. ____________

30. ____________

31. ____________

32. ____________

33. ____________

34. ____________

35. ____________

36. ____________

37. ____________

38. ____________

39. ____________

40. ____________

41. ____________

42. ____________

43. ____________

44. ____________

45. ____________

46. ____________

47. ____________

48. ____________

49. ____________

50. ____________

28. 0.3030030003 . . .

29. $-\sqrt{36}$

30. $0.24\overline{4}$

31. 0

32. $\dfrac{5\frac{2}{3}}{1\frac{1}{2}}$

33. $\pi \cong 3.14159\ldots$

34. $-\sqrt{64}$

35. $\dfrac{7}{0.2}$

36. $-\sqrt{10}$

Use Appendix D to find the following.

37. $\sqrt{42}$

38. $-\sqrt{31}$

39. $\sqrt{49}$

40. $\sqrt{-3}$

41. $\sqrt{72}$

42. $\sqrt{12}$

43. $\sqrt{-25}$

44. $-\sqrt{7}$

45. $\sqrt{63}$

46. $-\sqrt{81}$

47. $-\sqrt{14}$

48. $-\sqrt{45}$

49. $\sqrt{13}$

50. $\sqrt{75}$

9.2

Simplifying Radicals

To help simplify radical expressions, there are some rules that are necessary to have. Note the following example.

Example 1 Evaluate the following.

(a) $(\sqrt{16})^2 = (4)^2 = 16$

(b) $(\sqrt{49})^2 = (7)^2 = 49$

(c) $(\sqrt{100})^2 = (10)^2 = 100$

In general, we have the following rule.

RULE 9.2

$$(\sqrt{x})^2 = x$$

where x is nonnegative.

DO EXERCISES 1 AND 2.

What happens to the above rule if the exponent of 2 is under the radical sign, that is, $\sqrt{x^2} = ?$ Note the following example.

Example 2 Evaluate the following.

(a) $\sqrt{3^2} = \sqrt{9} = 3$

(b) $\sqrt{(-4)^2} = \sqrt{16} = 4$

(c) $\sqrt{(-8)^2} = \sqrt{64} = 8$

In general, we have the following rule.

RULE 9.3

$$\sqrt{x^2} = |x|$$

where x is a real number.

The absolute value is definitely needed in the Rule 9.3 to handle the cases where x is negative. Note that $\sqrt{(-3)^2} \neq -3$ but $\sqrt{(-3)^2} = |-3| = 3$.

Example 3 Simplify the following.

(a) $\sqrt{x^2y^2} = \sqrt{(xy)^2} = |xy|$

(b) $\sqrt{(t-4)^2} = |t-4|$

(c) $\sqrt{x^2 - 4x + 4} = \sqrt{(x-2)^2} = |x-2|$

DO EXERCISES 3 THROUGH 5.

LEARNING OBJECTIVES

After finishing Section 9.2, you should be able to:

- Simplify radical expressions involving square roots.

Evaluate the following if they exist.

1. $(\sqrt{a})^2$

2. $(\sqrt{-3})^2$

Evaluate the following if they exist.

3. $\sqrt{(-4)^2}$

4. $\sqrt{7^2}$

5. $\sqrt{x^2 + 2x + 1}$

6. Is $\sqrt{4 \cdot 25}$ the same as $\sqrt{4} \cdot \sqrt{25}$?

7. Does $\sqrt{(-9)(-36)}$ equal $\sqrt{-9} \cdot \sqrt{-36}$?

8. Does $|x \cdot y|$ equal $|x| \cdot |y|$?

Is there any relationship between $\sqrt{x \cdot y}$ and $\sqrt{x} \cdot \sqrt{y}$? Note the following example.

Example 4 Evaluate the following and note any comparisons.

(a) $\sqrt{4 \cdot 9} = \sqrt{36} = 6$ and $\sqrt{4} \cdot \sqrt{9} = 2 \cdot 3 = 6$

(b) $\sqrt{16 \cdot \frac{1}{4}} = \sqrt{4} = 2$ and $\sqrt{16} \cdot \sqrt{\frac{1}{4}} = 4 \cdot \frac{1}{2} = 2$

(c) $\sqrt{(-4)(-16)} = \sqrt{64} = 8$ and $\sqrt{-4} \cdot \sqrt{-16}$ is not defined since $\sqrt{-4}$ and $\sqrt{-16}$ do not exist as real numbers.

The following rule tells us that there is a relationship between $\sqrt{x \cdot y}$ and $\sqrt{x} \cdot \sqrt{y}$, but only if x and y are nonnegative.

RULE 9.4

$$\sqrt{x \cdot y} = \sqrt{x} \cdot \sqrt{y}$$

if x and y are nonnegative.

We might also note that

$$\sqrt{16 + 9} = \sqrt{25} = 5$$

and this is not the same as

$$\sqrt{16} + \sqrt{9} = 4 + 3 = 7$$

Thus, in general, we have

$$\sqrt{x + y} \neq \sqrt{x} + \sqrt{y}$$

DO EXERCISES 6 THROUGH 8.

Now we want to look at simplifying radicals. This process is somewhat similar to factoring a polynomial. In simplifying a radical, if we have a perfect square under the radical sign, then we may "factor" it out. Note the following.

Example 5 Simplify $\sqrt{12}$.

Note that $12 = 4 \cdot 3$ and 4 is a perfect square. Thus, we have

$$\sqrt{12} = \sqrt{4 \cdot 3} = \sqrt{2^2 \cdot 3}$$
$$= \sqrt{2^2} \cdot \sqrt{3} = 2\sqrt{3}$$

In Example 5 we used the following two rules.

1. $\sqrt{x \cdot y} = \sqrt{x} \cdot \sqrt{y}$ if x and y are nonnegative.
2. $\sqrt{x^2} = |x|$

Example 6 Simplify $\sqrt{180}$.

We see that $180 = 9 \cdot 20 = 3^2 \cdot 4 \cdot 5 = 3^2 \cdot 2^2 \cdot 5$. Thus,

$$\begin{aligned}\sqrt{180} &= \sqrt{3^2 \cdot 2^2 \cdot 5} = \sqrt{3^2} \cdot \sqrt{2^2} \cdot \sqrt{5} \\ &= 3 \cdot 2\sqrt{5} = 6\sqrt{5}\end{aligned}$$

Alternatively, we see that $180 = 36 \cdot 5 = 6^2 \cdot 5$ and

$$\begin{aligned}\sqrt{180} &= \sqrt{6^2 \cdot 5} \\ &= \sqrt{6^2} \cdot \sqrt{5} \\ &= 6\sqrt{5}\end{aligned}$$

DO EXERCISES 9 THROUGH 11.

Example 7 Simplify $\sqrt{8x^3}$.

Note that $8x^3 = 4 \cdot 2 \cdot x^2 \cdot x = 2^2 \cdot 2 \cdot x^2 \cdot x$. Thus,

$$\begin{aligned}\sqrt{8x^3} &= \sqrt{2^2 \cdot 2 \cdot x^2 \cdot x} \\ &= \sqrt{2^2} \cdot \sqrt{2} \cdot \sqrt{x^2} \cdot \sqrt{x} \\ &= 2|x|\sqrt{2x}\end{aligned}$$

In Example 7, the absolute value signs around the x are redundant since x has to be nonnegative for the original radical, $\sqrt{8x^3}$, to exist. If x were negative, then x^3 would also be negative and $\sqrt{8x^3}$ would not be defined. However, it does not hurt to leave the absolute value signs in, since in many problems they are necessary.

Example 8 Simplify $\sqrt{45a^3b^2}$.

Note that $45a^3b^2 = 9 \cdot 5 \cdot a^2 \cdot a \cdot b^2 = 3^2 \cdot 5 \cdot a^2 \cdot a \cdot b^2$. Thus,

$$\begin{aligned}\sqrt{45a^3b^2} &= \sqrt{3^2 \cdot a^2 \cdot b^2 \cdot 5a} \\ &= \sqrt{3^2} \cdot \sqrt{a^2} \cdot \sqrt{b^2} \cdot \sqrt{5a} \\ &= 3|a| \cdot |b|\sqrt{5a} \qquad (\text{or } 3a|b|\sqrt{5a})\end{aligned}$$

DO EXERCISES 12 THROUGH 14.

We have simplified expressions such as $\sqrt{x^2}$ and $\sqrt{x^3}$, but what happens if we have exponents that are larger than 3?

Example 9 Simplify $\sqrt{x^6}$.

We can do this two ways.

(a) $\sqrt{x^6} = \sqrt{(x^3)^2} = |x^3|$
Note the use of the exponential law—$(x^m)^n = x^{m \cdot n}$.

(b) $\sqrt{x^6} = \sqrt{x^2 \cdot x^2 \cdot x^2} = \sqrt{x^2} \cdot \sqrt{x^2} \cdot \sqrt{x^2} = |x|^3 = |x^3|$
Here, we used $x^m \cdot x^n = x^{m+n}$.

Simplify.

9. $\sqrt{50}$

10. $\sqrt{48}$

11. $\sqrt{200}$

Simplify.

12. $\sqrt{t^3}$

13. $\sqrt{18(x-3)^3}$

14. $\sqrt{16x^2y^2z^3}$

Simplify.

15. $\sqrt{3^5}$

16. $\sqrt{x^2y^4}$

17. $\sqrt{96t^7}$

18. $\sqrt{10x^3y^6}$

Example 10 Simplify $\sqrt{2^8x^5}$.

Note that $2^8x^5 = (2^4)^2x^4 \cdot x = (2^4)^2(x^2)^2x$. Thus,

$$\begin{aligned}\sqrt{2^8x^5} &= \sqrt{(2^4)^2(x^2)^2x} = \sqrt{(2^4)^2} \cdot \sqrt{(x^2)^2} \cdot \sqrt{x} \\ &= 2^4 \cdot |x^2|\sqrt{x} \\ &= 16x^2\sqrt{x}\end{aligned}$$

Since x^2 is always nonnegative, we have $|x^2| = x^2$.

DO EXERCISES 15 THROUGH 18.

DO SECTION PROBLEMS 9.2.

ANSWERS TO MARGINAL EXERCISES

1. a **2.** Does not exist. **3.** 4 **4.** 7 **5.** $|x + 1|$ **6.** Yes **7.** No **8.** Yes **9.** $5\sqrt{2}$ **10.** $4\sqrt{3}$ **11.** $10\sqrt{2}$ **12.** $t\sqrt{t}$ **13.** $3(x - 3)\sqrt{2(x - 3)}$ **14.** $4|xy|z\sqrt{z}$ **15.** $9\sqrt{3}$ **16.** $|x|y^2$ **17.** $4t^3\sqrt{6t}$ **18.** $x|y^3|\sqrt{10x}$

NAME COURSE/SECTION DATE

ANSWERS

SECTION PROBLEMS 9.2

Evaluate the following if they exist.

1. $(\sqrt{5})^2$
2. $(\sqrt{-5})^2$
3. $\sqrt{4^2}$
4. $\sqrt{(-4)^2}$
5. $\sqrt{a^2}$
6. $(\sqrt{2})^2$
7. $\sqrt{4x^2 + 4x + 1}$
8. $(\sqrt{x - y})^2$
9. Does $\sqrt{16 \cdot 9}$ equal $\sqrt{16} \cdot \sqrt{9}$?
10. Does $\sqrt{(-25)(-100)}$ equal $\sqrt{-25} \cdot \sqrt{-100}$?

Simplify.

11. $\sqrt{75}$
12. $-\sqrt{45}$
13. $\sqrt{49}$
14. $\sqrt{32}$
15. $\sqrt{28}$
16. $-\sqrt{x^2 + 6x + 9}$
17. $\sqrt{48}$
18. $\sqrt{8x^2}$
19. $\sqrt{16x^3y}$

1. ____
2. ____
3. ____
4. ____
5. ____
6. ____
7. ____
8. ____
9. ____
10. ____
11. ____
12. ____
13. ____
14. ____
15. ____
16. ____
17. ____
18. ____
19. ____

20. ______
21. ______
22. ______
23. ______
24. ______
25. ______
26. ______
27. ______
28. ______
29. ______
30. ______
31. ______
32. ______
33. ______
34. ______
35. ______
36. ______
37. ______
38. ______
39. ______
40. ______

20. $\sqrt{150}$

21. $-\sqrt{180}$

22. $\sqrt{99}$

23. $\sqrt{x^3y}$

24. $\sqrt{4x^2 - 8x + 4}$

25. $\sqrt{a^2b}$

26. $\sqrt{50x^2yz^3}$

27. $\sqrt{2^6}$

28. $\sqrt{128t^5}$

29. $-\sqrt{3^5}$

30. $\sqrt{7^3 \cdot 3^8}$

31. $\sqrt{7x^3y^7}$

32. $\sqrt{256y^4}$

33. $-\sqrt{7x^3y^7z^6}$

34. $\sqrt{6^2}$

35. $(\sqrt{6})^2$

36. $\sqrt{72}$

37. $\sqrt{a^3b^2}$

38. $\sqrt{3x^2 + 6x + 3}$

39. $\sqrt{x^7y^4}$

40. $\sqrt{6300}$

9.3

Operations with Radicals

We can perform various operations involving radicals. First, we will multiply two radicals together. In this process, we use the rules $\sqrt{x \cdot y} = \sqrt{x} \cdot \sqrt{y}$, where x and y are nonnegative, $(\sqrt{x})^2 = x$, and the distributive law.

Example 1 Multiply.

(a) $\sqrt{2} \cdot \sqrt{3} = \sqrt{2 \cdot 3} = \sqrt{6}$

(b) $\sqrt{5}(3 + \sqrt{2}) = (\sqrt{5})(3) + \sqrt{5} \cdot \sqrt{2}$
$= 3\sqrt{5} + \sqrt{10}$

DO EXERCISES 1 AND 2.

Example 2 Multiply.

(a) $\sqrt{3}(\sqrt{7} + \sqrt{3}) = \sqrt{3} \cdot \sqrt{7} + \sqrt{3} \cdot \sqrt{3}$
$= \sqrt{3 \cdot 7} + (\sqrt{3})^2$
$= \sqrt{21} + 3$

(b) $(\sqrt{3} - \sqrt{2})(\sqrt{3} + \sqrt{2}) = \sqrt{3} \cdot \sqrt{3} + \sqrt{3} \cdot \sqrt{2} - \sqrt{3} \cdot \sqrt{2} - \sqrt{2} \cdot \sqrt{2}$
$= (\sqrt{3})^2 + \sqrt{6} - \sqrt{6} - (\sqrt{2})^2$
$= 3 - 2$
$= 1$

Note that $\sqrt{6} - \sqrt{6}$ gave us zero. We will discuss adding and subtracting radicals later in this section.

DO EXERCISES 3 AND 4.

We can often simplify a radical after we have multiplied.

Example 3 Multiply and simplify.

(a) $\sqrt{2} \cdot \sqrt{10} = \sqrt{2 \cdot 10} = \sqrt{2 \cdot 2 \cdot 5}$
$= \sqrt{2^2 \cdot 5} = \sqrt{2^2} \cdot \sqrt{5} = 2\sqrt{5}$

(b) $\sqrt{6} \cdot \sqrt{30} = \sqrt{6 \cdot 30} = \sqrt{6 \cdot 6 \cdot 5}$
$= \sqrt{6^2 \cdot 5} = \sqrt{6^2} \cdot \sqrt{5} = 6\sqrt{5}$

DO EXERCISES 5 AND 6.

Example 4 Multiply and simplify.

(a) $\sqrt{3x} \cdot \sqrt{5x^2y^2} = \sqrt{3 \cdot 5 \cdot x \cdot x^2 \cdot y^2}$
$= \sqrt{x^2} \cdot \sqrt{y^2} \cdot \sqrt{15x}$
$= |x||y|\sqrt{15x}$ (or $x|y|\sqrt{15x}$)

LEARNING OBJECTIVES

After finishing Section 9.3, you should be able to:

- Add, subtract, and multiply certain radical expressions.

Multiply.

1. $\sqrt{7} \cdot \sqrt{5}$

2. $\sqrt{6}(\sqrt{5} - \sqrt{6})$

Multiply.

3. $\sqrt{x}(2 + 3\sqrt{y})$

4. $(\sqrt{x} + \sqrt{y})(\sqrt{x} - \sqrt{y})$

Multiply and simplify.

5. $\sqrt{2} \cdot \sqrt{14}$

6. $\sqrt{10} \cdot \sqrt{30}$

Multiply and simplify.

7. $\sqrt{ab} \cdot \sqrt{9a^2b}$

8. $\sqrt{3(x-1)} \cdot \sqrt{6x-6}$

Combine.

9. $4\sqrt{5} - 2\sqrt{5}$

10. $7\sqrt{8} + \sqrt{18} + 4\sqrt{20}$

11. $\sqrt{2} \cdot \sqrt{10} + \sqrt{45} - 3\sqrt{3} \cdot \sqrt{15}$

Combine.

12. $7\sqrt{3x} - 4\sqrt{3x} + 2\sqrt{3x}$

13. $\sqrt{12x^3y} - \sqrt{27xy^3}$

(b)
$$\begin{aligned}\sqrt{ab}(3\sqrt{a} - 2\sqrt{b}) &= 3\sqrt{ab} \cdot \sqrt{a} - 2\sqrt{ab} \cdot \sqrt{b} \\ &= 3\sqrt{a^2b} - 2\sqrt{ab^2} \\ &= 3\sqrt{a^2}\sqrt{b} - 2\sqrt{a}\sqrt{b^2} \\ &= 3|a|\sqrt{b} - 2|b|\sqrt{a} \qquad (\text{or } 3a\sqrt{b} - 2b\sqrt{a})\end{aligned}$$

DO EXERCISES 7 AND 8.

We can add or subtract radicals that have the same radicand by using the distributive law, but radicals with different radicands cannot be combined by addition or subtraction.

Example 5 Combine and/or simplify.

(a)
$$\begin{aligned}\sqrt{8} + 5\sqrt{18} &= \sqrt{4 \cdot 2} + 5\sqrt{9 \cdot 2} \\ &= \sqrt{4} \cdot \sqrt{2} + 5\sqrt{9} \cdot \sqrt{2} \\ &= 2\sqrt{2} + 5 \cdot 3\sqrt{2} \\ &= 2\sqrt{2} + 15\sqrt{2} \\ &= 17\sqrt{2}\end{aligned}$$

(b)
$$\begin{aligned}\sqrt{75} - 2\sqrt{18} + \sqrt{27} &= \sqrt{25 \cdot 3} - 2\sqrt{9 \cdot 2} + \sqrt{9 \cdot 3} \\ &= \sqrt{25} \cdot \sqrt{3} - 2\sqrt{9} \cdot \sqrt{2} + \sqrt{9} \cdot \sqrt{3} \\ &= 5\sqrt{3} - 6\sqrt{2} + 3\sqrt{3} \\ &= 8\sqrt{3} - 6\sqrt{2}\end{aligned}$$

DO EXERCISES 9 THROUGH 11.

Example 6 Combine and/or simplify.

(a)
$$\begin{aligned}\sqrt{x^3} - \sqrt{9x} &= \sqrt{x^2 \cdot x} - \sqrt{9} \cdot \sqrt{x} \\ &= \sqrt{x^2} \cdot \sqrt{x} - 3\sqrt{x} \\ &= x\sqrt{x} - 3\sqrt{x} \\ &= (x-3)\sqrt{x}\end{aligned}$$

(b)
$$\begin{aligned}\sqrt{45x^3y^2} - \sqrt{20xy^4} &= \sqrt{9 \cdot 5 \cdot x^2 \cdot x \cdot y^2} - \sqrt{4 \cdot 5 \cdot x \cdot (y^2)^2} \\ &= \sqrt{9} \cdot \sqrt{x^2} \cdot \sqrt{y^2} \cdot \sqrt{5x} - \sqrt{4} \cdot \sqrt{(y^2)^2} \cdot \sqrt{5x} \\ &= 3 \cdot |x||y|\sqrt{5x} - 2y^2\sqrt{5x} \\ &= (3|x||y| - 2y^2)\sqrt{5x} \quad [\text{or } (3x|y| - 2y^2)\sqrt{5x}]\end{aligned}$$

DO EXERCISES 12 AND 13.

DO SECTION PROBLEMS 9.3.

ANSWERS TO MARGINAL EXERCISES

1. $\sqrt{35}$ **2.** $\sqrt{30} - 6$ **3.** $2\sqrt{x} + 3\sqrt{xy}$ **4.** $x - y$ **5.** $2\sqrt{7}$
6. $10\sqrt{3}$ **7.** $3ab\sqrt{a}$ **8.** $3(x-1)\sqrt{2}$ **9.** $2\sqrt{5}$
10. $17\sqrt{2} + 8\sqrt{5}$ **11.** $-4\sqrt{5}$ **12.** $5\sqrt{3x}$ **13.** $(2x - 3y)\sqrt{3xy}$

NAME ____ COURSE/SECTION ____ DATE ____

SECTION PROBLEMS 9.3

Perform the indicated operation and simplify.

1. $\sqrt{3} \cdot \sqrt{7}$

2. $\sqrt{5} \cdot \sqrt{11}$

3. $\sqrt{3}(3 + \sqrt{2})$

4. $\sqrt{2}(\sqrt{5} - \sqrt{3})$

5. $\sqrt{5}(\sqrt{5} + \sqrt{7})$

6. $(\sqrt{2} + \sqrt{3})(\sqrt{2} - \sqrt{3})$

7. $(\sqrt{5} - \sqrt{2})(\sqrt{5} + 3)$

8. $(\sqrt{2} + \sqrt{3})^2$

9. $(\sqrt{5} - \sqrt{3})^2$

10. $(2\sqrt{2} - 3\sqrt{3})(2\sqrt{2} + 3\sqrt{3})$

11. $\sqrt{3} \cdot \sqrt{12}$

12. $\sqrt{15} \cdot \sqrt{6}$

13. $-\sqrt{2} \cdot \sqrt{10}$

14. $\sqrt{15} \cdot \sqrt{3} \cdot \sqrt{5}$

15. $\sqrt{x} \cdot \sqrt{8x}$

16. $\sqrt{2x - 2} \cdot \sqrt{9x - 9}$

17. $\sqrt{3xy} \cdot \sqrt{2x^3} \cdot \sqrt{6xy^3}$

18. $\sqrt{2a} \cdot \sqrt{10a^3} \cdot \sqrt{5ab^2}$

ANSWERS

1. ____
2. ____
3. ____
4. ____
5. ____
6. ____
7. ____
8. ____
9. ____
10. ____
11. ____
12. ____
13. ____
14. ____
15. ____
16. ____
17. ____
18. ____

19. ______

20. ______

21. ______

22. ______

23. ______

24. ______

25. ______

26. ______

27. ______

28. ______

29. ______

30. ______

31. ______

32. ______

33. ______

34. ______

35. ______

36. ______

19. $6\sqrt{3} - \sqrt{3}$

20. $2\sqrt{5} - 4\sqrt{5} + \sqrt{5}$

21. $4\sqrt{x} - 5\sqrt{x}$

22. $17\sqrt{ab} - 30\sqrt{ab} + \sqrt{ab}$

23. $2\sqrt{18} - \sqrt{8}$

24. $\sqrt{27} - 2\sqrt{48}$

25. $2\sqrt{20} - \sqrt{45} + 6\sqrt{5}$

26. $\sqrt{x^3} - \sqrt{x}$

27. $\sqrt{12x^3} + \sqrt{18x}$

28. $\sqrt{3} \cdot \sqrt{15} - \sqrt{5} + 2\sqrt{50}$

29. $\sqrt{2x} \cdot \sqrt{10x^2} - \sqrt{125x}$

30. $\sqrt{1} - \sqrt{0}$

31. $\sqrt{2x} \cdot \sqrt{8xy^2}$

32. $\sqrt{3}(2 - \sqrt{3})$

33. $(\sqrt{16} - \sqrt{7})(\sqrt{2} + 3)$

34. $3\sqrt{2x} + 4\sqrt{2x}$

35. $\sqrt{12} - 2\sqrt{27}$

36. $\sqrt{2x^3} - 6x\sqrt{8x}$

9.4

Rationalizing the Denominator

If an expression has a radical in the denominator, then we would like to be able to remove the radical from the denominator. This process is called *rationalizing the denominator.* In this section we will assume that all variables are nonnegative.

Example 1 Rationalize the denominator of $\frac{3}{\sqrt{2}}$.

We want to build the 2 in $\sqrt{2}$ up to a perfect square.

$$\frac{3}{\sqrt{2}} = \frac{3}{\sqrt{2}} \cdot 1 = \frac{3}{\sqrt{2}} \cdot \frac{\sqrt{2}}{\sqrt{2}} = \frac{3\sqrt{2}}{\sqrt{4}} = \frac{3\sqrt{2}}{2}$$

One reason we want to rationalize a denominator is that if we wanted to use $\sqrt{2} \cong 1.414$ and then calculate $\frac{3}{\sqrt{2}}$, we would have the somewhat messy division problem $\frac{3}{1.414}$. If we rationalize $\frac{3}{\sqrt{2}}$, however, and obtain $\frac{3\sqrt{2}}{2}$, then after substituting 1.414 for $\sqrt{2}$ we would obtain the somewhat easier calculation problem $\frac{3(1.414)}{2}$.

DO EXERCISE 1.

Example 2 Rationalize the denominators.

(a) $\frac{2}{\sqrt{6}} = \frac{2}{\sqrt{6}} \cdot \frac{\sqrt{6}}{\sqrt{6}} = \frac{2\sqrt{6}}{\sqrt{36}} = \frac{2\sqrt{6}}{6} = \frac{\sqrt{6}}{3}$

(b) $\frac{2}{\sqrt{12}} = \frac{2}{\sqrt{12}} \cdot \frac{\sqrt{3}}{\sqrt{3}} = \frac{2\sqrt{3}}{\sqrt{36}} = \frac{2\sqrt{3}}{6} = \frac{\sqrt{3}}{3}$

In part (b) of Example 3 we could have multiplied by $\sqrt{12}/\sqrt{12}$. However, it is a little easier to build 12 up to the perfect square 36 instead of to the perfect square 144.

DO EXERCISES 2 AND 3.

Example 3 Rationalize the denominators. Recall that all variables are nonnegative.

(a) $\frac{3}{\sqrt{x}} = \frac{3}{\sqrt{x}} \cdot \frac{\sqrt{x}}{\sqrt{x}} = \frac{3\sqrt{x}}{\sqrt{x^2}} = \frac{3\sqrt{x}}{x}$

(b) $\frac{1}{\sqrt{3t}} = \frac{1}{\sqrt{3t}} \cdot \frac{\sqrt{3t}}{\sqrt{3t}} = \frac{\sqrt{3t}}{\sqrt{(3t)^2}} = \frac{\sqrt{3t}}{3t}$

DO EXERCISES 4 AND 5.

LEARNING OBJECTIVES

After finishing Section 9.4, you should be able to:

- Rationalize the denominators in certain radical expressions.

Rationalize the denominator.

1. $\frac{2}{\sqrt{3}}$

Rationalize the denominators.

2. $\frac{2}{\sqrt{6}}$

3. $\frac{3}{\sqrt{18}}$

Rationalize the denominators. Assume all variables are nonnegative.

4. $\frac{\sqrt{x}}{\sqrt{y}}$

5. $\frac{1}{\sqrt{xy}}$

Simplify. Assume all variables are nonnegative.

6. $\sqrt{\dfrac{50x^2}{144}}$

7. $\dfrac{\sqrt{27x^3}}{\sqrt{3x}}$

Another type of rationalizing problem arises when we have a rational expression for a radicand, for example,

$$\sqrt{\frac{1}{2}} \quad \text{or} \quad \sqrt{\frac{4}{3x}}$$

Before we look at this type of problem, we give a quotient rule that is similar to the product rule $\sqrt{xy} = \sqrt{x} \cdot \sqrt{y}$ if x and y are nonnegative.

RULE 9.5

$$\sqrt{\frac{x}{y}} = \frac{\sqrt{x}}{\sqrt{y}}$$

if x and y are nonnegative and $y \neq 0$.

This rule is often helpful in simplifying certain radical expressions.

Example 4 Simplify. Recall that all variables are nonnegative.

(a) $\sqrt{\dfrac{3x^2}{16y^2}} = \dfrac{\sqrt{3x^2}}{\sqrt{16y^2}} = \dfrac{\sqrt{3} \cdot \sqrt{x^2}}{\sqrt{16} \cdot \sqrt{y^2}} = \dfrac{\sqrt{3}x}{4y}$

Absolute value signs are not needed around the x and the y since x and y are nonnegative.

(b) $\sqrt{\dfrac{175}{36}} = \dfrac{\sqrt{175}}{\sqrt{36}} = \dfrac{\sqrt{25 \cdot 7}}{6}$

$= \dfrac{\sqrt{25} \cdot \sqrt{7}}{6} = \dfrac{5\sqrt{7}}{6}$

(c) $\dfrac{\sqrt{8x^3y^3}}{\sqrt{2xy^2}} = \sqrt{\dfrac{8x^3y^2}{2xy^2}} = \sqrt{4x^2}$

$= \sqrt{4} \cdot \sqrt{x^2} = 2x$

DO EXERCISES 6 AND 7.

The following example shows two ways of simplifying radicals that have rational expressions for the radicand.

Example 5 Simplify.

(a) $\sqrt{\dfrac{1}{2}} = \dfrac{\sqrt{1}}{\sqrt{2}} = \dfrac{1}{\sqrt{2}} = \dfrac{1}{\sqrt{2}} \cdot \dfrac{\sqrt{2}}{\sqrt{2}} = \dfrac{\sqrt{2}}{2}$

or $\sqrt{\dfrac{1}{2}} = \sqrt{\dfrac{1}{2} \cdot \dfrac{2}{2}} = \sqrt{\dfrac{2}{4}} = \dfrac{\sqrt{2}}{\sqrt{4}} = \dfrac{\sqrt{2}}{2}$

(b) $\sqrt{\dfrac{4}{3x}} = \dfrac{\sqrt{4}}{\sqrt{3x}} = \dfrac{2}{\sqrt{3x}} = \dfrac{2}{\sqrt{3x}} \cdot \dfrac{\sqrt{3x}}{\sqrt{3x}} = \dfrac{2\sqrt{3x}}{3x}$

or $\sqrt{\dfrac{4}{3x}} = \sqrt{\dfrac{4}{3x} \cdot \dfrac{3x}{3x}} = \sqrt{\dfrac{4 \cdot 3x}{9x^2}} = \dfrac{\sqrt{4 \cdot 3x}}{\sqrt{9x^2}} = \dfrac{2\sqrt{3x}}{3x}$

(c) $\sqrt{\dfrac{x}{50}} = \dfrac{\sqrt{x}}{\sqrt{50}} = \dfrac{\sqrt{x}}{\sqrt{50}} \cdot \dfrac{\sqrt{2}}{\sqrt{2}} = \dfrac{\sqrt{2x}}{\sqrt{100}} = \dfrac{\sqrt{2x}}{10}$

or $\sqrt{\dfrac{x}{50}} = \sqrt{\dfrac{x}{50} \cdot \dfrac{2}{2}} = \sqrt{\dfrac{2x}{100}} = \dfrac{\sqrt{2x}}{\sqrt{100}} = \dfrac{\sqrt{2x}}{10}$

In (c) we could also multiply by $\dfrac{\sqrt{50}}{\sqrt{50}}$ instead of $\dfrac{\sqrt{2}}{\sqrt{2}}$.

DO EXERCISES 8 THROUGH 10.

DO SECTION PROBLEMS 9.4.

Simplify. Rationalize the denominators if necessary. Assume all variables are nonnegative.

8. $\sqrt{\dfrac{3}{5}}$

9. $\sqrt{\dfrac{y}{18x}}$

10. $\dfrac{\sqrt{16xy^2}}{\sqrt{3x^3}}$

ANSWERS TO MARGINAL EXERCISES

1. $\frac{2\sqrt{3}}{3}$ **2.** $\frac{\sqrt{6}}{3}$ **3.** $\frac{\sqrt{2}}{2}$ **4.** $\frac{\sqrt{xy}}{y}$ **5.** $\frac{\sqrt{xy}}{xy}$ **6.** $\frac{5x\sqrt{2}}{12}$
7. $3x$ **8.** $\frac{\sqrt{15}}{5}$ **9.** $\frac{\sqrt{2xy}}{6x}$ **10.** $\frac{4y\sqrt{3}}{3x}$

NAME | COURSE/SECTION | DATE | ANSWERS

SECTION PROBLEMS 9.4

Assume all variables are nonnegative. Rationalize the denominators.

1. $\frac{1}{\sqrt{5}}$ **2.** $\frac{2}{\sqrt{10}}$ **3.** $\frac{3}{\sqrt{12}}$ **4.** $\frac{3}{\sqrt{x}}$ **5.** $\frac{1}{\sqrt{a^3}}$

6. $\frac{1}{\sqrt{72}}$ **7.** $\frac{\sqrt{x}}{\sqrt{2x}}$ **8.** $\frac{1}{\sqrt{15}}$ **9.** $\frac{\sqrt{3}}{\sqrt{7}}$ **10.** $\frac{\sqrt{x}}{\sqrt{xy}}$

Simplify.

11. $\sqrt{\frac{x^2}{9y^2}}$ **12.** $\frac{\sqrt{75x^3}}{\sqrt{3xy^2}}$ **13.** $\frac{\sqrt{8x^3}}{\sqrt{2x}}$

14. $\sqrt{\frac{12y^3}{25z^2}}$ **15.** $\frac{\sqrt{50a^5b^2}}{\sqrt{2a^3}}$ **16.** $\sqrt{\frac{20x^5}{16x}}$

Simplify. Rationalize the denominator if necessary.

17. $\sqrt{\frac{1}{x}}$ **18.** $\sqrt{\frac{2}{3}}$ **19.** $\sqrt{\frac{3}{5}}$ **20.** $\sqrt{\frac{y}{z}}$

ANSWERS

1. ______
2. ______
3. ______
4. ______
5. ______
6. ______
7. ______
8. ______
9. ______
10. ______
11. ______
12. ______
13. ______
14. ______
15. ______
16. ______
17. ______
18. ______
19. ______
20. ______

21. ______

22. ______

23. ______

24. ______

25. ______

26. ______

27. ______

28. ______

29. ______

30. ______

31. ______

32. ______

33. ______

34. ______

35. ______

36. ______

37. ______

38. ______

39. ______

40. ______

21. $\sqrt{\frac{50}{63}}$

22. $\sqrt{\frac{68}{225}}$

23. $\sqrt{\frac{3x^2}{48}}$

24. $\frac{\sqrt{8x^4}}{\sqrt{5x^3}}$

25. $\sqrt{\frac{1}{15x}}$

26. $\sqrt{\frac{x}{6y}}$

27. $\frac{\sqrt{7x^3}}{\sqrt{14x}}$

28. $\sqrt{\frac{5}{12}}$

29. $\frac{\sqrt{2x^2}}{\sqrt{8x}}$

30. $\sqrt{\frac{x}{0.2}}$

31. $\frac{1}{\sqrt{10}}$

32. $\frac{1}{\sqrt{2x}}$

33. $\frac{1}{\sqrt{5}}$

34. $\frac{2}{\sqrt{8}}$

35. $\frac{y}{\sqrt{x}}$

36. $\frac{1}{\sqrt{x^3}}$

37. $\sqrt{\frac{3}{8}}$

38. $\sqrt{\frac{8x^2}{144}}$

39. $\sqrt{\frac{a^2}{b}}$

40. $\frac{\sqrt{150x^3y}}{\sqrt{50x^4}}$

NAME COURSE/SECTION DATE

ANSWERS

CHAPTER 9 REVIEW PROBLEMS

SECTION 9.1

Evaluate if possible.

1. $\sqrt{81}$

2. $-\sqrt{25}$

3. $\sqrt{-9}$

4. $-\sqrt{\frac{4}{9}}$

Express as a terminating decimal or find the repetend.

5. $\frac{1}{8}$

6. $\frac{3}{11}$

7. $\frac{5}{9}$

8. $\frac{7}{6}$

Identify as a rational or an irrational number.

9. $3.02\overline{02}$

10. $\sqrt{15}$

11. $1.020020002\ldots$

Use Appendix D to find the square roots.

12. $\sqrt{33}$

13. $\sqrt{-2}$

14. $\sqrt{83}$

15. $-\sqrt{16}$

16. $\sqrt{19}$

SECTION 9.2

Simplify.

17. $\sqrt{(-6)^2}$

18. $(\sqrt{-6})^2$

19. $\sqrt{72}$

20. $\sqrt{18x^4y^5}$

21. $\sqrt{4x^2 + 4x + 1}$

22. $\sqrt{\frac{4x^2}{9y^4}}$

1. ____________
2. ____________
3. ____________
4. ____________
5. ____________
6. ____________
7. ____________
8. ____________
9. ____________
10. ____________
11. ____________
12. ____________
13. ____________
14. ____________
15. ____________
16. ____________
17. ____________
18. ____________
19. ____________
20. ____________
21. ____________
22. ____________

23. ______

24. ______

25. ______

26. ______

27. ______

28. ______

29. ______

30. ______

31. ______

32. ______

33. ______

34. ______

35. ______

36. ______

37. ______

38. ______

39. ______

SECTION 9.3

Perform the indicated operation and simplify.

23. $\sqrt{10} \cdot \sqrt{2}$

24. $\sqrt{15} \cdot \sqrt{6}$

25. $\sqrt{21x^3} \cdot \sqrt{3x}$

26. $\sqrt{10}(\sqrt{5} + \sqrt{2})$

27. $(\sqrt{3} + \sqrt{5})(\sqrt{3} - \sqrt{5})$

28. $\sqrt{5} - 7\sqrt{5}$

29. $6\sqrt{20} + \sqrt{5} - \sqrt{45}$

30. $\sqrt{\frac{3}{16}} + \frac{\sqrt{12}}{2}$

SECTION 9.4

Assume all variables are nonnegative. Simplify and rationalize the denominator if necessary.

31. $\frac{3}{\sqrt{15}}$

32. $\frac{\sqrt{a}}{\sqrt{b}}$

33. $\frac{3}{\sqrt{7}}$

34. $\frac{1}{\sqrt{12}}$

35. $\frac{2}{\sqrt{3x}}$

36. $\sqrt{\frac{2}{3}}$

37. $\sqrt{\frac{2x^3}{3y^3}}$

38. $\frac{\sqrt{12x^4}}{\sqrt{3x^3}}$

39. $\sqrt{\frac{14}{5}}$

NAME ______ COURSE/SECTION ______ DATE ______

CHAPTER 9 TEST

Evaluate if possible.

1. $-\sqrt{64}$

2. $\sqrt{\frac{25}{81}}$

3. $\sqrt{-100}$

Use Appendix D to find the square roots.

4. $\sqrt{89}$

5. $-\sqrt{19}$

6. $\sqrt{5}$

Perform the indicated operation and/or simplify.

7. $\sqrt{28}$

8. $\sqrt{450}$

9. $\sqrt{18t^3}$

10. $\sqrt{5}(2 - \sqrt{5})$

11. $(\sqrt{3} - \sqrt{5})^2$

12. $\sqrt{14} \cdot \sqrt{6}$

13. $\sqrt{4y^2 + 8y + 4}$

14. $\sqrt{15ab^3} \cdot \sqrt{10a^2b}$

ANSWERS

1. ______
2. ______
3. ______
4. ______
5. ______
6. ______
7. ______
8. ______
9. ______
10. ______
11. ______
12. ______
13. ______
14. ______

15. ____________

16. ____________

17. ____________

18. ____________

19. ____________

20. ____________

21. ____________

22. ____________

23. ____________

24. ____________

25. ____________

15. $\sqrt{45} - 2\sqrt{20}$

16. $\sqrt{4x^3} + 3\sqrt{9x}$

17. $3\sqrt{2} - 5\sqrt{2} + 7\sqrt{2}$

18. $\sqrt{6} \cdot \sqrt{2} - \sqrt{27}$

19. $\sqrt{\dfrac{36a^2}{49b^4}}$

20. $\dfrac{\sqrt{75x^4y^2}}{\sqrt{3x}}$

21. $\dfrac{\sqrt{180}}{\sqrt{2}} - 3\sqrt{40}$

Rationalize the denominators. Assume all variables are nonnegative.

22. $\dfrac{3}{\sqrt{5}}$

23. $\dfrac{2}{\sqrt{8}}$

24. $\sqrt{\dfrac{3x}{2}}$

25. $\sqrt{\dfrac{5x}{7y}}$

CHAPTER 10

Quadratic Equations

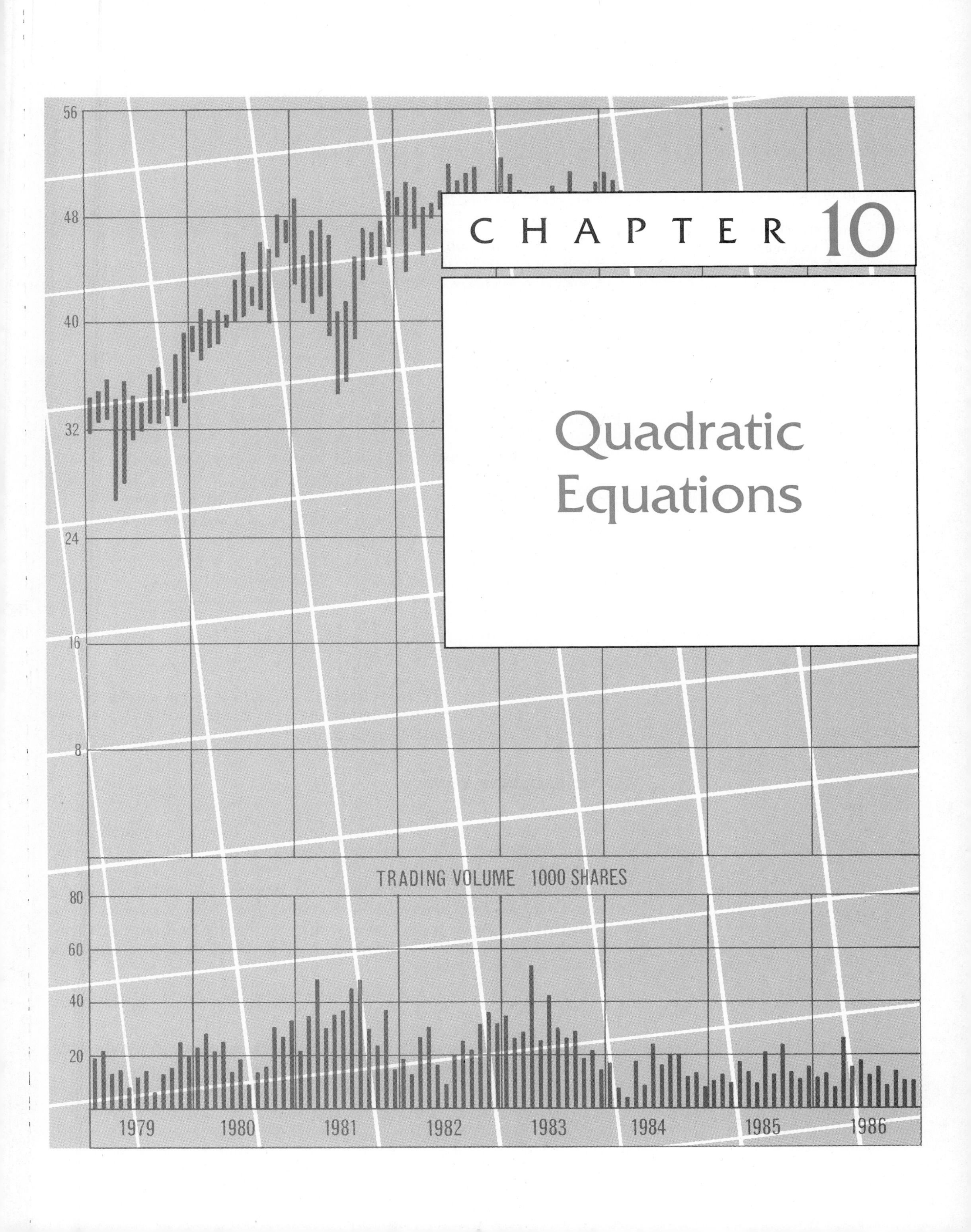

LEARNING OBJECTIVES

After finishing Section 10.1, you should be able to:

- Solve quadratic equations by factoring.
- Solve quadratic equations by taking principal square roots.

Solve by factoring.

1. $x^2 + x - 12 = 0$

2. $2x^2 = 7x - 3$

10.1

Solving by Taking Principal Square Roots

Recall that in Section 5.7 we looked at the general quadratic equation $ax^2 + bx + c = 0$ and solutions (or roots) to such equations. If we could factor $ax^2 + bx + c$, then we could find the solutions by applying the Law of Zero Products. An example will help refresh our memories.

Example 1 Solve $x^2 - 2x = 15$.

STANDARD FORM $\quad x^2 - 2x - 15 = 0$

FACTOR $\quad (x - 5)(x + 3) = 0$

LAW OF ZERO PRODUCTS $\quad$ (1) $x - 5 = 0$, or (2) $x + 3 = 0$

SOLVE $\quad$ (1) $x - 5 = 0$ gives us $x = 5$.
(2) $x + 3 = 0$ gives us $x = -3$.

CHECK

(1)

$x^2 - 2x$	15
$(5)^2 - 2(5)$	15
$25 - 10$	15
15	15

(2)

$x^2 - 2x$	15
$(-3)^2 - 2(-3)$	15
$9 + 6$	15
15	15

Unfortunately, there are many forms of $ax^2 + bx + c$ that cannot be factored, at least according to our definition of factoring. Since we will not be able to solve all quadratic equations by factoring, we must develop other methods.

DO EXERCISES 1 AND 2.

Now we want to look at a method for solving quadratic equations that do not have a first-degree term—such equations as $x^2 - 16 = 0$, $t^2 = 25$, $y^2 + 1 = 0$, and so on. Before we can look at this method, we need to look at equations of the form $|x| = a$ where a is nonnegative. Recall that $|x|$ is the distance that x is from the origin on a number line. Thus, a solution to the equation $|x| = a$ would be any number that is a distance a from the origin. Using the number line we see there are two such numbers, namely a and $-a$.

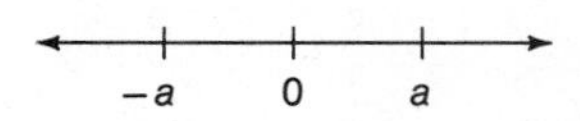

RULE 10.1 The equation $|x| = a$ (where a is nonnegative) has the solution $x = a$ or $x = -a$.

Example 2 Solve $|x| = 4$.

$|x| = 4$

$$(1)\quad x = 4 \qquad \text{or} \qquad (2)\quad x = -4$$

DO EXERCISE 3.

Example 3 Solve.

(a) $|x - 2| = 3$

$$(1)\quad x - 2 = 3 \qquad \text{or} \qquad (2)\quad x - 2 = -3$$
$$x = 5 \qquad\qquad x = -1$$

(b) $|2x + 3| = 5$

$$(1)\quad 2x + 3 = 5 \qquad \text{or} \qquad (2)\quad 2x + 3 = -5$$
$$2x = 2 \qquad\qquad 2x = -8$$
$$x = 1 \qquad\qquad x = -4$$

We will solve more equations of the form $|x| = a$ in Chapter 11.

DO EXERCISES 4 THROUGH 6.

If we have the equation $x^2 - 16 = 0$, then we can solve it by factoring.

$$x^2 - 16 = 0$$
$$(x - 4)(x + 4) = 0$$

Thus,

$$(1)\quad x = 4 \qquad \text{or} \qquad (2)\quad x = -4$$

We can also solve it by using the principal square root.

$x^2 - 16 = 0$

$x^2 = 16$ — We isolated the squared term on one side of the equality sign.

$\sqrt{x^2} = \sqrt{16}$ — We took the principal square root of both sides of the equation.

$|x| = 4$ — Recall that $\sqrt{x^2} = |x|$.

Thus,

$$(1)\quad x = 4 \qquad \text{or} \qquad (2)\quad x = -4$$

In general, we can solve any quadratic equation that does not have a first-degree term by taking the principal square root of both sides.

DO EXERCISE 7.

3. Solve $|x| = 10$.

Solve.

4. $|1 - t| = 3$

5. $|2x + 3| = 6$

6. Why is it necessary in $|x| = a$ to have a nonnegative; that is, what happens if a is negative?

7. Solve the equation $x^2 - 25 = 0$ using two different methods.

8. Solve $t^2 = 12$.

9. Solve $x^2 + 7 = 0$.

Example 4 Solve.

(a) $$x^2 - 49 = 0$$
$$x^2 = 49$$
$$\sqrt{x^2} = \sqrt{49}$$
$$|x| = 7$$

Thus,

$$(1)\quad x = 7 \qquad \text{or} \qquad (2)\quad x = -7$$

CHECK

(1)

$x^2 - 49$	0
$7^2 - 49$	0
$49 - 49$	0
0	0

(2)

$x^2 - 49$	0
$(-7)^2 - 49$	0
$49 - 49$	0
0	0

(b) $$t^2 = 8$$
$$\sqrt{t^2} = \sqrt{8}$$
$$|t| = 2\sqrt{2}$$

Thus,

$$(1)\quad t = 2\sqrt{2} \qquad \text{or} \qquad (2)\quad t = -2\sqrt{2}$$

CHECK

(1)

t^2	8
$(2\sqrt{2})^2$	8
$2^2 \cdot (\sqrt{2})^2$	8
$4 \cdot 2$	8
8	8

(2)

t^2	8
$(-2\sqrt{2})$	8
$(-2)^2 \cdot (\sqrt{2})^2$	8
$4 \cdot 2$	8
8	8

■ **DO EXERCISE 8.**

Example 5 Solve $y^2 + 1 = 0$.

$$y^2 + 1 = 0$$
$$y^2 = -1$$
$$\sqrt{y^2} = \sqrt{-1}$$

But $\sqrt{-1}$ is not a real number since the square root of a negative number does not exist as a real number. In this case, we say there is *no real solution* for the equation $y^2 + 1 = 0$. Thus, we see there are quadratic equations for which there are no real solutions.

■ **DO EXERCISE 9.**

Example 6 Solve $2x^2 - 9 = 0$.

$$2x^2 - 9 = 0$$
$$2x^2 = 9$$
$$x^2 = \frac{9}{2}$$
$$\sqrt{x^2} = \sqrt{\frac{9}{2}}$$
$$|x| = \frac{\sqrt{9}}{\sqrt{2}} = \frac{3}{\sqrt{2}}$$
$$= \frac{3}{\sqrt{2}} \cdot \frac{\sqrt{2}}{\sqrt{2}} = \frac{3\sqrt{2}}{2}$$

Thus,

$$(1)\quad x = \frac{3\sqrt{2}}{2} \qquad \text{or} \qquad (2)\quad x = -\frac{3\sqrt{2}}{2}$$

We could check these two solutions as we did in Example 4.

DO EXERCISE 10.

10. Solve $3y^2 - 8 = 0$.

Note that if we multiplied out the quadratic expression $(x - 1)^2 - 9$, we would have a first-degree term. That is,

$$(x - 1)^2 - 9 = x^2 - 2x + 1 - 9$$
$$= x^2 - 2x - 8$$

We cannot solve $x^2 - 2x - 8 = 0$ by taking principal square roots. However, when the equation is in the form $(x - 1)^2 - 9 = 0$, we can treat it as if it did not have a first-degree term and solve it by taking principal square roots.

Example 7 Solve $(x - 1)^2 - 9 = 0$.

$$(x - 1)^2 - 9 = 0$$
$$(x - 1)^2 = 9$$
$$\sqrt{(x - 1)^2} = \sqrt{9}$$
$$|x - 1| = 3$$

Thus,

$$(1)\quad x - 1 = 3 \qquad \text{or} \qquad (2)\quad x - 1 = -3$$
$$x = 4 \qquad\qquad\qquad x = -2$$

CHECK

(1)

$(x - 1)^2 - 9$	0
$(4 - 1)^2 - 9$	0
$(3)^2 - 9$	0
$9 - 9$	0
0	0

(2)

$(x - 1)^2 - 9$	0
$(-2 - 1)^2 - 9$	0
$(-3)^2 - 9$	0
$9 - 9$	0
0	0

DO EXERCISE 11.

11. Solve $(x - 3)^2 - 25 = 0$.

Solve.

12. $(y + 1)^2 = 8$

13. $(t - 1)^2 + 4 = 0$

Example 8 Solve $(x + 4)^2 = 5$.

$$(x + 4)^2 = 5$$
$$\sqrt{(x + 4)^2} = \sqrt{5}$$
$$|x + 4| = \sqrt{5}$$

Thus,

$$(1)\quad x + 4 = \sqrt{5} \qquad \text{or} \qquad (2)\quad x + 4 = -\sqrt{5}$$
$$x = -4 + \sqrt{5} \qquad\qquad x = -4 - \sqrt{5}$$

In general, we can solve any quadratic equation of the form $(x + k)^2 = r$ by taking principal square roots.

DO EXERCISES 12 AND 13.

DO SECTION PROBLEMS 10.1.

ANSWERS TO MARGINAL EXERCISES

1. $-4, 3$ **2.** $\frac{1}{2}, 3$ **3.** $10, -10$ **4.** $-2, 4$ **5.** $\frac{3}{2}, -\frac{9}{2}$ **6.** If a is negative, there are no solutions to $|x| = a$. **7.** $5, -5$ **8.** $2\sqrt{3}, -2\sqrt{3}$ **9.** No real solution **10.** $\frac{2\sqrt{6}}{3}, \frac{-2\sqrt{6}}{3}$ **11.** $8, -2$ **12.** $-1 + 2\sqrt{2}, -1 - 2\sqrt{2}$ **13.** No real solutions

NAME ______ COURSE/SECTION ______ DATE ______

ANSWERS

SECTION PROBLEMS 10.1

Solve by factoring.

1. $x^2 - 8x + 15 = 0$

2. $t^2 - 6t = 0$

3. $y^2 = 4y + 12$

4. $2x^2 - 8 = 0$

5. $6x^2 - 7x - 3 = 0$

6. $4t^2 + 4t + 1 = 0$

7. $x^2 + 5x - 14 = -20$

8. $6x^2 + 15x - 36 = 0$

9. $2x^2 = -5x + 12$

10. $3t^2 = 27$

Solve.

11. $|x| = 7$

12. $2|y| - 6 = 0$

13. $|a + 3| = 4$

14. $|3 - x| = 5$

15. $|2x - 1| = \frac{1}{2}$

16. $|x + 4| = -6$

Solve by factoring and by taking principal square roots.

17. $x^2 - 64 = 0$

18. $3y^2 - 12 = 0$

19. $(x - 1)^2 - 9 = 0$

20. $(y + 2)^2 - 1 = 0$

1. ______
2. ______
3. ______
4. ______
5. ______
6. ______
7. ______
8. ______
9. ______
10. ______
11. ______
12. ______
13. ______
14. ______
15. ______
16. ______
17. ______
18. ______
19. ______
20. ______

21. ______
22. ______
23. ______
24. ______
25. ______
26. ______
27. ______
28. ______
29. ______
30. ______
31. ______
32. ______
33. ______
34. ______
35. ______
36. ______
37. ______
38. ______
39. ______
40. ______
41. ______
42. ______

Solve by taking principal square roots.

21. $t^2 = 4$

22. $x^2 - 16 = 0$

23. $2y^2 - 18 = 0$

24. $x^2 + 25 = 0$

25. $t^2 - 7 = 0$

26. $x^2 - 12 = 0$

27. $3x^2 - 5 = 0$

28. $(x - 2)^2 - 36 = 0$

29. $x^2 - 16 = 0$

30. $x^2 = 64$

31. $t^2 - 5 = 0$

32. $y^2 + 7 = 0$

33. $(y + 3)^2 = 16$

34. $(x - 1)^2 - 12 = 0$

35. $(t + 5)^2 = 7$

36. $(x + 3)^2 + 9 = 0$

37. $\left(x - \frac{1}{2}\right)^2 = \frac{1}{4}$

38. $\left(y + \frac{3}{4}\right)^2 = \frac{7}{16}$

39. $\left(t + \frac{1}{3}\right)^2 + 1 = 0$

40. $(y - 10)^2 = 20$

41. $(x - 1.3)^2 - 1.44 = 0$

42. $(t - 3)^2 + 1 = 0$

10.2

Completing the Square

LEARNING OBJECTIVES

After finishing Section 10.2, you should be able to:

- Solve quadratic equations by completing the square.

Consider the equation $x^2 + 4x - 4 = 0$. We cannot solve this equation by factoring and, since it is not of the form $(x + k)^2 = r$, we cannot solve it by taking principal square roots. We note, however, that

$$\begin{aligned} x^2 + 4x - 4 &= x^2 + 4x + 4 - 8 \\ &= (x + 2)^2 - 8 \end{aligned}$$

and hence $x^2 + 4x - 4 = 0$ is the same equation as $(x + 2)^2 - 8 = 0$. Now, $(x + 2)^2 - 8 = 0$ can be written in the form $(x + k)^2 = r$ and can be solved by taking principal square roots.

A little terminology here is useful. Consider

$$x^2 + 4x - 4 = \underbrace{x^2 + 4x + 4}_{\text{A perfect trinomial square}} - 8 = (x + 2)^2 - 8$$

A *perfect trinomial square* is a second-degree trinomial that can be put in the form $(x + k)^2$. The advantage in having a perfect trinomial square in an equation is that we can then solve the equation by taking principal square roots.

At this point there are some fairly obvious questions to ask.

1. Can any quadratic equation $ax^2 + bx + c = 0$ be put in the form $(x + k)^2 = r$?
2. If so, how do we put it in the form $(x + k)^2 = r$?

The answer to the first question is yes, and the answer to the second question is by a process called *completing the square.*

Example 1 Put the equation $x^2 + 6x - 3 = 0$ in the form $(x + k)^2 = r$. To do this we have to complete the square on $x^2 + 6x$; that is, make $x^2 + 6x$ a perfect trinomial square. The steps are outlined below.

(a) Put the equation in the form $ax^2 + bx = c$.

$$x^2 + 6x - 3 = 0$$
$$x^2 + 6x = 3$$

(b) Obtain a coefficient of positive one for the x^2 term.

We already have this.

(c) Take $\frac{1}{2}$ the coefficient of the x term [this will be the value for k in $(x + k)^2 = r$], square it, and add this squared term to each side of the equation.

$$\left(\frac{1}{2}\right)(6) = 3 \qquad (\text{thus, } k = 3)$$
$$3^2 = 9$$
$$x^2 + 6x = 3$$
$$\underbrace{x^2 + 6x + 9}_{\text{A perfect trinomial square}} = 3 + 9$$

(d) Factor the left side so the equation will be in the form $(x + k)^2 = r$.

$$(x + 3)^2 = 12$$

Note

1. After putting the equation in the form $(x + 3)^2 = 12$, we solve the equation and obtain the two solutions $-3 + 2\sqrt{3}$ and $-3 - 2\sqrt{3}$.

2. In the process of completing the square, the reason for taking $\frac{1}{2}$ the coefficient of the x term, squaring it, and adding it (squared) to both sides of the equation is as follows. Suppose we want to put $x^2 + ax$ in the form $(x + k)^2$. Let us see what would have to be done.

   Note that $(x + k)^2 = x^2 + 2kx + k^2$. Thus for the two polynomials $x^2 + ax$ and $x^2 + 2kx + k^2$ to be equal, we need two things:

   (a) $a = 2k$ or $k = \dfrac{a}{2}$ (Hence, the $\frac{1}{2}$ the coefficient of the x term.)

   (b) k^2 has to be added to the polynomial $x^2 + ax$.

The steps for changing the equation $ax^2 + bx + c = 0$ to the form $(x + k)^2 = r$ are as follows.

STEP 1: Form $ax^2 + bx = c$

Put the quadratic equation in this form.

STEP 2: Coefficient of x^2

Obtain a coefficient of positive one for the x^2 term by multiplying each side of the equation by $1/a$.

STEP 3: Coefficient of x

Take $\frac{1}{2}$ the coefficient of the x term, square it, and add this squared term to both sides of the equation.

STEP 4: Form $(x + k)^2 = r$

Factor the left side so that the equation is in the form $(x + k)^2 = r$.

Example 2 Put the equation $2x^2 - 6x + 1 = 0$ in the form $(x + k)^2 = r$.

STEP 1: Form $ax^2 + bx = c$

$$2x^2 - 6x = -1$$

STEP 2: Coefficient of x^2

Multiply both sides by $\frac{1}{2}$ to obtain a coefficient of 1 for x^2.

$$\frac{1}{2}(2x^2) - \frac{1}{2}(6x) = \frac{1}{2}(-1)$$

$$x^2 - 3x = -\frac{1}{2}$$

STEP 3: Coefficient of x

Note $k = \frac{1}{2}(-3) = -\frac{3}{2}$.

$$\left(\frac{1}{2}\right)(-3) = -\frac{3}{2}$$

Square k.

$$\left(-\frac{3}{2}\right)^2 = \frac{9}{4}$$

Add k^2 to both sides.

$$\underbrace{x^2 - 3x + \frac{9}{4}}_{\text{A perfect trinomial square}} = -\frac{1}{2} + \frac{9}{4}$$

STEP 4: Form $(x + k)^2 = r$

$$\left(x - \frac{3}{2}\right)^2 = -\frac{2}{4} + \frac{9}{4}$$

$$\left(x - \frac{3}{2}\right)^2 = \frac{7}{4}$$

DO EXERCISES 1 AND 2.

Now that we can put any quadratic equation $ax^2 + bx + c = 0$ in the form $(x + k)^2 = r$, we can also solve any quadratic equation by completing the square and then taking principal square roots.

Example 3 Solve $x^2 - 8x + 5 = 0$.

(a) Complete the square.

Form $ax^2 + bx = c$ $\quad x^2 - 8x = -5$

Coefficient of x^2 $\quad$ The coefficient already is 1.

Coefficient of x $\quad \frac{1}{2}(-8) = -4$

Note that $k = -4$ and k^2 is $(-4)^2 = 16$. $\quad (-4)^2 = 16$

$$x^2 - 8x + 16 = -5 + 16$$

Form $(x + k)^2 = r$ $\quad (x - 4)^2 = 11$

(b) Solve by taking principal square roots.

$$\sqrt{(x - 4)^2} = \sqrt{11}$$

$$|x - 4| = \sqrt{11}$$

Thus,

(1) $x - 4 = \sqrt{11}$, $x = 4 + \sqrt{11}$ $\quad$ or $\quad$ (2) $x - 4 = -\sqrt{11}$, $x = 4 - \sqrt{11}$

Although it would be messy, we could check both of our solutions.

DO EXERCISE 3.

Example 4 Solve $(2x + 3)(x + 2) = 10$.

(a) Complete the square.

Form $ax^2 + bx = c$

$$2x^2 + 7x + 6 = 10$$

$$2x^2 + 7x = 4$$

Coefficient of x^2

$$\frac{1}{2}(2x^2) + \frac{1}{2}(7x) = \frac{1}{2}(4)$$

$$x^2 + \frac{7}{2}x = 2$$

By completing the square, put the following equations in the form $(x + k)^2 = r$.

1. $x^2 + 4x - 2 = 0$

2. $3x^2 = 9x + 6$

Solve by completing the square.

3. $x^2 - 6x + 2 = 0$

Solve by completing the square.

4. $2t^2 = -3t + 1$

5. $x^2 + 4x + 5 = 0$

Coefficient of x $\qquad \frac{1}{2}\left(\frac{7}{2}\right) = \frac{7}{4}$

Note that $k = \frac{7}{4}$ and k^2 is $(\frac{7}{4})^2 = \frac{49}{16}$. $\qquad \left(\frac{7}{4}\right)^2 = \frac{49}{16}$

$$x^2 + \frac{7}{2}x + \frac{49}{16} = 2 + \frac{49}{16}$$

Form $(x + k)^2 = r$ $\qquad \left(x + \frac{7}{4}\right)^2 = \frac{32}{16} + \frac{49}{16}$

$$\left(x + \frac{7}{4}\right)^2 = \frac{81}{16}$$

(b) Solve by taking principal square roots. $\qquad \sqrt{\left(x + \frac{7}{4}\right)^2} = \sqrt{\frac{81}{16}}$

$$\left|x + \frac{7}{4}\right| = \frac{9}{4}$$

Thus,

$$(1)\ x + \frac{7}{4} = \frac{9}{4} \quad \text{or} \quad (2)\ x + \frac{7}{4} = -\frac{9}{4}$$

$$x = \frac{2}{4} = \frac{1}{2} \qquad\qquad x = \frac{-16}{4} = -4$$

(c) CHECK

(1)

$(2x + 3)(x + 2)$	10
$[2(\frac{1}{2}) + 3](\frac{1}{2} + 2)$	10
$(1 + 3)(2\frac{1}{2})$	10
$(4)(\frac{5}{2})$	10
10	10

(2)

$(2x + 3)(x + 2)$	10
$[2(-4) + 3](-4 + 2)$	10
$(-8 + 3)(-2)$	10
$(-5)(-2)$	10
10	10

DO EXERCISES 4 AND 5.

DO SECTION PROBLEMS 10.2.

ANSWERS TO MARGINAL EXERCISES

1. $(x + 2)^2 = 6$ **2.** $\left(x - \frac{3}{2}\right)^2 = \frac{17}{4}$ **3.** $3 + \sqrt{7}, 3 - \sqrt{7}$
4. $\frac{-3 + \sqrt{17}}{4}, \frac{-3 - \sqrt{17}}{4}$ **5.** No real solution

NAME COURSE/SECTION DATE

ANSWERS

1. ________
2. ________
3. ________
4. ________
5. ________
6. ________
7. ________
8. ________
9. ________
10. ________
11. ________
12. ________
13. ________

SECTION PROBLEMS 10.2

Put the equation in the form $(x + k)^2 = r$.

1. $x^2 + 6x - 3 = 0$

2. $y^2 - 3y + 1 = 0$

3. $2x^2 - 8x + 3 = 0$

4. $3x^2 - 4x - 6 = 0$

Solve by completing the square.

5. $x^2 + 4x - 2 = 0$

6. $3x^2 = 9x + 6$

7. $2y^2 - 10y + 4 = 0$

8. $x^2 + x + 1 = 0$

9. $4x^2 + 12x + 9 = 0$

10. $-2x^2 + 4x + 5 = 0$

11. $\frac{t^2}{2} - t - 3 = 0$

12. $4y^2 - 100 = 0$

13. $x^2 + 6x - 3 = 0$

14. $2t^2 + t + 6 = 0$

15. $4t^2 + 25 = 20t$

16. $3x^2 + 4x - 3 = 0$

14. ______________

15. ______________

16. ______________

17. ______________

18. ______________

19. ______________

20. ______________

21. ______________

22. ______________

23. ______________

24. ______________

Solve by any method (factoring, principal square roots, or completing the square).

17. $x^2 - 7x + 12 = 0$

18. $t^2 = 3t - 5$

19. $9x^2 = -30x - 25$

20. $(x - 5)^2 - 16 = 0$

21. $2x^2 - 4x + 1 = 0$

22. $6x^2 - 7x - 20 = 0$

23. $3(x + 1)^2 - 12 = 0$

24. $2t^2 + 7t + 7 = 0$

10.3

Quadratic Formula

LEARNING OBJECTIVES

After finishing Section 10.3, you should be able to:

- Solve quadratic equations by using the quadratic formula.
- Solve additional rational equations.
- Solve equations that involve a square root.

We know that by completing the square we can find the solutions to the equation $ax^2 + bx + c = 0$. A natural question may be whether there is a general formula that will give us the solutions without having to complete the square each and every time. The answer to that question is that there is such a formula—the quadratic formula. We will now derive it by finding the solutions to $ax^2 + bx + c = 0$, $a \neq 0$.

1. Complete the square.

Form $ax^2 + bx = c$

$$ax^2 + bx = -c$$

Coefficient of x^2

Multiply each side by $\frac{1}{a}$.

$$\frac{1}{a}(ax^2) + \frac{1}{a}(bx) = \left(\frac{1}{a}\right)(-c)$$

$$x^2 + \frac{b}{a}x = -\frac{c}{a}$$

Coefficient of x

$$\frac{1}{2}\left(\frac{b}{a}\right) = \frac{b}{2a}$$

Note that $k = \frac{b}{2a}$ and k^2 is $\left(\frac{b}{2a}\right)^2 = \frac{b^2}{4a^2}$.

$$\left(\frac{b}{2a}\right)^2 = \frac{b^2}{4a^2}$$

$$\underbrace{x^2 + \frac{b}{a}x + \frac{b^2}{4a^2}}_{\text{A perfect trinomial square}} = \frac{b^2}{4a^2} - \frac{c}{a}$$

Form $(x + k)^2 = r$

$$\left(x + \frac{b}{2a}\right)^2 = \frac{b^2}{4a^2} - \frac{4ac}{4a^2}$$

$$= \frac{b^2 - 4ac}{4a^2}$$

2. Solve by taking principal square roots.

$$\sqrt{\left(x + \frac{b}{2a}\right)^2} = \sqrt{\frac{b^2 - 4ac}{4a^2}}$$

$$\left|x + \frac{b}{2a}\right| = \frac{\sqrt{b^2 - 4ac}}{\sqrt{4a^2}}$$

$$= \frac{\sqrt{b^2 - 4ac}}{2a}$$

Thus,

(1) $$x + \frac{b}{2a} = \frac{\sqrt{b^2 - 4ac}}{2a} \quad \text{or} \quad x = -\frac{b}{2a} + \frac{\sqrt{b^2 - 4ac}}{2a}$$

$$= \frac{-b + \sqrt{b^2 - 4ac}}{2a}$$

Solve by using the quadratic formula.

1. $x^2 - 2x - 5 = 0$

$$(2) \quad x + \frac{b}{2a} = -\frac{\sqrt{b^2 - 4ac}}{2a} \quad \text{or} \quad x = -\frac{b}{2a} - \frac{\sqrt{b^2 - 4ac}}{2a}$$
$$= \frac{-b - \sqrt{b^2 - 4ac}}{2a}$$

These two solutions to $ax^2 + bx + c = 0$ are normally combined and written as follows.

QUADRATIC FORMULA

$$x = \frac{-b \pm \sqrt{b^2 - 4ac}}{2a}$$

where one solution is $\dfrac{-b + \sqrt{b^2 - 4ac}}{2a}$ and the other solution is $\dfrac{-b - \sqrt{b^2 - 4ac}}{2a}$.

When we use the quadratic formula, we have to be careful that our equation is in standard form, and then note that a is the coefficient of x^2, b is the coefficient of x, and c is the constant term. We now list three steps we can follow to solve an equation by using the quadratic formula.

STEP 1: STANDARD FORM

Put the quadratic equation in standard form.

STEP 2: IDENTIFY

Identify a, b, and c.

STEP 3: SUBSTITUTE

Substitute a, b, and c in the quadratic formula and simplify to find our solutions.

Example 1 Solve $x^2 = 3x + 5$ by using the quadratic formula.

STEP 1: STANDARD FORM $\qquad x^2 - 3x - 5 = 0$

STEP 2: IDENTIFY $\qquad a = 1$, $b = -3$, and $c = -5$

STEP 3: SUBSTITUTE
$$x = \frac{-b \pm \sqrt{b^2 - 4ac}}{2a}$$
$$= \frac{-(-3) \pm \sqrt{(-3)^2 - 4(1)(-5)}}{2(1)}$$
$$= \frac{3 \pm \sqrt{9 + 20}}{2} = \frac{3 \pm \sqrt{29}}{2}$$

Thus, the two solutions are $\dfrac{3 + \sqrt{29}}{2}$ and $\dfrac{3 - \sqrt{29}}{2}$.

DO EXERCISE 1.

Example 2 Solve $2t^2 + t + 3 = 0$ by using the quadratic formula.

STEP 1: STANDARD FORM $\quad 2t^2 + t + 3 = 0$

STEP 2: IDENTIFY $\quad a = 2,\ b = 1,\ \text{and } c = 3$

STEP 3: SUBSTITUTE

$$t = \frac{-b \pm \sqrt{b^2 - 4ac}}{2a}$$

$$= \frac{-1 \pm \sqrt{1^2 - 4(2)(3)}}{2(1)}$$

$$= \frac{-1 \pm \sqrt{1 - 24}}{2}$$

$$= \frac{-1 \pm \sqrt{-23}}{2}$$

Since we have the square root of a negative number ($\sqrt{-23}$), which we know does not exist as a real number, we say there are *no real solutions* to $2t^2 + t + 3 = 0$.

DO EXERCISE 2.

Example 3 Solve $4x^2 + 20x + 25 = 0$.

STEP 1: STANDARD FORM $\quad 4x^2 + 20x + 25 = 0$

STEP 2: IDENTIFY $\quad a = 4,\ b = 20,\ c = 25$

STEP 3: SUBSTITUTE

$$x = \frac{-b \pm \sqrt{b^2 - 4ac}}{2a}$$

$$= \frac{-20 \pm \sqrt{(20)^2 - 4(4)(25)}}{2(4)}$$

$$= \frac{-20 \pm \sqrt{400 - 400}}{8}$$

$$= \frac{-20 \pm 0}{8}$$

Thus, the two solutions are

$$\frac{-20 + 0}{8} = -\frac{5}{2} \quad \text{and} \quad \frac{-20 - 0}{8} = -\frac{5}{2}$$

In Example 3, the two solutions are identical, and we often say in this case that we have a *double root* or a *root of multiplicity two.*

DO EXERCISES 3 AND 4.

Solve by using the quadratic formula.

2. $x^2 + 3x + 4 = 0$

Solve by using the quadratic formula.

3. $x^2 = -6x - 9$

4. $3t^2 - 2t - 2 = 0$

By using the discriminant, identify the type of solution for the following quadratic equations. (It is not necessary to solve the equation.)

5. $2x^2 - 10x + 3 = 0$

6. $4x^2 = 4x - 1$

7. $x^2 + 3x + 3 = 0$

You may have noticed from Examples 1, 2, and 3 or from Exercises 1, 2, 3, and 4 that there appear to be three possibilities for the solutions (roots) to $ax^2 + bx + c = 0$:

1. No real solution
2. A double root
3. Two distinct real solutions

This is in fact true, and the type of solution depends on $b^2 - 4ac$ (called the *discriminant*) in the quadratic formula

$$\frac{-b \pm \sqrt{b^2 - 4ac}}{2a}$$

The results are summarized as follows:

Discriminant	*Type of Solution*
1. $b^2 - 4ac < 0$	No real solution (since in this case we would have the square root of a negative number)
2. $b^2 - 4ac = 0$	A double root
3. $b^2 - 4ac > 0$	Two distinct real solutions

Example 4 Identify the type of solution for the following quadratic equations.

(a) $2x^2 - 3x + 5 = 0$
Here $a = 2$, $b = -3$, and $c = 5$. Thus,

$$b^2 - 4ac = (-3)^2 - 4(2)(5) = 9 - 40 < 0$$

and hence there is *no real solution*.

(b) $-x^2 + 6x + 2 = 0$
Here $a = -1$, $b = 6$, and $c = 2$. Thus,

$$b^2 - 4ac = (6)^2 - 4(-1)(2) = 36 + 8 > 0$$

and hence there are *two distinct real solutions*.

(c) $9x^2 - 6x + 1 = 0$
Here, $a = 9$, $b = -6$, and $c = 1$. Thus,

$$b^2 - 4ac = (-6)^2 - 4(9)(1) = 36 - 36 = 0$$

and hence there is a *double root*.

DO EXERCISES 5 THROUGH 7.

The quadratic formula can also be helpful in solving other types of equations. For example, quite often when we solve a rational equation it is necessary to solve a quadratic equation.

Example 5 Solve $\frac{1}{x} + \frac{1}{x-1} = 3.$

LCD

Multiply both sides by the LCD of $x(x-1)$.

$$x(x-1)\left(\frac{1}{x}\right) + x(x-1)\left(\frac{1}{x-1}\right) = 3x(x-1)$$

$$x - 1 + x = 3x^2 - 3x$$

SOLVE

Standard Form

$$0 = 3x^2 - 5x + 1$$

Identify

$$a = 3,\ b = -5,\ c = 1$$

Substitute

$$\begin{aligned} x &= \frac{-b \pm \sqrt{b^2 - 4ac}}{2a} \\ &= \frac{-(-5) \pm \sqrt{(-5)^2 - 4(3)(1)}}{2(3)} \\ &= \frac{5 \pm \sqrt{25 - 12}}{6} \\ &= \frac{5 \pm \sqrt{13}}{6} \end{aligned}$$

Thus, the two solutions are

$$\frac{5 - \sqrt{13}}{6} \quad \text{and} \quad \frac{5 + \sqrt{13}}{6}$$

(neither of which is a pole of the rational equation).

DO EXERCISES 8 AND 9.

Another type of equation we have not looked at yet is an equation that involves radicals, such as $\sqrt{x-4} = x - 6$. To solve such an equation, we must get rid of the square root, and the way to do this is to isolate the square root on one side of the equation and then square each side of the equation. Unfortunately, squaring both sides of an equation does not always give us an equivalent equation.

Example 6 If we square the equation $x = 3$, will we get an equivalent equation?

Note that the equation $x = 3$ has 3 for its only solution. If we square both sides of $x = 3$, we obtain the equation $x^2 = 9$, which has 3 and -3 for its solutions. Thus the equations $x = 3$ and $x^2 = 9$ are not equivalent. We do notice, however, that the solution to the equation $x = 3$ was also a solution to the equation we obtained when we squared each side. This will always be the case.

Solve.

8. $\frac{2}{x-3} = 1 + \frac{1}{x+1}$

9. $\frac{t}{t^2 + t - 2} = \frac{3}{t^2 - 1}$

Solve.

10. $\sqrt{x-1} = x - 7$

RULE 10.2 If we square both sides of an equation, then any solution(s) to the original equation will also be a solution to the new equation obtained by squaring both sides.

In Example 6, we notice that we picked up an extra solution when we squared each side. Often this will happen, and thus we have to be careful about checking our solutions when we solve an equation by squaring both sides.

RULE 10.3 If in the process of solving an equation it is necessary to square both sides of the equation, then we must always check our solution(s) in the original equation to see which solutions are valid.

Example 7 Solve $\sqrt{x-4} = x - 6$.

(a) Square each side to get rid of the square root.

$$(\sqrt{x-4})^2 = (x-6)^2$$
$$x - 4 = x^2 - 12x + 36$$

(b) Solve the resulting equation by factoring. We could also solve by using the quadratic formula. However, factoring is normally faster.

$$0 = x^2 - 13x + 40$$
$$0 = (x-5)(x-8)$$

Thus, (1) $x = 5$

or (2) $x = 8$

(c) Since we have squared each side in the solving process, we must check both solutions in the original equation $\sqrt{x-4} = x - 6$ to see whether they are valid solutions.

(1)

$\sqrt{x-4}$	$x-6$
$\sqrt{5-4}$	$5-6$
$\sqrt{1}$	-1
1	-1

(2)

$\sqrt{x-4}$	$x-6$
$\sqrt{8-4}$	$8-6$
$\sqrt{4}$	2
2	2

Thus, 8 is a solution to $\sqrt{x-4} = x - 6$, but 5 is not a solution.

DO EXERCISE 10.

Example 8 Solve $\sqrt{2y^2 - y + 3} - y = 1$.

Before we square, we must isolate the square root term. Otherwise we would not get rid of the square root when we square.

$$\sqrt{2y^2 - y + 3} = y + 1$$
$$(\sqrt{2y^2 - y + 3})^2 = (y+1)^2$$
$$2y^2 - y + 3 = y^2 + 2y + 1$$

Solve the resulting equation. (We will use the quadratic formula for practice even though $y^2 - 3y + 2$ can be factored.)

$$y^2 - 3y + 2 = 0$$
$$a = 1,\ b = -3,\ c = 2$$
$$y = \frac{-b \pm \sqrt{b^2 - 4ac}}{2a}$$
$$= \frac{-(-3) \pm \sqrt{(-3)^2 - 4(1)(2)}}{2(1)}$$
$$= \frac{3 \pm \sqrt{1}}{2} = \frac{3 \pm 1}{2}$$

Thus, the two solutions are

$$(1)\ y = \frac{4}{2} = 2 \qquad \text{and} \qquad (2)\ y = \frac{3-1}{2} = 1$$

Since we have squared each side in this solving process, we must check our solutions.

(1)

$\sqrt{2y^2 - y + 3} - y$	1
$\sqrt{2(2)^2 - 2 + 3} - 2$	1
$\sqrt{8 - 2 + 3} - 2$	1
$\sqrt{9} - 2$	1
$3 - 2$	1
1	1

(2)

$\sqrt{2y^2 - y + 3} - y$	1
$\sqrt{2(1)^2 - 1 + 3} - 1$	1
$\sqrt{2 - 1 + 3} - 1$	1
$\sqrt{4} - 1$	1
$2 - 1$	1
1	1

In this case, both solutions satisfy the original equation.

DO EXERCISE 11.

A final comment: we have seen various methods (such as factoring, completing the square, and the quadratic formula) for solving quadratic equations. Thus, when *given a choice,* which method do we use? From a strictly personal standpoint, I prefer to use factoring first (if, of course, the polynomial factors), then the quadratic formula (if the polynomial does not factor or I do not see how it factors), and, finally, completing the square.

DO SECTION PROBLEMS 10.3.

Solve.

11. $\sqrt{x + 4} = \sqrt{x^2 - 3x - 1}$

ANSWERS TO MARGINAL EXERCISES

1. $1 + \sqrt{6}, 1 - \sqrt{6}$ **2.** No real solution **3.** -3 **4.** $\frac{1 + \sqrt{7}}{3}, \frac{1 - \sqrt{7}}{3}$
5. Two distinct real solutions **6.** Double root **7.** No real solution
8. $\frac{3 + \sqrt{41}}{2}, \frac{3 - \sqrt{41}}{2}$ **9.** $1 + \sqrt{7}, 1 - \sqrt{7}$ **10.** 10 **11.** $5, -1$

NAME ____________________ COURSE/SECTION ____________ DATE ________

SECTION PROBLEMS 10.3

Solve.

1. $x^2 - 3x - 2 = 0$

2. $t^2 - 12t + 36 = 0$

3. $y^2 + 4y + 3 = 0$

4. $5x^2 - 1 = 2x$

5. $t^2 + 2t + 4 = 0$

6. $4x^2 + 25 = -20x$

7. $2x^2 = 6x$

8. $3x^2 - 8 = 0$

9. $2y^2 - 3y - 2 = 0$

10. $7x^2 - 2x + 1 = 0$

ANSWERS

1. ____________
2. ____________
3. ____________
4. ____________
5. ____________
6. ____________
7. ____________
8. ____________
9. ____________
10. ____________

11. ______________

12. ______________

13. ______________

14. ______________

15. ______________

16. ______________

17. ______________

18. ______________

19. ______________

20. ______________

11. $(x - 3)(x + 2) = 2x + 1$

12. $3(x^2 - 4) = -2x(x + 5)$

13. $6x^2 + 17x - 3 = 0$

14. $2y^2 - 3y + 4 = 0$

15. $9a^2 = 24a - 16$

16. $y^3 - 2y^2 - 5y = 0$

17. $2y^2 + 6y - 3 = 0$

18. $2x^2 = -5x - 2$

19. $y^2 - 6y + 10 = 0$

20. $y^2 = y - 1$

NAME COURSE/SECTION DATE

ANSWERS

Identify the type of solution (not necessary to solve).

21. $2x^2 - 3x - 4 = 0$

22. $4x^2 - 12x + 9 = 0$

23. $y^2 - 16y + 5 = 0$

24. $t^2 = 3t - 5$

25. $4a^2 = -2a - 25$

26. $-3x^2 - x + 7 = 0$

Solve.

27. $\frac{x}{x+1} = \frac{2}{x+2}$

28. $\frac{1}{x} - \frac{2}{x+3} = 4$

29. $\frac{x}{x^2 - x - 6} = \frac{2x}{x^2 - 4}$

30. $1 + \frac{1}{x} + \frac{x}{x+1} = 2$

31. $\sqrt{x+1} = x - 1$

32. $\sqrt{2x-5} + 4 = x$

21. ______
22. ______
23. ______
24. ______
25. ______
26. ______
27. ______
28. ______
29. ______
30. ______
31. ______
32. ______

33. ______________

34. ______________

35. ______________

36. ______________

37. ______________

38. ______________

39. ______________

40. ______________

33. $\sqrt{x+5} - \sqrt{4x-1} = 0$

34. $\sqrt{x^2+x+4} = x+1$

35. $(x+1)^2 - 9 = 0$

36. $y^2 + 2y + 5 = 0$

37. $5t^2 - 80 = 0$

38. $t^2 + 3t - 10 = 0$

39. $\sqrt{4t-3} - 2t = -3$

40. $\frac{1}{x} + \frac{3}{x-2} = 1$

10.4

Word Problems

LEARNING OBJECTIVES

After finishing Section 10.4, you should be able to:

- Solve additional word problems.
- Use the Pythagorean Theorem to solve certain word problems.

As usual, we save our favorite type of problem for last. Now that we can solve any quadratic equation, there are some additional word problems worth investigating.

Example 1 Larry Markham has a rectangular flower garden that measures 10 feet by 7 feet. If he wants to double the area of his flower garden by increasing the length and width by the same amount, what should the new dimensions be (to the nearest tenth of a foot)?

PICTURE

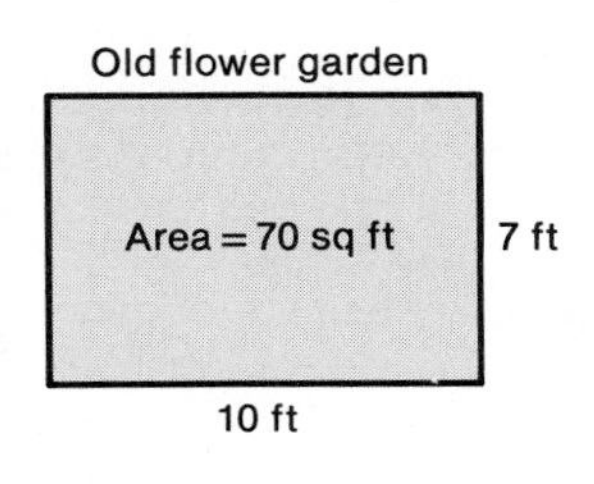

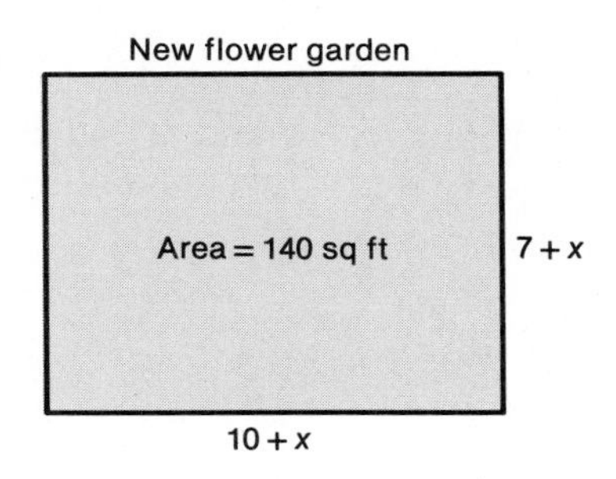

VARIABLE

Let x = amount (in feet) by which to increase the length and width.

EQUATION

We use the formula Area = (length)(width).

$$140 = (10 + x)(7 + x)$$
$$140 = 70 + 17x + x^2$$
$$0 = -70 + 17x + x^2$$

SOLVE

Use the quadratic formula. Note that $a = 1$, $b = 17$, $c = -70$.

$$x = \frac{-b \pm \sqrt{b^2 - 4ac}}{2a}$$
$$= \frac{-17 \pm \sqrt{(17)^2 - 4(1)(-70)}}{2(1)}$$
$$= \frac{-17 \pm \sqrt{289 + 280}}{2}$$
$$= \frac{-17 \pm \sqrt{569}}{2}$$
$$\cong \frac{-17 \pm 23.85}{2}$$

The two solutions are

$$\frac{-17 + 23.85}{2} \cong 3.4 \quad \text{and} \quad \frac{-17 - 23.85}{2}$$

(the latter is negative so we may disregard it since x must be positive). Thus, we would increase the length and width by approximately 3.4 feet each, giving us new dimensions of 13.4 feet and 10.4 feet.

DO EXERCISE 1.

For the next few problems, we will use the Pythagorean Theorem. Let us refresh our memories concerning this theorem.

We start with a *right* triangle (a triangle in which one of the angles measures 90°). The side opposite the 90°, or right, angle is called the *hypotenuse* and the other two sides are called the *legs.*

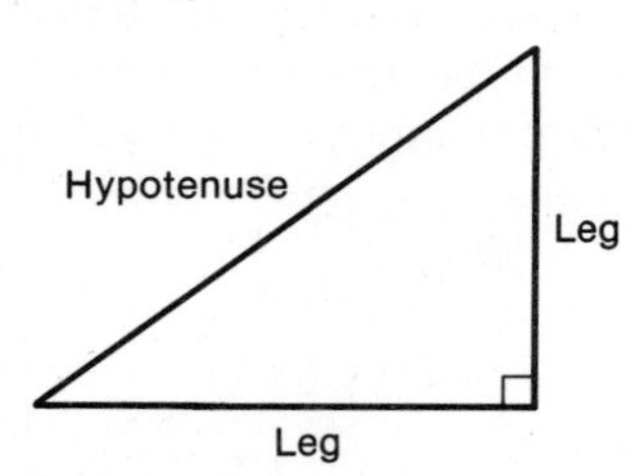

If we label the legs a and b and the hypotenuse c, then the Pythagorean Theorem states that the hypotenuse squared is equal to the sum of the squares of the legs.

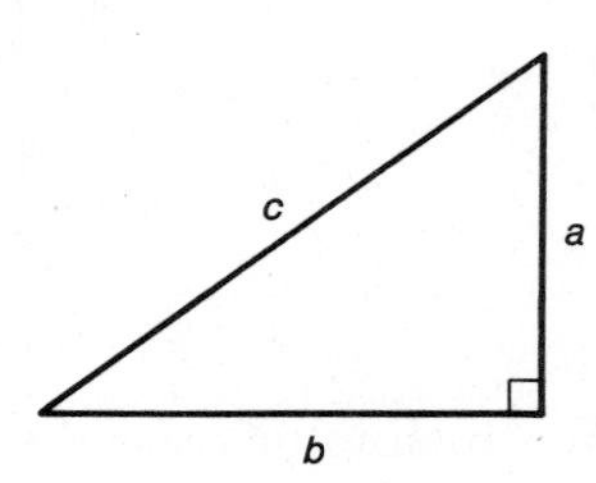

In symbols, the Pythagorean Theorem is written as follows.

$$c^2 = a^2 + b^2$$

Example 2 (a) Solve $c^2 = a^2 + b^2$ for c. (b) Solve $c^2 = a^2 + b^2$ for a.

(a) Take the principal square root of each side.

$$\sqrt{c^2} = \sqrt{a^2 + b^2}$$
$$|c| = \sqrt{a^2 + b^2}$$

Thus,

$$(1)\quad c = \sqrt{a^2 + b^2} \qquad \text{or} \qquad (2)\quad c = -\sqrt{a^2 + b^2}$$

However, since c is a distance and must be nonnegative, we may disregard $c = -\sqrt{a^2 + b^2}$ and we obtain

$$c = \sqrt{a^2 + b^2}$$

(Recall that, in general, $\sqrt{a^2 + b^2} \neq \sqrt{a^2} + \sqrt{b^2}$.)

(b) Isolate a^2.

$$a^2 = c^2 - b^2$$

Take the principal square root remembering that a cannot be negative.

$$\sqrt{a^2} = \sqrt{c^2 - b^2}$$

The solution is $a = \sqrt{c^2 - b^2}$.

DO EXERCISE 2.

1. A circle with a radius of 5 inches has an area of 25π square inches ($A = \pi r^2$). If you want to double the area of the circle, by how much (to the nearest hundredth of an inch) must you increase the radius?

2. Solve $c^2 = a^2 + b^2$ for b.

Example 3 Find side a in the following right triangle.

We can use the result from Example 2(b), and obtain

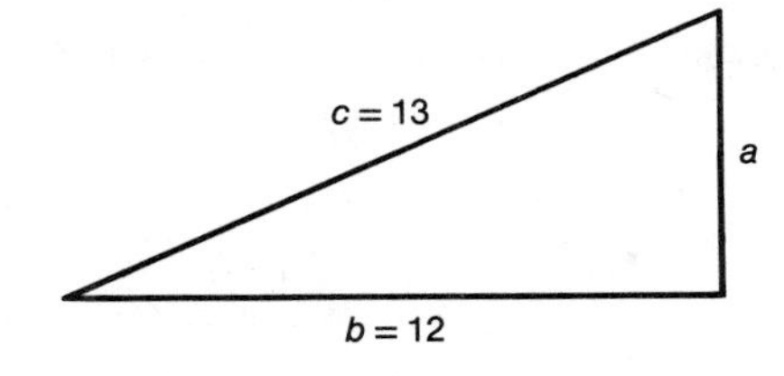

$$\begin{aligned} a &= \sqrt{c^2 - b^2} \\ &= \sqrt{(13)^2 - (12)^2} \\ &= \sqrt{169 - 144} \\ &= \sqrt{25} \\ &= 5 \end{aligned}$$

Example 4 Find x in the following right triangle.

The Pythagorean Theorem gives us

$$(17)^2 = x^2 + (x + 7)^2$$

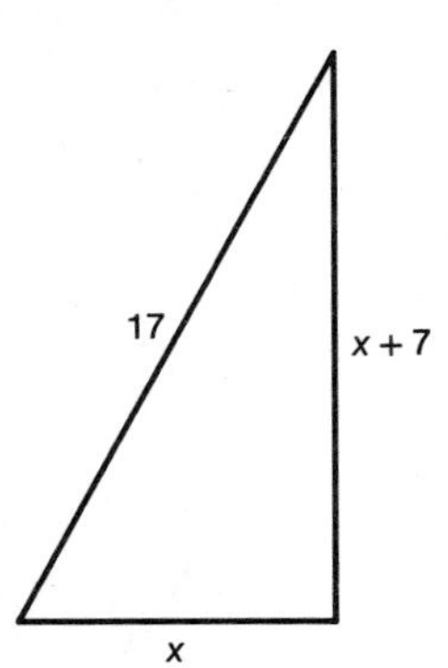

Solving for x (by factoring), we obtain

$$\begin{aligned} 289 &= x^2 + x^2 + 14x + 49 \\ 0 &= 2x^2 + 14x - 240 \\ &= 2(x^2 + 7x - 120) \\ &= 2(x - 8)(x + 15) \end{aligned}$$

Our two solutions are $x = 8$ and $x = -15$. Since x represents a distance, we may disregard -15 and obtain the one solution of $x = 8$. (Note that we could have used the quadratic formula to solve our equation.)

DO EXERCISES 3 AND 4.

The following example is a motion problem. Recall the motion formula $d = rt$ where d = distance, r = speed, and t = time.

Example 5 Buzz leaves home at 1:00 P.M. He walks east at 3 mph. At 2:00 P.M., David leaves the same house and walks south at 4.5 mph. What time is it when the men are 7.5 miles apart?

PICTURE

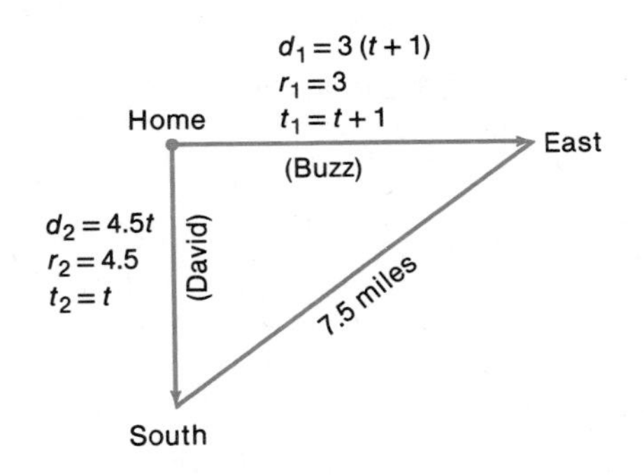

VARIABLE

Let t = David's time;
then $t + 1$ = Buzz's time.

3. The hypotenuse of a right triangle is 25 inches. Find the lengths of the legs of the triangle if one leg is 5 inches more than the other leg.

4. A twenty-foot ladder is leaning against a building. If the foot of the ladder is seven feet from the building, how far up the building does the ladder reach?

5. Joan leaves home at 9:00 A.M., traveling west at 30 mph. An hour later her friend Mary leaves home, traveling north at 25 mph. What time is it when they are 65 miles apart?

EQUATION

The motion formula for each trip gives us $d_1 = 3(t + 1)$ and $d_2 = 4.5t$. Also, the Pythagorean Theorem and the figure give us

$$(7.5)^2 = d_1^2 + d_2^2 = [3(t + 1)]^2 + (4.5t)^2$$

SOLVE

$$\begin{aligned}(7.5)^2 &= [3(t + 1)]^2 + (4.5t)^2\\ 56.25 &= 9(t^2 + 2t + 1) + 20.25t^2\\ 56.25 &= 9t^2 + 18t + 9 + 20.25t^2\\ 0 = 29.25t^2 + 18t - 47.25 &= 9(3.25t^2 + 2t - 5.25)\end{aligned}$$

We will use the quadratic formula on $3.25t^2 + 2t - 5.25 = 0$. Hence, $a = 3.25$, $b = 2$, $c = -5.25$ and

$$\begin{aligned}t &= \frac{-b \pm \sqrt{b^2 - 4ac}}{2a}\\ &= \frac{-2 \pm \sqrt{(2)^2 - 4(3.25)(-5.25)}}{2(3.25)}\\ &= \frac{-2 \pm \sqrt{4 + 68.25}}{6.5}\\ &= \frac{-2 \pm \sqrt{72.25}}{6.5}\\ &= \frac{-2 \pm 8.5}{6.5}\end{aligned}$$

Our two solutions are

$$t = \frac{-2 - 8.5}{6.5}$$

(which we may disregard because it is negative) and

$$t = \frac{-2 + 8.5}{6.5} = \frac{6.5}{6.5} = 1$$

Thus $t = 1$ hour, and it is 3:00 P.M. when the men are 7.5 miles apart.

DO EXERCISE 5.

DO SECTION PROBLEMS 10.4.

ANSWERS TO MARGINAL EXERCISES

1. $-5 + 5\sqrt{2} \cong 2.07$ in. **2.** $b = \sqrt{c^2 - a^2}$ **3.** 15 in. and 20 in.
4. $\sqrt{351} \cong 18.7$ ft **5.** 11:00 A.M.

NAME COURSE/SECTION DATE

ANSWERS

1. ____
2. ____
3. ____
4. ____
5. ____
6. ____
7. ____

SECTION PROBLEMS 10.4

1. The sum of two numbers is 6. The sum of their squares is 40. Find the numbers.

2. The sum of the squares of two numbers is 52. If the difference of the two numbers is 2, find the numbers.

3. An integer squared plus the next consecutive integer squared equals nine times the next consecutive integer minus 2. Find the three consecutive integers.

4. The sum of a number and its reciprocal is $\frac{29}{10}$. Find the number.

5. The area of a rectangle is 154 square inches. If the width is 3 inches less than the length, find the length.

6. If we want the length of a rectangle to be two inches more than the width and the area to be 80 square inches, find the dimensions of the rectangle.

7. The area of a rectangle is 12 sq ft and the perimeter is 14 ft. Find the dimensions of the rectangle. *Hint:*

$$(1)\quad 2l + 2w = 14 \qquad l + w = 7 \qquad w = 7 - l$$

$$(2)\quad A = lw \qquad A = l(7 - l)$$

8. ____________

9. ____________

10. ____________

11. ____________

12. ____________

13. ____________

14. ____________

8. The perimeter of a rectangle is 37 inches and the area is 84 square inches. Find the length and width.

Find x in the following.

9.

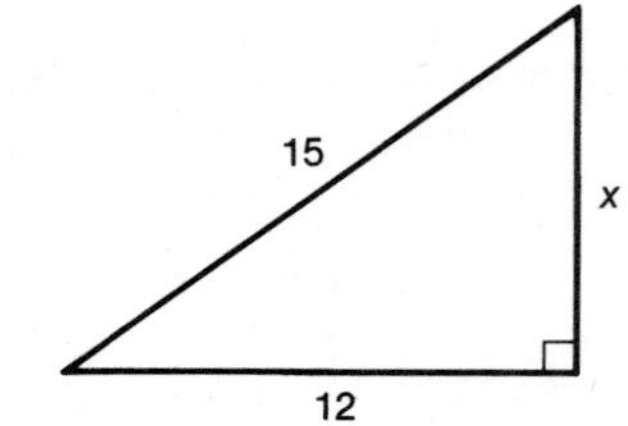

10.

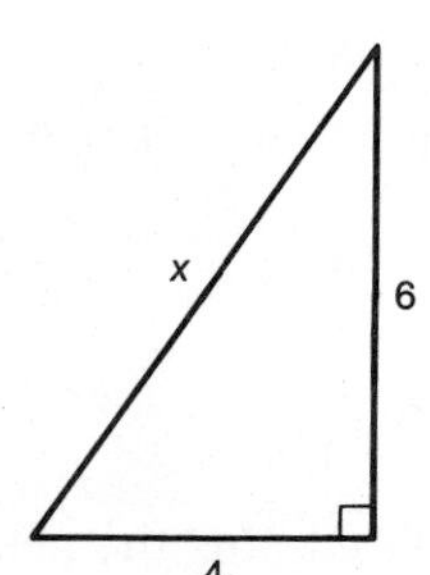

11.

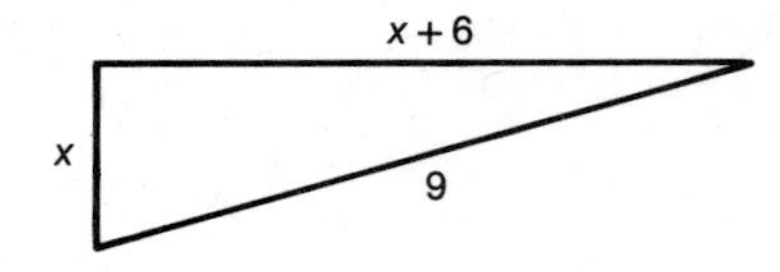

12.

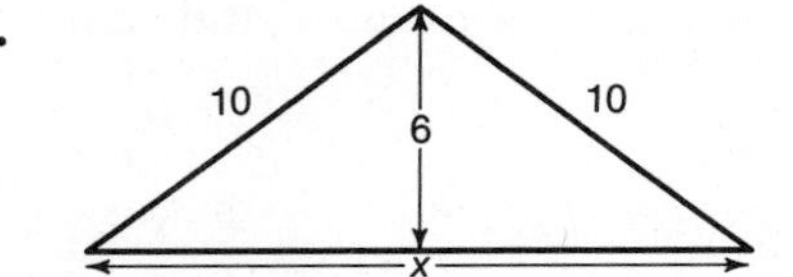

13. Cary Lowenstein and Gary Bother worked together and finished project ATB in 10 days. If Gary had done the project alone it would have taken him 2 more days than if Cary had done the project alone. If Gary had done project ATB by himself, how long would it have taken him?

14. If both the inlet and outlet pipes of a water storage tank are open, it takes 7 hours for the tank to fill. If it takes the outlet pipe 3 hours more to empty the tank than it does the inlet pipe to fill it, how long does it take the inlet pipe to fill the tank if the outlet pipe is closed?

NAME ______ COURSE/SECTION ______ DATE ______

ANSWERS

15. ______

16. ______

17. ______

18. ______

19. ______

20. ______

15. Louis has a certain number of dogs. If he squares the number he has and subtracts five times the number he has, the result is 14. How many dogs does Louis have?

16. If we have a rectangle with an area of 30 square inches and the length is twice the width minus $\frac{1}{2}$ inch, find the dimensions of the rectangle.

17. A ladder is leaning against a building. The ladder reaches 24 feet up the side of the building, and the foot of the ladder is 10 feet from the building. How long is the ladder?

18. The hypotenuse of a right triangle is 26. Find the lengths of the legs if one leg is 14 more than the other.

19. The area of a trapezoid $[A = h(B + b)/2]$ is 20 sq cm. The small base is 3 cm less than the large base and the height plus 1 is the same as the average of the small and large bases. Find the large base.

20. The height of a triangle is twice the base. If I increase the height and the base by 2, the area of the resulting triangle is 17 more than the area of the original triangle. Find the height and base of the original triangle.

21. ______________

22. ______________

23. ______________

24. ______________

25. ______________

21. The demand (q) of a certain product is related to the price (p) by the linear equation $q = 400 - 2p$ where q is in units and p is in dollars. If I want the revenue (R) from this product to be $18,200, what should I set the price at? *Hint:* Revenue = (price)(demand) or $R = pq$. Substitute for q in $R = pq$ what q is equal to in terms of p.

22. The demand (q) of product #103 is related to the price (p) of product #103 by the equation $q = 1000 - 20p$ where q is in units and p is in cents. What price should be set to generate $120.00 in revenue from this product? *Hint:* Be careful of the units. Note that $R = \$120$ is the same as $R = 12{,}000¢$.

23. Henry and Stan each ran 10 miles. If Henry ran 2 mph faster than Stan and finished 15 minutes ahead of Stan, how fast did each run?

24. Dena and Jeannie Novak each drove 30 miles. Dena drove 5 mph faster than Jeannie and finished 3 minutes ahead of Jeannie. How fast did Jeannie drive (to the nearest tenth of a mph)?

25. Set up the equation for the following problem. *Do not* solve it. At 3:00 A.M. a train leaves New York traveling west at 80 mph. At 3:30 A.M., another train leaves New York traveling south at 60 mph. At what time will they be 500 miles apart?

NAME

COURSE/SECTION

DATE

ANSWERS

1. ______

2. ______

3. ______

4. ______

5. ______

6. ______

7. ______

8. ______

9. ______

10. ______

11. ______

12. ______

13. ______

14. ______

15. ______

16. ______

17. ______

CHAPTER 10 REVIEW PROBLEMS

SECTION 10.1

Solve by factoring.

1. $x^2 - 4x + 3 = 0$

2. $y^2 + 2y - 15 = 0$

3. $2(x^2 - 4x) = x^2 - 15$

4. $12x^2 - 7x - 5 = 0$

Solve by taking principal square roots.

5. $2y^2 - 8 = 0$

6. $(x - 1)^2 = 49$

7. $(y + 2)^2 - 10 = 0$

8. $4(x + 1)^2 - 16 = 0$

9. $\dfrac{(y - 3)^2}{3} - 4 = 0$

SECTION 10.2

Solve by completing the square.

10. $y^2 = 8y - 16$

11. $x^2 - 7x + 2 = 0$

12. $2x^2 - 3x - 4 = 0$

13. $-\dfrac{y^2}{3} + 2y + 2 = 0$

SECTION 10.3

Solve by the quadratic formula.

14. $x^2 - 3x - 2 = 0$

15. $4t^2 + 12t + 9 = 0$

16. $2x^2 + 7x - 15 = 0$

17. $5x^2 - 8x + 3 = 0$

18. $16y^2 - 3y + 2 = 0$

19. $4t^2 + 9t - 5 = 0$

Solve by any method.

20. $x^2 - 5x - 6 = 0$

21. $t^2 - 6t + 3 = 0$

22. $8x^2 = 10x + 3$

23. $10y^2 - 3y - 4 = 0$

Solve.

24. $\sqrt{6 - x} = x$

25. $\sqrt{x^2 - 3x + 6} = 2 - x$

26. $\dfrac{3}{x^2 - 4} = \dfrac{2}{x^2 - 3x + 2}$

27. $\dfrac{1}{x - 2} - \dfrac{3}{x - 4} = \dfrac{1}{x + 5}$

SECTION 10.4

28. The area of a rectangle is 77 square inches and the perimeter is 36 inches. Find the dimensions of the rectangle.

18. ______

19. ______

20. ______

21. ______

22. ______

23. ______

24. ______

25. ______

26. ______

27. ______

28. ______

29. ______

30. ______

29. Find x in the triangle at the right.

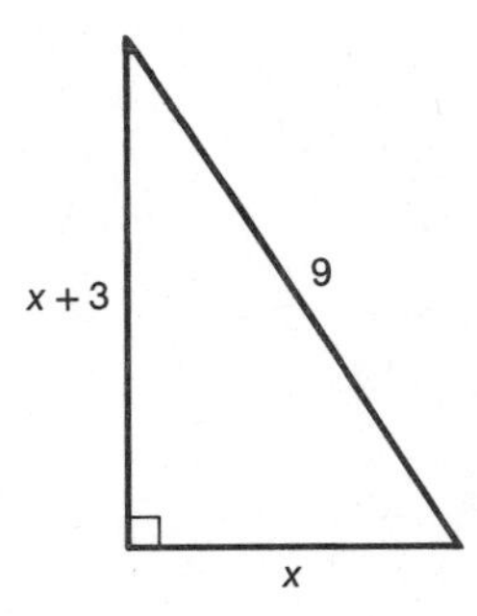

30. Find x in the triangle at the right.

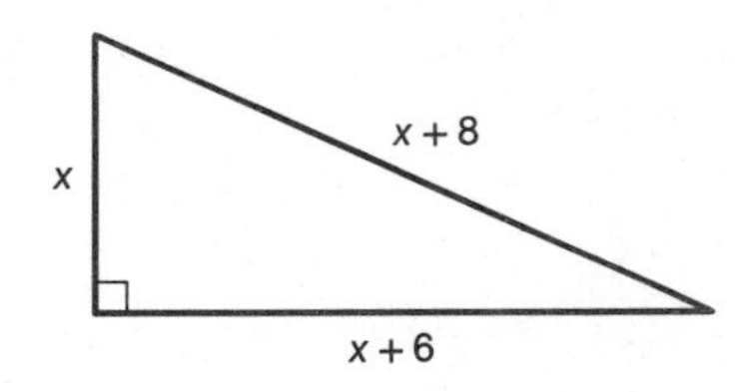

NAME COURSE/SECTION DATE

ANSWERS

1. ______
2. ______
3. ______
4. ______
5. ______
6. ______
7. ______

CHAPTER 10 TEST

1. Solve $2t^2 + 3t - 35 = 0$ by factoring.

2. Solve $x^2 - 81 = 0$ by principal square roots.

3. Solve $(y + 2)^2 - 3 = 0$ by principal square roots.

4. Solve $x^2 + 6x - 4 = 0$ by completing the square.

5. Solve $2t^2 = 3t + 4$ by completing the square.

Solve by any method.

6. $t^2 - 4t = -2$

7. $2x^2 - x + 5 = 0$

8. ______________

9. ______________

10. ______________

11. ______________

12. ______________

13. ______________

14. ______________

15. ______________

8. $3y^2 + 5y - 2 = 0$

9. $9x^2 + 4 = 12x$

10. $\sqrt{2x + 1} = x - 1$

11. $\dfrac{1}{t + 1} - \dfrac{t}{t - 2} = 1$

12. $\dfrac{1}{x^2 - 5x + 6} + \dfrac{x}{x - 2} = 0$

13. The hypotenuse of a right triangle is 13 feet. If one leg is twice the other leg plus 2, find the area of the right triangle.

14. The area of a triangle is 48 sq in. If the height is two-thirds of the base, find the height.

15. If both the inlet pipe and the outlet pipe of a water storage tank are open, it takes 10 hours for the tank to fill. It takes the outlet pipe 4 hours more to empty the tank than it does the inlet pipe to fill it. How long does it take the inlet pipe to fill the tank if the outlet pipe is closed?

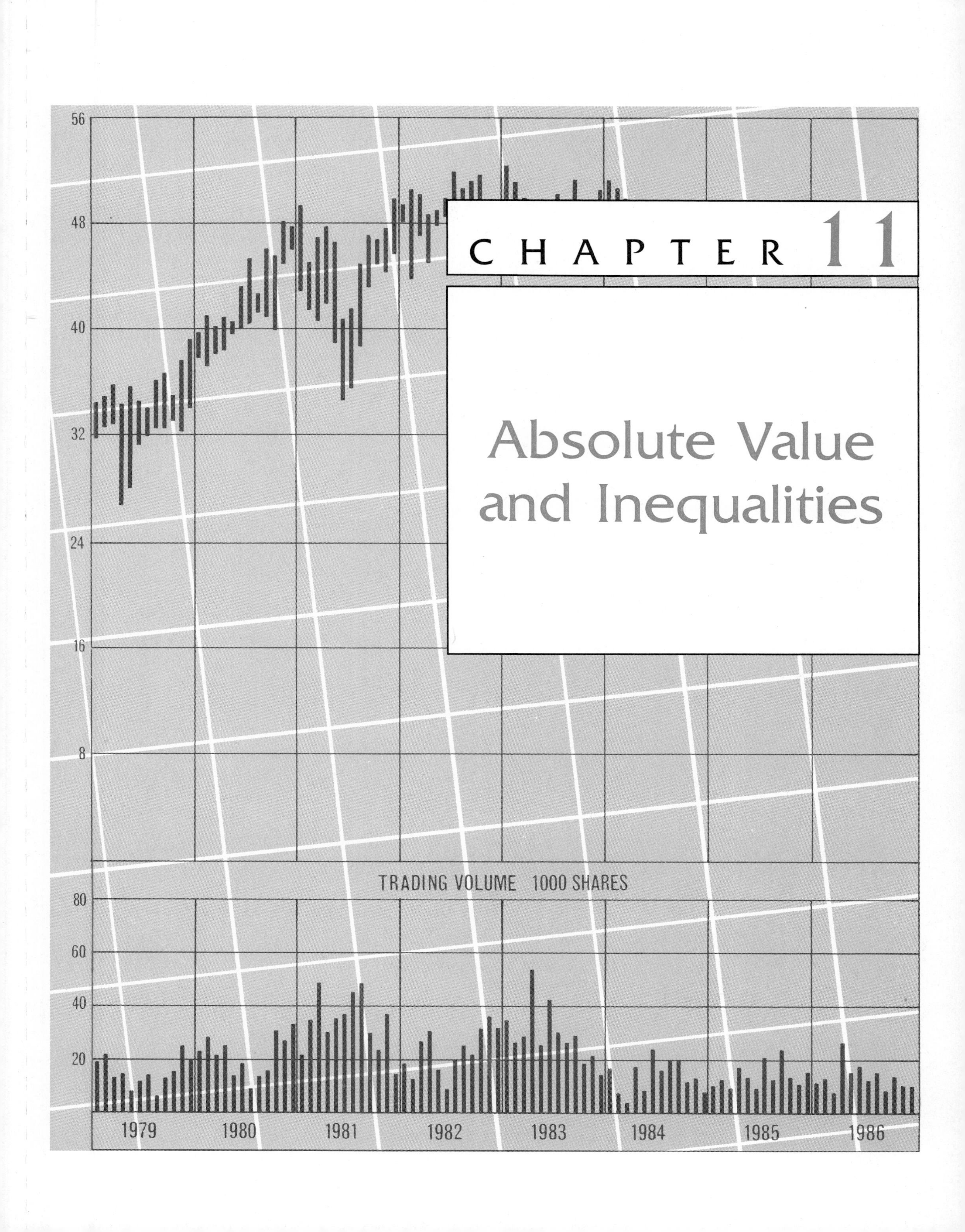

CHAPTER 11

Absolute Value and Inequalities

LEARNING OBJECTIVES

After finishing Section 11.1, you should be able to:

- Use set notation, graphical representation, or interval notation to represent inequalities.

Change the following English descriptions of a set into set notation.

1. The set of all real numbers less than 8

2. The set of all real numbers greater than or equal to $\frac{3}{4}$

3. The set of all real numbers between -2 and 3, including -2 but not including 3

4. Are the following two sets equal? $\{x \mid -1 \leq x \leq 4\}$ and $\{x \mid 4 \geq x \geq -1\}$

11.1

Introduction to Inequalities

Recall that the real numbers consist of all rational and irrational numbers. This means that there is a one-to-one correspondence between all the points on a number line and the set of all real numbers, since every real number corresponds to a point on the number line and every point on the number line is represented by a real number.

Now we want to discuss not a single real number, but intervals of real numbers. Such intervals arise naturally when we discuss the set of all real numbers less than or greater than a fixed number. To help in the discussion of such intervals (sometimes referred to as sets of real numbers), we will first introduce some notation.

DEFINITION 11.1

A. The symbol $<$ means "less than."

EXAMPLE: $\{x|x < 3\}$

The notation $\{x|x < 3\}$, called set notation, is read as "the set of all real numbers x such that x is less than 3."

B. The symbol $\leq$ means "less than or equal to."

EXAMPLE: $\{x|x \leq 3\}$

Note that 3 is included in the set $\{x|x \leq 3\}$ but 3 is not included in the set $\{x|x < 3\}$.

C. The symbol $>$ means "greater than."

EXAMPLE: $\{x|x > -2\}$

D. The symbol $\geq$ means "greater than or equal to."

EXAMPLE: $\left\{x|x \geq \frac{3}{2}\right\}$

We were first introduced to the symbols $<$ and $>$ in Chapter 1. These two symbols, as well as $\leq$ and $\geq$, are widely used in mathematics, and it is worth our time and trouble to become familiar with them.

Example 1 Change the following English descriptions of a set into set notation.

(a) The set of all real numbers greater than 3. $\{x|x > 3\}$

(b) The set of all real numbers less than or equal to -2. $\{x|x \leq -2\}$

(c) The set of all real numbers greater than -1 and less than or equal to 5. $\{x|-1 < x \leq 5\}$

DO EXERCISES 1 THROUGH 4.

If we want to depict the set of all real numbers less than (smaller than) 4, then one way to do it is by using set notation, as in $\{x|x < 4\}$. However, there are other ways. The next example looks at the *graphical representation* of various sets.

Example 2

Set Notation	*Graphical Representation*
(a) $\{x\|x < 3\}$	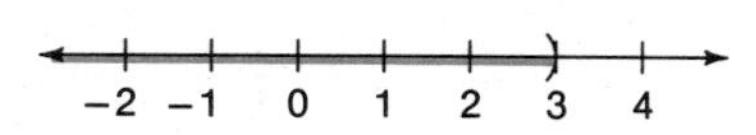

The parenthesis) on the graph indicates that 3 is not included in this set.

(b) $\{x\|x \geq -2\}$	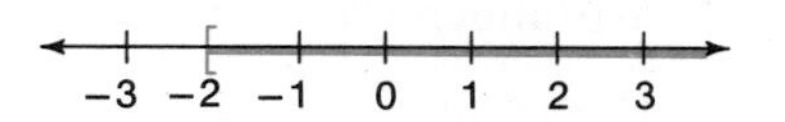

The bracket [on the graph indicates that 2 is included in this set.

(c) $\{x\|-3 < x \leq 2\}$	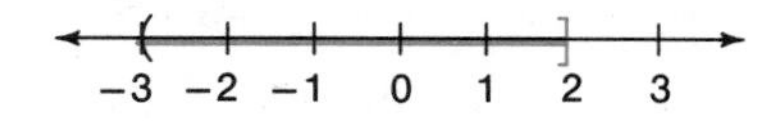

Once again, the parenthesis (means that -3 is not included in the set and the bracket] means that 2 is included in the set. Also, -3 is called a left-hand end point for this interval and 2 is called a right-hand end point.

DO EXERCISES 5 THROUGH 7.

We have another way to represent intervals of real numbers using what is called interval notation.

Example 3

Set Notation	*Graphical Representation*	*Interval Notation*
(a) $\{x\|x < 5\}$		$(-\infty, 5)$

The negative infinity symbol $-\infty$ means that there is no left-hand end point to our interval. In other words, we want to include *all* numbers smaller than 5. Note that 5 is a right-hand end point.

(b) $\{x\|x \geq -1\}$		$[-1, \infty)$

As in graphical representation, the [in interval notation means that we include the end point while) means we do not.

(c) $\{x\|-2 \leq x < 4\}$	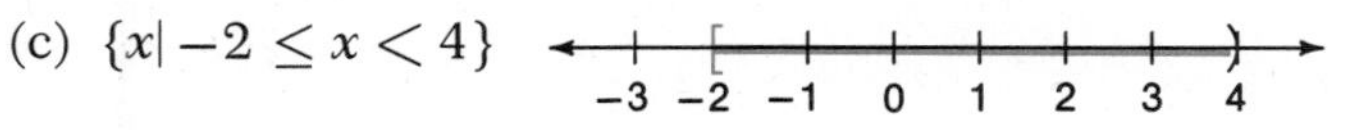	$[-2, 4)$
(d) $\left\{x\|0 \leq x \leq \frac{5}{2}\right\}$		$\left[0, \frac{5}{2}\right]$

DO EXERCISE 8.

We now want to look at the order property of the real numbers. This property was examined for whole numbers, fractions, and decimals in Chapter 1. First

Change the following from set notation to graphical representation or vice versa.

Set Notation	*Graphical Representation*
5. $\{x\|x > 3\}$	
6.	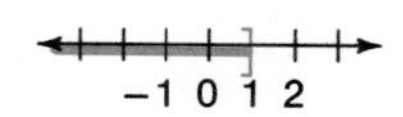
7. $\{x\|-1 \leq x \leq 4\}$	

8. Fill in the columns.

Set Notation	*Graphical Representation*	*Interval Notation*
(a) $\{x\|x < 3\}$		
(b)	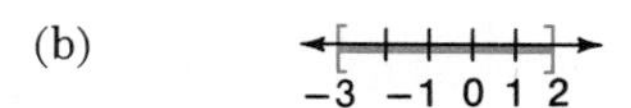	
(c)		$[-1, 4)$
(d)		$(-\infty, -2]$
(e) $\left\{x\|0 < x \leq \frac{5}{2}\right\}$		

9. Use Definition 11.2A to show that 1 is greater than -4.

10. Insert $<$, $>$, or $=$ between the following pairs of numbers.

(a) $-3, \quad -10$ (b) $\frac{7}{5}, \quad 1\frac{2}{5}$

(c) $0, \quad \frac{1}{5}$ (d) $\frac{1}{4}, \quad \frac{1}{3}$

(e) $\frac{5}{7}, \quad \frac{2}{3}$ (f) $0.109, \quad 0.110$

we give two definitions, one algebraic and one geometric, of how a real number x may be greater than another real number y.

DEFINITION 11.2

A. *Algebraic Definition:* x is greater than y if and only if $x - y$ is positive; or, in symbols, $x > y$ if and only if $x - y > 0$.

EXAMPLES:

$$4 > 3 \text{ since } 4 - 3 = 1 > 0$$
$$2 > -5 \text{ since } 2 - (-5) = 7 > 0$$
$$-1 > -3 \text{ since } -1 - (-3) = 2 > 0$$

B. *Geometric Definition:* x is greater than y if x is to the right of y on a number line.

EXAMPLES: Looking at the number line

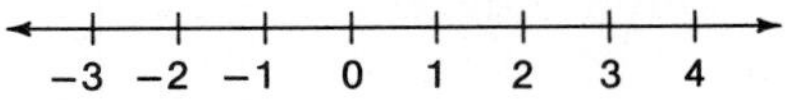

and using the definition, we see that

$$2 > -1 \qquad 3\frac{1}{3} > 2\frac{4}{5} \qquad 0 > -3$$

The order property of the real numbers says that we can compare the size of any two real numbers.

ORDER PROPERTY: Given two arbitrary real numbers, x and y, then exactly one of the following is true:

(1) $x > y$
(2) $x = y$
or (3) $x < y$

DO EXERCISES 9 AND 10.

DO SECTION PROBLEMS 11.1.

ANSWERS TO MARGINAL EXERCISES

1. $\{x \mid x < 8\}$ **2.** $\left\{x \mid x \geq \frac{3}{4}\right\}$ **3.** $\{x \mid -2 \leq x < 3\}$ **4.** Yes

5.
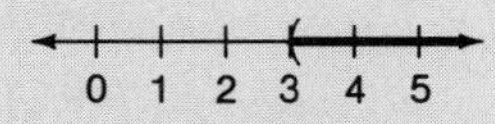

6. $\{x \mid x \leq 1\}$ **7.**
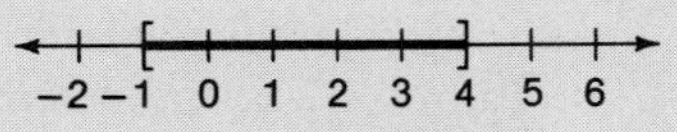

8. (a)
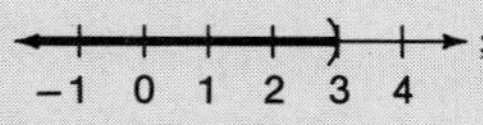

; $(-\infty, 3)$ (b) $\{x \mid -3 \leq x \leq 2\}$; $[-3, 2]$

(c) $\{x \mid -1 \leq x < 4\}$;

(d) $\{x \mid x \leq -2\}$;
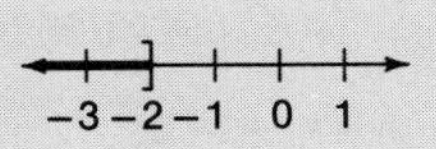

(e)
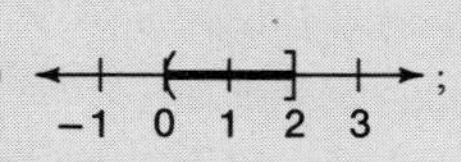

; $\left(0, \frac{5}{2}\right]$

9. $1 - (-4) = 5 > 0$ **10.** (a) $>$ (b) $=$ (c) $<$ (d) $<$ (e) $>$ (f) $<$

NAME COURSE/SECTION DATE

SECTION PROBLEMS 11.1

Change the English description to set notation.

1. The set of all real numbers less than -3
2. The set of all real numbers between 5 and 10, including 5 and 10
3. The set of all real numbers greater than or equal to 4
4. The set of all real numbers larger than 0 and smaller than or equal to 7
5. The set of all real numbers larger than -2
6. The set of all real numbers less than or equal to $\frac{3}{4}$
7. The set of all real numbers greater than 1 and less than or equal to 4
8. The set of all real numbers greater than or equal to -4 and less than or equal to 6.

Change the set notation into an English description.

9. $\{x|x > 3\}$
10. $\{x|x \leq -2\}$
11. $\{x|0 < x < 4\}$
12. $\{x|-1 \leq x < 5\}$
13. $\left\{x|x \geq \frac{2}{3}\right\}$
14. $\{x|2 \leq x \leq 10\}$

ANSWERS

1. ________
2. ________
3. ________
4. ________
5. ________
6. ________
7. ________
8. ________
9. ________
10. ________
11. ________
12. ________
13. ________
14. ________

15. ______________

16. ______________

17. ______________

18. ______________

19. ______________

20. ______________

21. ______________

22. ______________

Fill in the columns.

Set Notation	Graphical Representation	Interval Notation
15. $\{x \mid x > 2\}$		
16.	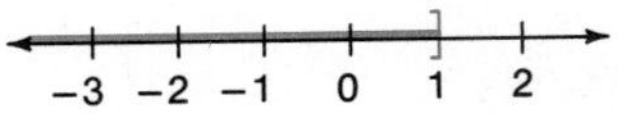	
17.		$[-2, \infty)$
18.		$[-1, 7)$
19. $\left\{x \mid 0 < x \le \frac{4}{3}\right\}$		
20.	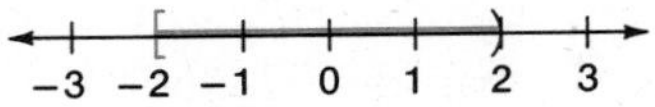	
21.	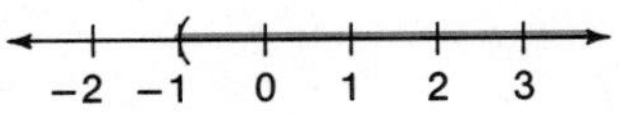	
22. $\{x \mid -1 \le x \le 2\}$		

NAME

COURSE/SECTION

DATE

ANSWERS

Set Notation	Graphical Representation	Interval Notation
23. $\left\{x \mid x > \frac{3}{4}\right\}$		
24.		$[-2, 2]$
25.		$(\infty, 2)$
26.	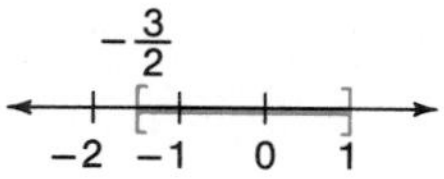	
27.		$(-2, 3)$
28.		$[-2, 0)$
29.	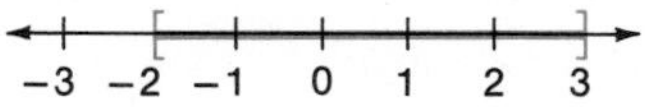	
30. $\{x \mid x \leq 2\}$		
31.	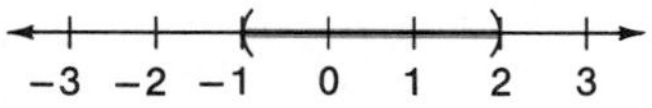	
32. $\{x \mid -2 < x < 3\}$		

23. ____________

24. ____________

25. ____________

26. ____________

27. ____________

28. ____________

29. ____________

30. ____________

31. ____________

32. ____________

33. ______

34. ______

35. ______

36. ______

37. ______

38. ______

39. ______

40. ______

41. ______

42. ______

43. ______

44. ______

45. ______

33. Give an algebraic definition for a real number x being less than a real number y.

34. Give a geometric definition for x is less than y.

Insert $<$, $>$, or $=$ between the following pairs of numbers.

35. $-3, 3$

36. $-5, -7$

37. $\frac{17}{20}, \frac{7}{8}$

38. $\frac{2}{\sqrt{4}}, 1$

39. $0.1102, 0.11019$

40. $0, -\frac{1}{2}$

41. $1.2, \frac{6}{5}$

42. $\frac{9}{12}, \frac{10}{15}$

43. $\frac{1}{9}, 0.11\overline{1}$

44. Show graphically that the set $\{x \mid -2 < x < 3\}$ and the set $\{x \mid 3 > x > -2\}$ are equal.

45. Show graphically that the set $\{x \mid 0 \leq x \leq 4\}$ and the set $\{x \mid 4 \geq x \geq 0\}$ are equal.

11.2

Solving Inequalities

LEARNING OBJECTIVES

After finishing Section 11.2, you should be able to:

- Solve inequalities using the addition and multiplication principles.

An inequality is just like an equation, except that we have one of the four inequality signs ($<$, $>$, $\geq$, or $\leq$) instead of an equality sign ($=$). The strategy we use in solving first-degree inequalities in one variable is the same strategy we used in solving first-degree equations in one variable. We start with an inequality where the solution is not obvious and change it to equivalent inequalities until we have an equivalent inequality for which the solution is obvious. Once again, there are two principles used for changing an inequality to an equivalent inequality: the addition principle and the multiplication principle. While the addition principle for inequalities is the same as that for equations, the multiplication principle for inequalities has an added twist that the multiplication principle for equations does not have. We shall discuss the addition principle first.

Note

1. $2 < 4$
2. If we add 3 to each side of the above inequality, we obtain $2 + 3 < 4 + 3$ or $5 < 7$, which is still a true statement.
3. If we add -3 to each side of the original inequality, we obtain $2 + (-3) < 4 + (-3)$ or $-1 < 1$, which is still a true statement.

The above remarks may be generalized as follows.

RULE 11.1 If $a < b$ and c is any real number, then $a + c < b + c$. This rule also holds for $\leq$, $>$, and $\geq$, and gives us the addition principle.

ADDITION PRINCIPLE If we add the same quantity to both sides of an inequality, then we obtain an equivalent inequality.

Example 1 Solve $x + 7 < -4$.

Just as in solving linear equations, we first want to isolate the x.

Add -7 to each side and simplify.

$$x + 7 + (-7) < -4 + (-7)$$
$$x < -11$$

Thus, $x + 7 < -4$ and $x < -11$ are equivalent inequalities and, hence, the solution to $x + 7 < -4$ is $\{x|x < -11\}$. The graphical representation is shown to the right.

−13 −12 −11 −10 −9

Example 2 Solve $y - 3 \geq -10$.

Add 3 to each side and simplify.

$$y - 3 + 3 \geq -10 + 3$$
$$y \geq -7$$

Solve. Demonstrate your solution on a number line.

1. $x - 3 \leq -2$

2. $t + \frac{3}{2} < 5$

Thus, the solution to $y - 3 \geq -10$ is $\{y | y \geq -7\}$. The graph is at the right.

−9 −8 −7 −6 −5 −4

■ **DO EXERCISES 1 AND 2.**

Example 3 Solve $3x - 4 \geq 2x + 3$.

Add $-2x$ to each side and simplify.

$$3x - 4 + (-2x) \geq 2x + 3 + (-2x)$$
$$x - 4 \geq 3$$

Add 4 to each side and simplify.

$$x - 4 + 4 \geq 3 + 4$$
$$x \geq 7$$

Thus, the solution is $\{x | x \geq 7\}$. The graph is at the right.

6 7 8 9 10

Example 4 Solve $2 < x + 5 \leq 7$.

This inequality says that 2 is less than $x + 5$, which in turn is less than or equal to 7.

Add -5 to all three "sides" and simplify.

$$2 + (-5) < x + 5 + (-5) \leq 7 + (-5)$$
$$-3 < x \leq 2$$

Thus, our solution is $\{x | -3 < x \leq 2\}$. The graph is at the right.

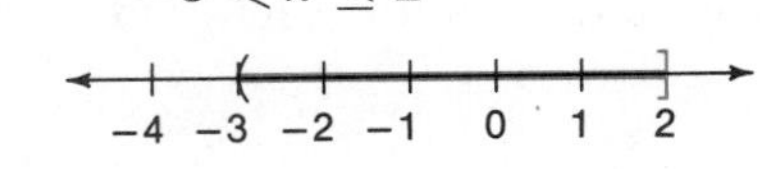

Solve. Demonstrate your solution on a number line.

3. $2x + 1 > x - 4$

4. $-2 < y - 4 < 3$

■ **DO EXERCISES 3 AND 4.**

Now we turn our attention to the multiplication principle.

Note

1. $2 < 4$
2. If we multiply each side of the inequality above by 3, we obtain

$$(2)(3) < (4)(3) \quad \text{or} \quad 6 < 12$$

which is a true statement.

3. If we multiply each side of the original inequality by -3, we obtain

$$(2)(-3) < 4(-3) \quad \text{or} \quad -6 < -12$$

which is not a true statement. We can make it a true statement if we change the "sense" of the inequality. In other words, it will be true if we change $<$ to $>$. Then we obtain $-6 > -12$ which is a true statement.

Thus, we have to break the multiplication principle for inequalities into two parts depending on whether the number by which we are multiplying both sides is positive or negative.

RULE 11.2

A. If $a < b$ and $c > 0$, then $ac < bc$.

B. If $a < b$ and $c < 0$, then $ac > bc$. Note that we have to change the sense of the inequality when we multiply by a negative number.

Rule 11.2 also holds for $>$, $\leq$, and $\geq$. If we have to change the sense of an inequality, then

1. $<$ becomes $>$.
2. $\leq$ becomes $\geq$.
3. $>$ becomes $<$.
4. $\geq$ becomes $\leq$.

MULTIPLICATION PRINCIPLE

A. If we multiply both sides of an inequality by a *positive* number, then we obtain an equivalent inequality.

B. If we multiply both sides of an inequality by a *negative* number *and change the sense of the inequality,* then we obtain an equivalent inequality.

Example 5 Solve $3x < 12$.

Multiply both sides by $\frac{1}{3}$ and simplify.

$$\frac{1}{3}(3x) < \frac{1}{3}(12)$$

$$x < 4$$

Thus, the solution is $\{x|x < 4\}$.

0 1 2 3 4

Example 6 Solve $-\frac{x}{4} \geq 2$.

Multiply both sides by -4 (hence we must change the sense of the inequality) and simplify.

$$(-4)\left(-\frac{x}{4}\right) \leq (-4)(2)$$

$$x \leq -8$$

Thus, the solution is $\{x|x \leq -8\}$. The graph is at the right.

−11 −10 −9 −8 −7 −6 −5

DO EXERCISES 5 THROUGH 7.

We can also use the addition and multiplication principles together to help solve inequalities.

Example 7 Solve $3x - 5 < 10$.

Add 5 to each side and simplify.

$$3x - 5 + 5 < 10 + 5$$

$$3x < 15$$

Multiply each side by $\frac{1}{3}$ and simplify.

$$\frac{1}{3}(3x) < \frac{1}{3}(15)$$

$$x < 5$$

Thus, the solution is $\{x|x < 5\}$. The graph is at the right.

3 4 5 6 7

DO EXERCISE 8.

Solve. Demonstrate your solution on a number line.

5. $5x \leq -15$

6. $-3x > 2$

7. $-\frac{2y}{3} < -1$

Solve. Demonstrate your solution on a number line.

8. $3x + 4 < -6$

Solve. Demonstrate your solution on a number line.

9. $-7 - 3t \geq 5$

Example 8 Solve $2 - t \geq -7$.

Add -2 to each side and simplify.

$$2 - t + (-2) \geq -7 + (-2)$$
$$-t \geq -9$$

Multiply each side by -1 (change the sense of the inequality) and simplify.

$$(-1)(-t) \leq (-1)(-9)$$
$$t \leq 9$$

Thus, the solution is $\{t | t \leq 9\}$. The graph is at the right.

6 7 8 9 10

DO EXERCISE 9.

Solve. Demonstrate your solution on a number line.

10. $2x - 4 \leq 4x + 1$

Example 9 Solve $-3x + 2 \geq -5x - 6$.

Add $5x$ to each side and simplify.

$$-3x + 2 + 5x \geq -5x - 6 + 5x$$
$$2x + 2 \geq -6$$

Add -2 to each side and simplify.

$$2x + 2 + (-2) \geq (-6) + (-2)$$
$$2x \geq -8$$

Multiply each side by $\frac{1}{2}$ and simplify.

$$\frac{1}{2}(2x) \geq \frac{1}{2}(-8)$$
$$x \geq -4$$

Thus, the solution is $\{x | x \geq -4\}$. The graph is at the right.

−5 −4 −3 −2 −1

DO EXERCISE 10.

And finally, here are two more examples.

Solve. Demonstrate your solution on a number line.

11. $-3(x + 4) < 2(x + 1) - 3$

Example 10 Solve $2(x - 5) \geq 4(x - 2) + 3$.

Remove parentheses and simplify.

$$2x - 10 \geq 4x - 8 + 3$$
$$2x - 10 \geq 4x - 5$$

Add $-4x$ to each side and simplify.

$$2x - 10 + (-4x) \geq 4x - 5 + (-4x)$$
$$-2x - 10 \geq -5$$

Add 10 to each side and simplify.

$$-2x - 10 + 10 \geq -5 + 10$$
$$-2x \geq 5$$

Multiply each side by $-\frac{1}{2}$ (change the sense of the inequality sign) and simplify.

$$\left(-\frac{1}{2}\right)(-2x) \leq \left(-\frac{1}{2}\right)(5)$$
$$x \leq -\frac{5}{2}$$

Thus, the solution is $\{x | x \leq -\frac{5}{2}\}$. The graph is at the right.

$-\frac{5}{2}$

−4 −3 −2 −1

DO EXERCISE 11.

Example 11 Solve $-3 < -3x - 1 \leq 2$.

Add 1 to all three sides and simplify.

$$-3 + 1 < -3x - 1 + 1 \leq 2 + 1$$
$$-2 < -3x \leq 3$$

Multiply all three sides by $-\frac{1}{3}$ (change the sense of both inequality signs) and simplify.

$$\left(-\frac{1}{3}\right)(-2) > \left(-\frac{1}{3}\right)(-3x) \geq \left(-\frac{1}{3}\right)(3)$$

$$\frac{2}{3} > x \geq -1$$

$$\text{or equivalently } -1 \leq x < \frac{2}{3}$$

Thus, the solution is $\{x \mid -1 \leq x < \frac{2}{3}\}$.

The graph is at the right.

-2 -1 0 $\frac{2}{3}$ 1 2

DO EXERCISES 12 AND 13.

DO SECTION PROBLEMS 11.2.

Solve. Demonstrate your solution on a number line.

12. $-4 \leq 2x + 3 < 1$

13. $\frac{2y}{3} - 1 < -\frac{1}{4} + \frac{1}{6}$

(*Hint:* First clear fractions by multiplying both sides by 12.)

ANSWERS TO MARGINAL EXERCISES

1. $\{x|x \leq 1\}$ **2.** $\left\{t|t < \frac{7}{2}\right\}$ **3.** $\{x|x > -5\}$ **4.** $\{y|2 < y < 7\}$
5. $\{x|x \leq -3\}$ **6.** $\left\{x|x < -\frac{2}{3}\right\}$ **7.** $\left\{y|y > \frac{3}{2}\right\}$ **8.** $\left\{x|x < -\frac{10}{3}\right\}$
9. $\{t|t \leq -4\}$ **10.** $\left\{x|x \geq -\frac{5}{2}\right\}$ **11.** $\left\{x|x > -\frac{11}{5}\right\}$
12. $\left\{x|-\frac{7}{2} \leq x < -1\right\}$ **13.** $\left\{y|y < \frac{11}{8}\right\}$

NAME

COURSE/SECTION

DATE

ANSWERS

1. ____
2. ____
3. ____
4. ____
5. ____
6. ____
7. ____
8. ____
9. ____
10. ____
11. ____

SECTION PROBLEMS 11.2

1. How do the two sets $\{x|x \geq 3\}$ and $\{x|x > 3\}$ differ?

2. Solve the inequality $-x \geq 10$ by using just the addition principle. (Using the multiplication principle, we would solve it as $(-1)(-x) \leq (-1)(10)$ or $x \leq -10$.)

Solve.

3. $x + 5 < -3$

4. $t - 7 \geq -2$

5. $y + \frac{2}{3} > -\frac{3}{4}$

6. $-3y < -9$

7. $2.1x \geq -0.63$

8. $2x - 10 \leq 2$

9. $7 - t > -1$

10. $\frac{x}{4} + 5 \geq -6$

11. $-3x - 4 \geq 2x + 6$

12. ______

13. ______

14. ______

15. ______

16. ______

17. ______

18. ______

19. ______

20. ______

21. ______

22. ______

23. ______

24. ______

12. $3 - 5x < 2x + 6$

13. $\frac{x}{3} - \frac{1}{4} < -2$

14. $3x - 5 \leq 10$

15. $-2x + 1 > -3$

16. $2y - 5 < -y + 6$

17. $\frac{t}{2} - \frac{1}{3} \geq \frac{1}{4}$

18. $3 - 2x \geq -5$

19. $\frac{2x - 3}{4} < 1$

20. $\frac{y}{-2} \geq -3$

21. $\frac{7 - x}{4} < 3$

22. $6x - 5 \leq -10x - 4$

23. $3(1 - 2y) < -2(y + 3) - 4$

24. $3(y - 1) - 2(y + 3) \leq 4y$

NAME COURSE/SECTION DATE

25. $-\frac{2}{3}t + \frac{1}{4} > 2t - \frac{4}{3}$

26. $-4(x + 1) - 3x < 0$

27. $2t - 4 > 5t + 5$

28. $-4 < 2x + 1 < 0$

29. $3 \leq \frac{-x - 4}{2} < 5$

30. $-2x + 7 = -3$

31. $-2 < 3x \leq 6$

32. $0 \leq 1 - x \leq 4$

33. $-2 \leq 2x + 4 \leq 2$

34. $-2(1 - t) > 3t - 4$

35. $\frac{x}{3} - \frac{3}{4} < -x + \frac{1}{6}$

36. $-3(y + 5) - y \geq -2(1 - y)$

37. $\frac{1}{2} < 2x - 1 < 3$

ANSWERS

25. ____________

26. ____________

27. ____________

28. ____________

29. ____________

30. ____________

31. ____________

32. ____________

33. ____________

34. ____________

35. ____________

36. ____________

37. ____________

38. ______

39. ______

40. ______

41. ______

42. ______

43. ______

44. ______

45. ______

46. ______

47. ______

48. ______

49. ______

50. ______

38. $\frac{4-3x}{-2} \geq -5$

39. $6a - 1 = 11$

40. $x - \pi > 0$

41. $2x - 5 \leq 4$

42. $\frac{t}{-2} + 1 > -3$

43. $\frac{3y}{2} - \frac{1}{4} \leq y$

44. $4 \leq 2x \leq 5$

45. $-3 < 3t - 1 < 6$

46. $0 \leq 1 - 2y < 5$

47. $\frac{2x}{3} - \frac{x+1}{4} \leq \frac{1}{6}$

48. $3(x + 1) - 2(2x + 5) > 3x$

49. $-4(1 - 2y) > -y + 5$

50. $\frac{2-6x}{-3} \leq -2$

11.3

Absolute Value in Equations and Inequalities

We gave a geometric definition of absolute value in Chapter 2. We begin this section by stating it again.

DEFINITION 11.3 The absolute value of x, denoted $|x|$, is the distance x is from the origin on a number line.

EXAMPLES: $|4| = 4$ $\quad |6 - 11| = |-5| = 5$

$|-3| = 3$ $\quad \left|-\frac{5}{2}\right| = \frac{5}{2}$

There is also an equivalent algebraic definition, which we now give. We should be familiar with both definitions. For the work in this section, however, we will use the geometric definition.

DEFINITION 11.4 The absolute value of x is given by

$$|x| = \begin{cases} x \text{ if } x \geq 0 \\ -x \text{ if } x < 0 \end{cases}$$

The algebraic definition of absolute value is split into two parts, and which part we use depends on whether we are taking the absolute value of a nonnegative number or a negative number.

Example 1

(a) $|5| = 5$ Since $5 \geq 0$, we use the top part of the definition which says $|x| = x$.

(b) $|-7| = -(-7) = 7$ Since $-7 < 0$, we use the bottom part of the definition which says $|x| = -x$.

(c) $-|-5| = -(5) = -5$

(d) $|8 + (-12)| = |-4| = 4$

(e) $|8| + |-12| = 8 + 12 = 20$

Also, recall that

$$|x \cdot y| = |x| \cdot |y|$$

but in general

$$|x + y| \neq |x| + |y|$$

(See Examples 1(d) and 1(e).)

DO EXERCISES 1 THROUGH 6.

LEARNING OBJECTIVES

After finishing Section 11.3, you should be able to:

- Solve equations of the form $|x| = a$.
- Solve inequalities of the form $|x| < a$ and $|x| > a$.

Evaluate using either the geometric or algebraic definition of absolute value.

1. $|24|$

2. $|-17|$

3. $-|-3|$

4. $|7 - 12|$

5. $|7| - |12|$

6. Are the following true or false?
(a) $|x \cdot y| = |x| \cdot |y|$
(b) $|x - y| = |x| - |y|$
(c) $|x^2| = x^2 = |x|^2$
(d) $|x - y| = |y - x|$
(e) $|-x| = -|x|$
(f) $|x + y| = |x| + |y|$

Solve.

7. $|x| = 5$

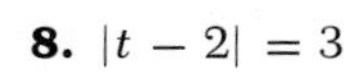

8. $|t - 2| = 3$

We looked at the equation $|x| = a$ (where a is nonnegative) and solutions to this equation in Chapter 10. We would like to look at this equation again.

Suppose we want to solve the equation $|x| = 2$. We realize that $|x|$ means the distance x is from the origin on a number line, so the equation $|x| = 2$ tells us that we are looking for points that are a distance of 2 units from the origin. If we look at the number line

−3 −2 −1 0 1 2 3

we see that there are two points, 2 and −2, that are a distance of two units from the origin. Thus, the equation $|x| = 2$ has two solutions, namely $x = 2$ or $x = -2$.

If we consider the arbitrary nonnegative number a and the number line

$-a$ 0 a

then the following proposition is fairly obvious.

PROPOSITION 11.1 The equation $|x| = a$ $(a \geq 0)$ has two solutions

$$(1)\quad x = a \qquad \text{or} \qquad (2)\quad x = -a$$

If the number a is negative $(a < 0)$, then the equation $|x| = a$ has no solution since $|x| \geq 0$ for all real numbers x. That is, $|x|$ can never be less than zero. For example, the equation $|x| = -5$ has no solution.

Example 2 Solve $|x| = 4$.

From Proposition 11.1, we have the two solutions (1) $x = 4$ or (2) $x = -4$.

Example 3 Solve $|y - 1| = 3$.

From Proposition 11.1, we have the two solutions

$$(1)\quad y - 1 = 3 \qquad \text{or} \qquad (2)\quad y - 1 = -3$$
$$y = 4 \qquad\qquad\qquad y = -2$$

DO EXERCISES 7 AND 8.

Example 4 Solve $|3 - x| - 2 = 7$.

Before we can use Proposition 11.1, we must isolate the absolute value.

(a) Add 2 to each side of the equation and simplify.

$$|3 - x| - 2 + 2 = 7 + 2$$
$$|3 - x| = 9$$

(b) From Proposition 11.1, we have

$$(1)\quad 3 - x = 9$$
$$-x = 6$$
$$x = -6$$

or

$$(2)\quad 3 - x = -9$$
$$-x = -12$$
$$x = 12$$

(c) Check

(1)

$\lvert 3 - x \rvert - 2$	7
$\lvert 3 - (-6) \rvert - 2$	7
$\lvert 9 \rvert - 2$	7
$9 - 2$	7
7	7

(2)

$\lvert 3 - x \rvert - 2$	7
$\lvert 3 - 12 \rvert - 2$	7
$\lvert -9 \rvert - 2$	7
$9 - 2$	7
7	7

Example 5 Solve $|2x| = -4$.

There is no solution because an absolute value cannot be less than 0.

DO EXERCISES 9 THROUGH 11.

Now we want to solve the inequality $|x| \leq 2$ by using the geometric definition of absolute value. We recall that $|x|$ is the distance x is from the origin, so $|x| \leq 2$ means that we want all points that are a distance less than or equal to 2 from the origin. If we look at the number line

−3 −2 −1 0 1 2 3

we see that all points to the right of the origin up to and including 2 are a distance less than or equal to 2 from the origin.

−3 −2 −1 0 1 2 3

However, all points to the left of the origin down to and including -2 are also a distance less than or equal to 2 from the origin.

−3 −2 −1 0 1 2 3

Thus, the solution to $|x| \leq 2$ is all points between -2 and 2 (including -2 and 2). The complete graphical representation is

−3 −2 −1 0 1 2 3

The interval notation is $\{x | -2 \leq x \leq 2\}$.

If we consider the arbitrary nonnegative number a and the number line

$-a$ 0 a

then the following proposition is fairly obvious.

> *PROPOSITION 11.2* The inequality $|x| \leq a$ $(a \geq 0)$ has as its solution $\{x | -a \leq x \leq a\}$. Similarly, the inequality $|x| < a$ $(a > 0)$ has as its solution $\{x | -a < x < a\}$.

If the number a is negative $(a < 0)$, then the inequality $|x| \leq a$ has no solution since $|x|$ can never be less than a negative number. For example, the inequality $|x| \leq -7$ has no solution.

Solve.

9. $|-1 - y| = -2$

10. $\left|\frac{-x - 2}{3}\right| = 4$

11. $|3t| + 6 = 14$

12. Solve $|x| < \frac{7}{2}$.

Example 6 Solve $|x| < \frac{4}{3}$.

From Proposition 11.2, we have the solution $\{x | -\frac{4}{3} < x < \frac{4}{3}\}$. The graph is at the right.

DO EXERCISE 12.

13. Solve $|2x - 3| < 4$.

Example 7 Solve $|2x + 3| \leq 3$.

From Proposition 11.2, we have

$-3 \leq 2x + 3 \leq 3$

$-6 \leq 2x \leq 0$ (We added -3 to all sides.)

$-3 \leq x \leq 0$ (We multiplied all sides by $\frac{1}{2}$.)

Thus, the solution is $\{x | -3 \leq x \leq 0\}$. The graph is at the right.

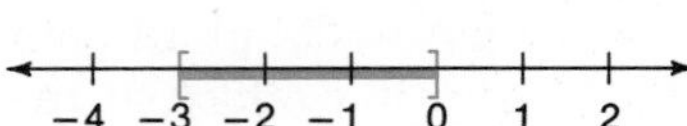

DO EXERCISE 13.

Solve.

14. $|1 - 2x| - 3 \leq 2$

15. $|t| \leq -3$

16. Is the set $\{t | 11 > t > -3\}$ the same as the set $\{t | -3 < t < 11\}$?

Example 8 Solve $|1 - t| + 3 < 2$. (We must isolate the absolute value before we can apply Proposition 11.2.)

Isolate the absolute value by adding -3 to both sides.

$|1 - t| + 3 + (-3) < 2 + (-3)$

$|1 - t| < -1$

We see that $|1 - t| < -1$ has no solution because an absolute value cannot be less than 0.

Example 9 Solve $|-t + 4| < 7$.

From Proposition 11.2, we have

$-7 < -t + 4 < 7$

$-11 < -t < 3$ (We added -4 to all sides.)

$11 > t > -3$ (We multiplied all sides by -1 and changed sense of inequalities.)

Thus, the solution is $\{t | 11 > t > -3\}$. The graph is at the right.

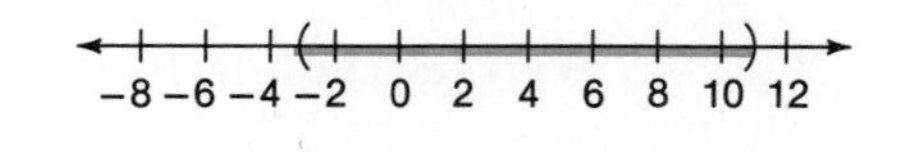

DO EXERCISES 14 THROUGH 16.

Now we want to look at the inequality $|x| \geq 2$. Once again, using the geometric definition, $|x| \geq 2$ means we want all points whose distance from the origin is greater than or equal to 2. Certainly all points to the right of and including 2 are in this category.

−3 −2 −1 0 1 2 3

All points to the left of and including −2 are also in this category.

−3 −2 −1 0 1 2 3

Thus, the solution to $|x| \geq 2$ includes all points to the right of and including 2 ($\{x|x \geq 2\}$) and all points to the left of and including -2 ($\{x|x \leq -2\}$). That is, the solution is

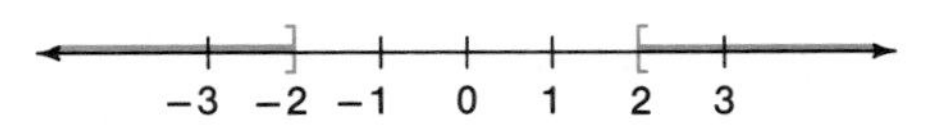

or, in interval notation, $\{x|x \geq 2 \text{ or } x \leq -2\}$. Note that the two inequalities $x \geq 2$ and $x \leq -2$ may not be combined into one three-sided inequality.

If we consider an arbitrary nonnegative number a and the number line

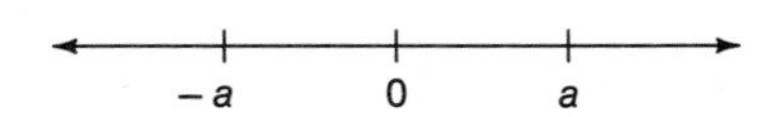

then the following proposition is fairly obvious.

PROPOSITION 11.3 The inequality $|x| \geq a$ $(a \geq 0)$ has as its solution

$$\{x|x \geq a \text{ or } x \leq -a\}$$

Similarly, the inequality $|x| > a$ $(a \geq 0)$ has as its solution

$$\{x|x > a \text{ or } x < -a\}$$

If the number a is negative ($a < 0$), then the inequality $|x| \geq a$ has the set of all real numbers as its solution since the absolute value of any real number is nonnegative and, hence, is automatically greater than a negative number. For example, the solution to $|x| \geq -3$ is all real numbers.

Example 10 Solve $|x| > 6$.

From Proposition 11.3, we have the solution

$$\{x|x > 6 \quad \text{or} \quad x < -6\}$$

The graph is at the right.

−6 −4 −2 0 2 4 6

DO EXERCISE 17.

Example 11 Solve $|2x - 4| \geq 4$.

From Proposition 11.3, we have

$$(1) \quad 2x - 4 \geq 4$$
$$2x \geq 8$$
$$x \geq 4$$

$$\text{or} \quad (2) \quad 2x - 4 \leq -4$$
$$2x \leq 0$$
$$x \leq 0$$

17. Solve $|x| \geq \frac{5}{2}$.

Solve.

18. $|t-3|>5$

19. $|3y-2|\le 4$

20. $\dfrac{|-x+1|}{3}>2$

21. $\left|7x-\frac{1}{4}\right|\ge -2$

Thus, the solution is $\{x|x\ge 4$ or $x\le 0\}$. The graph is at the right.

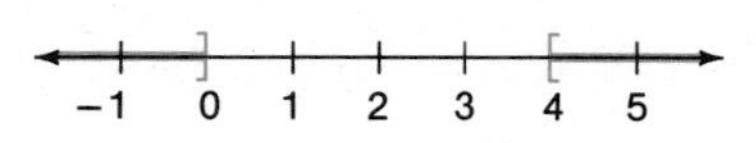

Example 12 Solve $|2-t|-3>-2$.

Isolate the absolute value by adding 3 to each side.

$$|2-t|-3+3>-2+3$$
$$|2-t|>1$$

From Proposition 11.3, we have

$$(1)\quad 2-t>1$$
$$-t>-1$$
$$t<1$$

$$\text{or}\quad (2)\quad 2-t<-1$$
$$-t<-3$$
$$t>3$$

Thus, the solution is $\{t|t<1$ or $t>3\}$. The graph is at the right.

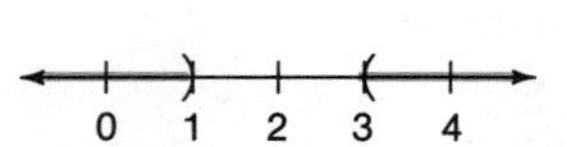

DO EXERCISES 18 THROUGH 21.

As a brief review of the three propositions in this section, we present the following chart.

Equation/Inequality	*Solution* *For* $a\ge 0$	*For* $a<0$
1. $\|x\|=a$	(1) $x=a$ or (2) $x=-a$	No solution
2. $\|x\|\le a$	$-a\le x\le a$	No solution
3. $\|x\|\ge a$	(1) $x\ge a$ or (2) $x\le -a$	All real numbers

DO SECTION PROBLEMS 11.3.

ANSWERS TO MARGINAL EXERCISES

1. 24 **2.** 17 **3.** −3 **4.** 5 **5.** −5 **6.** (a) T (b) F (c) T (d) T (e) F (f) F **7.** 5, −5 **8.** 5, −1 **9.** No solution **10.** 10, −14 **11.** $\frac{8}{3}, -\frac{8}{3}$ **12.** $-\frac{7}{2}<x<\frac{7}{2}$ **13.** $-\frac{1}{2}<x<\frac{7}{2}$ **14.** $-2\le x\le 3$ **15.** No solution **16.** Yes **17.** $x\ge\frac{5}{2}$ or $x\le -\frac{5}{2}$ **18.** $t>8$ or $t<-2$ **19.** $-\frac{2}{3}\le y\le 2$ **20.** $x>7$ or $x<-5$ **21.** Any real number

NAME COURSE/SECTION DATE

SECTION PROBLEMS 11.3

Solve.

1. $|2 - y| = 3$
2. $|3x| = 6$
3. $|2a + 1| = 7$
4. $|x - 1| = -2$
5. $3|t - 1| = 6$
6. $|-2x - 4| + 1 = 2$
7. $|7x + 1| = -2$
8. $\left|\frac{3x + 1}{2}\right| = 4$
9. $|1 - t| = 3$
10. $2|2x - 5| = 10$
11. $|-2 - 3y| - 2 = 4$
12. $\left|1 - \frac{x}{3}\right| = 7$
13. $|x| < 3$
14. $|3y| \leq 6$
15. $|2m + 3| \leq 4$
16. $|x - 3| + 4 < 5$
17. $\left|1 - \frac{x}{2}\right| \leq 3$
18. $|2x - 3| \leq -2$
19. $|y + 3| - 2 \leq -1$
20. $2|x + 3| - 5 = -2$
21. $|t| > 4$
22. $\left|\frac{x}{3}\right| \geq 1$
23. $|2y + 1| > -3$
24. $2|x - 2| \leq -4$

ANSWERS

1. ________
2. ________
3. ________
4. ________
5. ________
6. ________
7. ________
8. ________
9. ________
10. ________
11. ________
12. ________
13. ________
14. ________
15. ________
16. ________
17. ________
18. ________
19. ________
20. ________
21. ________
22. ________
23. ________
24. ________

25. $|z - 1| \geq 3$

26. $4|2a + 1| \geq 12$

27. $|3 - 2x| > 2$

28. $|1 - t| \geq 4$

29. $\left|\frac{2 - 3x}{3}\right| \leq 4$

30. $\left|\frac{c}{3}\right| = 2$

31. $|0.4n - 1.2| > 2.4$

32. $\frac{1}{2}|2x + 6| = 3$

33. $|2 - t| < 4$

34. $|2x - 1| > -3$

35. $|1 - 3t| < -2$

36. $\left|\frac{1 - x}{2}\right| > 3$

37. $\frac{|2t - 1|}{2} - 1 \leq 4$

38. $1 - 6x > 7$

39. $|1 - 6x| > 7$

40. $|x| = \pi$

41. $|3 - 2t| = 4$

42. $|3x + 1| - 3 = 5$

43. $|3x| \leq -4$

44. $\left|\frac{1 - t}{-2}\right| < 1$

45. $|3.1x - 4.2| > -2$

46. $|x + 6| + 3 \leq 4$

47. $\frac{|1 - 3t|}{4} > 2$

48. $|x| > 0$

49. $\left|\frac{1 - 3x}{2}\right| = 1$

50. $|-2 - 3t| \leq 4$

25. ____________

26. ____________

27. ____________

28. ____________

29. ____________

30. ____________

31. ____________

32. ____________

33. ____________

34. ____________

35. ____________

36. ____________

37. ____________

38. ____________

39. ____________

40. ____________

41. ____________

42. ____________

43. ____________

44. ____________

45. ____________

46. ____________

47. ____________

48. ____________

49. ____________

50. ____________

NAME | COURSE/SECTION | DATE | ANSWERS

CHAPTER 11 REVIEW PROBLEMS

SECTION 11.1

Translate the English description into set notation.

1. The set of all real numbers greater than 10.

2. The set of all real numbers less than or equal to -3.

3. The set of all real numbers between 0 and 5, including 0 but not including 5.

Fill in the columns.

Set Notation	*Graphical Representation*	*Interval Notation*
4. $\{x \mid x > 3\}$		
5.	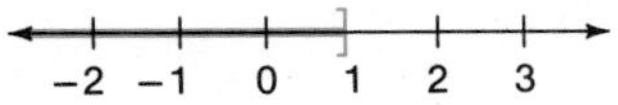	
6.		$(-\infty, -1]$
7. $\left\{x \,\middle|\, -\frac{3}{2} < x < \frac{5}{2}\right\}$		

SECTION 11.2

Solve.

8. $2 - t \leq 4$

9. $3x + 1 > -6$

10. $1.3 + 2t < -2.7$

ANSWERS

1.

2.

3.

4.

5.

6.

7.

8.

9.

10.

11. ______
12. ______
13. ______
14. ______
15. ______
16. ______
17. ______
18. ______
19. ______
20. ______
21. ______
22. ______
23. ______
24. ______
25. ______
26. ______
27. ______
28. ______
29. ______
30. ______
31. ______
32. ______
33. ______
34. ______
35. ______

11. $-2t + 1 < -3$

12. $3 - 4x \geq -5$

13. $2(x + 3) < 5$

14. $7y - 3 \leq 2y + 5$

15. $6 - 3x > 2x + 4$

16. $-3(1 - x) < 2(x + 4)$

17. $\frac{x}{3} - \frac{1}{4} \geq \frac{5}{6}$

18. $\frac{2x - 3}{4} < -1$

19. $-\frac{t}{2} - \frac{3}{4} < t + \frac{1}{4}$

20. $\frac{1.1 - 0.5y}{-3} \geq -2.1$

21. $x - 2 > 0$

SECTION 11.3

Solve.

22. $|x| = 3$

23. $|-3x| = 9$

24. $|3y| = 2$

25. $|x - 3| = 7$

26. $\left|\frac{t}{4}\right| = 7$

27. $|2.3x| = -4$

28. $|x| \leq 2.3$

29. $|2y + 1| - 3 \geq 2$

30. $3|2x + 4| \leq 5$

31. $\left|\frac{t - 1}{4}\right| > 3$

32. $|x + 5| \leq -2$

33. $3|y + 1| - 4 > 2$

34. $|1 - 2x| < 3$

35. $\left|\frac{y + 1}{3}\right| > 4$

NAME COURSE/SECTION DATE

CHAPTER 11 TEST

Fill in the columns.

	Set Notation	*Graphical Representation*	*Interval Notation*
1.	$\{x \mid x > 2\}$		
2.		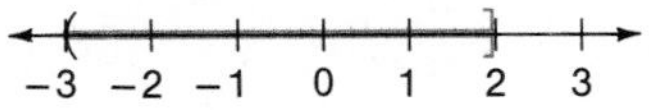	
3.			$(-\infty, 3]$

4. Evaluate: (a) $|-25|$ (b) $|1.45|$ (c) $-|-2|$

Solve.

5. $2t + 4 > 5$

6. $2 - x \leq -4$

ANSWERS

1. ______

2. ______

3. ______

4. (a) ______

(b) ______

(c) ______

5. ______

6. ______

7. ____________

8. ____________

9. ____________

10. ____________

11. ____________

12. ____________

13. ____________

14. ____________

15. ____________

7. $\frac{y}{2} - \frac{1}{3} \geq \frac{3}{4}$

8. $2(1 - 3x) + 3 < -5x - (2x + 1)$

9. $|t - 3| = 1$

10. $|3x - 1| \leq 6$

11. $|2 - y| > 2$

12. $\left|\frac{t - 2}{3}\right| + 1 = 2$

13. $|3 - a| < -2$

14. $2|x + 1| - 1 > 3$

15. $-2 \leq 1 - 2x \leq 5$

NAME COURSE/SECTION DATE

ANSWERS

1. ______
2. ______
3. ______
4. ______
5. ______
6. ______
7. ______
8. ______
9. ______
10. ______
11. ______
12. ______

FINAL EXAMINATION

CHAPTER 1

Perform the indicated operation.

1. $\frac{7}{10} \cdot \frac{15}{2} \cdot \frac{3}{14}$

2. $\frac{7}{15} + \frac{1}{18}$

3. $1\frac{1}{4} \div 2\frac{2}{3}$

4. $10\frac{1}{4} - 5\frac{3}{5}$

5. 0.002×1.4

6. Divide 0.00578 by 1.7.

CHAPTER 2

Calculate and simplify.

7. $-3 - (-4) + 5$

8. $(-2)(-3) - (5)(-2)$

9. $2\frac{1}{4} - 1\frac{1}{3} - 2\frac{1}{5}$

10. $\frac{(2^{-2})^3 \cdot 2^4}{(2^2 \cdot 2^{-3})^2}$

11. $2x - (x - y) - 2y$

12. Evaluate $b^2 - 4ac$ when $a = 2$, $b = -3$, and $c = -2$.

13. ____________

14. ____________

15. ____________

16. ____________

17. ____________

18. ____________

19. ____________

20. ____________

21. ____________

22. ____________

23. ____________

CHAPTER 3

13. Solve $-2x + 5 = -7$.

14. Solve $\frac{3x}{4} - \frac{2}{3} = \frac{x}{6}$.

15. Solve $-3(y + 4) - 2y = 2(1 - y) + 4$.

16. Solve $N = PQ + R$ for Q.

CHAPTER 4

17. A 150-foot rope is cut into three pieces. The second piece is three times the length of the first piece and the third piece is twice the length of the second piece. How long is each piece?

18. Helen mixes a 40% alcohol solution with a 70% alcohol solution to obtain 15 pints of a solution that is 55% alcohol. How much of the 40% alcohol solution did she use in her mixture?

CHAPTER 5

19. Multiply $2x^2 - x + 3$ by $2x + 1$.

20. Factor $2y^2 - 8$.

21. Factor $12x^2 + 8x - 15$.

22. Solve $x^2 + x - 12 = 0$.

23. A rectangle has a length that is 5 feet more than the width. If the area of the rectangle is 500 square feet, find the length.

NAME COURSE/SECTION DATE

CHAPTER 6

24. Graph $3x - 2y = 6$.

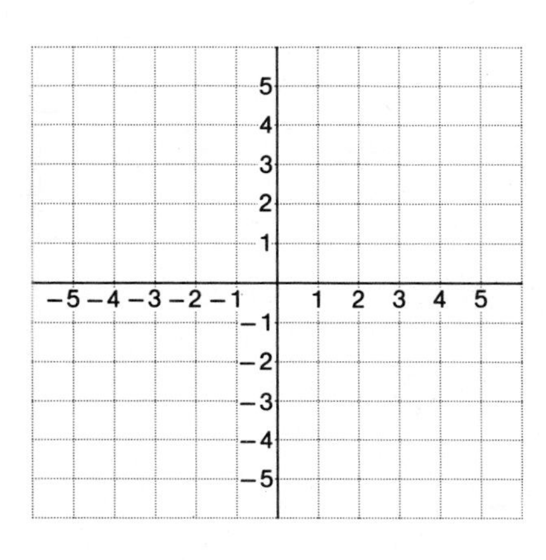

25. Graph $y = x^2 - 2$.

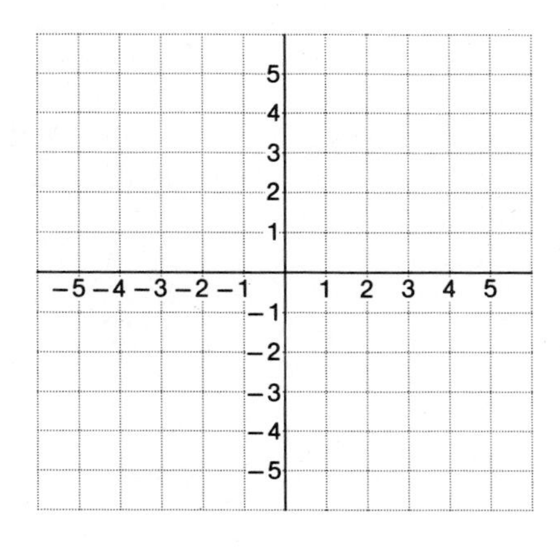

26. Find the slope of the straight line $-2x + 5y = 10$.

27. Find the equation of the line with slope -3 that goes through the point $(-1, 4)$.

28. Find the equation of the line that goes through the points $(2, -5)$ and $(-4, -7)$.

CHAPTER 7

29. Solve. $\begin{Bmatrix} 2x - y = 4 \\ -3x + y = -2 \end{Bmatrix}$

30. Solve. $\begin{Bmatrix} 4s - 2t = 5 \\ -2s + t = -2 \end{Bmatrix}$

31. Solve. $\begin{Bmatrix} 2a - 3b = 4 \\ -5a - 7b = -4 \end{Bmatrix}$

ANSWERS

24. See graph.

25. See graph.

26. ______

27. ______

28. ______

29. ______

30. ______

31. ______

32. ______

33. ______

34. ______

35. ______

36. ______

37. ______

32. At 1:00 a car leaves St. Louis going east at 40 mph. At 2:00 another car leaves St. Louis going west at 45 mph. At what time will they be 210 miles apart?

33. Pat's car gets 20 mpg using regular gas and 24 mpg using gasohol. On a recent trip of 1020 miles, Pat used a total of 45 gallons of gas. How many gallons of regular gas did she use on this trip?

CHAPTER 8

34. Multiply and reduce: $\dfrac{x-3}{x^2-x-2} \cdot \dfrac{x^2+3x+2}{x^2-3x}$

35. Divide and reduce: $\dfrac{4a^2b}{c^3} \div \dfrac{6b^3}{c^2a}$

36. Add: $\dfrac{1}{x^2-3x+2} + \dfrac{x}{x^2-4}$

37. Simplify: $\dfrac{1-\dfrac{1}{t^2}}{1+\dfrac{1}{t}}$

NAME COURSE/SECTION DATE

ANSWERS

38. ______

39. ______

40. ______

41. ______

42. ______

43. ______

44. ______

38. Solve: $\dfrac{3}{2y-1} - \dfrac{4}{y+2} = 0$

39. John can do a job in $2\frac{1}{2}$ hours and Hank can do the same job in 3 hours. If they work together, how long will it take to do the job?

CHAPTER 9

40. Simplify: $\sqrt{45x^3y^2}$

41. Combine: $2\sqrt{18} - 4\sqrt{8}$

42. Rationalize the denominator for $\dfrac{4}{\sqrt{2x}}$.

CHAPTER 10

43. Solve $x^2 - 4x - 3 = 0$ by completing the square.

44. Solve $2y^2 = 3y - 4$ by the quadratic formula.

45. ______________

46. ______________

47. ______________

48. ______________

49. ______________

50. ______________

45. Solve $\frac{1}{x} - \frac{3}{x + 2} = \frac{1}{x + 1}$.

46. Solve $\sqrt{3x + 1} = x - 1$.

CHAPTER 11

Solve.

47. $2(1 - x) - 3x < x + 5$

48. $|2x| = 5$

49. $|1 - 2t| \leq 3$

50. $\left|\frac{3y + 1}{2}\right| \geq 4$

Appendix A

Unit Conversion

BRITISH AMERICAN UNITS

Linear Measurements

12 inches (in.) = 1 foot (ft)
3 feet = 1 yard (yd)
5280 feet = 1760 yards = 1 mile (mi)

Volume Measurements

Liquid
16 fluid ounces (fl oz) = 1 pint (pt)
2 cups = 1 pint
2 pints = 1 quart (qt)
4 quarts = 1 gallon (gal)

Dry
2 pints = 1 quart
8 quarts = 1 peck (pk)
4 pecks = 1 bushel (bu)

Weight Measurements

16 ounces (oz) = 1 pound (lb)
2000 pounds = 1 ton (t)

Time Measurements

60 seconds (sec) = 1 minute (min)
60 minutes = 1 hour (hr)
24 hours = 1 day
7 days = 1 week
365 days = 1 year (yr)
10 years = 1 decade
100 years = 1 century

METRIC UNITS

Linear Measurements

1 kilometer (km) = 1000 meters (m)
1 hectometer (hm) = 100 m
1 dekameter (dam) = 10 m
1 meter = 1 m
1 decimeter (dm) = $\frac{1}{10}$ m (or 1 m = 10 dm)
1 centimeter (cm) = $\frac{1}{100}$ m (or 1 m = 100 cm)
1 millimeter (mm) = $\frac{1}{1000}$ m (or 1 m = 1000 mm)

Volume Measurements

1 kiloliter (kL) = 1000 liters (L)
1 hectoliter (hL) = 100 L
1 dekaliter (daL) = 10 L
1 liter = 1 L
1 deciliter (dL) = $\frac{1}{10}$ L (or 1 L = 10 dL)
1 centiliter (cL) = $\frac{1}{100}$ L (or 1 L = 100 cL)
1 milliliter (mL) = $\frac{1}{1000}$ L (or 1 L = 1000 mL)

Weight Measurements	1 kilogram (kg) = 1000 grams (g)
	1 hectogram (hg) = 100 g
	1 dekagram (dag) = 10 g
	1 gram = 1 g
	1 decigram (dg) = $\frac{1}{10}$ g (or 1 g = 10 dg)
	1 centigram (cg) = $\frac{1}{100}$ g (or 1 g = 100 cg)
	1 milligram (mg) = $\frac{1}{1000}$ g (or 1 g = 1000 mg)
	1 metric ton = 1000 kg

Appendix B

Geometric Formulas

DEFINITION B.1 The *perimeter* of a rectangle is the distance around the edge of the rectangle. In general, the *perimeter* of a polygon (an n-sided closed figure where n is a whole number) is the distance around the outside edge of the polygon.

GEOMETRIC FORMULA 1 The perimeter p of a square with side s is given by $p = 4s$.

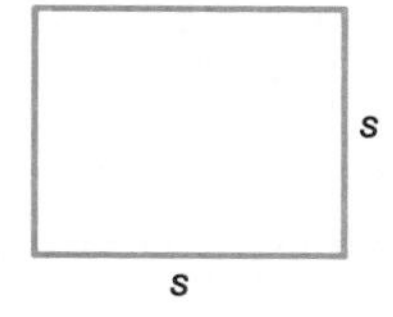

GEOMETRIC FORMULA 2 The perimeter of a rectangle is given by $p = 2L + 2W$.

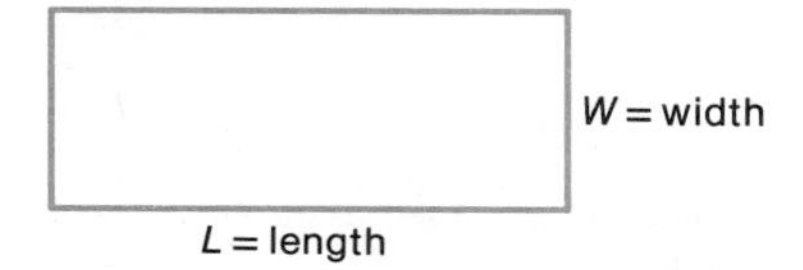

DEFINITION B.2 *One square inch* (sq in.) is the area of a square that is one inch on each side. There are similar definitions for one square foot, one square centimeter, one square meter, and so on, where the units change accordingly.

To change from one square unit to another square unit, we need the following conversion factors.

1 sq ft = (12 in.)(12 in.)
= 144 sq in.

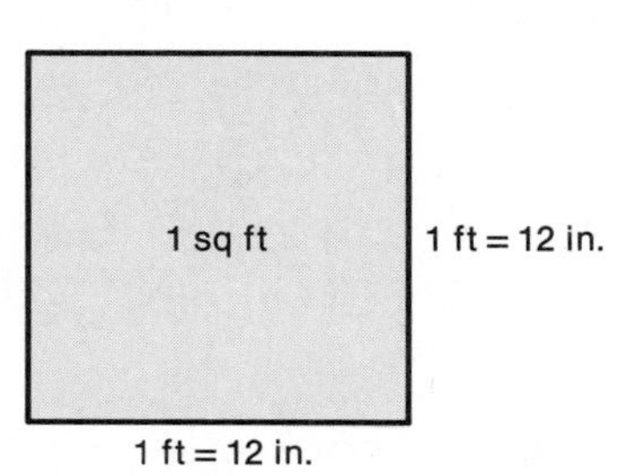

1 sq yd = (3 ft)(3 ft)
= 9 sq ft

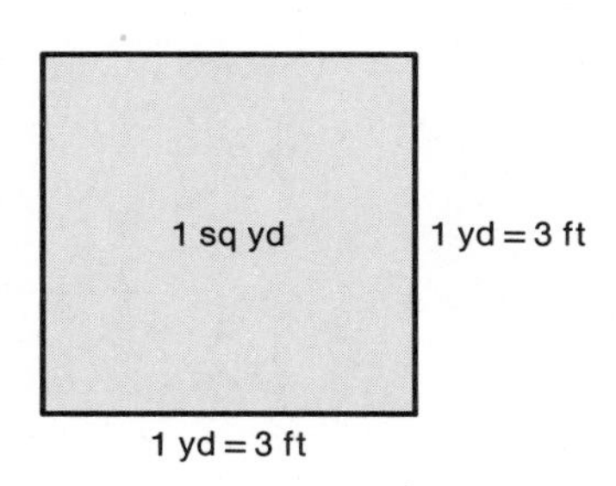

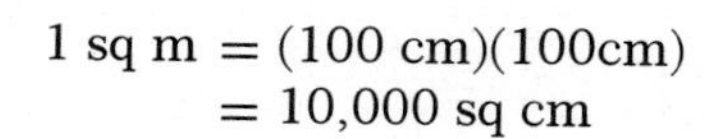

$$1 \text{ sq m} = (100 \text{ cm})(100\text{cm}) \\ = 10{,}000 \text{ sq cm}$$

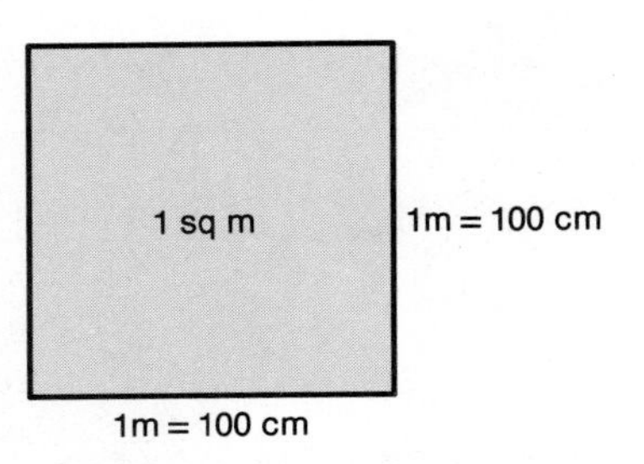

Other conversion factors in square units are obtained similarly.

GEOMETRIC FORMULA 3 The area of a rectangle is given by $A = L \cdot W$.

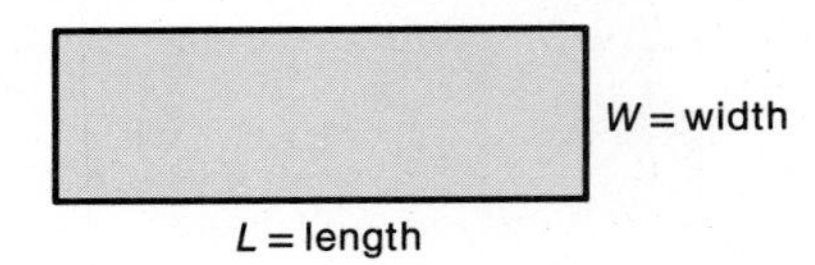

DEFINITION B.3 A *parallelogram* is a four-sided polygon that has two pairs of parallel sides.

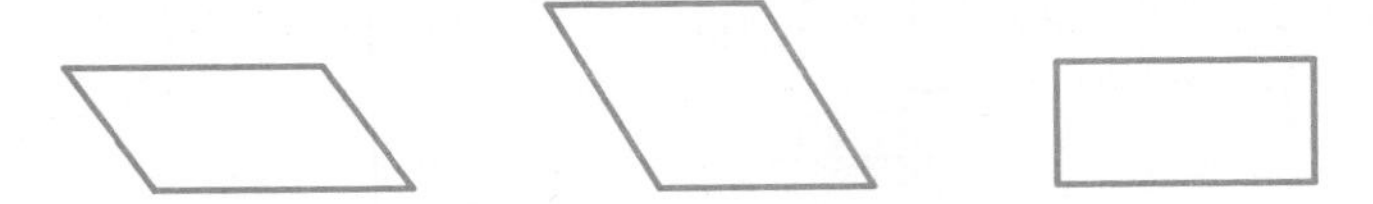

GEOMETRIC FORMULA 4 The area of a parallelogram with base b and height h is given by $A = b \cdot h$.

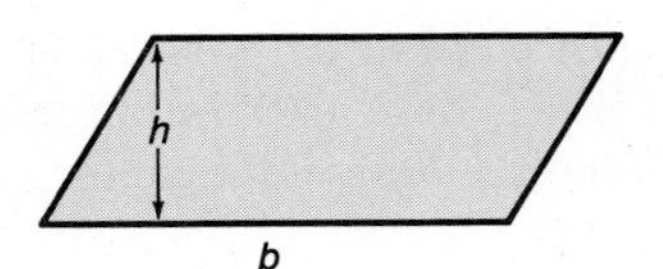

DEFINITION B.4 A triangle is a three-sided polygon.

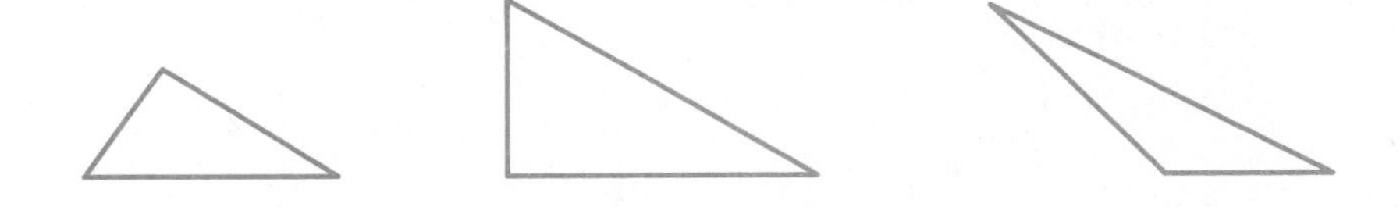

GEOMETRIC FORMULA 5 The area of a triangle is $A = \frac{b \cdot h}{2}$, where b = base, h = height.

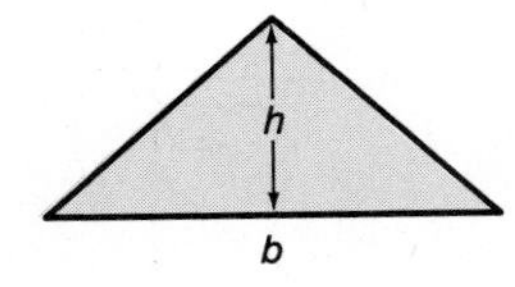

To find the height that is associated with a particular base (any side of the triangle may be considered the base), you start with the vertex opposite the base and then take the shortest distance (perpendicular to the base) from the vertex to the base as the height.

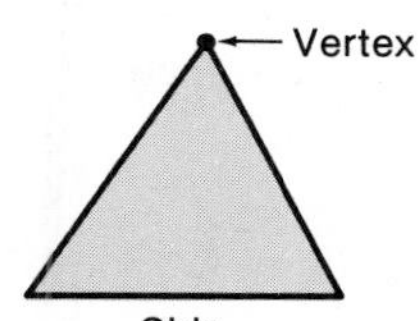

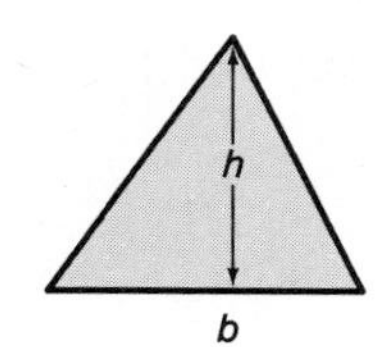

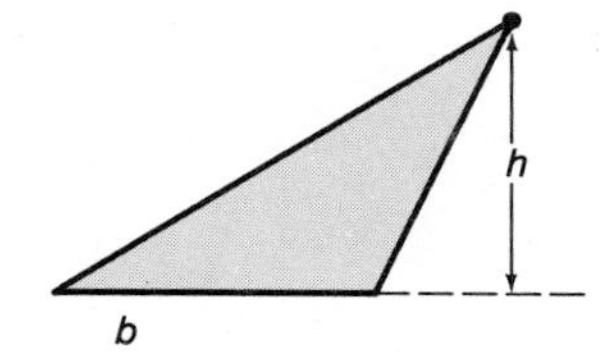

GEOMETRIC FORMULA 6 The sum of the angles of any triangle is 180 degrees (180°).

For example, the angle labeled α in the following triangle is $180° - (120° + 25°) = 180° - 145° = 35°$.

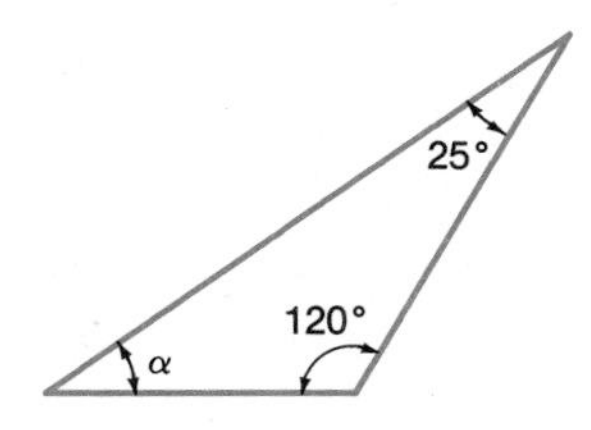

DEFINITION B.5 A trapezoid is a four-sided polygon with one set of parallel sides (the bases).

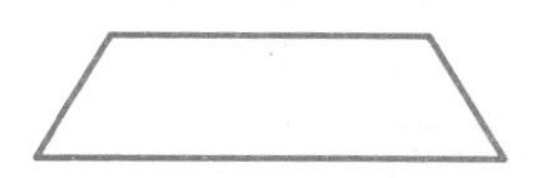
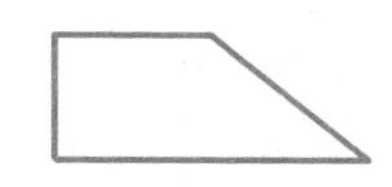
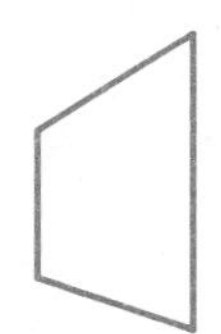

GEOMETRIC FORMULA 7 The area of a trapezoid is given by $A = \dfrac{h(B + b)}{2}$ where

B = large base

b = small base

h = height

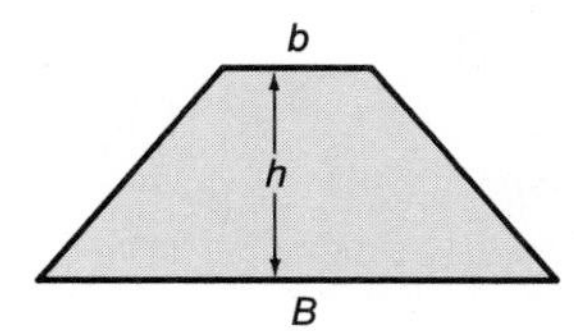

Note that the bases of a trapezoid must be parallel, and that the height is the shortest perpendicular distance between the bases.

DEFINITION B.6

r = radius, the distance from the center of the circle to the outer rim
d = diameter, the distance across the circle if you go through the center of the circle
C = circumference, the distance around the rim of the circle

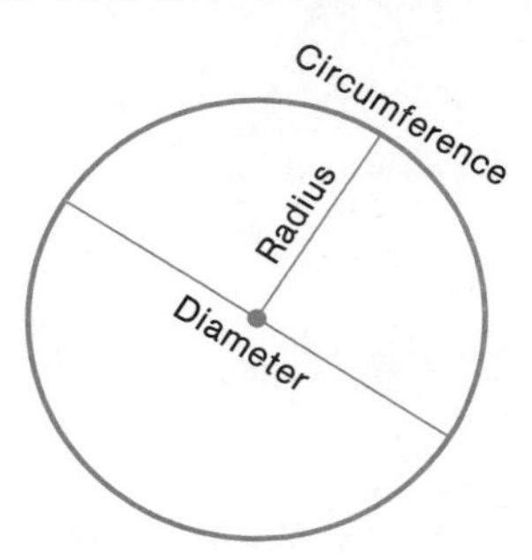

GEOMETRIC FORMULA 8 The diameter of a circle is twice the radius, $d = 2r$; or the radius is one-half the diameter, $r = d/2$.

GEOMETRIC FORMULA 9 The circumference of a circle is given by $C = 2\pi r$ where π is a constant and has the following commonly used approximations: $\pi \cong 3.14$, $\pi \cong 3.1416$, $\pi \cong \frac{22}{7}$.

GEOMETRIC FORMULA 10 The area of a circle is given by $A = \pi r^2$ where $r^2 = r \cdot r$.

DEFINITION B.7 One cubic inch (cu in.) is the volume of a cube that is one inch on a side. There are similar definitions for one cubic foot, one cubic centimeter, and so on, where the units change accordingly.

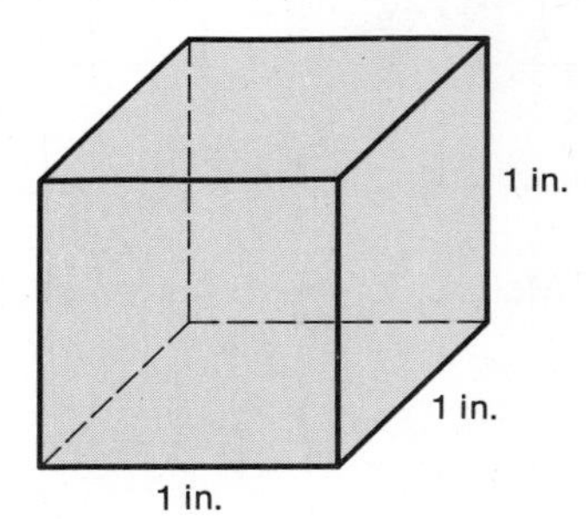

GEOMETRIC FORMULA 11 The volume of a rectangular solid is given by the formula $V = LWH$ where L = length, W = width, and H = height.

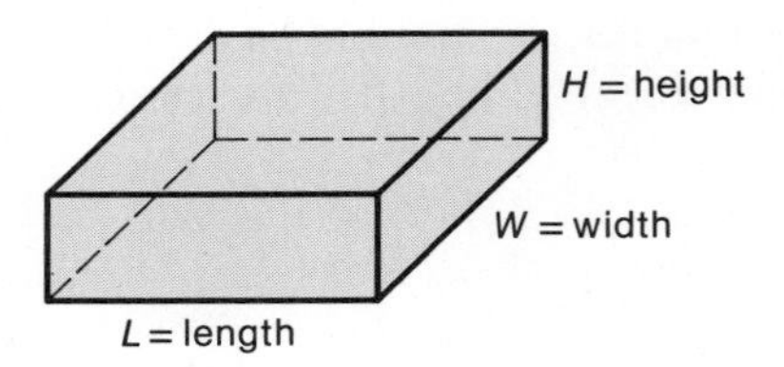

GEOMETRIC FORMULA 12 The volume of a cylinder is given by $V = \pi r^2 h$ where r = radius and h = height.

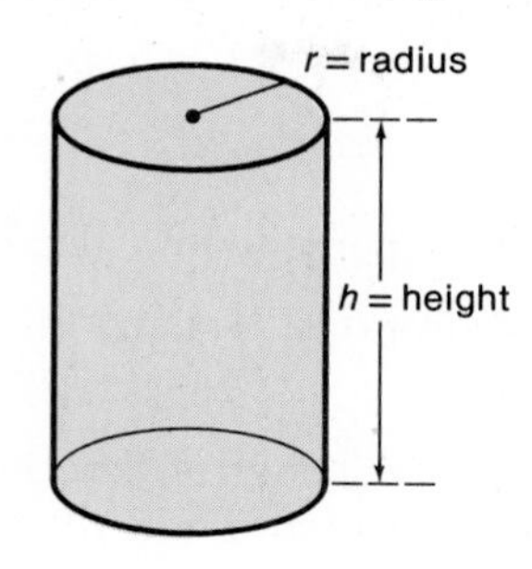

Appendix C

Percent

RULE C.1 To change a percent to a decimal or a fraction, divide the percent by 100 and drop the % sign; that is, $x\% = \frac{x}{100}$.

RULE C.2 To change a decimal or fraction to a percent, multiply the decimal or fraction by 100 and add the % sign.

PERCENT FORMULA 1 The basic formula for percent problems is $P = R \cdot B$ where P = percentage, R = rate (or percent), and B = base.

1. The formula $P = R \cdot B$ can be put into words as follows:

$$\underbrace{\text{The percent}}_{P}\ \underbrace{\text{is the result of}}_{=}\ \underbrace{\text{taking a percent (rate)}}_{R}\ \underbrace{\text{of}}_{\cdot}\ \underbrace{\text{the base}}_{B}$$

2. From remark 1 we notice that the base is always what we take a percent (rate) of.
3. We will use the words "rate" and "percent" interchangeably.
4. The rate (percent) R always has a percent sign (%) after it. The base B and the percentage P are never used with a percent sign.
5. Before the rate R can be substituted in the percent formula, it must be changed from a percent to a decimal (or fraction).

The general procedural steps involved in solving a percent problem are as follows.

STEP 1: IDENTIFY

Identify the rate, base, and percentage in the problem. One of these three will be the unknown quantity that we are looking for.

STEP 2: SELECT

Select the appropriate percent formula (eventually there will be three such formulas). The formula to be used depends on whether P, B, or R is the unknown quantity.

STEP 3: SOLVE

Substitute in the appropriate percent formula and solve for the unknown quantity.

PERCENT FORMULA 2 When the unknown quantity in a percent problem is the rate, we use the formula $R = \frac{P}{B}$.

PERCENT FORMULA 3 When the unknown quantity in a percent problem is the base, we use the formula $B = \frac{P}{R}$.

Appendix D

Table of Square Roots (1 through 100)

x	$\sqrt{x}$	x	$\sqrt{x}$
1	1	51	7.141
2	1.414	52	7.211
3	1.732	53	7.280
4	2	54	7.348
5	2.236	55	7.416
6	2.449	56	7.483
7	2.646	57	7.550
8	2.828	58	7.616
9	3	59	7.681
10	3.162	60	7.746
11	3.317	61	7.810
12	3.464	62	7.874
13	3.606	63	7.937
14	3.742	64	8
15	3.873	65	8.062
16	4	66	8.124
17	4.123	67	8.185
18	4.243	68	8.246
19	4.359	69	8.307
20	4.472	70	8.367
21	4.583	71	8.426
22	4.690	72	8.485
23	4.796	73	8.544
24	4.899	74	8.602
25	5	75	8.660
26	5.099	76	8.718
27	5.196	77	8.775
28	5.292	78	8.832
29	5.385	79	8.888
30	5.477	80	8.944
31	5.568	81	9
32	5.657	82	9.055
33	5.745	83	9.110
34	5.831	84	9.165
35	5.916	85	9.220
36	6	86	9.274
37	6.083	87	9.327
38	6.164	88	9.381
39	6.245	89	9.434
40	6.325	90	9.487
41	6.403	91	9.539
42	6.481	92	9.592
43	6.557	93	9.644
44	6.633	94	9.695
45	6.708	95	9.747
46	6.782	96	9.798
47	6.856	97	9.849
48	6.928	98	9.899
49	7	99	9.950
50	7.071	100	10

Answers

Section Problems 1.1

1. Natural number, whole number, and fraction **3.** Fraction **5.** Fraction **7.** 3 is the numerator and 8 is the denominator. **9.** 5 is the numerator and 1 is the denominator. **11.** Yes, n cannot be zero. **13.** (Answers may vary.) **15.** 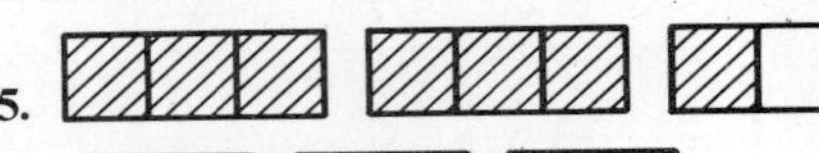

17. (Answers may vary.) **19.**

Section Problems 1.2

1. 1, 2, 3, 4, 6, 8, 12, 24 **3.** 1, 2, 3, 4, 6, 8, 9, 12, 18, 24, 36, 72 **5.** 1, 3, 5, 7, 15, 21, 35, 105 **7.** $14 = 2 \times 7$ **9.** $9 = 3 \times 3$ **11.** 17 **13.** $24 = 2 \cdot 2 \cdot 2 \cdot 3$ **15.** $8 = 2 \cdot 2 \cdot 2$ **17.** 53

19. $\frac{1}{3}$ **21.** $\frac{15}{28}$ **23.** 2 **25.** $\frac{20}{7}$ **27.** $\frac{2}{3}$ **29.** $\frac{14}{5}$ **31.** $\frac{10}{9}$ **33.** $\frac{7}{5}$ **35.** $\frac{1}{3}$ **37.** $\frac{1}{90}$

39. 3 **41.** 4 **43.** $\frac{2}{5}$ **45.** $\frac{3}{2}$ **47.** $\frac{24}{35}$ **49.** 5 **51.** 0 **53.** $\frac{3}{10}$ **55.** $\frac{8}{3}$ **57.** $\frac{1}{6}$

Section Problems 1.3

1. 10 **3.** 54 **5.** 20 **7.** $\frac{25}{12}$ **9.** $\frac{17}{8}$ **11.** $\frac{2}{9}$ **13.** $\frac{1}{10}$ **15.** $\frac{4}{21}$ **17.** $\frac{1}{10}$ **19.** $\frac{29}{21}$

21. $\frac{7}{10}$ **23.** $\frac{11}{10}$ **25.** $\frac{35}{12}$ **27.** $\frac{8}{15}$ **29.** $\frac{323}{420}$ **31.** $\frac{2}{3}$ **33.** $\frac{5}{24}$ **35.** $\frac{41}{36}$ **37.** 1

39. $\frac{11}{20}$ **41.** $\frac{3}{20}$ **43.** $\frac{77}{51}$ **45.** $\frac{179}{60}$

Section Problems 1.4

1. $\frac{7}{4}$ **3.** $\frac{11}{3}$ **5.** $\frac{99}{8}$ **7.** $\frac{14}{9}$ **9.** $6\frac{2}{3}$ **11.** $4\frac{1}{4}$ **13.** $4\frac{2}{3}$ **15.** $20\frac{5}{11}$ **17.** $\frac{9}{14}$

19. $\frac{16}{5}$ **21.** $7\frac{2}{3}$ **23.** $\frac{121}{15}$ or $8\frac{1}{15}$ **25.** $\frac{67}{12}$ **27.** $13\frac{1}{4}$ **29.** $\frac{461}{15}$ or $30\frac{11}{15}$ **31.** $\frac{24}{5}$

33. $\frac{2}{5}$ **35.** $\frac{61}{8}$ **37.** $\frac{41}{12}$ **39.** $\frac{23}{28}$ **41.** $\frac{28}{5}$ **43.** $\frac{293}{36}$ or $8\frac{5}{36}$

Section Problems 1.5

1. $\frac{43}{100}$ **3.** $\frac{1234}{10{,}000}$ **5.** $2\frac{1364}{10{,}000}$ or $\frac{21{,}364}{10{,}000}$ **7.** 3.1 **9.** 0.014 **11.** 49.33 **13.** 0.1321

15. 1.422 **17.** 3.013 **19.** 39.026 **21.** 11.902 **23.** (a) 6,440 (b) 6,400 (c) 6,000 (d) 10,000 **25.** (a) 15,540 (b) 15,500 (c) 16,000 (d) 20,000 **27.** (a) 0.2 (b) 0.15 (c) 0.155 (d) 0.1547 **29.** (a) 0.5 (b) 0.53 (c) 0.527 (d) 0.5265 **31.** (a) 39,830 (b) 39,800 (c) 40,000 (d) 40,000

Section Problems 1.6

1. 100.8 **3.** 0.086 **5.** 0.331 **7.** 0.2376 **9.** 16.125 **11.** 0.228 **13.** 0.151 **15.** 0.0936
17. 1.04 **19.** 53.721 **21.** 0.256 **23.** $0.41\overline{6}$ **25.** 0.16 **27.** $1.\overline{3}$ **29.** 1.75

Section Problems 1.7

1. $4 \cdot 2$ **3.** $2 \cdot 4$ **5.** $3 \cdot 5 + 2 \cdot 9$ **7.** $5 + 5 + 5$ **9.** $1 + 1 + 1 + 1 + 1$

11. $3 + 3 + 8 + 8 + 8 + 8$ **13.** 6 **15.** 9 **17.** 5 **19.** 4 **21.** 10 **23.** 14 **25.** $\frac{1}{3}$

27. 6 **29.** $\frac{3}{2}$ **31.** $\frac{1}{1.8}$ **33.** 6^4 **35.** $\left(\frac{1}{3}\right)^5$ **37.** $2^3 \cdot 4^5$ **39.** $4 \cdot 4 \cdot 4$ **41.** $\frac{1}{5} \cdot \frac{1}{5} \cdot \frac{1}{5} \cdot \frac{1}{5}$

43. $2 \cdot 2 \cdot 2 \cdot 3 \cdot 3$ **45.** 3^6 **47.** $\left(\frac{1}{2}\right)^{10}$ **49.** Cannot simplify **51.** 3^8 **53.** 7^{10} **55.** 6^8

Section Problems 1.8

1. Distributive **3.** Commutative (multiplication) **5.** Associative (addition) **7.** Distributive
9. Commutative (multiplication) **11.** Answers will vary. **13.** Answers will vary.
15. $2(3 + 4) = 2 \cdot 3 + 2 \cdot 4$ **17.** $4(2 + 3 + 5) = 4 \cdot 2 + 4 \cdot 3 + 4 \cdot 5$ **19.** $<$ **21.** $>$
23. $<$ **25.** $=$ **27.** $<$ **29.** Because subtraction is not associative

Chapter 1 Review Problems

1. 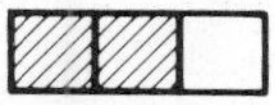 **2.**  **3.** 7 is the numerator; 4 is the denominator **4.** Every whole number is a fraction but not vice versa. **5.** A quotient of two whole numbers a/b except $b \neq 0$ **6.** $3 \cdot 7$ **7.** $2 \cdot 2 \cdot 5 \cdot 5$ **8.** $\frac{2}{7}$ **9.** 8 **10.** 6 **11.** $\frac{5}{34}$

12. $\frac{1}{21}$ **13.** 0 **14.** 12 **15.** 90 **16.** $\frac{17}{12}$ **17.** $\frac{17}{20}$ **18.** $\frac{8}{15}$ **19.** $\frac{17}{8}$ **20.** $3\frac{4}{5}$

21. 22 **22.** $\frac{5}{4}$ **23.** $\frac{38}{3}$ **24.** $2\frac{8}{15}$ **25.** $\frac{9}{5}$ **26.** $\frac{33}{7}$ **27.** 32 **28.** $\frac{14}{1000}$ **29.** $\frac{136}{100}$

30. 0.006 **31.** 14.21 **32.** 27.046 **33.** 9.868 **34.** 14.991 **35.** 3.1625 **36.** (a) 48,740 (b) 48,700 (c) 49,000 (d) 50,000 **37.** (a) 0.197 (b) 0.20 (c) 0.2 **38.** 0.0276 **39.** 1.6
40. 0.00009 **41.** 4.7 **42.** 0.375 **43.** 0.417 **44.** $2 \cdot 4$ **45.** $5 \cdot 8$ **46.** $15 - 4 = 11$ since $11 + 4 = 15$ **47.** $24 \div 8 = 3$ since $3 \times 8 = 24$ **48.** $\frac{1}{12}$ **49.** $\frac{2}{9}$ **50.** 9^4 **51.** $5^2 \cdot 6^5$

52. 9^7 **53.** 2^{18} **54.** 3^7 **55.** Distributive **56.** Commutative **57.** Associative **58.** $<$
59. $>$

Chapter 1 Test

1. $\frac{6}{7}$ **2.** $\frac{13}{30}$ **3.** $8\frac{7}{15}$ **4.** $\frac{1}{8}$ **5.** $7\frac{7}{12}$ **6.** $\frac{11}{36}$ **7.** $\frac{15}{2}$ **8.** $\frac{17}{2}$ **9.** 14.474 **10.** 0.07

11. 17.419 **12.** 0.24 **13.** 180 **14.** (a) 3^{12} (b) 2^7 **15.** 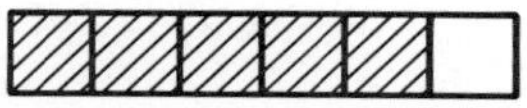 **16.** (a) 12.482
(b) 12.48 (c) 12.5 (d) 12 **17.** (a) $>$ (b) $=$ (c) $>$ **18.** Commutative (multiplication)
19. Associative (addition) **20.** Distributive

Section Problems 2.2

1. -2 **3.** -7 **5.** 0 **7.** 2 **9.** -5 **11.** -5 **13.** -8 **15.** 2 **17.** -13 **19.** -8
21. 7 **23.** 3 **25.** 7 **27.** 199 **29.** -303 **31.** -202 **33.** -1082 **35.** -82 **37.** 89
39. -46 **41.** -155 **43.** 132 **45.** 111

Section Problems 2.3

1. -8 **3.** 10 **5.** -1 **7.** -5 **9.** -17 **11.** 4 **13.** 1 **15.** 16 **17.** 16 **19.** 3
21. -8 **23.** 9 **25.** 13 **27.** 3 **29.** 0 **31.** -33 **33.** -7 **35.** -46 **37.** -12
39. 5 **41.** -10 **43.** 29 **45.** -10 **47.** -534 **49.** -9 **51.** -111 **53.** 307 **55.** 575

Section Problems 2.4

1. -30 **3.** 20 **5.** -24 **7.** -40 **9.** -2 **11.** -2 **13.** 0 **15.** -5 **17.** -10
19. 21 **21.** 0 **23.** -6 **25.** 7488 **27.** -4 **29.** -11 **31.** 5 **33.** 9 **35.** 39
37. 23 **39.** 14 **41.** 25 **43.** 8 **45.** -5 **47.** -2 **49.** 4

Section Problems 2.5

1. -1.8 **3.** -9.75 **5.** $-\frac{8}{7}$ **7.** $\frac{23}{20}$ **9.** $-\frac{3}{8}$ **11.** -8.1 **13.** -2.3 **15.** -2 **17.** $\frac{1}{2}$

19. $\frac{-7}{60}$ **21.** $\frac{2}{15}$ **23.** 3.25 **25.** 0.415 **27.** 6 **29.** $\frac{-2}{5}$ **31.** $-\frac{1}{2}$ **33.** $\frac{5}{3}$ **35.** -7

37. $\frac{8}{15}$ **39.** 0 **41.** $-\frac{9}{20}$ **43.** $\frac{13}{24}$ **45.** $-4\frac{5}{12}$ **47.** -51.85 **49.** -1.2

Section Problems 2.6

1. $3 \cdot 3 \cdot 3 \cdot 3 \cdot 3 \cdot 3$ **3.** $\left(-\frac{2}{3}\right)\left(-\frac{2}{3}\right)\left(-\frac{2}{3}\right)\left(-\frac{2}{3}\right)$ **5.** 7^3 **7.** $(1.2)^4$ **9.** $x^3 + y^2$ **11.** 125

13. $\frac{9}{16}$ **15.** $\frac{16}{5}$ **17.** 2.8561 **19.** $-\frac{1}{27}$ **21.** 140.608 **23.** -6 **25.** 2 **27.** -35

Section Problems 2.7

1. $\frac{1}{81}$ **3.** $\frac{1}{16}$ **5.** 25 **7.** $\frac{1}{2.25}$ or $\frac{4}{9}$ **9.** 4^6 **11.** $\frac{1}{4}$ **13.** $\frac{1}{36}$ **15.** $\frac{1}{y^6}$ **17.** 8 **19.** $\frac{1}{y^6}$

21. 1 **23.** $\frac{1}{y^8}$ **25.** x^6 **27.** y^6 **29.** $\frac{1}{x^6}$ **31.** $\frac{1}{4x^6}$ **33.** $\frac{1}{x^2}$ **35.** $\frac{27x^3}{y^3}$ **37.** 1

39. $\frac{4}{x^4y^5}$ **41.** 1 **43.** $\frac{1}{7x}$ **45.** 1 **47.** $-\frac{1}{(-5)^5}$ **49.** x^3 **51.** 2 **53.** 1 **55.** y^4

Section Problems 2.8

1. Commutative **3.** Associative **5.** Distributive **7.** Commutative **9.** Distributive

11. $3x$; $4y$ **13.** $0.5y$; $-1.4x$ **15.** $\frac{7}{2}$; -3 **17.** $3x + 12$ **19.** $7x + (-7y)$ or $7x - 7y$

21. $-28x + 4y + 20$ **23.** $-6x + \frac{9}{5}$ **25.** $x + (-2y) + \left(-\frac{5}{2}\right)$ or $x - 2y - \frac{5}{2}$

27. $-7.2x + 9$ **29.** $3(x + 4)$ **31.** $-5[1 + (-5a)]$ or $-5(1 - 5a)$ **33.** $-3(2x - y + 6)$
35. $-2[2x + (-3y) + 4]$ or $-2(2x - 3y + 4)$ **37.** $-3a(2b + c - 1)$ **39.** $-3(x + 4)$
41. $7xy(x - 2y)$

Section Problems 2.9

1. $3x$ and $2x$ **3.** x^2y and $-2x^2y$ **5.** $6x + (-8)$ or $6x - 8$ **7.** $-x + (-2)$ or $-x - 2$
9. $x + 4y$ **11.** $9x$ **13.** $a^2 + 5a + (-b)$ or $a^2 + 5a - b$ **15.** $4m^2 + 3m + (-3mn) + (-n)$ or $4m^2 + 3m - 3mn - n$ **17.** $x + 8y$ **19.** $11R$ **21.** $-5x + (-11)$ or $-5x - 11$
23. $-0.24x + 2.448y$ **25.** $2x + (-10\pi) + 5$ or $2x - 10\pi + 5$ **27.** -3 **29.** $-3x + 8$
31. $2x + (-y)$ or $2x - y$ **33.** $-2a + 3b$ **35.** $-3x + 17$ **37.** $-5x + (-12)$ or $-5x - 12$
39. $5x^2 + (-2x) + (-2)$ or $5x^2 - 2x - 2$

Section Problems 2.10

1. (a) 3 (b) -3 (c) -13 (d) 4 **3.** -44 **5.** 204 **7.** 83 **9.** -268 **11.** 10.4 sq ft

13. 3750 watts **15.** 43.96 in. **17.** $36\frac{2}{3}$°C **19.** 86 lb **21.** 6400 ft **23.** (a) -14 (b) 31

25. 7 **27.** 14 **29.** 25 **31.** -13 **33.** (a) 15 (b) -29

Chapter 2 Review Problems

1. 9 **2.** -7 **3.** -38 **4.** -385 **5.** 545 **6.** -9 **7.** -8 **8.** -5 **9.** 11 **10.** 1
11. -1 **12.** -788 **13.** 204 **14.** 3 **15.** 18 **16.** 716 **17.** -62 **18.** 10 **19.** 2

20. -12 **21.** -40 **22.** 0 **23.** 19 **24.** 6 **25.** 6 **26.** -3 **27.** 16 **28.** $-\frac{63}{10}$

29. $\frac{50}{7}$ **30.** $\frac{5}{4}$ **31.** $\frac{5}{16}$ **32.** $\frac{33}{16}$ **33.** 3.7 **34.** -8.47 **35.** 0.9 **36.** 474

37. -5.126 **38.** $\left(\frac{1}{3}\right)\left(\frac{1}{3}\right)\left(\frac{1}{3}\right)\left(\frac{1}{3}\right)$ **39.** $xxxx + yy$ **40.** $(-5)^4$ **41.** $x^2 + (2y)^3 + z^1$

42. 64 **43.** 32 **44.** $-\frac{1}{16}$ **45.** 26 **46.** 22 **47.** $-6\frac{2}{5}$ **48.** -20 **49.** $\frac{1}{9}$ **50.** 64

51. $\frac{1}{x^2}$ **52.** $\frac{1}{2^{10}}$ **53.** 1 **54.** $\frac{36y^4}{x^6}$ **55.** $\frac{1}{49}$ **56.** $\frac{1}{9}$ **57.** $\frac{1}{x^7}$ **58.** 3 **59.** $4x^4y^6$

60. $\frac{4y^8}{x^8}$ **61.** Commutative **62.** Distributive **63.** Associative **64.** Associative

65. Commutative **66.** Distributive **67.** $6x + 12$ **68.** $-4x + (-8y) + 20$ or $-4x - 8y + 20$
69. $-3[2x + (-3)]$ or $-3(2x - 3)$ **70.** $2[2x + (-3y) + 6]$ or $2(2x - 3y + 6)$ **71.** $3s[t + (-3)]$ or $3s(t - 3)$ **72.** $-2x + y$ **73.** $3y$ **74.** $4x + (-13y)$ or $4x - 13y$ **75.** a **76.** -16
77. 54 sq in. **78.** 41 **79.** -10 **80.** 50 **81.** -10°C

Chapter 2 Test

1. -2 **2.** -10 **3.** 2 **4.** 15 **5.** 3 **6.** 380 **7.** $-\frac{11}{3}$ **8.** $-\frac{1}{10}$ **9.** 2.46 **10.** $\frac{9}{20}$

11. $-3\frac{5}{12}$ **12.** $\frac{1}{2}$ **13.** $-4\frac{1}{5}$ **14.** 39 **15.** $3x - 6y + 15$ **16.** $-3(2x - 5y)$

17. $-5x + y$ **18.** $7a + 5$ **19.** $-5t^2 + 6t + 13$ **20.** $\frac{16}{81}$ **21.** $\frac{1}{49}$ **22.** $-\frac{1}{32}$ **23.** $\frac{1}{x}$

24. 3 **25.** $\frac{9x^4}{y^8}$ **26.** $\frac{8x^9}{y^9}$ **27.** (a) 1 (b) -15 **28.** 63 **29.** 4 **30.** (a) Commutative (b) Distributive (c) Associative

Section Problems 3.1

1. No—has two variables **3.** Not linear—exponent other than 1; that is, $1/s = s^{-1}$. **5.** Linear **7.** Linear **9.** Yes **11.** No **13.** Yes **15.** Answers will vary. **17.** Answers will vary. **19.** Answers will vary. **21.** Answers will vary. **23.** If we add the same quantity to both sides of an equation, we obtain an equivalent equation. **25.** Equations that have exactly the same solutions

Section Problems 3.2

1. 1 **3.** 15 **5.** $\frac{1}{4}$ **7.** $\frac{3}{2}$ **9.** 7 **11.** 3 **13.** -5 **15.** 7 **17.** -3 **19.** 6 **21.** 4.4

23. -2 **25.** $\frac{1}{12}$ **27.** -80 **29.** -3.7 **31.** $-\frac{6}{5}$ **33.** $-\frac{1}{12}$ **35.** $-\frac{1}{4}$

Section Problems 3.3

1. 2 **3.** -5 **5.** 3 **7.** 15 **9.** 20 **11.** 7 **13.** -7 **15.** -3 **17.** 2 **19.** -6

21. 40 **23.** 1 **25.** 30 **27.** -3 **29.** 12 **31.** $-\frac{5}{2}$ **33.** 60.25 **35.** 6

Section Problems 3.4

1. 7 **3.** 1 **5.** 2 **7.** 2 **9.** -1 **11.** -2 **13.** 2 **15.** 5 **17.** $\frac{6}{5}$ **19.** 8 **21.** No solution **23.** $-\frac{5}{2}$ **25.** Infinitely many solutions **27.** $-\frac{1}{4}$ **29.** 0 **31.** 7.4 **33.** Infinitely many solutions

Section Problems 3.5

1. -1 **3.** -4 **5.** $\frac{-4}{5}$ **7.** $\frac{-9}{10}$ **9.** Infinitely many solutions **11.** 13 **13.** -18.3

15. 3 **17.** $\frac{4}{5}$ **19.** -23 **21.** -2 **23.** $\frac{1}{4}$ **25.** -13 **27.** $\frac{5}{14}$ **29.** -3 **31.** $\frac{15}{4}$ **33.** $\frac{-4}{5}$

Section Problems 3.6

1. $h = \frac{2A}{b}$ **3.** $R = \frac{I}{PT}$ **5.** $h = \frac{2A}{B + b}$ **7.** $d = rt$ **9.** $D^2 = \frac{2gHP}{NL}$ **11.** $l = \frac{p - 2w}{2}$

13. $C = \frac{5}{9}(F - 32°)$ **15.** $x = \frac{5y + 10}{3}$ **17.** $x = \frac{-y}{2} - \frac{3}{2}$ **19.** $a = \frac{5b}{2} + \frac{3}{2}$

21. $x = \frac{y}{2} - \frac{3}{2}$ **23.** $t = 2 - \frac{2s}{5}$ **25.** $s = 1 - \frac{t}{3}$ **27.** $h = \frac{A - 2\pi r^2}{2\pi r}$ **29.** $P = \frac{Q}{Q - 1}$

31. $X = \frac{A}{M + N}$ **33.** $y = \frac{2ax - 4}{7c + 3b}$

Chapter 3 Review Problems

1. No—two variables **2.** Yes **3.** No—two variables or exponent other than 1 **4.** No

5. Answers will vary. **6.** -12 **7.** -4 **8.** -12 **9.** $-\frac{5}{4}$ **10.** -2.8 **11.** 3 **12.** $\frac{11}{3}$

13. 5 **14.** 6 **15.** -18 **16.** -6 **17.** -2 **18.** No solution **19.** Infinitely many solutions **20.** Infinitely many solutions **21.** -5 **22.** -2.2 **23.** $\frac{9}{5}$ **24.** $-\frac{5}{4}$ **25.** $\frac{47}{26}$

26. 2.85 **27.** $\frac{5}{9}$ **28.** $P = \frac{A - N}{Q}$ **29.** $C = \frac{5}{9}(F - 32°)$ **30.** $d = \frac{Ve}{c}$ **31.** $P = \frac{I}{RT}$

32. $h = \frac{A}{b}$ **33.** $x = \frac{5y}{2} + 5$ **34.** $y = \frac{2x}{5} - 2$

Chapter 3 Test

1. (a) No—2 variables (b) No—2 variables (c) No—2 variables **2.** No **3.** Answers will vary.

4. -1 **5.** -3 **6.** -4 **7.** -1 **8.** $-\frac{11}{2}$ **9.** -8 **10.** -6 **11.** $\frac{5}{12}$ **12.** 2

13. -2 **14.** Infinitely many solutions **15.** 6.4 **16.** $-\frac{1}{3}$ **17.** -4 **18.** $M = \frac{CQ}{100}$

19. $l = \frac{p - 2w}{2}$ **20.** $y = \frac{2x}{3} + \frac{4}{3}$

Section Problems 4.1

1. $x - 7$ **3.** $4x$ **5.** $\frac{x}{-3}$ **7.** $\frac{x}{y}$ or $\frac{y}{x}$ **9.** $2(n + 10)$ **11.** $\frac{3}{5}x$ **13.** $4x + 5$ **15.** $\frac{x}{2} - 8$

17. $4(x + 2)$ **19.** $\frac{2}{3}x + 4$ **21.** $x + y = 17$ **23.** $x + 3 = 4x - 7$ **25.** $\frac{3}{4}x = -3$

27. $x + 1 - 2x = -4x + 6$ **29.** $\frac{x}{2} + \frac{3}{4} = -\frac{5}{6} + x$ **31.** $\frac{7 - 2x}{4}$ **33.** $6 - 2x = 4x + 1$

35. $-2x = \frac{x}{2} + \frac{1}{3}$

Section Problems 4.2

1. Length $= 12\frac{1}{2}$ in.; width $= 5$ in. **3.** 6 **5.** 8 **7.** 35°, 70°, 75° **9.** 40

11. Car—$8400; pickup—$7800 **13.** Preparation—3 hr, installation—6 hr, cleanup—1 hr **15.** 20 **17.** 4 ft by 16 ft **19.** 10.5¢ **21.** Dan—6, Donald—12, Doug—4 **23.** 10 lb, 16 lb, 20 lb

Section Problems 4.3

1. $250 **3.** $80,000 **5.** 300 **7.** 120 adults **9.** $1800 **11.** 41.76 sq ft **13.** 1400 **15.** 20% **17.** $3600

Section Problems 4.4

1. 30 and 36 **3.** 4 and 8 **5.** 11.1% **7.** 5 and 36 **9.** 5 and 6 **11.** 20 min **13.** $600 **15.** 17 and 22

Section Problems 4.5

1. 13 nickels, 12 dimes **3.** 6.25 pt of 70% solution, 18.75 pt of 50% solution **5.** $5000 at 5%, $7500 at 6% **7.** 26 and 32 **9.** $13\frac{1}{3}$ gal of 40% solution, $6\frac{2}{3}$ gal of 70% solution **11.** 0.4 L of 36% solution, 1.6 L of 16% solution **13.** 9 pt of 20% solution, 6 pt of 45% solution **15.** $32

Section Problems 4.6

1. $11\frac{1}{4}$ ft from 60-lb weight **3.** 150 lb **5.** 3.4 yr **7.** 90 **9.** 12% **11.** 9.5 hr **13.** 7, 8, and 9 **15.** 4 ft from the fulcrum on the right side

Chapter 4 Review Problems

1. $3x + 7$ **2.** $3(x + 7)$ **3.** $2x - 4y$ **4.** $\frac{x}{y} + 6$ **5.** $\frac{2x + 5}{3} = 6$ **6.** 4 in. by 11 in. **7.** 32°, 64°, and 84° **8.** Joe = $4.50, Sue = $4.00, Helen = $6.00 **9.** 194 **10.** 180 **11.** Approx. 10% **12.** 60 **13.** $160 **14.** 14, 17, and 20 **15.** 40 and 42 **16.** 9 **17.** 6 nickels, 7 dimes, 5 quarters **18.** $\frac{10}{17}$ **19.** $7\frac{1}{2}$ pt of 20% solution and $2\frac{1}{2}$ pt of 60% solution **20.** 8 gal **21.** 6 ft from the 120-lb weight **22.** $4\frac{4}{5}$ ft from the fulcrum **23.** $68,900 **24.** $4\frac{5}{6}$ yr **25.** 36, 38 **26.** 28, 29, and 30 **27.** 165 sq m **28.** 22 and 30 **29.** 18 **30.** 12 dimes, 16 quarters **31.** $7\frac{1}{2}$ ft from the fulcrum **32.** $3\frac{2}{3}$ ft to the right of the fulcrum **33.** $2\frac{2}{3}$ yr **34.** 12 Cuban and 8 American

Chapter 4 Test

1. 12 and 13 **2.** \$45 **3.** 5 and 7 **4.** 20 oz of \$1.30/oz tea, 12 oz of \$1.70/oz tea **5.** 30 ft by 40 ft **6.** $9\frac{1}{7}$ ft from 150-lb weight **7.** $\frac{6}{5}$ **8.** $2\frac{1}{5}$ yr **9.** \$5,000 **10.** 15

Section Problems 5.1

1. Yes **3.** No—negative exponent **5.** Yes **7.** No—negative exponent **9.** $-x^2$; $-3x$; 4 **11.** $\frac{t}{2}$; -5 **13.** $0.2y^2$; $\frac{-y}{4}$; y **15.** -1; -3; 4 **17.** $\frac{1}{2}$; -5 **19.** -1; -3; -1.4 **21.** 4; 2; 0; polynomial has degree 4 **23.** 4; 4; 5; polynomial has degree 5 **25.** 7; 2; 0; polynomial has degree 7 **27.** $-2x^5 + x^4 + 3x^3 - x + 1$ **29.** $-t^4 - t^3 + 3t^2 + t - 4$ **31.** Answers will vary. **33.** Answers will vary. **35.** Answers will vary. **37.** Answers will vary. **39.** $-3x + 3$ **41.** $3xy + y^2 - xy^2$ **43.** $y - 12$

Section Problems 5.2

1. $-x^2 - x - 1$ **3.** $3t^3 - 2t^2 + 6$ **5.** $4x^3 + 2x^2 + 6x + 2$ **7.** $x^3 - 2x^2 + 5x - 2$ **9.** $-x^4 + x^3 + 2x - 5$ **11.** $3x^2 + 2x$ **13.** $-x^3 + x^2 + 4x - 4$ **15.** $2a^2b - 7ab + 2b^2 - 1$ **17.** $2x^2 - 14x - 2$ **19.** $4s^2 + 2s$

Section Problems 5.3

1. $-21x^3$ **3.** $-12t^3$ **5.** $-\frac{3}{2}x^3y^4$ **7.** $-8t^3 + 12t^2 - 4t$ **9.** $-3x^3 + 6x^2 - 15x$ **11.** $-\frac{3}{2}x^3y + 2x^3y^2 - 3x^2y$ **13.** $2ac + 2ad - bc - bd$ **15.** $2x^3 - 6xy + yx^2 - 3y^2$ **17.** $y^3 - 2y^2 - y + 2$ **19.** $3x^4 - 7x^3 + 5x^2 - 13x + 4$ **21.** $x^2 + x - 20$ **23.** $y^2 - 7y + 12$ **25.** $x^2 - 25$ **27.** $2t^2 + 7t - 4$ **29.** $6t^2 - 13t - 5$ **31.** $42x^2 - 27x + 3$ **33.** $42x^2 + 15x - 3$ **35.** $2t^2 - 3t - 20$ **37.** $3t^2 - 25t + 42$ **39.** $x^3 + 4x^2 - 9x - 36$ **41.** $x^3 + 3x^2 + 3x + 1$ **43.** $2s^3 + 2st^2 - ts^2 - t^3$ **45.** $-12y^3 - 26y^2 + 10y$ **47.** $1 - b - 6b^2$ **49.** $0.14t^2 - 3.72t - 5.6$

Section Problems 5.4

1. $x^2 - 16$ **3.** $4y^2 - 1$ **5.** $s^2 - 9$ **7.** $4y^2 + 4y + 1$ **9.** $y^2 - 2y + 1$ **11.** $4t^2 - 12t + 9$ **13.** $36x^2 - 36x + 9$ **15.** $t^2 - 2.6t + 1.69$ **17.** $y^2 - 6y + 9$ **19.** $3x^2 - 11x + 10$ **21.** $8x^3 + 12x^2 + 6x + 1$ **23.** $s^3 + 3s^2 - 4$ **25.** $x^4 - 8x^2 + 16$ **27.** $4s^2 - 9$ **29.** $y^3 - y^2 - 4y + 4$

Section Problems 5.5

1. $4(x - 2y + 3)$ **3.** $ab^2(3a - 4)$ **5.** $(b - 4)(a + 7)$ **7.** $x(x^2 - 3x + 9)$ **9.** $(y - 6)(y + 6)$ **11.** $(t - 4)(t + 4)$ **13.** $(7y - 6)(7y + 6)$ **15.** $3(y - 3)(y + 3)$ **17.** $2t(t - 2)(t + 2)$ **19.** $(2 - x)(2 + x)$ **21.** $xy(x - y)(x + y)$ **23.** $(x - 1)(x + 1)(y + 3)$ **25.** $2(t - 5)(t + 5)$ **27.** $3(y - 4)(y + 4)$ **29.** Not factorable

Section Problems 5.6

1. $(x-5)(x-4)$ **3.** $(x+7)(x-2)$ **5.** $(s+6)(s+2)$ **7.** Doesn't factor
9. $(x-2)(x-5)$ **11.** $x(x-1)(x+1)$ **13.** $(y+2)(y+1)$ **15.** $(x-3)(x+3)$
17. $(2y+3)(y+4)$ **19.** $(2s-3)(s+2)$ **21.** $t(s-3)(s+3)$ **23.** $3(2x-1)(x+4)$
25. $(3y-2)^2$ **27.** $(5x+1)(3x-1)$ **29.** $(2t-3)(t+1)$ **31.** $2ab(a-2)(a+2)$
33. $(6t-1)(2t-5)$ **35.** $(9x-5)(x+4)$ **37.** Doesn't factor **39.** $3(x+5)(x-5)$
41. $2(3x^2-7x+5)$ **43.** $(3p-2)(2p-1)$ **45.** $(4t-5)(3t+1)$

Section Problems 5.7

1. 2, 5 **3.** 2, -2 **5.** -1, -6 **7.** 3, 5 **9.** 3 **11.** $\frac{3}{2}, -\frac{5}{3}$ **13.** $\frac{1}{3}, -2$ **15.** 4, -3

17. $-3, -\frac{1}{2}$ **19.** $\frac{3}{2}$ **21.** -1, 6 **23.** $\frac{1}{3}, 2$ **25.** No solutions by factoring **27.** 4, -2

29. 0, 7

Section Problems 5.8

1. -2 or 6 **3.** $b=5$, $B=10$, $h=6$ **5.** 0 and 1 or 8 and 9 **7.** 90 ft **9.** 5 in. **11.** 5, 6 or -6, -5 **13.** 1.5 in.

Chapter 5 Review Problems

1. No—$\sqrt{x}$ **2.** Yes **3.** No—variable in denominator **4.** x^4; $-x^3$; $3x^2$; $7x$; -5 **5.** $-x^2y^2$; $-3xy^2$; $1.4y$ **6.** 1; -1; 3; 7; -5 **7.** 0.2; $-\frac{1}{4}$; 7 **8.** 5; 3; 2; 0; polynomial has degree 5

9. 3; 6; polynomial has degree 6 **10.** $-x^5+6x^4+3x+2$ **11.** $-1+t+t^2+3t^4-7t^7$
12. Answers will vary. **13.** $3x^3-3x^2+8x-5$ **14.** $-t^4+7t^3+9t^2-t+2$
15. $5a^2b^2+4ab-a^3+b^3$ **16.** $-y^3+6y^2-2y$ **17.** $-12x^3$ **18.** $24x^3y^3$
19. $-28a^3b^2+4a^2b^3-12ab^3$ **20.** $x^3+x^2-10x+8$ **21.** $ac-ad+bc-bd$
22. $y^2-3y-10$ **23.** $t^2-11t+30$ **24.** $2x^2+3x-20$ **25.** $48y^2-22y-15$
26. $12x^2+32x+5$ **27.** x^2-9 **28.** $s^2+10s+25$ **29.** $4t^2-12t+9$ **30.** $100x^2-49$
31. x^3+3x^2+3x+1 **32.** $(x-7)(x+7)$ **33.** $-2t(2t+5)$ **34.** $ab(a-b)$
35. $(y-3)(x+4)$ **36.** $(x-4)(x+4)(x-3)(x+3)$ **37.** $(y-3)(y-1)$
38. $(x+5)(x+2)$ **39.** Doesn't factor **40.** $(2x+3)(3x+2)$ **41.** $2(3x-4)(2x+5)$

42. $(9a+4)(2a+3)$ **43.** -4, 3 **44.** 1, 5 **45.** $-\frac{1}{2}, -\frac{3}{5}$ **46.** $-\frac{7}{6}, \frac{5}{7}$ **47.** 5 or -2

48. 3, 5 or -5, -3 **49.** 8, 10

Chapter 5 Test

1. $2a^2-6a+3$ **2.** $-x^3+4x^2-2x-7$ **3.** $2m^3+2mn+m^2n+n^2$

4. $2t^4-4t^3-3t^2-4t-5$ **5.** $6x^2-13x-5$ **6.** $9a^2+12a+4$ **7.** (a) 3 (b) $-x^3$; $\frac{7}{2}x^2$; x; -1.5 (c) -1; $\frac{7}{2}$; 1; -1.5 **8.** $-t^5+6t^3+2t^2-t+4$ **9.** $-3x(2a-3b)$

10. $(t-5)(t+3)$ **11.** $(2y-3)(y+4)$ **12.** $3(x-2)(x+2)$ **13.** $(3x+4)(x+1)$

14. $(6b+5)(2b-1)$ **15.** $-5, 2$ **16.** $-\frac{1}{2}, -2$ **17.** $-2, 2$ **18.** $-1, \frac{2}{3}$ **19.** 14 and 16 or -2 and 0 **20.** Length = 3 ft, width = 16 ft

Section Problems 6.1

1. Yes **3.** No—exponent other than 1 **5.** Yes **7.** Yes **9.** $x - 2y = -1$

11. $2a - 4b = 7$ **13.** $\frac{x}{2} - \frac{y}{3} = -\frac{1}{4}$ **15.** $2y = 5$ **17.** (b), (d) **19.** (b) **21.** 4

23. -17 **25.** $-\frac{5}{2}$ **27.** $\frac{4}{3}$ **29.** Answers will vary. **31.** Answers will vary. **33.** Answers will vary.

Section Problems 6.2

1. II **3.** III **5.** III **7.** IV **9–17.** See graphs. **19.** $P = (3, 4)$ **21.** $R = (1, -2)$

23. $T = (0, 3)$ **25.** $V = \left(-1\frac{1}{2}, -1\frac{1}{2}\right)$ **27.** $Y = (-3, 0)$ **29.** $A = \left(3\frac{1}{2}, -3\right)$

31. $C = \left(4\frac{1}{2}, 2\frac{1}{3}\right)$ **33.** $E = (-3, -3)$ **35.** Answers will vary. **37.** Answers will vary.
39. The x-coordinate will always be zero.

Section Problems 6.3

1.

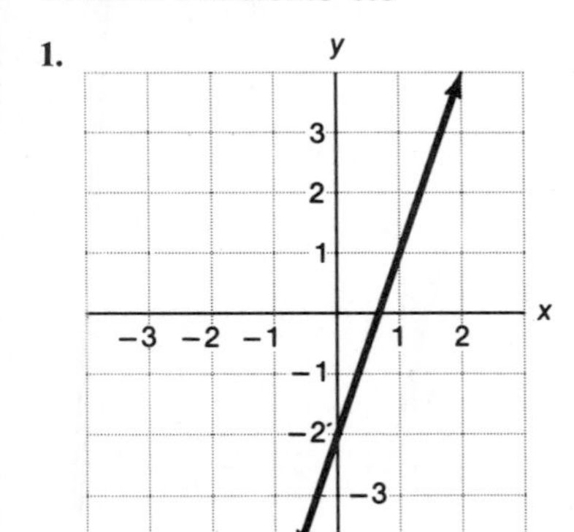

3.

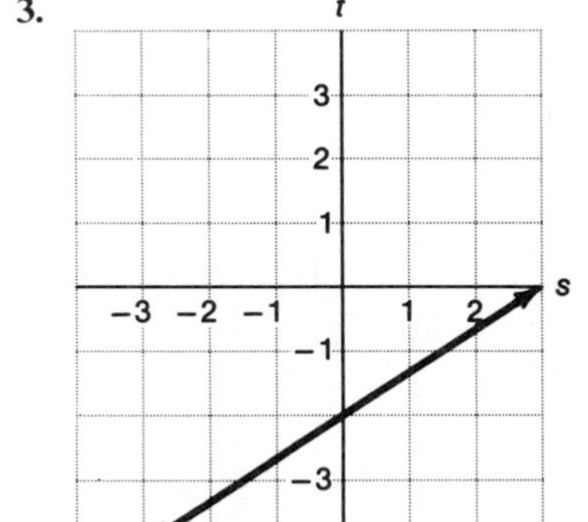

5.

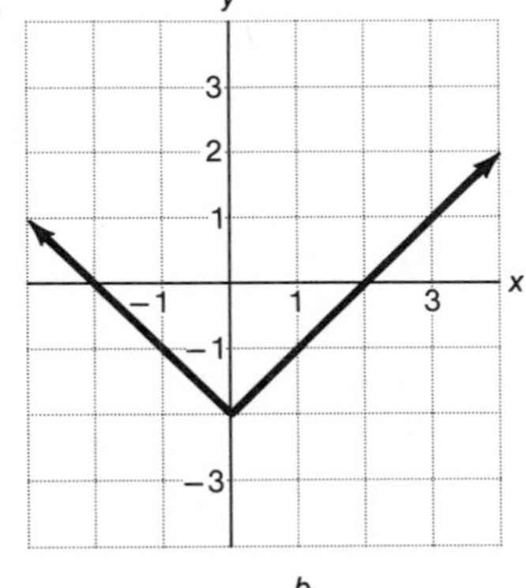

7.

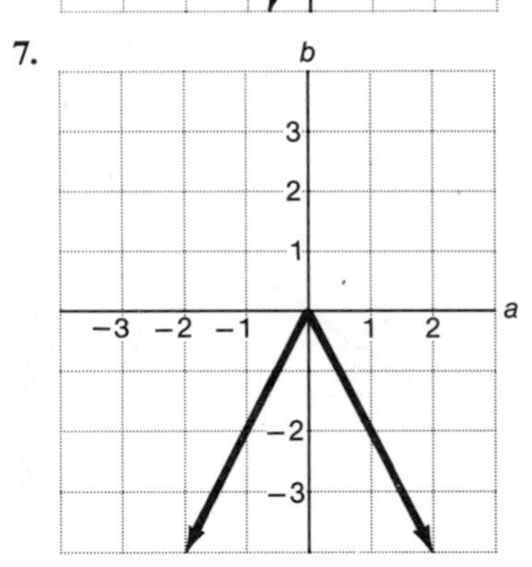

9.

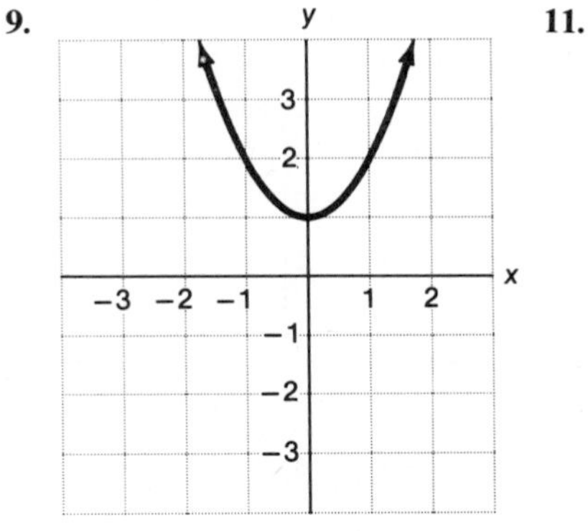

11.

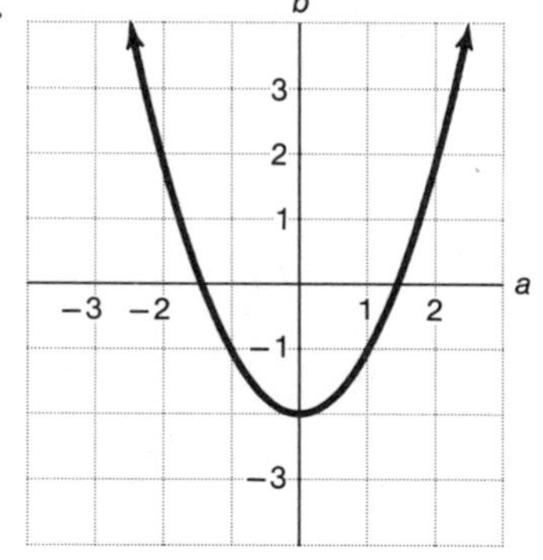

13.

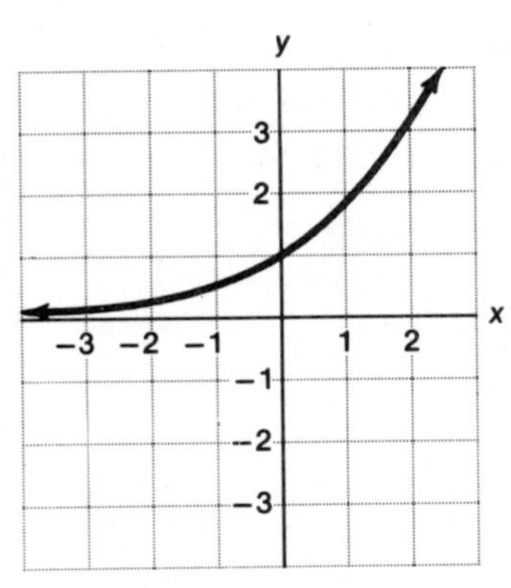

15.

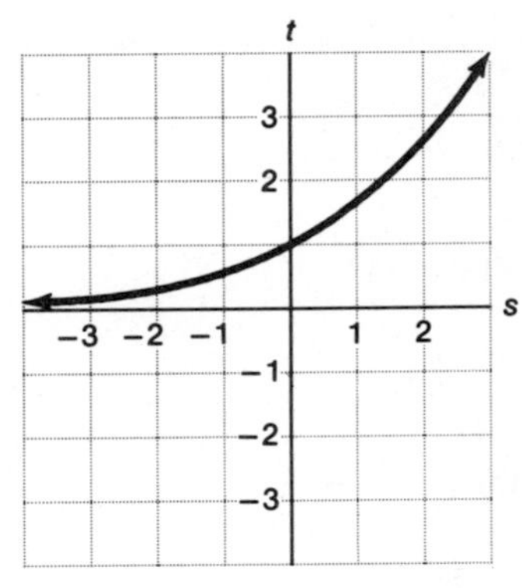

17.

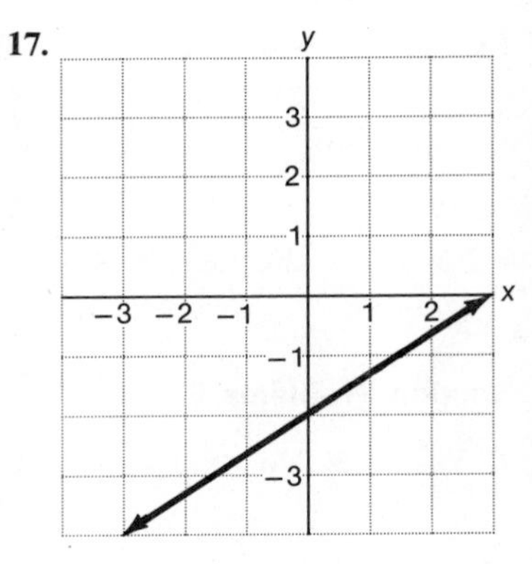

19.

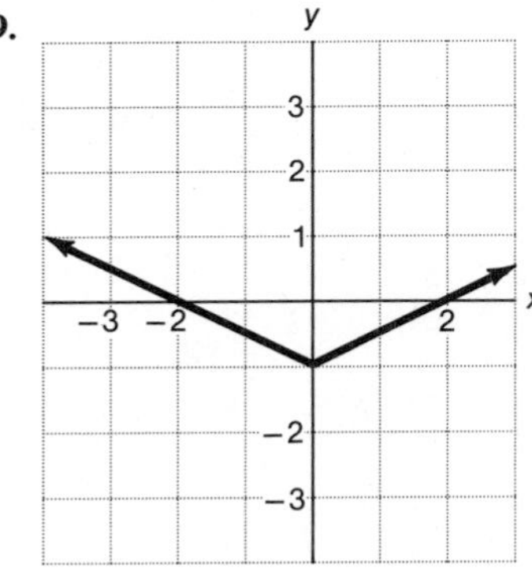

Section Problems 6.4

1. Yes **3.** No—three variables **5.** Yes **7.** $3x - 4y = -7$ **9.** $x = 1.7$ **11.** $3m - n = 4$

13. x-intercept $= \left(\frac{8}{3}, 0\right)$; y-intercept $= (0, -2)$ **15.** s-intercept $= (6, 0)$; t-intercept $= (0, -2)$

17. a-intercept $= (-5, 0)$, b-intercept $= \left(0, \frac{10}{7}\right)$ **19.** No x-intercept; y-intercept $= (0, 4)$

21.

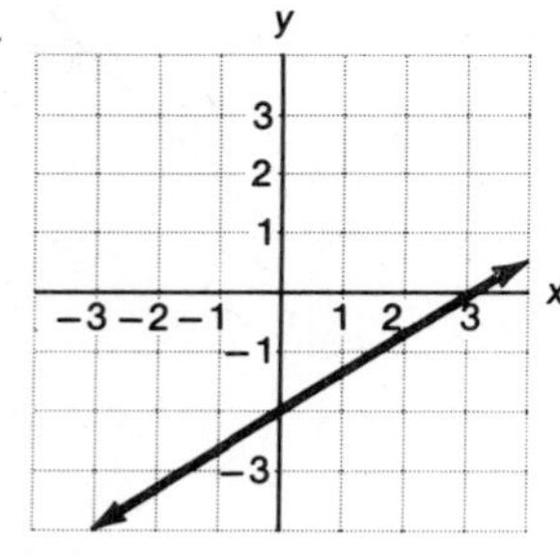

23.

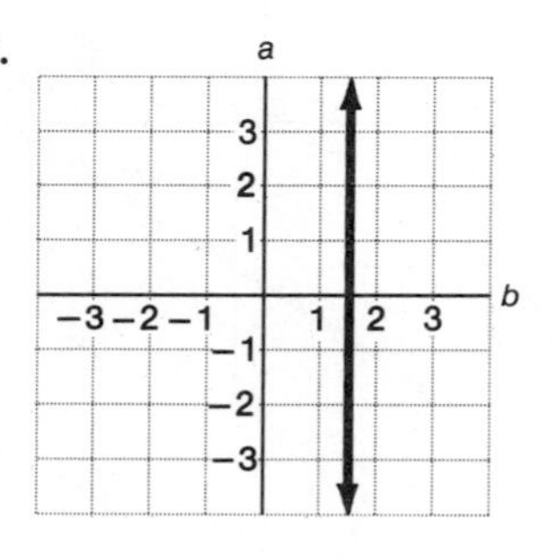

25.

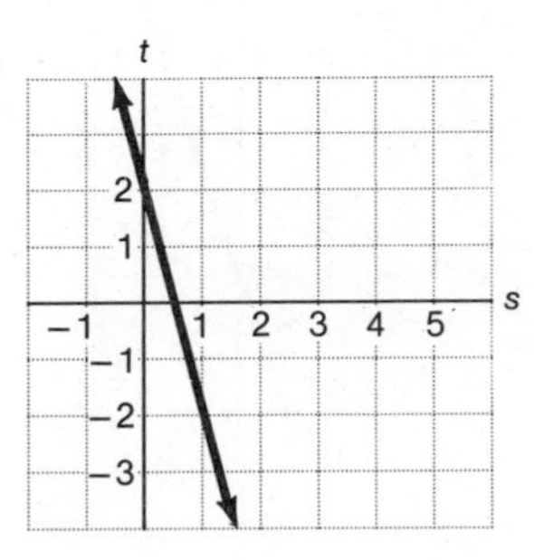

27.

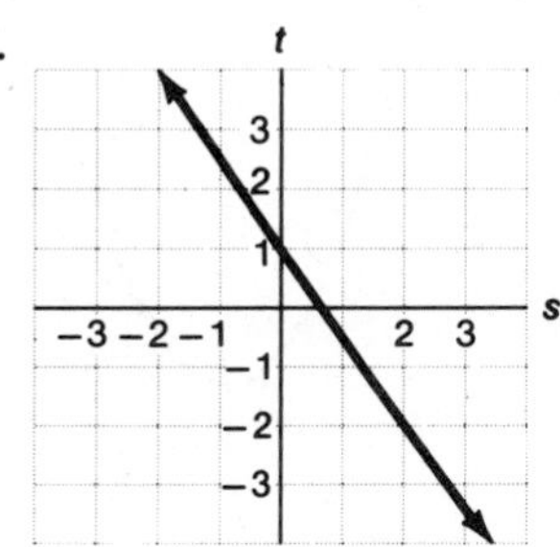

29.

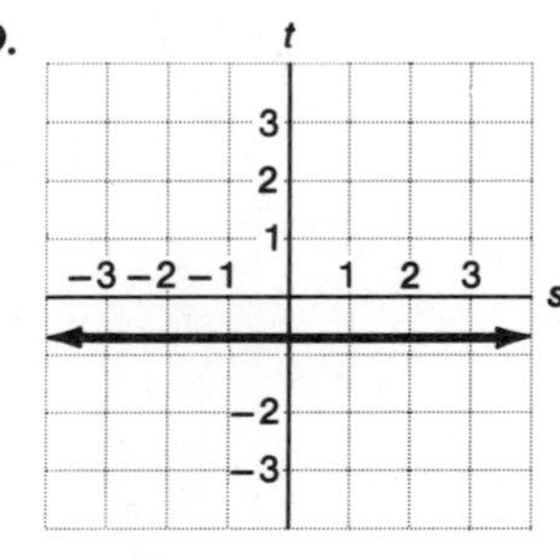

31.

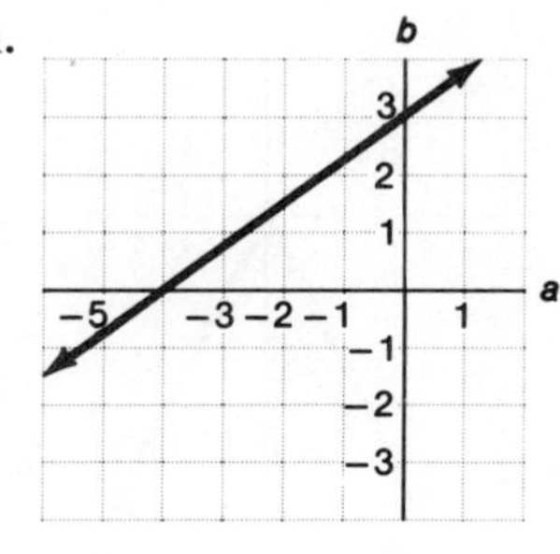

33.

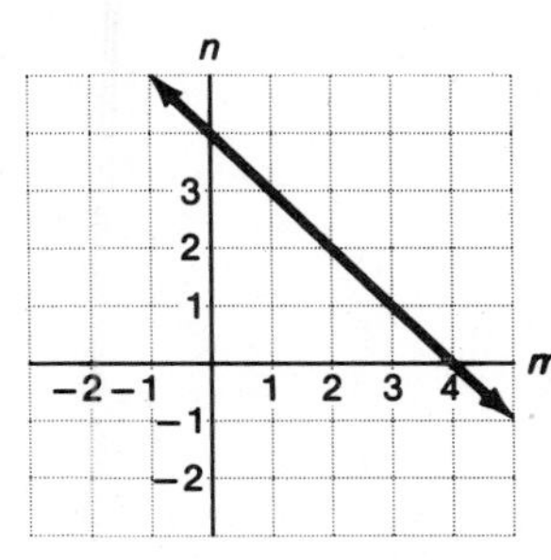

35. All horizontal lines (except $y = 0$) **37.** All lines that go through the origin

Section Problems 6.5

1. $\frac{2}{5}$ **3.** 4 **5.** Slope undefined **7.** -12 **9.** 0 **11.** 1 **13.** 0 **15.** -1

17. Undefined **19.** $\frac{7}{2}$ **21.** $-\frac{5}{6}$ **23.** 0 **25.** Undefined **27.** $\frac{1}{7}$

29.

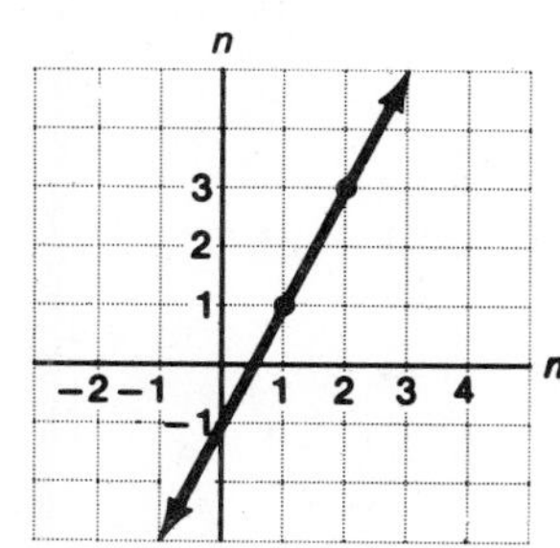

31.

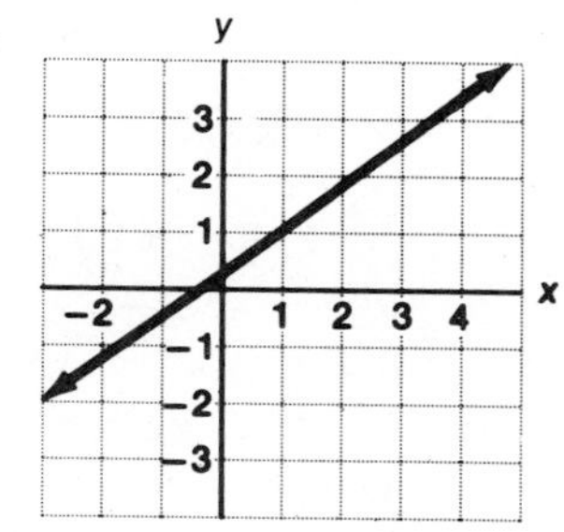

33.

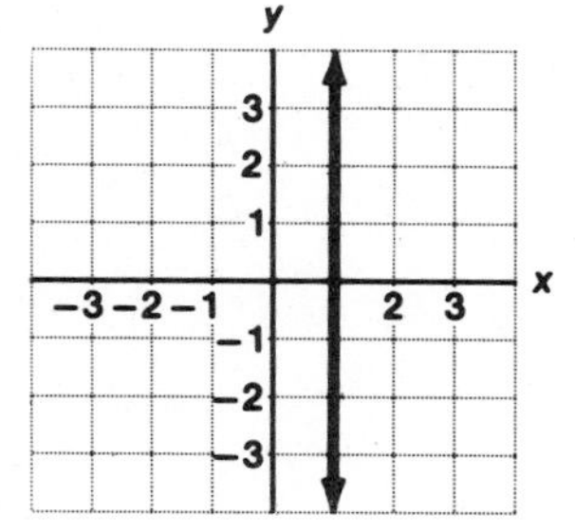

35. As we go from left to right along the x-axis, the line goes upward. **37.** The absolute value of the x-coordinate tells us how far from the y-axis the point is, and the absolute value of the y-coordinate tells us how far from the x-axis the point is.

Section Problems 6.6

1. $y = 2x - 6$ **3.** $y = \frac{5}{2}$ **5.** $b = \frac{1}{3}a - \frac{7}{3}$ **7.** $y = \frac{3}{2}x + 3$ **9.** $m = -3$; $(0, 1)$ **11.** No slope, no vertical axis intercept **13.** $m = 0$; $(0, -2)$ **15.** $m = \frac{5}{7}$; $(0, -2)$ **17.** $m = \frac{1}{3}$; $(0, 2)$

19. $m = \frac{2}{3}$; $\left(0, -\frac{5}{3}\right)$ **21.** $y = \frac{3}{2}x + 2$ **23.** $y = -1$ **25.** $x = 9$ **27.** $y = 4x - 6$

29. $y = -\frac{3}{2}x - 12$ **31.** $y = \frac{2}{3}x - \frac{11}{3}$ **33.** $x = -1$ **35.** $y = 4$ **37.** $y = \frac{2}{9}x + \frac{13}{9}$

39. $y = \frac{3}{5}x - \frac{14}{5}$ **41.** $\frac{3}{5}$ **43.** $y = 2x + 1$ **45.** $y = -\frac{5}{2}x + 7$ **47.** $y = 5$

49. $y = -\frac{5}{3}x + 24$ **51.** $y = \frac{4}{3}x - 1$

Section Problems 6.7

1. $2\frac{4}{5}$ yr **3.** (a) $y = 300 + .06x$ (b) \$2500 in sales **5.** \$39 **7.** (a) $c = 20q + 100$ (b) 35

9. (a) $R = 80x$, $C = 14{,}000 + 45x$, $P = 35x - 14{,}000$ (b) 400 **11.** (a) $P = 12x + 2600$ (b) \$6,200

Chapter 6 Review Problems

1. No—exponent other than 1 **2.** No—three variables **3.** Yes **4.** (a), (d)

5. $-3x + \frac{1}{4}y = -1$ **6.** $4x - 8y = 5$ **7.** Answers will vary. **8.** 19 **9.** IV **10.** III

11. I **12.** II

13–18.

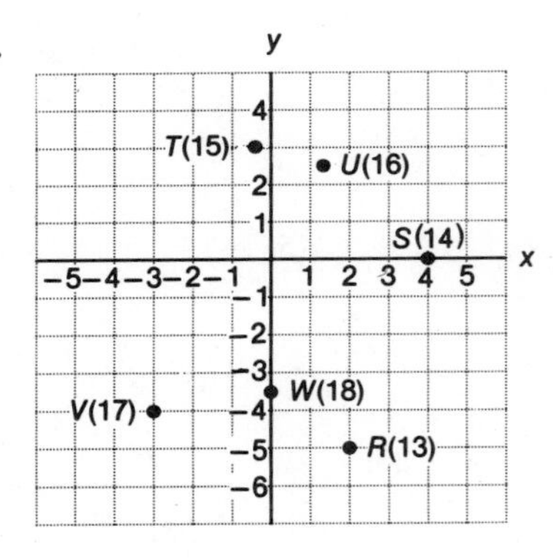

19. $P = (4, 2)$ **20.** $Q = (-3, 0)$ **21.** $R = \left(-1\frac{1}{2}, -2\frac{1}{2}\right)$ **22.** $S = \left(0, 4\frac{1}{3}\right)$ **23.** $T = \left(3\frac{1}{2}, -2\right)$

24.

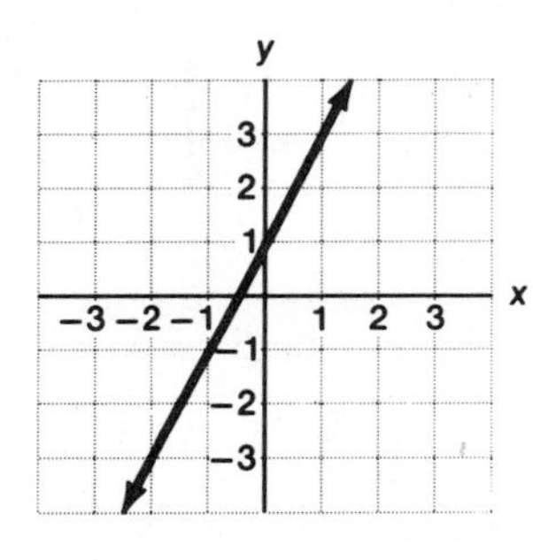

25.

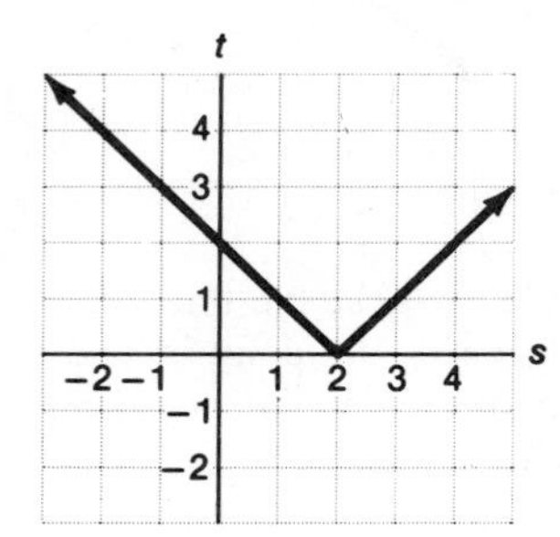

26.

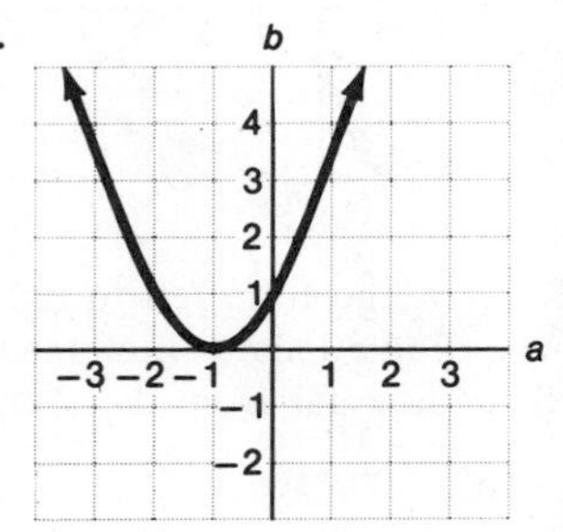

27. $-3x + y = -4$ **28.** $-2a + b = 14$ **29.** x-intercept $= (-4, 0)$; y-intercept $= (0, 3)$

30. a-intercept $= \left(\frac{8}{3}, 0\right)$; b-intercept $= (0, -2)$

31.

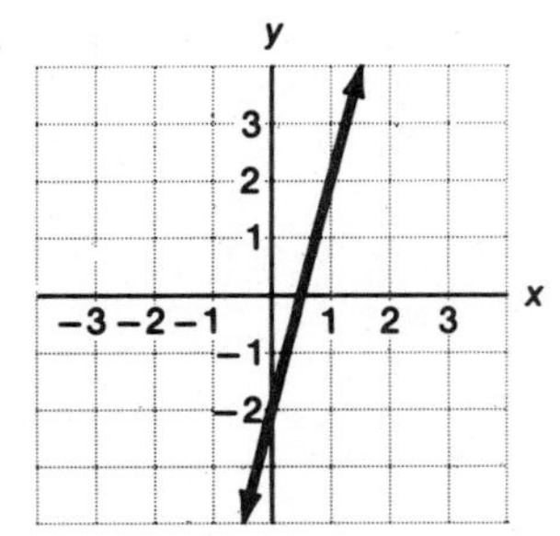

32.

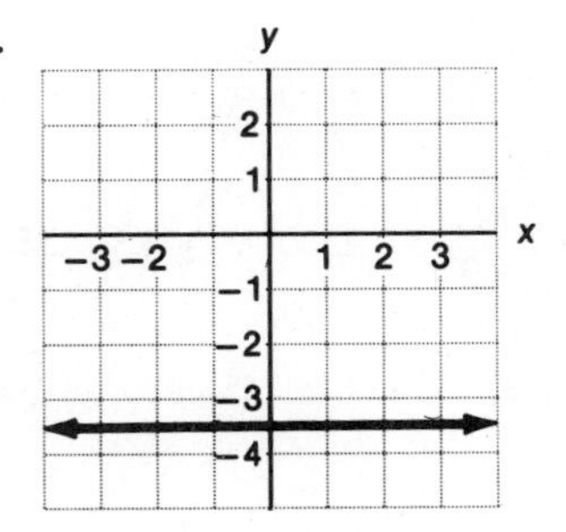

33.

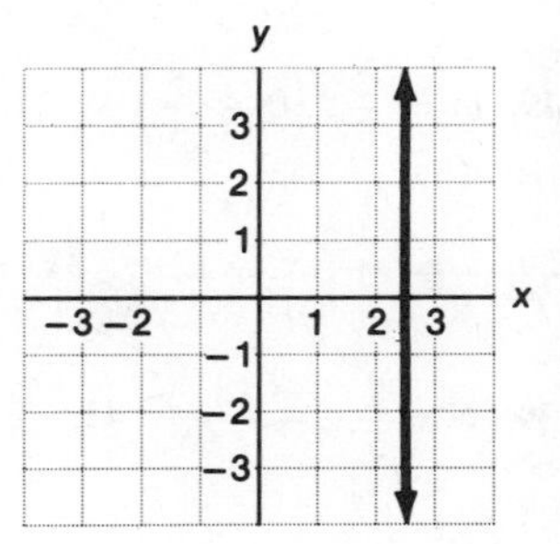

34. $-\frac{3}{5}$ **35.** 1 **36.** 3 **37.** No slope **38.** $\frac{5}{2}$ **39.** 0 **40.** $-\frac{2}{9}$ **41.** Undefined

42–44.

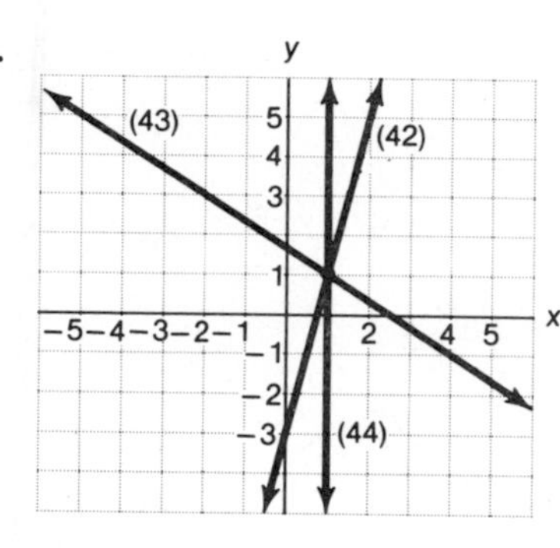

45. $y = x + 5$ **46.** $t = s - 5$ **47.** $y = 7$

48. $m = \frac{3}{2}$; $(0, -4)$ **49.** $m = \frac{2}{5}$; $(0, -2)$ **50.** No slope; no t-intercept **51.** $y = -\frac{1}{3}x - 1$

52. $y = 5$ **53.** $x = 3$ **54.** $y = 1.2x - 1.6$ **55.** $y = 2x - 1$ **56.** $y = -\frac{7}{3}x + \frac{1}{3}$

57. $x = 1$ **58.** $y = 5$ **59.** $y = -3x + 11$ **60.** $y = \frac{3}{2}x + 7$ **61.** (a) $V = 12{,}000 - 2400t$

(b) \$4000 (c) $4\frac{1}{6}$ yr **62.** (a) 25 (b) 16 or 17 $\left(\text{actually } 16\frac{2}{3}\right)$

Chapter 6 Test

1. (a) No—3 variables (b) Yes (c) Yes (d) No—absolute value (e) No—xy term **2.** $n = 1$

3. $P = (3, 2)$; $Q = (-3, 1)$; $R = (0, -2)$; $S = \left(-\frac{5}{2}, -\frac{3}{2}\right)$; $T = (1, 0)$

4.

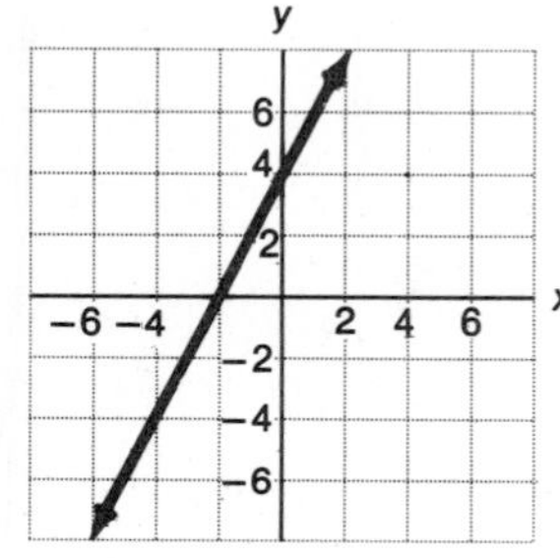

5.

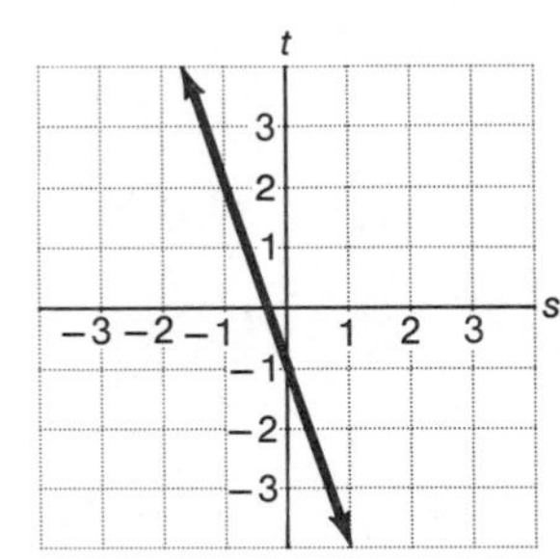

6.

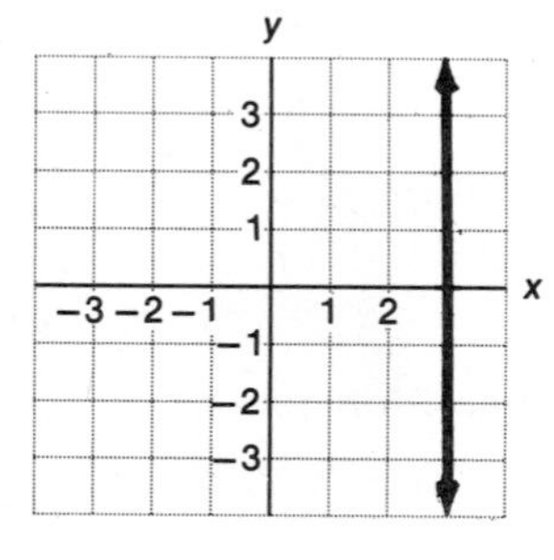

7.

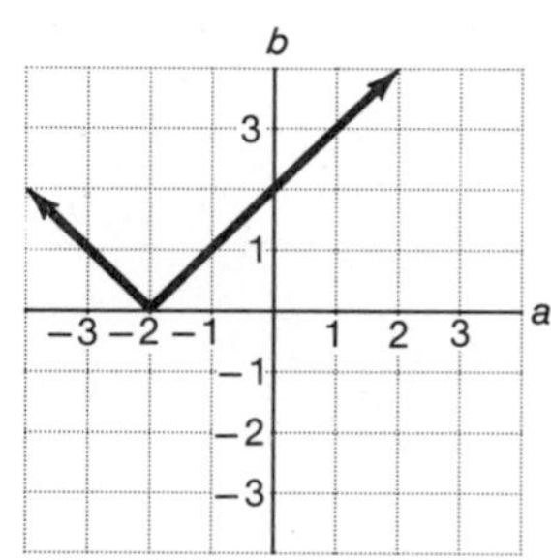

8. $m = \frac{3}{2}$; $(0, -4)$ **9.** $m = \frac{2}{3}$; y-intercept $= (0, 4)$;

x-intercept $= (-6, 0)$

10.

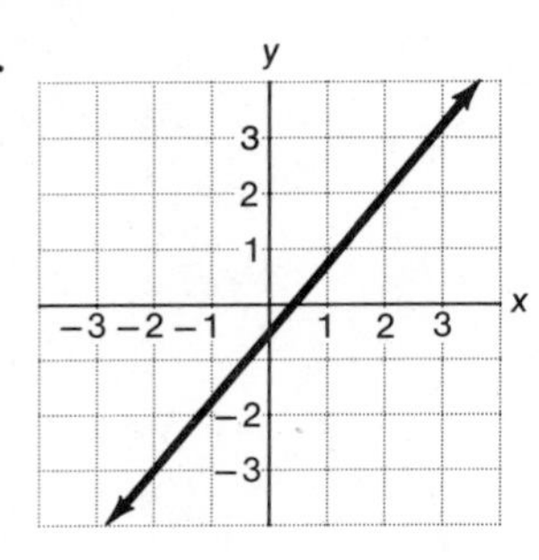

11. L_1—no slope; slope of $L_2 = \frac{5}{6}$ 12. $y = -3x + 7$

13. $y = 4x - 13$ 14. $y = -\frac{3}{2}x + 5$

15. (a) $V = 25{,}000 - 6000t$ (b) $2\frac{1}{2}$ yr

Section Problems 7.1

1. (a) No (b) Yes (c) No 3. (a) Yes (b) Yes (c) Yes 5. No solution 7. Unique
9. (a) Different slopes (b) Same slope, different y-intercept (c) Same slope, same y-intercept

Section Problems 7.2

1. (2, 1) 3. $\left(-\frac{3}{2}, 3\right)$ 5. $(-1, 3)$ 7. $(-3, 4)$ 9. $(-2, 5)$ 11. (3, 1) 13. $(-1, 2)$ 15. 2 and 2 17. 2 19. 2 dimes and 6 nickels

Section Problems 7.3

1. $\left(\frac{13}{5}, \frac{6}{5}\right)$ 3. (5, 9) 5. $(5, -2)$ 7. (5, 6) 9. $(4, -1)$ 11. Infinitely many solutions

13. (14, 23) 15. $(3, -1)$ 17. $(0, -4)$ 19. (14, 3) 21. $(-4, -6)$ 23. $(5, -2)$

25. $\left(-\frac{38}{7}, -\frac{43}{7}\right)$ 27. $\left(\frac{34}{5}, -\frac{1}{5}\right)$ 29. $\left(-\frac{11}{8}, \frac{1}{8}\right)$ 31. $\frac{7}{2}$ and $\frac{23}{2}$ 33. 32 oz of 50¢ kind and 88 oz of 80¢ kind 35. 4 L 37. 8 in. by 12 in. 39. 7 gal of white lightning and 12 gal of blue lightning

Section Problems 7.4

1. $(1, -3)$ 3. $(-1, -7)$ 5. $(5, -3)$ 7. (2, 2) 9. No solution 11. $(-1, -3)$

13. $(-4, 6)$ 15. No solution 17. (7, 2) 19. $\left(\frac{7}{3}, \frac{5}{3}\right)$ 21. $\left(\frac{26}{43}, -\frac{2}{43}\right)$ 23. 25

25. $\left(\frac{9}{5}, \frac{18}{5}\right)$ 27. 145 29. 8 hamburgers, 6 french fries 31. $(-4, 2)$ 33. (11, 6)

35. (0, 2) 37. $\left(\frac{3}{5}, \frac{2}{5}\right)$ 39. $(-3, -11)$ 41. $(-9, -8)$ 43. (1, 2) 45. $\left(\frac{13}{22}, \frac{5}{22}\right)$

Section Problems 7.5

1. 170 mph 3. 4:30 5. $22\frac{1}{2}$ min $\left(\text{or } \frac{3}{8} \text{ hr}\right)$ 7. 420 mi 9. 300 mph 11. 20 oz of $1.00 and 10 oz of $.70 kind

Chapter 7 Review Problems

1. No **2.** Yes **3.** Unique, no solution, or infinitely many solutions **4.** (1, 0) **5.** No solution **6.** Infinitely many solutions **7.** (2, −2) **8.** 2 and −3 **9.** (4, 3) **10.** $\left(\frac{10}{3}, \frac{11}{3}\right)$

11. Infinitely many solutions **12.** $\left(\frac{5}{2}, \frac{23}{2}\right)$ **13.** $\left(\frac{4}{5}, -\frac{17}{5}\right)$ **14.** (1, −1) **15.** No solution

16. (22, 14) **17.** (26, 18) **18.** $\left(\frac{13}{5}, \frac{19}{5}\right)$ **19.** $\left(3, \frac{13}{2}\right)$ **20.** $\left(\frac{26}{25}, \frac{32}{25}\right)$ **21.** 10 and −8

22. 9 pt of 40% solution, 3 pt of 60% solution **23.** 500 mi east of Fargo and 480 mi west

24. 15 min $\left(\text{or } \frac{1}{4} \text{ hr}\right)$ **25.** 12 and 7 **26.** 23 and 45 **27.** 13 of primer and 17 of overcoat

Chapter 7 Test

1. Yes **2.** (−2, 3) **3.** (3, −1) **4.** (−1, −4) **5.** (−3, 5) **6.** (4, 5) **7.** No solution **8.** (18, 12) **9.** Unique solution means lines intersect; no solution means parallel distinct lines; infinitely many solutions means same line. **10.** 2 and −1 **11.** 9 L of 40% solution, 3 L of 60% solution **12.** $9000 **13.** 4:00 **14.** Susan finds 25 and Larry finds 30. **15.** Answers will vary.

Section Problems 8.1

1. Yes **3.** Yes **5.** No—numerator and denominator not polynomials **7.** Yes **9.** 0 **11.** 5, −4 **13.** 0, 4, −4 **15.** 3 **17.** None **19.** 0, 1, 3

Section Problems 8.2

1. $\frac{5}{12}$ **3.** $\frac{3}{14}$ **5.** $\frac{2}{3}$ **7.** $\frac{63}{55}$ **9.** $\frac{a}{b}$ **11.** $\frac{6}{5a^2}$ **13.** $\frac{xy^2}{4z}$ **15.** $\frac{4}{3}$ **17.** $\frac{x-3}{1-4z}$

19. −1 **21.** $-\frac{3}{5}$ **23.** $\frac{x(x-1)}{x+1}$ **25.** $\frac{y-3}{y+1}$ **27.** $t-4$ **29.** $-\frac{n+3}{n}$ **31.** $\frac{y}{y+1}$

33. Cannot be reduced **35.** $\frac{1}{t-5}$

Section Problems 8.3

1. $\frac{12}{35}$ **3.** $\frac{25}{6}$ **5.** $\frac{1}{y^2}$ **7.** $\frac{-3x(y-1)}{y(x+1)}$ **9.** $\frac{x^2}{(x+1)(x+4)}$ **11.** 0 **13.** $\frac{-3(x-1)}{x+6}$

15. $\frac{(2x+1)(x+9)}{x}$ **17.** $\frac{x(3-x)}{x+2}$ **19.** $\frac{2y-1}{(y+3)(y-1)}$ **21.** $\frac{3}{2}$ **23.** $2b$ **25.** $\frac{1}{12y}$

27. 0 **29.** $\frac{4x^2y}{z}$ **31.** $\frac{x(x+3)}{x+4}$ **33.** $\frac{x}{x-1}$ **35.** $\frac{a-2}{a+1}$ **37.** $\frac{(a-2)(a-1)}{(a-3)(a+2)}$ **39.** $x(x+2)^2$

Section Problems 8.4

1. $\frac{3x+2}{x-1}$ **3.** 1 **5.** $\frac{7-t}{t}$ **7.** $\frac{x+2}{x+3}$ **9.** $\frac{-t+1}{t+3}$ **11.** 140 **13.** $x(x-3)$

15. $x(x-1)^2$ **17.** $(x-1)^2(x+4)^2$ **19.** $x(x-1)$ **21.** $-2(t)(t+1)$ **23.** $x(x^2-3x)$
25. $21x(x-4)$

Section Problems 8.5

1. $\frac{9}{35}$ **3.** $\frac{1}{45}$ **5.** $\frac{x+3}{x+4}$ **7.** $\frac{5}{x-5}$ **9.** $\frac{a^2-a+1}{a-1}$ **11.** $\frac{7t-3}{(t-1)(t+1)}$ **13.** $\frac{-t+23}{(t-5)(t+1)}$

15. $\frac{a^2+a+1}{a^2}$ **17.** $\frac{8y-1}{y(y-1)}$ **19.** $\frac{2t-1}{(t-1)(t+1)}$ **21.** $\frac{2y^2+9y-2}{(y+2)(y-2)(y+4)}$ **23.** $\frac{4x-3}{x(x-1)}$

25. $\frac{x^2-x-4}{(x-4)(x-1)}$ **27.** $\frac{2t^2+5t-21}{(t-2)(t-3)(t+1)}$ **29.** $\frac{y}{y+3}$ **31.** $\frac{(x+1)(x-1)}{x^2}$

33. $\frac{1}{(t-3)(t-2)}$ **35.** $\frac{x^2+17x+2}{(x-1)^2(x+4)^2}$ **37.** $\frac{3x^2+8x+4}{(2x+3)(x+1)(x-1)}$ **39.** $\frac{x}{x-1}$ **41.** -1

43. $\frac{x-3}{x+3}$

Section Problems 8.6

1. 6 **3.** 2, -2 **5.** 2, -1 **7.** $-\frac{23}{4}$ **9.** 8, -1 **11.** $\frac{7}{8}$ **13.** 2, -6 **15.** $-\frac{28}{3}$ **17.** No solution (3 is a pole) **19.** $\frac{19}{4}$ **21.** $\frac{7}{5}$ **23.** -1 **25.** $\frac{9}{4}$ **27.** 3 **29.** $\frac{5}{2}$ (note that -1 is a pole) **31.** $\frac{24}{37}$ **33.** $\frac{25}{t^2}$ **35.** $T=\frac{I}{PR}$ **37.** $Q=\frac{P}{P-1}$ **39.** $x=\frac{AC}{a-BC}$

Section Problems 8.7

1. 1 or -3 **3.** 4 or -1 **5.** 570 mph and 30 mph **7.** $1\frac{4}{11}$ hr **9.** 50 tables **11.** $2\frac{13}{21}$ hr

13. \$6000 for a pickup; \$7500 for a car **15.** 55 mph for Saul; 45 mph for William **17.** 8′ by 15′
19. 70 mph

Section Problems 8.8

1. 80 mi **3.** 150,000 gal **5.** \$1.88 **7.** $44\frac{4}{9}$ min **9.** $4\frac{4}{5}$ hr **11.** 40 cm **13.** $5\frac{2}{5}$ hr

15. 8 people **17.** 62.5 mph **19.** 300 cycles per second **21.** 120 rpm **23.** 45 **25.** 4 hr

Chapter 8 Review Problems

1. Yes **2.** No—denominator not a polynomial **3.** 4, -4 **4.** None **5.** $\frac{x}{3}$ **6.** $\frac{x+3}{x-1}$

7. $\frac{t-1}{t^2(t+1)(t+2)}$ **8.** $\frac{3y(x-1)}{x(y+2)}$ **9.** $\frac{3(x-1)}{x(x+2)}$ **10.** $\frac{2(x-4)(x+1)}{x-2}$ **11.** $3a^2b^2c^4$ **12.** 0

13. $\frac{2(x+2)}{x(x-6)}$ 14. $\frac{-1}{x-4}$ 15. $\frac{a+4}{a-5}$ 16. 60 17. $(x-3)(x+2)^2(x+1)$

18. $\frac{6x+8}{(x+3)(x-2)}$ 19. $\frac{y^3-y^2+2y-3}{y(y-1)}$ 20. $\frac{6}{x(x-1)(x+2)}$ 21. $\frac{4x-5}{(x-2)(x+1)^2}$

22. $\frac{x^2-1}{x^2+1}$ 23. $\frac{35}{3}$ 24. $\frac{6}{5}$ 25. -6 26. $P = \frac{s}{(1+i)^2}$ 27. $C = \frac{AB}{1+A}$ 28. $\frac{1}{2}$ or $\frac{1}{3}$

29. 20 30. $1\frac{1}{59}$ hr 31. 26 32. $3\frac{3}{13}$ hr 33. $2\frac{2}{3}$ mi 34. 2.8 hr

Chapter 8 Test

1. $\frac{3-2a}{2b(1+b)}$ 2. $\frac{x+5}{x-1}$ 3. $\frac{-12s}{t(t-1)}$ 4. $\frac{(x-3)(x-4)}{2}$ 5. $\frac{3b}{2}$ 6. $\frac{2(b-1)}{3}$

7. $\frac{x(x-2)}{(x+2)(x+4)}$ 8. $\frac{t^2+7t-3}{(t-1)(t+4)}$ 9. $\frac{4y^2-3y-3}{(y+1)(y-3)}$ 10. $\frac{5a-7}{(a+3)(a-1)(a-2)}$

11. $\frac{5t-2}{3t-1}$ 12. 1 13. No solution 14. 2 15. $\frac{3}{2}$ or -2 16. $\frac{x}{(x+1)(2x-1)}$

17. $R = \frac{R_1R_2}{R_1+R_2}$ 18. -10 19. 2.4 hr 20. 70

Section Problems 9.1

1. 5 and -5 3. $\frac{1}{6}$ and $-\frac{1}{6}$ 5. -10 7. $\frac{3}{4}$ 9. Doesn't exist as a real number 11. Doesn't exist as a real number 13. $\frac{9}{8}$ 15. 0.875 17. $0.\overline{45}$ 19. $0.\overline{6}$ 21. $0.\overline{1}$ 23. 0.0625

25. Irrational 27. Irrational 29. Rational 31. Rational 33. Irrational 35. Rational
37. 6.481 39. 7 41. 8.485 43. Doesn't exist as a real number 45. 7.937 47. -3.742
49. 3.606

Section Problems 9.2

1. 5 3. 4 5. $|a|$ 7. $|2x+1|$ 9. Yes 11. $5\sqrt{3}$ 13. 7 15. $2\sqrt{7}$ 17. $4\sqrt{3}$
19. $4|x|\sqrt{xy}$ 21. $-6\sqrt{5}$ 23. $|x|\sqrt{xy}$ 25. $|a|\sqrt{b}$ 27. 2^3 or 8 29. $-9\sqrt{3}$
31. $|xy^3|\sqrt{7xy}$ 33. $-|xy^3z^3|\sqrt{7xy}$ 35. 6 37. $a|b|\sqrt{a}$ 39. $x^3y^2\sqrt{x}$

Section Problems 9.3

1. $\sqrt{21}$ 3. $3\sqrt{3}+\sqrt{6}$ 5. $5+\sqrt{35}$ 7. $5+3\sqrt{5}-\sqrt{10}-3\sqrt{2}$ 9. $8-2\sqrt{15}$ 11. 6
13. $-2\sqrt{5}$ 15. $2x\sqrt{2}$ 17. $6x^2y^2\sqrt{x}$ 19. $5\sqrt{3}$ 21. $-\sqrt{x}$ 23. $4\sqrt{2}$ 25. $7\sqrt{5}$
27. $2x\sqrt{3x}+3\sqrt{2x}$ 29. $(2x-5)\sqrt{5x}$ 31. $4x|y|$ 33. $4\sqrt{2}+12-\sqrt{14}-3\sqrt{7}$ 35. $-4\sqrt{3}$

Section Problems 9.4

1. $\frac{\sqrt{5}}{5}$ 3. $\frac{\sqrt{3}}{2}$ 5. $\frac{\sqrt{a}}{a^2}$ 7. $\frac{\sqrt{2}}{2}$ 9. $\frac{\sqrt{21}}{7}$ 11. $\frac{x}{3y}$ 13. $2x$ 15. $5ab$ 17. $\frac{\sqrt{x}}{x}$

19. $\frac{\sqrt{15}}{5}$ 21. $\frac{5\sqrt{14}}{21}$ 23. $\frac{x}{4}$ 25. $\frac{\sqrt{15x}}{15x}$ 27. $\frac{x\sqrt{2}}{2}$ 29. $\frac{\sqrt{x}}{2}$ 31. $\frac{\sqrt{10}}{10}$ 33. $\frac{\sqrt{5}}{5}$

35. $\frac{y\sqrt{x}}{x}$ 37. $\frac{\sqrt{6}}{4}$ 39. $\frac{a\sqrt{b}}{b}$

Chapter 9 Review Problems

1. 9 **2.** -5 **3.** Doesn't exist **4.** $-\frac{2}{3}$ **5.** 0.125 **6.** $0.\overline{27}$ **7.** $0.\overline{5}$ **8.** $1.1\overline{6}$

9. Rational **10.** Irrational **11.** Irrational **12.** 5.745 **13.** Doesn't exist **14.** 9.110 **15.** -4 **16.** 4.359 **17.** 6 **18.** Doesn't exist **19.** $6\sqrt{2}$ **20.** $3x^2y^2\sqrt{2y}$ **21.** $|2x + 1|$

22. $\frac{2|x|}{3y^2}$ **23.** $2\sqrt{5}$ **24.** $3\sqrt{10}$ **25.** $3x^2\sqrt{7}$ **26.** $5\sqrt{2} + 2\sqrt{5}$ **27.** -2 **28.** $-6\sqrt{5}$

29. $10\sqrt{5}$ **30.** $\frac{5\sqrt{3}}{4}$ **31.** $\frac{\sqrt{15}}{5}$ **32.** $\frac{\sqrt{ab}}{b}$ **33.** $\frac{3\sqrt{7}}{7}$ **34.** $\frac{\sqrt{3}}{6}$ **35.** $\frac{2\sqrt{3x}}{3x}$ **36.** $\frac{\sqrt{6}}{3}$

37. $\frac{x\sqrt{6xy}}{3y^2}$ **38.** $2\sqrt{x}$ **39.** $\frac{\sqrt{70}}{5}$

Chapter 9 Test

1. -8 **2.** $\frac{5}{9}$ **3.** Doesn't exist **4.** 9.434 **5.** -4.359 **6.** 2.236 **7.** $2\sqrt{7}$ **8.** $15\sqrt{2}$

9. $3t\sqrt{2t}$ **10.** $2\sqrt{5} - 5$ **11.** $8 - 2\sqrt{15}$ **12.** $2\sqrt{21}$ **13.** $2|y + 1|$ **14.** $5ab^2\sqrt{6a}$

15. $-\sqrt{5}$ **16.** $(2x + 9)\sqrt{x}$ **17.** $5\sqrt{2}$ **18.** $-\sqrt{3}$ **19.** $\frac{6|a|}{7b^2}$ **20.** $5x|y|\sqrt{x}$ **21.** $-3\sqrt{10}$

22. $\frac{3\sqrt{5}}{5}$ **23.** $\frac{\sqrt{2}}{2}$ **24.** $\frac{\sqrt{6x}}{2}$ **25.** $\frac{\sqrt{35xy}}{7y}$

Section Problems 10.1

1. 3, 5 **3.** 6, -2 **5.** $\frac{3}{2}, -\frac{1}{3}$ **7.** $-2, -3$ **9.** $\frac{3}{2}, -4$ **11.** 7, -7 **13.** 1, -7

15. $\frac{3}{4}, \frac{1}{4}$ **17.** 8, -8 **19.** 4, -2 **21.** 2, -2 **23.** 3, -3 **25.** $\sqrt{7}, -\sqrt{7}$

27. $\frac{\sqrt{15}}{3}, -\frac{\sqrt{15}}{3}$ **29.** 4, -4 **31.** $\sqrt{5}, -\sqrt{5}$ **33.** 1, -7 **35.** $-5 + \sqrt{7}, -5 - \sqrt{7}$

37. 1, 0 **39.** No real solution **41.** 2.5, 0.1

Section Problems 10.2

1. $(x + 3)^2 = 12$ **3.** $(x - 2)^2 = \frac{5}{2}$ **5.** $-2 + \sqrt{6}, -2 - \sqrt{6}$ **7.** $\frac{5 + \sqrt{17}}{2}, \frac{5 - \sqrt{17}}{2}$

9. $-\frac{3}{2}$ **11.** $1 + \sqrt{7}, 1 - \sqrt{7}$ **13.** $-3 + 2\sqrt{3}, -3 - 2\sqrt{3}$ **15.** $\frac{5}{2}$ **17.** 3, 4 **19.** $-\frac{5}{3}$

21. $\frac{2 + \sqrt{2}}{2}, \frac{2 - \sqrt{2}}{2}$ **23.** 1, −3

Section Problems 10.3

1. $\frac{3 + \sqrt{17}}{2}, \frac{3 - \sqrt{17}}{2}$ **3.** −1, −3 **5.** No real solution **7.** 0, 3 **9.** $2, -\frac{1}{2}$

11. $\frac{3 + \sqrt{37}}{2}, \frac{3 - \sqrt{37}}{2}$ **13.** $-3, \frac{1}{6}$ **15.** $\frac{4}{3}$ **17.** $\frac{-3 + \sqrt{15}}{2}, \frac{-3 - \sqrt{15}}{2}$ **19.** No real solution **21.** Two distinct real roots **23.** Two distinct real roots **25.** No real solution **27.** $\sqrt{2}, -\sqrt{2}$ **29.** 0, 4 (−2 is a pole) **31.** 3 **33.** 2 **35.** 2, −4 **37.** 4, −4 **39.** 3

Section Problems 10.4

1. $3 + \sqrt{11}, 3 - \sqrt{11}$ **3.** 5, 6, and 7 **5.** 14 in. **7.** 3 ft by 4 ft **9.** 9 **11.** $\frac{-6 + 3\sqrt{14}}{2}$

13. $(11 + \sqrt{101})$ days **15.** 7 **17.** 26 ft **19.** $6\frac{1}{2}$ cm **21.** $70 or $130 **23.** 10 mph—Henry, 8 mph—Stan **25.** $6400x^2 + 3600\left(x - \frac{1}{2}\right)^2 = 250{,}000$

Chapter 10 Review Problems

1. 1, 3 **2.** 3, −5 **3.** 3, 5 **4.** $1, -\frac{5}{12}$ **5.** 2, −2 **6.** 8, −6 **7.** $-2 + \sqrt{10}, -2 - \sqrt{10}$

8. 1, −3 **9.** $3 + 2\sqrt{3}, 3 - 2\sqrt{3}$ **10.** 4 **11.** $\frac{7 + \sqrt{41}}{2}, \frac{7 - \sqrt{41}}{2}$ **12.** $\frac{3 + \sqrt{41}}{4}, \frac{3 - \sqrt{41}}{4}$

13. $3 + \sqrt{15}, 3 - \sqrt{15}$ **14.** $\frac{3 + \sqrt{17}}{2}, \frac{3 - \sqrt{17}}{2}$ **15.** $-\frac{3}{2}$ **16.** $\frac{3}{2}, -5$ **17.** $1, \frac{3}{5}$ **18.** No real solution **19.** $\frac{-9 + \sqrt{161}}{8}, \frac{-9 - \sqrt{161}}{8}$ **20.** 6, −1 **21.** $3 + \sqrt{6}, 3 - \sqrt{6}$

22. $-\frac{1}{4}, \frac{3}{2}$ **23.** $\frac{4}{5}, \frac{1}{2}$ **24.** 2 **25.** −2 **26.** 7 (2 is a pole) **27.** $\frac{-1 + \sqrt{7}}{3}, \frac{-1 - \sqrt{7}}{3}$

28. 7 in. by 11 in. **29.** $\frac{-3 + \sqrt{153}}{2}$ **30.** $2 + 4\sqrt{2}$

Chapter 10 Test

1. $\frac{7}{2}, -5$ **2.** 9, −9 **3.** $-2 + \sqrt{3}, -2 - \sqrt{3}$ **4.** $-3 + \sqrt{13}, -3 - \sqrt{13}$ **5.** $\frac{3 + \sqrt{41}}{4}$,

$\frac{3 - \sqrt{41}}{4}$ **6.** $2 + \sqrt{2}, 2 - \sqrt{2}$ **7.** No real solution **8.** $\frac{1}{3}, -2$ **9.** $\frac{2}{3}$ **10.** 4 **11.** 0

12. $\frac{3 + \sqrt{5}}{2}, \frac{3 - \sqrt{5}}{2}$ **13.** 30 sq ft **14.** $4\sqrt{2}$ in. **15.** $(-2 + 2\sqrt{11})$ hr

Section Problems 11.1

1. $\{x \mid x < -3\}$ **3.** $\{x \mid x \geqslant 4\}$ **5.** $\{x \mid x > -2\}$ **7.** $\{x \mid 1 < x \leqslant 4\}$ **9.** The set of all real numbers greater than 3 **11.** The set of all real numbers greater than 0 and less than 4 **13.** The set of all real numbers greater than or equal to $\frac{2}{3}$ **15.** 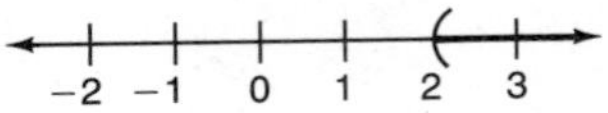

$(2, \infty)$

17. $\{x \mid x \geqslant -2\}$ 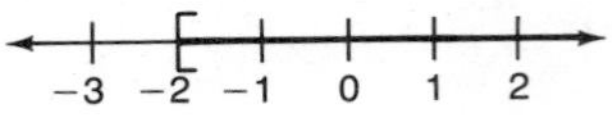

19. 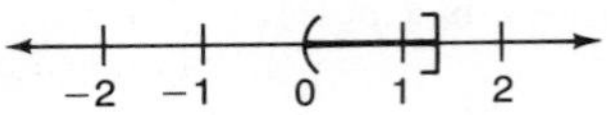

$\left(0, \frac{4}{3}\right]$ **21.** $\{x \mid x > -1\}$ $(-1, \infty)$ **23.** 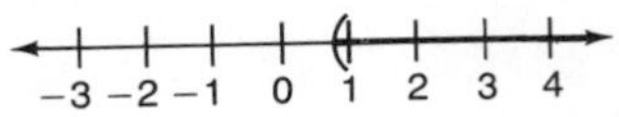

$\left(\frac{3}{4}, \infty\right)$ **25.** $\{x \mid x < 2\}$ 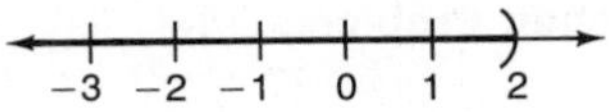

27. $\{x \mid -2 < x < 3\}$ 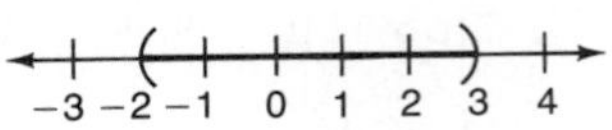

29. $\{x \mid -2 \leqslant x \leqslant 3\}$ $[-2, 3]$ **31.** $\{x \mid -1 < x < 2\}$ $(-1, 2)$ **33.** Answers will vary. **35.** $<$ **37.** $<$ **39.** $>$ **41.** $=$ **43.** $=$ **45.** Answers may vary.

Section Problems 11.2

1. 3 is in $\{x \mid x \geqslant 3\}$, but 3 is not in $\{x \mid x > 3\}$. **3.** $x < -8$ **5.** $y > -\frac{17}{12}$ **7.** $x \geqslant -0.3$

9. $t < 8$ **11.** $x \leqslant -2$ **13.** $x < -\frac{21}{4}$ **15.** $x < 2$ **17.** $t \geqslant \frac{7}{6}$ **19.** $x < \frac{7}{2}$ **21.** $x > -5$

23. $y > \frac{13}{4}$ **25.** $t < \frac{19}{32}$ **27.** $t < -3$ **29.** $-14 < x \leqslant -10$ **31.** $-\frac{2}{3} < x \leqslant 2$

33. $-3 \leqslant x \leqslant -1$ **35.** $x < \frac{11}{16}$ **37.** $\frac{3}{4} < x < 2$ **39.** $a = 2$ **41.** $x \leqslant \frac{9}{2}$ **43.** $y \leqslant \frac{1}{2}$

45. $-\frac{2}{3} < t < \frac{7}{3}$ **47.** $x \leqslant 1$ **49.** $y > 1$

Section Problems 11.3

1. $-1, 5$ **3.** $3, -4$ **5.** $3, -1$ **7.** No solution **9.** $-2, 4$ **11.** $\frac{4}{3}, -\frac{8}{3}$ **13.** $-3 < x < 3$

15. $-\frac{7}{2} \leqslant m \leqslant \frac{1}{2}$ **17.** $-4 \leqslant x \leqslant 8$ **19.** $-4 \leqslant y \leqslant -2$ **21.** $t > 4$ or $t < -4$ **23.** All real numbers **25.** $z \geqslant 4$ or $z \leqslant -2$ **27.** $x < \frac{1}{2}$ or $x > \frac{5}{2}$ **29.** $-\frac{10}{3} \leqslant x \leqslant \frac{14}{3}$ **31.** $n > 9$ or

$n < -3$ **33.** $-2 < t < 6$ **35.** No solution **37.** $-\frac{9}{2} \leqslant t \leqslant \frac{11}{2}$ **39.** $x < -1$ or $x > \frac{4}{3}$

41. $-\frac{1}{2}, \frac{7}{2}$ **43.** No solution **45.** All real numbers **47.** $t < -\frac{7}{3}$ or $t > 3$ **49.** $-\frac{1}{3}, 1$

Chapter 11 Review Problems

1. $\{x \mid x > 10\}$ **2.** $\{x \mid x \leqslant -3\}$ **3.** $\{x \mid 0 \leqslant x < 5\}$ **4.** −1 0 1 2 3 4 5; $(3, \infty)$

5. $\{x \mid x \leqslant 1\}$; $(-\infty, 1]$ **6.** $\{x \mid x \leqslant -1\}$; −3 −2 −1 0 1 2 3

7. $-\frac{3}{2}$ $\frac{5}{2}$ −2 −1 0 1 2 3; $\left(-\frac{3}{2}, \frac{5}{2}\right)$ **8.** $t \geqslant -2$ **9.** $x > -\frac{7}{3}$ **10.** $t < -2$

11. $t > 2$ **12.** $x \leqslant 2$ **13.** $x < -\frac{1}{2}$ **14.** $y \leqslant \frac{8}{5}$ **15.** $x < \frac{2}{5}$ **16.** $x < 11$ **17.** $x \geqslant \frac{13}{4}$

18. $x < -\frac{1}{2}$ **19.** $t > -\frac{2}{3}$ **20.** $y \geqslant -10.4$ **21.** $x > 2$ **22.** 3, −3 **23.** 3, −3

24. $\frac{2}{3}, -\frac{2}{3}$ **25.** −4, 10 **26.** 28, −28 **27.** No solution **28.** $-2.3 \leqslant x \leqslant 2.3$ **29.** $y \geqslant 2$ or $y \leqslant -3$ **30.** $-\frac{17}{6} \leqslant x \leqslant -\frac{7}{6}$ **31.** $t > 13$ or $t < -11$ **32.** No solution **33.** $y > 1$ or $y < -3$ **34.** $-1 < x < 2$ **35.** $y > 11$ or $y < -13$

Chapter 11 Test

1. −2 −1 0 1 2 3 4 5; $(-2, \infty)$ **2.** $\{x \mid -3 < x \leqslant 2\}$; $(-3, 2]$ **3.** $\{x \mid x \leqslant 3\}$;
−2 −1 0 1 2 3 4 5 **4.** (a) 5 (b) 1.45 (c) −2 **5.** $t > \frac{1}{2}$ **6.** $x \geqslant 6$ **7.** $y \geqslant \frac{13}{6}$

8. $x < -6$ **9.** 2, 4 **10.** $-\frac{5}{3} \leqslant x \leqslant \frac{7}{3}$ **11.** $y < 0$ or $y > 4$ **12.** −1, 5 **13.** No solution

14. $x > 1$ or $x < -3$ **15.** $-2 \leqslant x \leqslant \frac{3}{2}$

Final Examination

1. $\frac{9}{8}$ **2.** $\frac{47}{90}$ **3.** $\frac{15}{32}$ **4.** $4\frac{13}{20}$ **5.** 0.0028 **6.** .0034 **7.** 6 **8.** 16 **9.** $-\frac{77}{60}$ **10.** 1

11. $x - y$ **12.** 25 **13.** 6 **14.** $\frac{8}{7}$ **15.** −6 **16.** $Q = \frac{N - R}{P}$ **17.** 15, 45, and 90 ft

18. $7\frac{1}{2}$ pt **19.** $4x^3 + 5x + 3$ **20.** $(y - 2)(y + 2)$ **21.** $(6x - 5)(2x + 3)$ **22.** −4, 3

23. 25 ft

24.

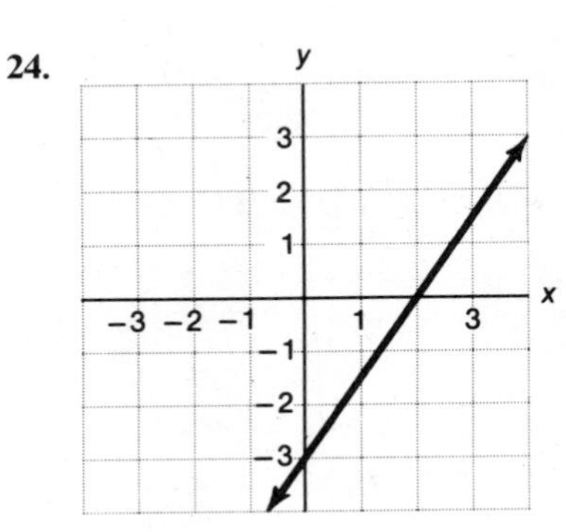

25.

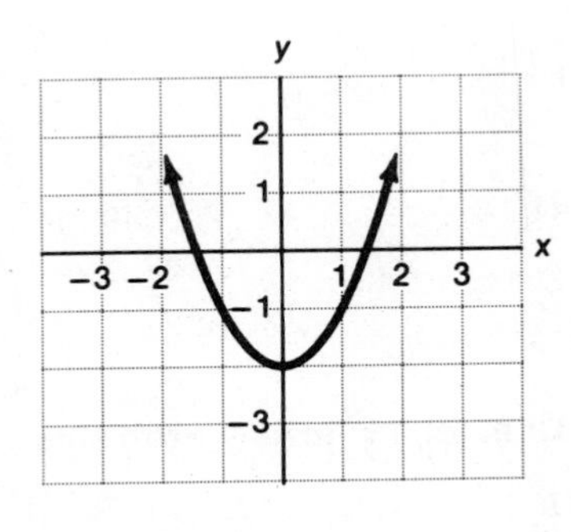

26. $\frac{2}{5}$ **27.** $y = -3x + 1$ **28.** $y = \frac{1}{3}x - \frac{17}{3}$ **29.** $(-2, -8)$ **30.** No solution

31. $\left(\frac{40}{29}, -\frac{12}{29}\right)$ **32.** 4:00 **33.** 15 gal **34.** $\frac{x + 2}{x(x - 2)}$ **35.** $\frac{2a^3}{3b^2c}$ **36.** $\frac{x^2 + 2}{(x - 2)(x - 1)(x + 2)}$

37. $\frac{t - 1}{t}$ **38.** 2 **39.** $1\frac{4}{11}$ hr **40.** $3x|y|\sqrt{5x}$ **41.** $-2\sqrt{2}$ **42.** $\frac{2\sqrt{2x}}{x}$ **43.** $2 + \sqrt{7}$, $2 - \sqrt{7}$ **44.** No real solution **45.** $\frac{-1 + \sqrt{7}}{3}$, $\frac{-1 - \sqrt{7}}{3}$ **46.** 5 **47.** $x > -\frac{1}{2}$

48. $\frac{5}{2}, -\frac{5}{2}$ **49.** $-1 \leqslant t \leqslant 2$ **50.** $y \geqslant \frac{7}{3}$ or $y \leqslant -3$

Index

2 3 4 5 6 7 8 9 0